ESTATES GAZETTE LAW REPORTS

Estates Gazette Law Reports

2010

Volume 3

Edited by
HH Judge Hazel Marshall QC

Routledge
Taylor & Francis Group
LONDON AND NEW YORK

EG Books

EG Books is an imprint of Routledge

2 Park Square, Milton Park, Abingdon, Oxon OX14 4RN

711 Third Avenue, New York, NY 10017

Routledge is an imprint of the Taylor & Francis Group, an informa business

British Library Cataloguing-in-Publication Data
A catalogue record for this book is available from the British Library

Library of Congress Cataloging-in-Publication Data
A catalog record for this book is available from the Library of Congress

ISBN: 978-0-08-096939-8
ISSN: 0951-9289

CONTENTS

Cases reported in "Estates Gazette" [2010] issues 35-50

SUPPLEMENT

Cases not reported in "Estates Gazette"

TABLE OF CASES

Titles of cases shown in bold type
Names of parties reversed shown in ordinary type

INDEX OF SUBJECT MATTER

Hildron Finance Ltd *v* Sunley Holdings Ltd

CHANCERY DIVISION

[2010] EWHC 1681 (Ch)

6 July 2010

HENDERSON J

Estates Gazette 9 October 2010

[2010] 40 EG 104

Contract — Frustration — Sale of block of flats — Purchaser agreeing to sell long lease of porter's flat if no longer needed as such and to pay overage to vendor — Sale of flat prevented by enfranchisement of block under Leasehold Reform, Housing and Urban Development Act 1993 — Whether agreement to be construed so as to require overage payment — Whether agreement frustrated

The respondent purchased the freehold of a block of flats from the appellant in March 1986. All the flats save one were let on long leases that reserved a service charge and provided for the landlord to employ a resident porter; the remaining flat was occupied by the porter. The parties entered into an overage agreement under which, should the porter's flat cease to be needed for a resident porter at any time during the following 21 years, the respondent would sell a new long lease of that flat on the open market, with vacant possession, and pay 50% of the net proceeds of sale to the appellant after the deduction of certain items.

In 2004, a majority of tenants served an initial notice to acquire the freehold of the block pursuant to the right to collective enfranchisement introduced by the Leasehold Reform, Housing and Urban Development Act 1993. That notice was registered and, by virtue of section 19(1) of the Act, it became unlawful for the respondent to sell a long lease of the porter's flat because it formed part of the common parts that the tenants were entitled to acquire under section 2(1)(b). By 2007, the flat was no longer needed for a resident porter such that the obligation to sell it would otherwise have been triggered.

The enfranchisement price for the block was determined; this included a value of £200,000 attributable to the porter's flat. The appellant claimed that it was entitled to a payment either under the overage agreement or as damages for its breach. The respondent obtained a declaration in the county court that the appellant was not entitled to any payment. The appellant appealed. It contended that the court should seek to give effect to the underlying objectives of the overage agreement in the light of the changed circumstances. On that approach, it should be awarded 50% of the net sum received for the porter's flat.

***Held*: The appeal was dismissed. When imputing an intention to the parties to an agreement in respect of unforeseen circumstances, a certain amount of reconstruction is required, but that must be firmly grounded in the words that the parties actually used in their original factual matrix. The overage agreement could not be construed so as to require a payment to the appellant on the enfranchisement of the block. Although, in a general sense, its purpose was to unlock and share the increased capital value of the flat if it were no longer required to accommodate a porter, the steps to be taken in that event were narrowly and prescriptively formulated and were predicated on the assumption that the flat could be sold on the open market, with a new long lease on the standard form, for a price of which the appellant could approve. A sale of the freehold interest in the flat as part of a disposal of the entire block, compelled by statute and at a price that had to be ascertained according to the criteria of the 1993 Act, rather than by the exposure of the flat to a real as opposed to notional open market, had been outside the parties' contemplation at the date of the agreement. Moreover, there was no causative connection between the porter's vacation of the flat and the events that led to the enfranchisement of the block. The points of distinction between the agreed steps set out in the agreement and the steps that had led to the respondent receiving £200,000 for the freehold interest in the flat were so numerous and significant that it was impossible to impute a common intention to the parties that the agreement should be modified to accommodate them.**

It was not possible to construe the agreement in such a way that it applied to the events that had occurred. Nor would it be reasonable to construe it as imposing an absolute liability on the respondent even where performance had become impossible. The correct solution was to treat the contract as having been discharged by frustration without fault on either side.

The following cases are referred to in this report.

Attorney General of Belize v *Belize Telecom Ltd* [2009] UKPC 10; [2009] 2 All ER 1127; [2009] 2 All ER (Comm) 1; [2009] 1 WLR 1988
Baily v *De Crespigny* (1869) LR 4 QB 180
Bromarin AB v *IMD Investments Ltd* [1999] STC 301
Casson v *Ostley PJ Ltd* [2001] EWCA Civ 1013; [2003] BLR 147
Hildron Finance Ltd v *Greenhill Hampstead Ltd* [2008] 1 EGLR 179, LT
Investors Compensation Scheme Ltd v *West Bromwich Building Society (No 1)*; *Investors Compensation Scheme Ltd* v *Hopkin & Sons*; *Alford* v *West Bromwich Building Society*; *Armitage* v *West Bromwich Building Society* [1998] 1 WLR 896; [1998] 1 All ER 98, HL
Mediterranean Salvage & Towage Ltd v *Seamar Trading & Commerce Inc (The Reborn)* [2009] EWCA Civ 531; [2010] 1 All ER (Comm) 1
Slipper v *Tottenham & Hampstead Junction Railway Co* (1867) LR 4 Eq 112, Ch
Sudbrook Trading Estate Ltd v *Eggleton* [1983] 1 AC 444; [1982] 3 WLR 315; [1982] 3 All ER 1; (1982) 44 P&CR 153; [1983] 1 EGLR 47; (1982) 265 EG 215, HL

This was an appeal by the appellant, Sunley Holdings Ltd, from a decision of Mr Recorder Antonio Bueno QC, sitting in Central London County Court, allowing a claim by the respondent, Hildron Finance Ltd, for a declaration that it was not liable to make any payment under an overage agreement made between the parties.

Alexander Hill-Smith (instructed by John May Law) appeared for the appellant; Stuart Hornett (instructed by Charles Russell LLP, of Cheltenham) represented the respondent.

Giving judgment, **HENDERSON J** said:

[1] The issue on this appeal is whether the appellant, Sunley Holdings Ltd (Sunley), is entitled to receive an overage payment pursuant to the terms of an agreement that it entered into on 25 March 1986 (the porter's flat agreement) when it sold the freehold interest in a large block of flats on Hampstead High Street, London NW3, known as Greenhill, to Hildron Finance Ltd (Hildron). Hildron is the respondent

to the appeal, which is brought by Sunley (with permission granted by Floyd J) against the relevant parts of the judgment and order of Mr Recorder Antonio Bueno QC, dated 10 July 2009, in Central London County Court.

[2] At the time of the sale in 1986, the porter's flat in Greenhill (flat 9) was occupied by the resident porter employed by the landlord, Sunley. All the other flats (some 136 in all) were let on long leases reserving a ground rent and service charge. It was the understanding of both Sunley and Hildron, as recorded in a recital to the porter's flat agreement, that the leases all provided for the landlord, at its discretion, to employ a porter and other employees for the purpose of performing the landlord's covenants and to make available, either in the block itself or elsewhere, accommodation for such porter or other employees. The important point is that the landlord was at liberty to provide the services of a non-resident porter or porters, although there had in fact been a resident porter living in a flat in Greenhill at all times since the block was built in the 1930s.

[3] The purpose of the porter's flat agreement in broad terms was to regulate future dealings with the porter's flat by Hildron and its successors in title for a period of 21 years and to require Hildron to sell a new long lease of the flat on the open market with vacant possession if there came a time when the flat was no longer required as accommodation for a resident porter or employee. The net proceeds of any such sale were then to be divided equally between Sunley and Hildron, after deducting from the gross proceeds the sum of £10,500 together with the reasonable legal costs and surveyor's fees incurred by Hildron in connection with the negotiation and grant of the new lease.

[4] The present dispute concerns the effect on the porter's flat agreement of the collective enfranchisement provisions contained in Part I of the Leasehold Reform, Housing and Urban Development Act 1993 (the 1993 Act). On 8 November 2004, a requisite majority of qualifying tenants served an initial notice on Hildron pursuant to section 13 of the 1993 Act, claiming to exercise the right to collective enfranchisement in respect of the entire block. The proposed nominee purchaser was a company called Greenhill Hampstead Freehold Ltd. By a counternotice dated 20 January 2005, Hildron admitted the tenants' right to enfranchise and put forward counter-proposals for the price to be paid. Meanwhile, the initial notice was registered at HM Land Registry by the participating tenants pursuant to section 97(1) of the 1993 Act, which provides that such a notice may be registered "as if it were an estate contract". The effect of the registration was to impose certain constraints on the rights of Hildron as the freeholder to deal with the premises specified in the notice, as set out in section 19. Although Sunley has at times disputed the point, it was common ground at the hearing below, and was again common ground by the time of the hearing of the appeal, that the effect of section 19 would have been to prevent the grant of any new long lease of the porter's flat by Hildron because the flat formed part of the common parts that the tenants were entitled to acquire under section 2(1)(b) of the 1993 Act. By virtue of section 19(1), any purported grant of such a lease would have been void: the relevant words are "any transaction shall be void to the extent that it purports to effect… any such grant of a lease as is mentioned in paragraph (a)…".

[5] It follows, and is also now common ground, that once the initial notice had been registered, section 19 made it unlawful for Hildron to comply with the express terms of the porter's flat agreement in the event that the porter's flat was no longer required to accommodate a porter or other employee, and if Hildron had purported to grant a new long lease of the flat in such circumstances the lease would have been void.

[6] The qualifying tenants and Hildron were unable to agree on the price to be paid for the freehold of Greenhill and the matter was referred to the leasehold valuation tribunal (LVT) to determine: see LON/ENF/1486/05. In its decision dated 12 July 2006, the LVT determined the total value of the landlords' interests in the property (including the agreed value of the interests of certain intermediate landlords) as £2,298,172. Of this sum, the amount attributable to the porter's flat was £50,000. In reaching this conclusion, the LVT accepted Hildron's submissions, buttressed by expert evidence, that a willing buyer in the market place would not have been purchasing the porter's flat as a stand-alone investment but as part of a complex investment in the entire block, and that the service charge provisions in the leases did not allow for the recovery of a notional market rent by the landlord in respect of the porter's flat.

[7] Hildron then appealed to the Lands Tribunal (LT); and one of the issues in respect of which permission to appeal was given was the amount to be attributed to the porter's flat. The appeal was heard by HH Judge Reid QC and Mr Norman Rose FRICS on 12 and 13 November 2007, and on 10 January 2008 they gave their decision*. They held that the total price payable by the nominee purchaser was £2,835,255 and that the sum attributable to the porter's flat was £200,000. It was agreed between the expert valuers on both sides that the open market value of the porter's flat was £300,000 as at the valuation date of 20 January 2005 (the date of Hildron's counternotice, which the LT held in agreement with the LVT to be the correct date to take). The LT differed from the LVT in holding that the later tranches of long leases contained clear provisions that entitled the lessor to recover a notional rent for the porter's flat as part of the service charge, but they agreed that no notional rent was recoverable under the "old leases" that comprised the first tranche. They also expressly found that it was open to the lessor to discharge its obligations in respect of portering services by the provision of non-resident porters and thus to sell the porter's flat: see [72] of the decision. Nevertheless, they declined to place a value on the flat equivalent to its agreed open market value but instead settled on the figure of £200,000. For present purposes, their reasoning appears sufficiently from [76] of the decision:

> In our view, the value put on the flat by [Hildron] is over-optimistic. The sale of the flat or its reduction into possession so as to let it would create other problems in relation to the provision of portering services. It is unlikely that any purchaser would regard the flat as a stand-alone investment for which it would be prepared to pay the full open market price as though it were a one-off retail sale. In addition, were the flat to be retained as a porter's flat, there would be a shortfall on the notional rent because of the terms of the old leases. As against that, the figures proposed by Mr Maunder Taylor [the expert witness for the nominee purchaser] seem to us to be too low. Even allowing for the "bulk discount" point and the difficulty created by the terms of the old leases as to recovering the notional rent, we take the view that a purchaser would attribute a greater value than £150,000 to the flat. In our judgment, the appropriate figure is one of £200,000. If we are wrong in concluding that the sale of the porter's flat is not prohibited, our opinion as to the value that a purchaser would attribute to the right to receive a notional rent under the more recent leases amounting, we understand, to some two-thirds of the total would be £100,000.

[8] Following the decision of the LT, Hildron was still unwilling to execute a voluntary transfer of the freehold to the nominee purchaser. One of the reasons for this was that Hildron and Sunley were unable to agree whether the overage provisions in the porter's flat agreement had been triggered, and the restriction against the registered title that had been entered to protect Sunley's rights under the agreement remained in place. Eventually, on 23 May 2008, the nominee purchaser issued a Part 8 claim form in Central London County Court seeking a vesting order pursuant to section 24 of the 1993 Act. The defendants to the claim were Hildron and Sunley. Apart from a vesting order, the claimant asked for declarations that Sunley was not entitled to any payment from Hildron before giving its consent to a transfer of the freehold to the claimant and that Sunley was obliged to give its unconditional consent to such a transfer.

[9] In due course, Hildron made a Part 20 claim against Sunley, seeking declarations to the same effect as those sought by the nominee purchaser, and Sunley filed a defence and counter-claim asking for declarations that Hildron was in breach of the porter's flat agreement and also in breach of another agreement entered into at the same time as the porter's flat agreement, namely the car-parking agreement. These were the live issues that the learned recorder had to determine when the case eventually came on for trial before him on 17 and 18 February 2009. By that stage, a vesting order was no longer sought

* Editor's note: Reported at [2008] 1 EGLR 179

A and the nominee purchaser had dropped out of the proceedings because, on 20 January 2009, the freehold had been transferred by Hildron to the nominee purchaser, with Sunley's consent, subject to an agreed retention by Hildron from the purchase price of £150,000 that was paid into a joint account to abide the result of Sunley's claim in the Part 20 proceedings.

[10] It is unnecessary for me to say anything further about the car-parking agreement. The recorder rejected Sunley's claim founded on that agreement and Sunley has not appealed against that part of his decision.

[11] With regard to the porter's flat agreement, there is one B important fact that I have not yet mentioned. On 4 September 2006, almost 20 months after it had served its counternotice under the 1993 Act, Hildron granted an assured shorthold tenancy of flat 9 for a term of one year to a Mr Helge Kostka, an employee of Deutsche Bank. It is clear that on the grant of this tenancy, if not before, flat 9 was no longer required as accommodation for a porter or any other employee of Hildron's, with the consequence that the initial triggering event under the porter's flat agreement was satisfied. As I have pointed out, however, it would have been unlawful for Hildron to market a new long lease of the flat because of the provisions of section 19 of the 1993 Act. There is no suggestion that the grant of the assured shorthold tenancy infringed C section 19. The critical question that the recorder had to determine was whether the supervening illegality, which made it impossible to perform the porter's flat agreement in accordance with its express terms, relieved Hildron from any obligation thereunder or whether Sunley was still entitled to rely on it in the changed circumstances where the flat was in effect sold by Hildron, pursuant to the provisions of the 1993 Act, for a consideration of £200,000.

Porter's flat agreement

[12] I will now set out the relevant provisions of the porter's flat agreement. It recited that:

(1) by a transfer of even date, Greenhill had been transferred by D Sunley to Hildron on the terms therein mentioned;

(2) Greenhill comprised a block of some 137 flats, with six purpose-built garages and ancillary grounds for the enjoyment and benefit of the flat occupiers;

(3) all the flats (with the exception of flat 9) were then subject to long-term leases reserving ground rent and service charge;

(4) the leases each provided for the landlord, at its discretion, to employ a porter and other employees for the purpose of performing the landlord's covenants and to make accommodation available for the porter or other employee either in the block or elsewhere;

E (5) flat 9 was currently occupied by the resident porter employed by the freeholder in accordance with the terms of the leases; and

(6) in respect of future dealings with flat 9, the parties had agreed to enter into the deed.

[13] The operative clauses then provided as follows:

> 1. HILDRON for itself and its successors in title as aforesaid HEREBY COVENANTS with Sunley for a period of 21 years from the date hereof that
>
> (a) if flat No 9 is no longer required as accommodation for a porter or other employee of the landlord under the provisions of the said leases it will forthwith sell on the open market and with vacant possession a new lease of flat No 9 in the form and to the effect of the standard form of block lease (a specimen F copy of which is annexed hereto) at a price equal to the highest market price reasonably negotiable such price to be first approved in writing by Sunley (such approval not to be unreasonably withheld)
>
> (b) it shall not grant a lease or tenancy of or otherwise dispose of any interest in flat No 9 except in accordance with the provisions of this Deed
>
> (c) until such time as a new lease is sold in accordance with the provisions of this Deed Hildron shall not use nor permit or suffer flat No 9 to be used otherwise than as accommodation for a porter or other employee of the landlord under the provisions of the said leases
>
> (d) on the grant of the new lease as hereinbefore provided Hildron will pay to Sunley a sum equal to 50% of the net sale proceeds (the net sale proceeds being the gross price achieved for such new lease as approved by Sunley under Clause 1(a) of this Deed less the sum of £10,500.00 together with Hildron's reasonable legal costs and surveyors fees in connection with the negotiation and grant of such new lease in so far as such costs and fees are not made the G responsibility of the purchaser).
>
> 2. HILDRON HEREBY FURTHER COVENANTS with Sunley that it will forthwith in conjunction with its application to HM Land Registry to register the transfer hereinbefore recited lodge this Deed at HM Land Registry so that its contents may be noted in the register of title number 461978 and will make an application in Form 75 to register a restriction against title number 461978 in respect of this Deed.
>
> 3. HILDRON HEREBY FURTHER COVENANTS with Sunley that during the term specified in Clause 1 hereof it will not dispose of its interest in the said property known as Greenhill aforesaid unless simultaneously with such disposal the person persons or body acquiring such interest shall have entered into a Deed with Sunley in the terms (*mutatis mutandis*) of this Deed. H

[14] The reason why the sum of £10,500 is deductible from the gross sale price of the new lease under clause 1(d) is that this sum represented the price actually paid by Hildron for the porter's flat in 1986. Evidence was given at the trial by Mr Paul Alexander, whose company had been the managing agent of Greenhill for the past 23 years and who had acted for Hildron when it bought the freehold from Sunley in 1986. He explained that the price for the porter's flat of £10,500 had been based on its investment value, calculated as seven years' purchase of a notional yearly rent at the time of £1,505. It should be noted that the porter's flat agreement contained no provision to index-link or otherwise update the J figure of £10,500 during the 21-year life of the agreement.

[15] Against this background, the basic commercial purpose of the porter's flat agreement is, in my judgment, reasonably clear. Hildron had paid Sunley no more than the investment value of the flat, based on a notional yearly rent of £1,505. Part but not all of this notional rent could be recovered by the landlord through the service charge mechanism in the leases that were not "old leases". The position with regard to the old leases was, at best, uncertain. The flat had always been occupied by a resident porter employed by the landlord, but the parties were in agreement that it would be open to the landlord in the future to provide accommodation for a porter outside the block. As it K happens, the correctness of this assumption has recently been confirmed by the LT in its decision of 10 January 2008 (see [7] above), but nothing turns on this: what matters is the contemporary understanding of the parties, as recorded in recital (4). Accordingly, a time might well come when the flat would no longer be required for a resident porter and when the landlord would be free to sell it on the open market. In those circumstances, a bargain was struck. Hildron (or its successors in title) would be obliged, as soon as the flat became available, to sell it on the open market with vacant possession for the highest price that it could reasonably negotiate. The interest to be sold would be a new long lease in the standard form used for the block. The price would L be subject to the prior written approval of Sunley, such approval not to be unreasonably withheld. On completion of the sale, Hildron would pay Sunley 50% of the net proceeds of sale (defined as the gross price achieved less the specified deductions, including the £10,500 already paid by Hildron in 1986). The agreement was to remain in force for 21 years, until 25 March 2007. Meanwhile, Hildron was not to grant any lease or tenancy of the flat, or otherwise dispose of any interest in the flat, except in accordance with the provisions of the deed, nor was it to use the flat, or permit its use, otherwise than as tied accommodation for a porter or other employee under the provisions of the leases. The rights of Sunley under the agreement were also to be protected by the M entry of a restriction against the registered title, and if Hildron disposed of the freehold it was obliged to ensure that a deed in similar terms was entered into between its successor in title and Sunley.

[16] It may be noted that Hildron was under no obligation to take steps to accommodate a porter outside the block, and if it chose not to do so the initial triggering event in clause 1(a) would never occur. Once the triggering event had occurred, however, the consequences were immediate and mandatory. The requisite sale had to be put in hand forthwith, Sunley had to approve the price and the proceeds of sale then had to be divided in the specified manner.

Judgment of the recorder

[17] Sunley's pleaded case, in its defence and counter-claim settled by Mr Gerard van Tonder of counsel, was to the following effect. By January 2007 at the latest, flat 9 was no longer required as accommodation for the porter or any other employee of Hildron. Hildron was accordingly obliged to comply with clause 1(a) of the porter's flat agreement and in breach of its obligation had failed to sell the flat. Sunley had thereby suffered loss and damage to be calculated as 50% of the net sale proceeds that would have been achieved on a sale of the flat. The point that it would have been prima facie unlawful for Hildron to sell the flat was obliquely recognised in para 14 of the defence, which said that Sunley relied on Article 1 of the First Protocol to the European Convention on Human Rights to avoid the so-called *Slipper* principle: see *Slipper* v *Tottenham & Hampstead Junction Railway Co* (1867) LR 4 Eq 112. The *Slipper* principle, as I understand it, is that where leasehold property is acquired under compulsory powers conferred by an Act of parliament, any requirement for the lessor's consent is taken away by operation of the statute. As amplified in his skeleton argument for the trial, counsel's contention appears to have been that the invalidation by the 1993 Act of the sale required by the porter's flat agreement would amount to an unlawful deprivation of Sunley's possessions that could not be justified in the public interest, with the consequence, so it was said, that section 19 of the 1993 Act should be construed as though it did not invalidate the grant of a long lease of the flat. I confess that I find the argument difficult to follow, and I am not surprised that it was rejected by the recorder.

[18] In his skeleton argument for the trial, Mr van Tonder seemed to have departed from his pleaded case that the damages recoverable by Sunley should be based on a retrospective valuation of the flat to be carried out by a jointly instructed expert. He submitted, rather, that the effect of section 19 was to require the flat to be sold to the nominee purchaser instead of on the open market, and that Hildron could easily perform the substance of its obligation under the porter's flat agreement by paying to Sunley 50% of the net sale proceeds received from the nominee purchaser. The sum claimed by way of damages was therefore £94,750 (that is, one-half of £200,000 after the deduction of £10,500 before division).

[19] These submissions, too, were rejected by the recorder. He pointed out that the price of £200,000 paid for the flat, following the decision of the LT, was less than the agreed open market value of the flat with vacant possession. In addition, it was paid for the freehold interest in the flat and not for a long lease, as envisaged by the porter's flat agreement. He thought that this latter point was probably of little significance, but continued:

> 26. However, Hildron has a much more formidable argument. Thus, it contends that once the s13 notice was served (on 8 November 2004) compliance with Clause 1 of the Porter's Flat Agreement became impossible, for these reasons:
>
> (1) first, s19(1)(a) of the 1993 Act prohibited Hildron from granting a long lease and would have rendered any transaction purporting to sell Flat 9 void; and
>
> (2) secondly, it would have been legally and/or practically impossible to sell a lease of the flat after a unilateral notice protecting the s13 notice was registered by the residents on 12 November 2004. Hildron, therefore, contends that the effect in law was to discharge the covenant, on the ground that performance, albeit arising after the contract was entered into, would have been illegal…
>
> 27. Sunley did not contend that there was any breach by Hildron of Clause 3 of the Porter's Flat Agreement (not to sell the Property without requiring the purchaser to enter into the same covenants); it concedes that, when the transfer of the property finally took place in early 2009, the Porter's Flat Agreement had expired by effluxion of time, on 24 March 2007, with the result that any possible obligation had lapsed; and it acknowledges that even if this agreement had not expired, Hildron would still have been prima facie relieved from all liability, as assignments made in compliance with statutory requirements cannot result in breaches of covenant…
>
> 28. However, Sunley argued further that, as a matter of construction and so as to give effect to the overage agreement, Clause 1 of the Porter's Flat Agreement should be construed as requiring Hildron "to perform the substance of its obligations" by paying 50% of what it received for the flat (less deductions), as leasehold enfranchisement (whether under the 1993 Act or otherwise) was not even on the horizon when this agreement was entered into, in March 1986.
>
> 29. The situation, as it appears to me, is clear. The application of the so-called *Slipper* principle does not arise, as the Porter's Flat Agreement had already expired when the freehold was finally transferred to Greenhill [that is, the nominee purchaser] and, as I see it, there is insufficient evidence to support the contention that Flat 9 was "no longer… required" for a porter during the 21 year life of the covenant. If I am wrong, *Slipper* is determinative of the position and there is no conceptual basis, as I see it, upon which the application of this principle can be avoided. Mr van Tonder expressly disavowed any reliance upon the existence of an independent implied term supporting the manner in which this covenant should be construed, and rightly so.
>
> 30. Mr Hornett's contention, simply put, was that Mr van Tonder was effectively inviting the Court to re-write Clause 1 of the Porter's Flat Agreement in a manner which was wholly different to anything the parties can reasonably have contemplated and that such a step would do extreme violence to elementary principles of contract law. I agree.

[20] It can be seen, therefore, that the recorder rejected Sunley's arguments based on the construction of the porter's flat agreement and held that counsel's argument amounted to an impermissible rewriting of the agreement between the parties, in a way that went far beyond anything that they could reasonably have contemplated.

[21] Rather puzzlingly, the recorder also seems to have doubted (in [29] of his judgment) whether there was enough evidence to support the contention that the initial trigger condition in clause 1(a) of the porter's flat agreement had been satisfied. Mr Alexander had given uncontradicted evidence that flat 9 ceased to be used as the porter's flat in September 2006, and (as I have already said) that it was let on an assured shorthold tenancy dated 4 September 2006. Thus, there can be no doubt, and it was common ground before me, that the initial trigger condition was indeed satisfied.

(1) Sunley's primary argument on the appeal

[22] Since the hearing below, there has been a change of counsel on Sunley's side and Sunley is now represented by Mr Alexander Hill-Smith. The primary submission that he advanced in support of the appeal is an argument of construction. He submitted, in short, that the court should seek to give effect to the underlying commercial purpose of the porter's flat agreement in circumstances that neither party could have foreseen when it was entered into. In 1986, there was no legal impediment to the marketing and granting of a new long lease if the porter's flat ceased to be used to accommodate a porter. Further, as matters then stood, that would have been the obvious way to realise the maximum capital value from a disposal of the flat and would have placed it in the same position as all the other flats in the block. There was no legislative equivalent to the 1993 Act in existence in 1986 and nobody could have foreseen the statutory scheme that parliament would subsequently enact, or the prohibitions that would be contained in section 19 of the 1993 Act. In those circumstances, said Mr Hill-Smith, the task of the court is to consider the underlying contractual objectives and to apply them in the light of the new circumstances. It is only if such an exercise is impossible that the court should hold that the original agreement has been discharged by frustration.

[23] In support of this submission, Mr Hill-Smith relied on the guidance given by the Court of Appeal on the approach to construction where there are unforeseen circumstances in *Bromarin AB* v *IMD Investments Ltd* [1999] STC 301. The leading judgment was given by Chadwick LJ, with whom Potter and Beldam LJJ agreed. The case concerned two agreements in identical terms dated 19 March 1990, by each of which the defendant, IMD, bought the shares in a company with accrued capital losses. Bromarin was one of the vendors. Under the law as it then stood, IMD would be able to make use of the losses as a deduction against future chargeable gains by transferring assets pregnant with gains to the purchased company, which could then dispose of those assets and set the accrued losses against the chargeable gain arising on the disposal. The sale price of the shares consisted of an immediate cash sum of £10,000, payable on completion, and a secondary consideration that was payable over a period of five years and that consisted (in very broad terms) of one-half of any tax saved.

A As one would expect, the detailed provisions for the ascertainment and computation of the secondary consideration were complex and involved a number of defined terms. One of those terms was "Losses", which was defined as meaning "the amount of capital losses allowable to the Company [that is, the purchased company] as at the date hereof [that is, 19 March 1990] for deduction from any chargeable gains accruing to it for the purposes of corporation tax".

[24] The effect of the contractual scheme was that, in the absence of further provision, no secondary consideration would become payable "unless (i) the purchaser chose to transfer an asset or assets into the company; (ii) the asset was subsequently disposed of by the B company, which was under the control of the purchaser; and (iii) a chargeable gain accrued on that disposal": see p307b-c. Accordingly, as Chadwick LJ pointed out, the amount and the timing of payments of secondary consideration would depend on decisions taken by the purchaser over which the vendors had no control. In order to protect the vendors against the possibility that the purchaser might choose to defer indefinitely the realisation of chargeable gains in the company, and thus the payment of secondary consideration, a further clause was introduced that imposed a new payment obligation on the expiry of five years, consisting of an amount equal to half of the corporation tax on "deemed chargeable gains". The "deemed chargeable gains" were C defined as being the difference between the company's chargeable gains accruing to it over the period of the agreement up to and including 31 March 1995 and the amount of the losses (defined as above).

[25] There would have been no difficulty in applying these provisions to the events that happened if there had not been a material and unforeseen change in the law during the five-year period. By virtue of new provisions enacted in 1993, the ability to use the pre-entry losses of a capital loss company purchased by another group was brought to an end and the relevant transitional provisions had the effect that the losses allowable to the purchased companies as at the date of the agreements in March 1990 ceased to be allowable in respect of chargeable gains D accruing after 16 March 1993. No assets had been introduced into the purchased companies before the latter date. Against that background, Bromarin claimed that it was still entitled to be paid the sale price for the shares in the company it had sold to IMD. IMD disputed liability on the basis that the losses had not been used and that there was no possibility of using them after the change in the law. At first instance, Park J decided the case in favour of IMD. Adopting a purposive construction, in reliance on the now well-known principles that had recently been enunciated by Lord Hoffmann in *Investors Compensation Scheme Ltd* v *West Bromwich Building Society (No 1)* [1998] 1 WLR 896, at pp912-913, he construed the defined term "Losses", in the context of E the provision relating to the secondary consideration, as meaning only losses that would be allowable as a deduction from chargeable gains as at 31 March 1995. He considered that the parties could not have intended the term to bear its defined meaning in circumstances that they had not foreseen.

[26] Chadwick LJ then continued, at p310e:

> The difficulty with that approach is that it is commonplace that problems of construction, in relation to commercial contracts, do arise where the circumstances which actually exist at the time when the contract falls to be construed are not circumstances which the parties foresaw at the time when they made the agreement. If the parties have foreseen the circumstances which F actually arise, they will normally, if properly advised, have included some provision which caters for them. What that provision may be will be a matter of negotiation in the light of an appreciation of the circumstances for which provision has to be made.
>
> It is not, to my mind, an appropriate approach to construction to hold that, where the parties contemplated event "A", and they did not contemplate event "B", their agreement must have been taken as applying only in event "A" and cannot apply in event "B". The task of the court is to decide, in the light of the agreement that the parties made, what they must have been taken to have intended in relation to the event, event "B", which they did not contemplate. That is, of course, an artificial exercise, because it requires there to be attributed to the parties an intention which they did not have (as a matter of fact) because they did not appreciate the problem which needed to be addressed. But it is an exercise which the courts have been willing to undertake for as long as commercial contracts have come before them for construction. It is an exercise G which requires the court to look at the whole agreement which the parties made, the words which they used and the circumstances in which they used them, and to ask what should reasonable parties be taken to have intended by the use of those words in that agreement, made in those circumstances, in relation to this event which they did not in fact foresee.

[27] Chadwick LJ went on to apply those principles and held that there was no warrant for departing from the defined meaning of "Losses". There was no basis for giving the words of the definition a different meaning, and the risk of a change in the law was to be borne by the purchaser that had control over the timing of disposals.

[28] I have referred to the factual background in *Bromarin* at some H length, partly because it is helpful to place the guidance given by Chadwick LJ in its context but also because Mr Hill-Smith sought to draw some support from a similarity between the position in that case and the facts of the present case. He pointed out, correctly, that both cases were concerned with entitlement to a secondary consideration under a contract payable after a significant period of time had elapsed from the date of the contract, and that in both cases there was a significant change in the law between the date of the contract and the relevant date on which the issue arose. That is true, but I do not think that these similarities help Mr Hill-Smith's argument. On the contrary, the actual decision of the Court of Appeal in *Bromarin* provides a J salutary reminder of the need to construe a contract in the factual matrix in which it was actually made and of the danger of making a question-begging assumption about what the parties would have intended in circumstances that they could not foresee.

[29] Mr Hill-Smith also submitted that where unforeseen events have occurred, the court may need to work creatively in order to give effect to the parties' underlying contractual intentions. He referred in this context to the observations of Sedley LJ in *Casson* v *Ostley PJ Ltd* [2001] EWCA Civ 1013; [2003] BLR 147, in [29] to [32]:

> 29. In the present case, as in most cases on the interpretation of contracts which reach this court, the words fall short of the facts. It may be axiomatic that K we are to deduce the parties' intentions from the words they have used, but the intention itself is in most such cases a fiction. Occasionally, it is true, something which has been agreed on has just been poorly expressed and can be elucidated; but far more often the parties have simply and understandably not even thought about the event which has now caused a problem. No more than a legislature can they be expected to anticipate every eventuality; but when the unexpected happens, as it regularly does, they and the courts have only the now insufficient words on the page to fall back on.
>
> 30. What is the court then to do? It may not simply make the contract which it believes the parties would have made if they had thought about the issue. It must keep in focus those agreed purposes which are evident. It must give what effect it can to the words on the page. But since, *ex hypothesi*, the words on L the page do not fit the facts, the court has to work creatively; and consistency requires it to do so by adopting and observing principles in lawyers' language, rules of construction. …
>
> 31. …
>
> 32. "Construction" has two meanings, one derived from the verb to construe, the other from the verb to construct. It may be as well to admit that under the guise of the first, the courts in cases like this are doing the second. We mitigate the uncovenanted effect of literalism not by nakedly writing a new contract for the parties but by construing the words according to principles which enable the contract, in effect, to be reconstructed. It is a very reasonable stopping place on the road that runs between second-guessing parties who have simply contracted incautiously and leaving a party at the mercy of unconsidered words. M

[30] These observations were made in the context of a case where it was necessary to construe an exemption clause in a building contract and to decide whether it excluded loss or damage from fire caused by the negligence of the defendant or its subcontractors. The court held, for a number of reasons, that the clause did not provide exemption in such circumstances. The judgments were *extempore*, and neither of the other two members of the court, Schiemann LJ and Sir Murray Stuart-Smith, expressly associated themselves with Sedley LJ's observations. Nevertheless, I find the observations helpful, and I do not think that they conflict with Chadwick LJ's description of the exercise that the court is required to perform. There is inevitably a considerable degree

of reconstruction involved in imputing an intention to the parties in respect of circumstances that they did not foresee. The point made by Chadwick LJ is that any such reconstruction has to be firmly grounded in the words that they actually used in their original factual matrix.

[31] Relying on these authorities, Mr Hill-Smith submitted that the purpose of the porter's flat agreement was to quantify the entitlement of Sunley to overage once the trigger event had occurred during the 21-year period. The agreement provided for the following steps to be taken following the trigger event:

(i) the grant of a long lease;

(ii) the marketing of the lease;

(iii) the sale of the lease at a price to be approved by Sunley; and

(iv) the division of the proceeds of sale.

The overall purpose of these steps, he submitted, was to allow the porter's flat to be realised at something approaching market value. As a consequence of the service of the initial notice, the steps could no longer take place in literal compliance with the terms of the contract. In particular, because of section 19 of the 1993 Act, it was no longer possible to grant and sell a new long lease of the flat. Nevertheless, the freehold of the flat was sold through the statutory machinery of the 1993 Act, which uses the terms "vendor" and "purchaser" in connection with the process. The grant of a new lease was not of fundamental importance to the agreement, but was rather considered by the parties, in the circumstances prevailing in 1986, to be the appropriate way of unlocking the value contained in the flat. In the changed context of the collective enfranchisement process, the equivalent unlocking was achieved by obtaining the determination of the LT of the price to be paid for the freehold interest in the flat. Accordingly, the grant of the lease should not be treated as a critical part of the contractual process.

[32] The marketing and sale of the lease, with Sunley's approval of the price, was also no longer possible once the initial notice had been registered at the Land Registry, but their equivalent, in the light of the changed circumstances, was the sale by Hildron to the participating tenants. This is the sale that has taken place. It is true that Sunley did not approve the sale, but Sunley is entitled to and does waive the requirement for its consent. Further, the price payable under the 1993 Act is based on the open market value of the flat, subject to certain specified assumptions: see para 3(1) of Schedule 6, which provides that:

> the value of the freeholder's interest in the specified premises is the amount which… that interest might be expected to realise if sold on the open market by a willing seller… on the following assumptions…

It is thus essentially equivalent to the open market sale envisaged by clause 1 of the porter's flat agreement.

[33] The relevant price was then fixed by the LT at £200,000 on the assumption that the landlord would be free to accommodate the porter elsewhere and dispose of the flat with vacant possession: see [76] of its decision, quoted in [7] above.

[34] Hildron's counter-argument, advanced as it was below by Mr Stuart Hornett, may be summarised as follows. The triggering event under the porter's flat agreement was not simply that the flat should no longer be required as accommodation for the porter. Before Sunley could become entitled to receive any payment, there had also to be a sale of a new lease pursuant to the terms of the agreement, and only then would Sunley be entitled to a sum equal to half of the net sale proceeds. Subclauses 1(a) and (d) were inextricably linked, and if the lease were not sold, Sunley could not be entitled to anything under clause 1(d). The failure to sell the flat under clause 1(a) might in theory give rise to a claim in damages, which is how Sunley's case is still pleaded. However, the complete answer to any such claim is that Hildron was prohibited, by section 19 of the 1993 Act, from complying with its obligations under clause 1(a). The court is not free to ignore the express words of the contract when ascertaining its general commercial purpose: see *Bromarin*. To construe the agreement in the radical way contended for by Sunley would be to ignore the bargain that the parties actually made and to substitute a quite different one:

(a) the disposal of the flat as part of the enfranchisement process is not a sale on the open market as required by clause 1(a);

(b) the disposal of the flat as part of a freehold disposal of the entire block is not an isolated sale of the flat on a long lease with vacant possession, as again required by clause 1(a);

(c) the valuation of the flat by the LT is not "a price equal to the highest market price reasonably obtainable" within the meaning of clause 1(a); and

(d) the valuation methodology is different.

[35] Furthermore, submitted Mr Hornett, the enfranchisement process could have led to several different outcomes in respect of the price attributable to the porter's flat. The parties might have reached agreement on the value to be attributed to it. Alternatively, Hildron might have decided not to appeal against the decision of the LVT, in which case it would have received only £50,000. Moreover, even on the appeal to the LT, the price that was determined for the flat was £100,000 less than its agreed open market value.

[36] More generally, Mr Hornett submitted that this is a case in which no underlying purpose can be discerned separate from the actual provisions agreed between the parties. Sunley has been unable to offer any precise formulation of the underlying purpose for which it contends and is merely asking the court to impute retrospectively to the parties a supposed common intention derived from the events that have actually happened. Unlike many overage agreements, the porter's flat agreement is short, simple and relatively unsophisticated. There is no proper basis for assuming that the parties had intended it to have any wider application than its terms expressly envisaged.

[37] I have not found this an easy question, and I was initially attracted by Mr Hill-Smith's skilful submissions. On reflection, however, I have come to the clear view that Sunley is asking the court to go several steps too far in its reconstruction of the porter's flat agreement. It is true that, in a general sense, the evident purpose of the agreement was to unlock and share the increased capital value of the flat if it were no longer required to accommodate a porter. However, the agreed steps to be taken in that event were both narrowly and prescriptively formulated. They were predicated on the assumption that the flat could then be sold in the open market with a new long lease in the standard form. The only sale ever envisaged by the parties was a separate, self-contained sale of the porter's flat for a price that Sunley would be able to approve. I am satisfied that a sale of the freehold interest in the flat, as part of a disposal of the entire block, was well outside the contemplation of the parties in 1986; and the further fact that such a sale was then brought about by the compulsory operation of the 1993 Act, rather than by a voluntary decision of the landlord, does not make the position any better. On the contrary, it introduces the further point that the price attributable to the flat was now something outside Sunley's control, to be ascertained by application of the criteria in Schedule 6 to the 1993 Act and not by the exposure of the flat to the real, as opposed to a notional, open market.

[38] Taken cumulatively, the points of distinction between the agreed steps set out in the porter's flat agreement and the steps that have in fact led to the receipt by Hildron of £200,000 for the freehold interest in the flat are, in my view, so many and so significant that it is impossible to impute a common intention to the parties that the porter's flat agreement should be modified so as to accommodate them. I have already noted several of the points of distinction in my summary of Mr Hornett's argument. A further one, to which he rightly drew attention, is that there would be a chronological dislocation. The agreement envisages that the open market sale of the new lease is to take place as soon as possible after the triggering event in clause 1(a). Yet the service of the initial notice that ultimately compelled Hildron to transfer the freehold took place on 8 November 2004, the best part of two years before the occurrence of the trigger event, and the valuation date adopted by the LVT and the LT was the date of service of Hildron's counternotice in January 2005, still some 20 months before the trigger event. This dislocation reflects the more general point that there is no causative connection between the vacation of the flat by the porter and

the events that led to the enfranchisement of the block and the receipt by Hildron of the £200,000.

[39] For these reasons, I would decline Sunley's invitation to construe the porter's flat agreement in such a way that it applies to the events that have happened.

(2) Other arguments

[40] Mr Hill-Smith put Sunley's case in various alternative ways, but the conclusion that I have reached on his primary construction argument means that none of his other arguments has any prospect of success. One argument was based on the well-established principle that where the machinery provided in a contract breaks down and it is not essential to the achievement of the underlying contractual purpose, the court will substitute its own machinery: see *Sudbrook Trading Ltd* v *Eggleton* [1983] 1 AC 444*. However, there is clearly no scope for the operation of this principle if I am correct in my view that the machinery laid down in the porter's flat agreement was essential to its operation. Another argument was based on the implication of a term to cater for the unforeseen circumstances that arose. Again, however, there can be no scope for the implication of a term, whether the exercise is properly to be regarded as one of contractual interpretation, as Lord Hoffmann said in *Attorney General of Belize* v *Belize Telecom Ltd* [2009] UKPC 10; [2009] 1 WLR 1988, in [19], or whether a more traditional test of necessity is applied, the need for which has been reaffirmed by the Court of Appeal, after *Belize*, in *Mediterranean Salvage & Towage Ltd* v *Seamar Trading & Commerce Inc (The Reborn)* [2009] EWCA Civ 531†, in [15], *per* Sir Anthony Clarke MR, with whose judgment Rix and Carnwath LJJ agreed. If the question is properly regarded as being one of construction, the argument can add nothing to Sunley's argument on the first question. If the test is one of necessity, it comes nowhere near being satisfied on the facts. The porter's flat agreement provided a perfectly clear and simple mechanism, which was unforeseeably invalidated by the collective enfranchisement process. If, as I have held, it is not possible to discern a broader underlying purpose, there can be no basis for implying a term to salvage the contract. The true position is simply that the contract became impossible to perform in the manner envisaged and intended by the parties.

[41] In these circumstances, I accept Hildron's submission that the porter's flat agreement was discharged by a species of frustration: see *Chitty on Contracts* (30th ed), vol I, in para 23-022 and *Baily* v *De Crespigny* (1869) LR 4 QB 180. In that case, a lessor covenanted that he and his assigns would not build on a piece of land adjoining the demised premises. When a railway company, in the exercise of powers derived from a subsequent statute, compulsorily acquired the land and built a station on it, the Court of Queen's Bench (Sir Alexander Cockburn CJ, Lush, Hannen and Hayes JJ) held that the lessor was discharged from his covenant by the subsequent Act of parliament, which put it out of his power to perform it. As Hannen J, delivering the judgment of the court, said, at pp186-187:

> The legislature by compelling him to part with his land to a railway company, whom he could not bind by any stipulation, as he could an assignee chosen by himself, has created a new kind of assign, such as was not in the contemplation of the parties when the contract was entered into. To hold the defendant responsible for the acts of such an assignee is to make an entirely new contract for the parties.

[42] It is also worth quoting a passage from earlier in Hannen J's judgment, at p185:

> There can be no doubt that a man may by an absolute contract bind himself to perform things which subsequently become impossible, or to pay damages for the non-performance, and this construction is to be put upon an unqualified undertaking, where the event which causes the impossibility was or might have been anticipated and guarded against in the contract, or where the impossibility arises from the act or default of the promissor.
>
> But where the event is of such a character that it cannot reasonably be supposed to have been in the contemplation of the contracting parties when the contract was made, they will not be held bound by general words which, though large enough to include, were not used with reference to the possibility of the particular contingency which afterwards happens.

It is accepted by Sunley that the parties could not in 1986 have foreseen the way in which the 1993 Act would subsequently operate and, in my judgment, it would plainly be unreasonable to construe clause 1(a) of the porter's flat agreement as imposing an absolute liability on Hildron even where performance became impossible. The right solution, therefore, is to treat the contract as discharged by frustration without fault on either side.

[43] It remains for me to notice briefly one further argument that Mr Hill-Smith included in his skeleton argument but did not advance with any enthusiasm at the hearing. It depended on some views expressed by Professor Sir Guenter Treitel in his book on *Frustration and Force Majeure* (2nd ed), in paras 15-007 to 15-009, where he said that the decision in *Baily* appeared to produce an unjust result because the lessor was able to retain the compensation that he received for the compulsory purchase and was not required to pay over any part of it to the tenant. Professor Treitel commented that the decision might appear to be a logical deduction from the rule that frustration operates automatically, and pointed out that in some systems of law such a result can be avoided, at least to some extent, by applying the principle of conditional discharge or surrogate benefit. He recognised, however, that English law "does not directly recognise any such principle" (see para 15-008), and went on to say (in para 15-009) that an English court could "prevent a party from so profiting only by holding that a contract has *not* been frustrated". Although this discussion is of great interest, and any views expressed by Professor Treitel deserve the utmost respect, Mr Hill-Smith realistically accepted that, sitting as a first-instance judge, I was bound to apply the general rule of English law that frustration operates automatically, and I could not properly use the threat of frustration as a way of compelling Hildron to accept altered performance of its bargain with Sunley.

Conclusion

[44] For these reasons, Sunley's appeal will be dismissed.

Appeal dismissed.

---

* Editor's note: Also reported at [1983] 1 EGLR 47

† Editor's note: Reported at [2010] 1 All ER (Comm) 1

COUNCIL TAX

R (on the application of Goremsandu) *v* Harrow London Borough Council

ADMINISTRATIVE COURT

[2010] EWHC 1873 (Admin)

26 July 2010

HH JUDGE ANTHONY THORNTON QC, sitting as a deputy High Court judge

Estates Gazette 13 November 2010

[2010] 45 EG 94

Council tax — Liability — House in multiple occupation (HMO) — Regulation 2 of Council Tax (Liability for Owners) Regulations 1992 — House let to group of tenants under series of annual shorthold furnished tenancies — Landlord storing unwanted furniture in conservatory — Tenants making separate payments in respect of rent — One tenant ceasing to be tenant but remaining in occupation — Whether property an HMO within Class C of regulation 2 — Whether landlord liable for council tax as owner of house

The appellant owned a bungalow in the respondent local authority's area. In October 1999, she let it on an annual shorthold tenancy to a group of four tenants on a furnished basis. The tenants did not want to use the furniture and the appellant agreed to store it in the conservatory. Successive tenancies were granted until January 2002, when the subsisting tenancy was surrendered and a new one granted to only three of the tenants. The fourth, P, remained in occupation until 2005 with the appellant's agreement. The tenancies specified a rent of £1,200 per month for the property, although in practice each tenant paid his share separately to the appellant; after P ceased to be a tenant, he paid a lesser amount. The last tenancy ended in December 2007.

The respondents decided that the storage of furniture in the conservatory rendered the bungalow a house in multiple occupation (HMO) during the period of the tenancies; the inhabitants had had a licence to occupy part only of the dwelling, or had not been liable to pay rent or a licence fee for the dwelling as a whole, within the definition in Class C of regulation 2 of the Council Tax (Liability for Owners) Regulations 1992. The respondents concluded that, under section 8 of the Local Government Finance Act 1992, the appellant, rather than the tenants, was liable for council tax for that period and billed her for outstanding council tax of £11,035, less sums already paid by the tenants. The appellant paid that tax.

Her appeal against the respondents' decision was dismissed by the valuation tribunal, which found that the bungalow had been an HMO by virtue of: (i) the arrangement with the conservatory; (ii) the arrangement whereby each tenant made a separate payment of rent; and (iii) the presence of P as a non-tenant. The appellant appealed.

***Held*: The appeal was allowed. (1) Under the series of shorthold tenancies throughout the relevant period, the tenants had paid rent in return for the exclusive possession of the entire bungalow, including the conservatory, and the exclusive use of all the furniture provided by the appellant. Although the storage in the conservatory of the unwanted furniture prevented access to that area, the conservatory remained part of the house demised to the tenants under their furnished tenancy and the only items stored in it were those for which the tenants were paying rent. They continued to have the right to use the furniture or to store it elsewhere and use the conservatory. If the conservatory was locked and they had no key, they could have requested a key at any time. The tribunal had erred in disregarding the terms of the lease. Since the bungalow was occupied by tenants who, by the unvaried terms of their tenancy, were entitled to occupy the whole of it, including the conservatory, it was not an HMO within the first limb of the definition in Class C. (2) The arrangements for the separate payment of rent by each tenant did not render the property an HMO within the second limb of Class C. The rent charges did not give rise to a licence whereby the tenants occupied part only of the dwelling, nor were they paid in respect of part only of it. The tenants were liable under the tenancy agreements to pay rent in respect of the house as a whole, albeit that each wrote a separate cheque for his part of the rent. They were jointly liable for the entire rent under the terms of the lease, so that if one tenant defaulted in paying his share, the appellant could have claimed that share from the others. The separate payments were an arrangement of convenience that did not affect or diminish the tenants' overall liability for the rent in respect of the house as a whole. (3) The bungalow had not been an HMO by virtue of P's continued occupation after January 2002. P lived in the bungalow as a whole under either a sublease or a licence from the three tenants and was liable to pay either rent or a licence fee in respect of the dwelling, including the conservatory. It was immaterial that this sum differed from that paid by the other three tenants. (4) Thus, the appellant was not liable for council tax during the relevant period. The respondents were obliged to repay the sums that she had paid to them, with interest at the judgment rate, and were not entitled to set those sums against future or unpaid council tax liabilities.**

No cases are referred to in this report.

This was an appeal by the appellant, Katia Goremsandu, from a decision of London (North West) Valuation Tribunal dismissing an appeal against a decision of the respondents, Harrow London Borough Council, as to the appellant's liability under the Council Tax (Liability for Owners) Regulations 1992 to pay council tax as the owner of a house in multiple occupation.

The appellant appeared in person; Adrian Davis (instructed by the legal department of Harrow London Borough Council) represented the respondents.

Giving judgment, **HH JUDGE ANTHONY THORNTON QC** said:

Introduction

[1] This judgment relates to the appeal brought by the appellant, Mrs Katia Goremsandu, against the decision of the London (North West) Valuation Tribunal (VT) dated 23 June 2009. The respondents to the appeal are the London borough of Harrow (LBH). The appeal is brought pursuant to regulation 32(1) of the Council Tax (Alteration of Lists and Appeals) Regulations 1993. This permits appeals as of right on a question of law from the VT to the Administrative Court and, on appeal, the Administrative Court may confirm, vary, set aside, revoke

or remit the decision of the VT or make any order that the VT could have made.

[2] The VT was itself hearing an appeal brought by Mrs Goremsandu from a decision of the LBH dated 13 February 2009 that had decided that Mrs Goremsandu had been liable to pay council tax on a house that was owned by her, 85A Whitchurch Lane, Edgware, Middlesex HA8 6LN (the house), from 3 April 2002 until 18 February 2008 on the basis that that property, which had been let to three tenants under a series of annual shorthold tenancies since 1 January 2002, was, during that period, a house in multiple occupation (HMO) and, as such, Mrs Goremsandu was liable to pay council tax as its owner. Until that date, LBH had treated the house as being let to the tenants under a single tenancy that covered the entire house, so that, on that basis, the tenants were liable for council tax. The tenants had paid some, but not all, of the sum chargeable for council tax in the relevant period and the effect of LBH's decision was that Mrs Goremsandu was liable to pay the balance due over the entire period, a figure of £11,035.79, less the sums previously paid by the tenants plus bailiff's costs and other disbursements. The precise sum that Mrs Goremsandu has already paid towards this HMO liability was not proved in evidence so that, if she succeeds in this appeal, the precise sum that LBH will have to repay her will need to be established or agreed.

[3] LBH raised two preliminary procedural issues that, if successful, would prevent the appeal from being brought. The first was that the appeal was brought out of time, since it was filed in the Administrative Court on 28 July 2009, 35 days after LBH received the decision of the VT on 24 June 2009, the decision being dated 23 June 2009. The relevant regulation provides that Mrs Goremsandu had 28 days in which to bring her appeal from the date she was notified of the VT's decision.

[4] There was no evidence as to when Mrs Goremsandu received the VT's decision, but since LBH had received its copy of the decision on the day following the decision being published, it is to be inferred that it was sent out to the parties by the VT by first-class post on 23 June 2009. If two clear days is allowed, being the normal period provided for by the rules of the Supreme Court between dispatch of a notice by post and its presumed receipt, Mrs Goremsandu is to be taken to have received the VT's decision on 25 June 2009 and should have filed her notice of appeal on 22 July 2009, four weeks later. Thus, the notice was filed six days late on 28 July 2009. No evidence was provided by Mrs Goremsandu to explain the delay in lodging her appeal, and until counsel representing LBH filed his skeleton around four weeks before the hearing of this appeal, no point had been taken by LBH that the appeal had been lodged out of time.

[5] I was asked to extend time for lodging the notice of appeal by Mrs Goremsandu, who represented herself at the hearing as she had before the VT. The period of the extension requested, between 22 July 2009 and 28 July 2009, is a period of four working days. LBH could point to no prejudice that they had suffered by this delay and had taken no point about it until it was referred to in their counsel's skeleton a few days before the hearing of the appeal. In fairness to LBH, counsel was not instructed to oppose an application to extend time. Since Mrs Goremsandu was representing herself and had a highly arguable question of law to put forward on this appeal, I granted her the extension of time that she sought.

[6] LBH also pointed to a failure by Mrs Goremsandu to identify the question of law arising on this appeal, as required by the Council Tax (Liability for Owners) Regulations 1992 (the 1992 Regulations). Although it is the case that Mrs Goremsandu had failed to identify with any precision the relevant questions of law in the statement that the 1992 Regulations require to be served by an appellant who is appealing a council tax issue, her appeal documentation identified what could clearly be identified as questions of law, namely the true meaning and effect of the relevant regulation, whether that regulation was correctly applied to the facts found by the VT and whether the VT had erred in law in failing to give effect to the provisions of the relevant tenancy agreements under which the tenants had occupied the house.

Relevant facts

[7] Mrs Goremsandu has owned the house for many years. It is a detached bungalow with a conservatory attached to the rear. It is located in a small garden and was let out on annual furnished shorthold tenancies by Mrs Goremsandu from 21 October 1999 until the last of these tenancies that she granted on 1 January 2007 expired on 31 December 2007. The first three tenancies were granted to four tenants, Messrs Abraham, Fletcher, Herrity and Parravani. The third of these tenancies was surrendered on 1 January 2002 and a new shorthold tenancy was granted to Messrs Abraham, Fletcher and Herrity. When the group of four tenants moved into the house, they did not want to use the furniture that was let to them with the furnished shorthold tenancy and they agreed with Mrs Goremsandu that this furniture would be stored in the conservatory, which they also did not want to use. The glass door at the rear of the bungalow was locked, as was the door from the conservatory into the garden, solely to ensure the security of the furniture. This arrangement continued from 21 October 1999 until 31 December 2001 and from 1 January 2002 without interruption until the three remaining tenants moved out on around 1 February 2008, having occupied the house pursuant to six successive shorthold tenancies in identical terms. The last of these tenancies expired on 31 December 2007, and the tenants had then held over for a short period before vacating the house.

[8] Mrs Goremsandu had built a new two-room extension at the back of the bungalow that connected with the back of the conservatory. This extension was completed at around the time that the three tenants moved out and Mrs Goremsandu and her son then moved into the bungalow with its extension. It would appear that Mrs Goremsandu took possession of the rear extension on around 15 January 2008, half-way through the four-week period during which the three tenants held over by agreement. Their last tenancy agreement came to an end on 31 December 2007 and, with the agreement of Mrs Goremsandu, they held over until 1 February 2008. The VT made no finding as to whether the rear extension was first occupied by Mrs Goremsandu before the three tenants moved out or soon afterwards. However, if Mrs Goremsandu began to occupy the extension prior to the three tenants moving out, the longest period that the house was occupied by the three tenants in the bungalow simultaneously with the rear extension being occupied by Mrs Goremsandu was 14 days at the very end of the three tenants' period of occupancy of the house.

[9] Throughout their occupation of the house from 1999, the four or three tenants were billed for council tax by LBH and such payments as they made were paid by them to LBH. It was only after the three tenants had left that LBH learnt of the arrangement whereby Mrs Goremsandu's furniture was stored in the locked conservatory and, based on that knowledge and statements from two of the tenants, LBH decided that the house had been occupied under an HMO arrangement since 1 April 2002. This, if correct, had the effect of making Mrs Goremsandu liable for council tax rather than the three tenants, who had previously been sent demands for the entire council tax liability relating to the occupancy of the house.

Applicable law

[10] The liability for the payment of council tax is imposed on the resident or owner in a hierarchy of liability unless the property is an HMO: section 6 of the Local Government Finance Act 1992. If it is an HMO, the residence requirements are displaced by the provisions of regulation 2 of the 1992 Regulations as amended by SI 1993/151 and SI 1995/620. These impose liability on the owner of the dwelling. The material provisions of this collection of SIs is as follows:

> The following are the classes of chargeable dwellings prescribed for the purposes of section 8(1) of the [LGFA]
>
> Houses in multiple occupation, etc
>
> Class C a dwelling which
>
> (a) was originally constructed or subsequently adapted for occupation by persons who do not constitute a single household; or
>
> (b) is inhabited by a person who, or by two or more persons each of whom either
>
> (i) is a tenant of, or has a licence to occupy, part only of the dwelling; or

A (ii) has a licence to occupy, but is not liable (whether alone or jointly with other persons) to pay rent or a licence fee in respect of the dwelling as a whole.

Decision of the VT

[11] The decision of the VT contained the following relevant findings of fact.

[12] "(1) … the 30 March 2004 letter from Mr Parravani, in addition to the letter from the appellant, dated 18 December 2005, indicate that each tenant was responsible for a separate rent charge in respect of the appeal dwelling."

[13] The first letter referred to was written by Mr Parravani to LBH B in connection with the payment of housing benefit in a letter written from 85A Whitchurch Lane. This states that he wanted to appeal against a recently imposed housing benefit restriction of £61.68 per week given that he was paying £69.24 per week, which he considered to be a reasonable rent and very good value for such a bungalow as he was then occupying. Mr Parravani had ceased to be a tenant of Mrs Goremsandu on 31 December 2001, but there was evidence that he remained in occupation of the house in a different capacity until some time in 2005, when he vacated the house. This was provided by Mr Parravani in this letter when giving details of the bungalow that he was occupying. These details were of the bungalow as a whole. Thus, C this evidence suggests that Mr Parravani was a subtenant or licensee of the three tenants following the change of tenants from the group of four to the group of three, and that his changed status had been accepted as such by Mrs Goremsandu. His subtenancy or licence clearly extended to the whole of the house, and any payment that he made as rent was made in reduction of the three tenants' liability to pay Mrs Goremsandu £1,200 per month. Thus, his tenancy or licence did not relate to part only of the bungalow and his rent payment to the three tenants, or to Mrs Goremsandu on their behalf, clearly related to the house as a whole. It follows that this letter is not evidence of the house being an HMO at any time during which Mr Parravani occupied it after 31 December 2001.

D [14] The second letter referred to was written by Mrs Goremsandu and was addressed to "Whom it may concern", and a copy of this letter was sent to LBH on 18 December 2005. It read:

> This letter is to confirm that Peter Fletcher remains a tenant at the below address and continues to pay £300 rent per calendar month to myself. There has been no break in the tenancy since 1999 including from 3 September 2005 until the present date. Other permanent tenants include Kieran Herrity and Abe Abraham who still live at the property and have done so since 1999. Gary Parravani left the property on 10 October 2005.

E [15] That letter confirms that Messrs Fletcher, Herrity and Abraham were tenants pursuant to a tenancy that had been in existence since 1999. They were, in other words, joint tenants under a single tenancy agreement. As is common with a group of single joint tenants, each appears to have entered into an arrangement with Mrs Goremsandu, the landlord, to pay her directly his "share" of the rent. However, the formal tenancy was to be found in the tenancy agreement, which provided for a single rent per month for the bungalow of £1,200, for which each of the three joint tenants was jointly and severally liable. Thus, the payment of £300 per month referred to was a payment that related to Mr Fletcher's entitlement to occupy the entire bungalow.

F [16] The conclusion to be drawn is that each tenant was responsible, at least in the first instance, for a separate rent charge in respect of the appeal house, but, given the definition of an HMO, that did not make the house a multiple occupation dwelling since the rent charge related to the tenancy and joint occupation of the whole of the bungalow, including the conservatory.

[17] "(2) the plans of the appeal dwelling indicate that both the conservatory and subsequent extension of the house were capable of occupation along with the rest of the house. However, the Tribunal notes that the appellant restricted access to this part of the dwelling, that she did not allow the tenants to use the extension area and the conservatory area before that was constructed as part of their occupation of the subject property."

[18] This finding relates to the conservatory and the rear extension. G It is a confused finding because the first sentence refers to the "subsequent extension of the house", whereas the second sentence refers to the appellant restricting access to use the "extension area" and the "conservatory area before that was constructed". It is clear, however, that the VT was referring to the restriction on the tenants' access to the conservatory throughout their tenancy that the finding states was imposed by Mrs Goremsandu and to the additional restriction on their access to the extension once it had been constructed.

[19] It also appears from this finding that the VT was erroneously concluding that Mrs Goremsandu had restricted access to the extension area for an appreciable period of time and that both areas had been H used by Mrs Goremsandu to store her furniture. It is clear from Mrs Goremsandu's evidence to the VT, supported by the evidence of each of the three tenants, that the construction of the extension was completed only when, or only a few days before, the three tenants vacated the bungalow on 1 February 2008. Mrs Goremsandu's evidence was recorded as part of LBH's case, which repeated the information that she had given to LBH on 20 February 2008 to the effect that she had completed a two-room rear extension on 15 January 2008. This date was accepted by LBH at the hearing of this appeal as being the date on which the extension was completed. The error made by the VT arose because the presenting officer, in opening LBH's case, stated J that the completion date had been 15 January 2002, repeating an error made in a customer counter receipt by the note-taker when noting a discussion between Mrs Goremsandu and an enquiry counter-clerk in February 2008. Thus, there is only minimal evidential value to be drawn from the fact that Mrs Goremsandu first occupied the newly completed extension on 15 January 2008 since the tenants moved out on 1 February 2008.

[20] The crucial finding was that Mrs Goremsandu restricted access to and did not allow the tenants to use the conservatory. The VT then found that, in consequence, the tenants had a licence to occupy only part of the dwelling. This finding was based on the evidence of the three K tenants set out in written form. Although this was not expressly referred to, the VT must have rejected the evidence of Mrs Goremsandu. The decision recorded her evidence as being that the conservatory area was used for the storage of her property. It was not a living space but merely contained decorative plants. Mrs Goremsandu also gave evidence that the three tenants were occupying the property pursuant to the series of tenancy agreements and that they were the named tenants in those agreements. The agreements covered the entire bungalow, including the conservatory.

[21] The three tenants' evidence may be summarised as follows:

L (a) Mr Herrity: Mr Herrity, in a letter to LBH dated 20 February 2009, stated that between January 2002 and the middle of 2007 he, Mr Fletcher and Mr Abraham had a joint tenancy agreement as one household. He personally collected the money to pay the bills, including council tax, and paid that tax to LBH on time. He also stated:

> I also wish to mention that we had access to all parts of the house including the back room which was used as a storage area. We could access the storage room via all internal doors which were unlocked.

The VT dismissed this evidence because Mr Herrity, in using the expression "back room" did not detail specifically whether this was M the conservatory area or some other part of the property. However, this is a wholly unreasonable conclusion to draw from the use of that expression. Mr Herrity had occupied the bungalow for many years, during which time there was no extension attached to it because this had not been built during his period of occupancy. The bungalow, as is clear from other evidence before the VT, comprised four bedrooms, a lounge, a large kitchen and two toilets, with showers and a sauna. The conservatory was attached to the back of the bungalow and was not a living area but was used for plants. This is the only possible area that Mr Herrity could be referring to by his expression "back room", as is confirmed by the layout of the bungalow and his subsequent reference to "the storage room". It follows that this evidence, from the tenant who appears to have been responsible for the administration of the tenants' tenancy obligations, provides clear and firm support for Mrs Goremsandu's evidence.

(b) Mr Abraham: Mr Abraham, as recorded in a customer counter receipt dated 14 July 2008, stated that the conservatory was not accessible to the tenants because furniture was stored there and the conservatory was locked.

(c) Mr Fletcher: Mr Fletcher, as recorded in a note of a telephone conversation with him, stated that the conservatory was used only to store the landlord's belongings. It remained locked throughout the tenancy and the tenants had no access. He confirmed this evidence in answers given to a questionnaire e-mailed to him by LBH on 13 May 2009, save that he added that the tenants had no access to the key to the conservatory.

[22] This potentially conflicting evidence had to be considered along with the terms of the leases. These were in common form and were drafted as shorthold tenancies. The landlord was stated to be Mrs Katia Goremsandu and the tenants Mr Herrity, Mr Abraham and Mr Fletcher. The house that was let was stated to be "85A Whitchurch Lane, Edgware, Middlesex, HA8 6LN". The rent was stated to be "£1,200 per month renewable". The tenancy was a furnished tenancy and the agreement provided that the landlord agreed to let and the tenant agreed to take the property and contents for the term and at the rent payable. The tenants were jointly liable for the entire rent, so that if any tenant defaulted in paying Mrs Goremsandu his share of the overall rent she could have claimed that part of the rent from either or both of the other tenants. The tenants were also required to leave the contents at the end of the tenancy in approximately the same places in which they were positioned at its commencement. These contents were those listed in the inventory. These terms were dismissed by the VT as providing any indication of who occupied the property. The decision recorded its view that these terms did not, in this instance, provide definitive evidence of whether the subject property fell within the definition of an HMO set out in the relevant legislation. However, no explanation was given as to why the terms of the lease could be disregarded in this way and why it was open to the VT to find that the tenants had only a licence to occupy part of the building.

[23] The finding also recorded that Mrs Goremsandu restricted the tenants from using the conservatory. This finding is clearly correct in respect of the factual position, whereby the tenants could not use the conservatory while it remained clogged with the furniture that they had rented but did not want to use and while it remained locked with no access to the key. However, there was no evidence that Mrs Goremsandu had imposed a variation of the terms of the tenancy agreements on the tenants or that she had unilaterally torn up those agreements and imposed on the tenants a licence in different terms. Thus, the factual description of the state of the conservatory had to be considered against the terms of the tenancy agreements. It was an error for the VT to find, without explanation or evidence, that the storage of furniture in the conservatory had created a situation in which the tenancy agreements were discarded or rendered of no effect as a result or that these agreements had been replaced by licences to occupy the house less the conservatory.

[24] "(3) the tenants were liable to pay an individual rent charge for their occupation of the subject dwelling."

[25] The three tenants were subject to a contractual liability to pay £1,200 per month to Mrs Goremsandu. Each had agreed with the others that they would pay a proportion of that sum, and their respective sums were collected by Mr Herrity and forwarded by him to her. Mr Parravani's rent payment was presumably paid in this way until he ceased to be a tenant, whereupon he continued to occupy the entire house paying the same weekly or monthly figure in the same way as before. The only difference was that he was no longer a tenant of Mrs Goremsandu and was no longer directly liable to her for his payments since he no longer had a contractual relationship with her. Out of convenience, and presumably in agreement with the three tenants, he paid Mrs Goremsandu his "rent", presumably through Mr Herrity. This rent was the sum that he actually owed to the three tenants for his continued shared occupation of the entire house.

[26] It follows that the individual rent charge paid each month by each tenant did not alter each tenant's status in the house. Each occupied the entire house, as did Mr Parravani until he moved out of the house some time in 2005.

Discussion

[27] Notwithstanding the VT's conclusion that the three tenants were not tenants but were licensees of part of the house, the starting point in considering whether the house was an HMO during the period 3 April 2002 to 18 February 2008 must be the series of shorthold tenancies in existence during all but the last six weeks of this period. The parties successively entered into six of these tenancies in identical terms in that period, they paid what they described as rent to Mrs Goremsandu and, in return for that rent, they were provided with exclusive occupancy of the entire bungalow including the conservatory, and to exclusive use of all the furniture provided by Mrs Goremsandu as described in the inventory. In fact, this arrangement had started in October 1999, and all that had happened in January 2002 was that the original group of four tenants was replaced by the existing group less Mr Parravani, who continued to occupy the entire bungalow and to pay "rent" as the subtenant or licensee of the three remaining tenants.

[28] The evidence presented to the VT shows that the furniture in the conservatory consisted of furniture provided at the outset of the first tenancy by Mrs Goremsandu to the tenants as part of their furnished tenancy. This furniture was of course owned by her. The furniture was not needed by the tenants so it was stored in the conservatory, which remained part of the house that had been demised to the tenants under the terms of their lease. The lease gave the tenants exclusive possession of the entire house, including the conservatory. However, so much furniture was stored in the conservatory that no one, including the tenants, could gain access into that area. The door into the conservatory was or may have been locked and the key was or may have been inaccessible the evidence is contradictory on this factual detail. However, the furniture was still being hired by the tenants as part of their furnished tenancy, they continued to have the right to use it or, if they could find other storage facilities for it, to store it elsewhere and use the conservatory as a conservatory rather than as a storage area. If the conservatory was locked and they had no access to the key, they could at any time have called for the key in order to make use of any of the furniture or to relocate the furniture and use the conservatory.

[29] This conclusion arising from an analysis of the facts as found in conjunction with the terms of the lease is confirmed by Mr Herrity's evidence and it is consistent with a careful reading of the evidence of both Mr Abraham and Mr Fletcher. They stated that the conservatory was inaccessible and that it was full of Mrs Goremsandu's furniture, both of which statements are factually correct. They also stated that the conservatory door was locked with no apparent access to the key, which is possible. However, they were not asked, and expressed no opinion, whether they were still paying rent for the furniture and whether they could have insisted on using it and, subject to clearing it, of using the conservatory. They never addressed the relationship between the demise and rent provided for by the tenancy agreements and their rights over the furniture and the conservatory since they were never asked about these matters.

[30] These matters must, however, be considered as part of any consideration of the statutory provisions defining an HMO. The bungalow could have been an HMO during the relevant period only if one or other of these two statutory definitions applied:

(1) the bungalow was inhabited by a person who, or by two or more persons each of whom, was a tenant of or had a licence to occupy part only of the dwelling; or

(2) the bungalow was inhabited by a person who had a licence to occupy it, but who was not liable (whether alone or jointly with other persons) to pay rent or a licence fee in respect of the dwelling as a whole.

[31] The bungalow, including the conservatory, was occupied by the three tenants, who, by the unvaried terms of their tenancy, were entitled to occupy the entire house, including the conservatory. Thus, the first

limb is not satisfied. The three tenants were also liable to pay rent in respect of the house as a whole, albeit that each paid a separate cheque for part of the rent. This was an arrangement of convenience that did not affect or diminish their overall liability for the rent in respect of the house as a whole. Thus, the second limb is not satisfied.

[32] The VT found that Mr Parravani's occupancy created an HMO. Between 1 January 2003 and sometime in 2005, he was occupying the bungalow, but not as a joint tenant of Mrs Goremsandu. He was however: see his letter to LBH dated 30 March 2004:

> currently paying £69.85 rent… for the bungalow and believe that this not only to be a reasonable rent but very good value for such a bungalow. … I enclose some details of the house…

Details of the bungalow are provided by him in the letter. Although these do not include the word "conservatory", the details he provided were not intended to be definitive (the letter says "some details" were enclosed). These details were, however, intended to detail the premises he was occupying, and it is clear from that description that he was occupying the entire house and garden, including the conservatory since it was an integral part of the house.

[33] Thus, Mr Parravani was "inhabiting" the entire bungalow under either a sublease or a licence and was liable to pay either rent or a licence fee in respect of the dwelling as a whole. The fact that this was a different sum to that paid by any of the three tenants is immaterial. On his own admission, Mr Parravani was paying £69.85 in rent (or a licence fee) to enable him to occupy the entire house, including the conservatory.

Error of law by the VT

[34] The VT had erred in law in not giving effect to the terms of the tenants' shorthold tenancies and in not applying the statutory definition of an HMO despite setting it out in its decision. In particular, it considered that individual rent charges gave rise to "multiple occupation", which is the wrong test to apply. The statutory test that it should have applied is whether the rent charges gave rise to a licence, whereby they occupied only part only of the dwelling, or whether they were paid in respect of part only of the dwelling. On analysis, the individual rent charges were all paid to allow each tenant to occupy the entire house, as provided for in the tenancy agreement, and each payment related to the occupation of the house as a whole.

[35] The VT also considered that the fact that Mrs Goremsandu's furniture was kept locked in the conservatory meant that the tenants did not occupy the conservatory. That factual finding was not sufficient because the statutory test that had to be applied was concerned with whether the tenant was a "tenant of part only of the dwelling". The tenants remained tenants of the conservatory even if factually they were unable to use or readily gain access to it. The only items of furniture that were stored in the conservatory were items for which the tenants were paying rent, and since the terms of the tenancy had not been varied, they were in law entitled to call for the key at any time and, if they chose to do so, to exclude Mrs Goremsandu from the conservatory and also to exclude her furniture from that area so long as they stored or used the furniture in another location. They were therefore tenants of the whole dwelling, which included the conservatory.

Conclusion

[36] The appeal will be allowed. Mrs Goremsandu is entitled to the following declarations:

(1) 85A Whitchurch Lane, HA8 6LN was not an HMO between 3 April 2002 until 18 February 2008 for the purposes of regulation 2 of the 1992 Regulations; and

(2) Mrs Goremsandu is not liable for council tax for that house between those dates and is entitled to be repaid by LBH all sums paid by her for or on account of such council tax liability and any bailiff's fees or other charges or disbursements referable to such council tax with interest at 6% from the date(s) of payment until repayment.

[37] LBH did not establish that the house was an HMO from 15 January 2008 to 1 February 2008, the period of a potential overlap between Mrs Goresmandu's occupation of the new extension and the departure of the three tenants. The evidence was equally consistent with Mrs Goremsandu moving her belongings into the new extension in that period and not actually occupying it. Thus, this period is to be treated in the same way as the other periods in question. Since LBH have wrongfully claimed and have been wrongfully paid council tax payments and related charges, it must repay the sums wrongfully paid and may not simply set those sums against future or unpaid council tax liabilities. LBH should also pay interest at the judgment rate of 6% from the dates of the individual wrongfully demanded payments until the date of repayment.

[38] The parties should attempt to agree the terms of the order to give effect to this judgment. Mrs Goremsandu is entitled to the relatively small sum in costs that she is entitled to as a successful litigant in person, which should be agreed or summarily assessed when the judgment is handed down in open court. If the terms of the order can be agreed, the attendance of the parties at the handing down of this judgment is excused unless either party wants to attend. Attendance will be necessary if the terms of the order cannot be agreed.

Appeal allowed.

EASEMENTS

A
HXRUK II (CHC) Ltd *v* Heaney

CHANCERY DIVISION: LEEDS DISTRICT REGISTRY

9 September 2010

HH JUDGE PETER LANGAN QC, sitting as a deputy High Court judge

Estates Gazette 6 November 2010

[2010] 44 EG 126

Easements — Right of light — Remedies — Mandatory injunction or damages — Extension to commercial building interfering with
B ***right of light enjoyed by neighbouring listed building — Whether small award of damages sufficient to compensate for infringement of right — Whether oppressive to grant mandatory injunction requiring removal of offending parts of extension***

In 2007, the claimant purchased a five-storey commercial property in the centre of Leeds for £18.75m. It had the benefit of a planning permission for redevelopment and extension, including the construction of two additional floors at the sixth- and seventh-floor levels. The property was subject to an easement of light in favour of the defendant's nearby office building, a Grade II-listed Victorian property that the defendant had restored. The price paid by the claimant included a
C negotiated reduction of £350,000 to allow for claims by the defendant for interference with his right of light.

Negotiations and correspondence between the parties did not produce an agreed resolution of the light issue. In September 2008, works were begun on erecting the steel frame for the sixth and seventh floors in the claimant's building. In November 2008, the defendant threatened injunction proceedings but these did not materialise. The claimant's works were completed by July 2009.

In August 2009, the claimant brought proceedings for declaratory relief, seeking a ruling that a small award of damages was an appropriate remedy for the admitted actionable interference with
D the defendant's right of light. The defendant counterclaimed for a mandatory injunction requiring the claimant to demolish the offending parts of the development. The claimant contended that the injury to the defendant's rights could be compensated by a money award and that a mandatory injunction would be oppressive, both by reason of the injurious effect it would have on the claimant and by reason of the defendant's conduct, including his failure to issue a claim form, such that the claimant had been obliged to do so in order to introduce some certainty. It relied on the fact that the loss of adequately lit space in the defendant's building had been calculated at less than 1% of its net lettable area and the conventional "book value" damages for the loss
E of light would be £80,000, only 2% of the £4m value of the building.

***Held*: The claim was dismissed; the counterclaim was allowed. An actionable interference with an easement may be compensated by an award of damages, rather than a mandatory injunction, only if: (i) the injury to the dominant owner's rights is small; (ii) it is capable of being estimated in money; (iii) it can adequately be compensated by a small money payment; and (iv) it would be oppressive to the servient owner to grant an injunction. In order to avoid an injunction, the servient owner must show that all four criteria are satisfied. The injury to the defendant's rights lay close to the margin but could not be characterised as small for the purpose of the first criterion. The figures advanced by the claimant had to**
be taken into account, as did the fact that good light was relatively F
more important in a person's home than in industrial or commercial premises. However, other relevant factors included the character of the defendant's building, the commitment and investment that he had put into its restoration and the fact that he operated several businesses within it. Taking those matters into account, as a matter of impression as much as of analysis, the reduced flow of light constituted damage of a kind for which the defendant should not have to accept a money payment. The claimant's infringement had been neither inadvertent nor necessary. It had been committed with a view to profit; the claimant could easily, although less profitably,
have reduced the dimensions of the sixth and seventh floors. The G
court should not sanction what had been done by compelling the defendant to take monetary compensation, which he did not want. Accordingly, a mandatory injunction should be granted.

***Per curiam*: Where an award of damages is appropriate, this may take the form either of "book value" damages for nuisance at common law or of an award of damages in lieu of an injunction under section 50 of the Senior Courts Act 1981, based on the figure that the party with the right could fairly have demanded, in a hypothetical negotiation between the parties, for relaxing that right.**

The following cases are referred to in this report. H

AMEC Developments Ltd v *Jury's Hotel Management (UK) Ltd* (2001) 82 P&CR 22; [2001] 1 EGLR 81; [2001] 07 EG 163

Carr-Saunders v *Dick McNeil Associates Ltd* [1986] 1 WLR 922; [1986] 2 All ER 888; (1986) 53 P&CR 14; [1986] 2 EGLR 181; 279 EG 1359

Gafford v *Graham* (1999) 77 P&CR 73; [1999] 3 EGLR 75; [1999] 41 EG 159, CA

Jacklin v *Chief Constable of West Yorkshire* [2007] EWCA Civ 181

Jaggard v *Sawyer* [1995] 1 WLR 269; [1995] 2 All ER 189; [1995] 1 EGLR 146; [1995] 13 EG 132, CA

Pell Frischmann Engineering Ltd v *Bow Valley Iran Ltd* [2009] UKPC 45; [2010] BLR 73

Shelfer v *City of London Electric Lighting Co Ltd (No 1)*; *Meux's Brewery Co*
v *City of London Electric Lighting Co* [1895] 1 Ch 287, CA J

Regan v *Paul Properties DPF No 1 Ltd* [2006] EWCA Civ 1391; [2007] Ch 135; [2006] 3 WLR 1131; [2006] 3 EGLR 94; [2006] 46 EG 210

Tamares (Vincent Square) Ltd v *Fairpoint Properties (Vincent Square) Ltd* [2007] EWHC 212 (Ch); [2007] 1 WLR 2167; [2007] 1 EGLR 26; [2007] 14 EG 106

Wrotham Park Estate Co Ltd v *Parkside Homes Ltd* [1974] 1 WLR 798; [1974] 2 All ER 321; (1973) 27 P&CR 296; 229 EG 617, Ch

This was the hearing of a claim by the claimant, HXRUK II (CHC) Ltd, for declaratory relief as to the appropriateness of damages as a remedy for an infringement of a right of light; and a counterclaim by
the defendant, Marcus Heaney, for a mandatory injunction requiring the K
removal of structures that infringed that right.

Andrew Francis (instructed by Shoosmiths) appeared for the claimant; Stephen Bickford-Smith (instructed by Pinsent Masons LLP, of Leeds) represented the defendant.

Giving judgment, **HH JUDGE PETER LANGAN QC** said:

Introduction

[1] This action is concerned with rights of light. Unusually, the claimant is the servient not the dominant owner. The claimant brought the proceedings with a view to obtaining declarations as to its freedom

from liability to the defendant, while the substantive relief in the form of an injunction or damages is asked for in the counterclaim.

[2] The following matters are not in dispute. (1) An easement of light exists in favour of windows in the defendant's building, the former head office of the Yorkshire Penny Bank. (2) The claimant, by the addition of two floors to its nearby building, formerly called block A, Cloth Hall Court and now called Toronto Square, has committed an actionable interference with the access of light to the defendant's building. (3) The defendant is entitled to a remedy in respect of that interference. As to (3), until a late stage in the proceedings the claimant maintained that the defendant's conduct had been such as to debar him from obtaining any relief from the court, but that contention was not pursued in closing submissions.

[3] The litigation has thus become focused on the question of remedy. There are two areas in which the court has to come to a decision. (1) The defendant seeks a mandatory injunction requiring the claimant to cut back the offending works. The claimant says that the appropriate remedy is an award of damages and the matters that would have been raised in support of the "no remedy" case remain pertinent to this question of injunction or damages. (2) If there is to be an award of damages, there is a dispute as to how the damages are to be quantified. These two issues are interrelated because the amount of any damages that might be awarded is one of the considerations that will be relevant to the question of whether an injunction should be granted.

[4] The leading textbook on the subject is Bickford-Smith and Francis, *Rights of Light: The Modern Law* (2nd ed), 2007. I have been fortunate to have had the assistance of the co-authors as counsel in this case, Mr Andrew Francis, for the claimant, and Mr Stephen Bickford-Smith, for the defendant. I am grateful to them both for their exceptionally helpful written and oral submissions.

Narrative

[5] The buildings with which I am concerned are situated in Infirmary Street, which is in the centre of Leeds. They are familiar to many people, myself included, who commute to work by bus. As one stands at the bus stops on the north side of Infirmary Street, one can see on the other side of the street: (i) to one's right, with a frontage to Infirmary Street, the old bank building; (ii) to one's left, the claimant's building; and (iii) between the two properties, Toronto Street, a private road and pavement with (outside the claimant's building) four mature trees. The light with which this case is concerned comes into windows on the left-hand elevation of the bank building that faces the claimant's building across Toronto Street.

[6] The defendant purchased the bank building, the address of which is 2 Infirmary Street, in November 2003. In his witness statement, the defendant described his property as "a striking Victorian building… grade II listed… featuring impressive stone carvings and window apertures along with a very striking turret". The defendant went on to say:

> When I purchased the building it was part let as offices with a vacant banking hall and [was] in need of restoration. Over the years I have spent approximately 3 million pounds restoring the building to its former glory, re-instating the original detail, re-letting the impressive revamped offices, and created a conference and banqueting venue, with what will be residential accommodation on the top floor.
>
> 2 Infirmary Street contains five floors of mixed use accommodation. The total net internal areas of the building are approximately 32,000 square feet. There are various office tenants with a range of lease terms, with the predominant occupier being Aspire, which is my conference and banqueting company.

[7] Cloth Hall Court was constructed as a complex of four office buildings in or around 1980. The building with which I am concerned was known as block A. It was constructed in an L-shape, with the long arm of the L facing onto Infirmary Street and the short arm facing onto Toronto Street, opposite the side elevation of the defendant's building. As it faced the defendant's building, block A comprised a ground floor and five upper floors, with the fifth and top floor having a mansard roof.

[8] In March 2007, the then owners of block A, Cloth Hall Realty Ltd (CHRL), obtained planning permission from Leeds City Council for the redevelopment of block A. The main elements of the redevelopment were to be: (i) the refurbishment of the accommodation up to fourth-floor level; (ii) the demolition and reconstruction of the fifth floor; (iii) the construction of sixth and seventh floors; (iv) the addition of an extension, to be integrated with the remainder of the building, within the elbow of the L, and running from ground- to seventh-floor levels.

[9] Highcross Strategic Advisers Ltd (HSAL) is a company that, on behalf of a fund known as Highcross Fund 2, concentrates its interest on the industrial and office sectors of the property market throughout the UK. HSAL, according to the evidence of Mr Christopher Mills, the development director, seeks "to purchase commercial properties that have the potential to offer significant added value through active management or development". The claimant is a Jersey-registered company, which is part of Highcross Fund 2, and was the vehicle used for the purchase of block A. That property is the claimant's only asset.

[10] In May 2007, CB Richard Ellis (CBRE), a firm of commercial property agents based in Leeds, brought Cloth Hall Court to the attention of HSAL as a potential investment. It seems that the idea must have been immediately attractive to HSAL because due diligence got under way quickly and solicitors were instructed as early as 7 June 2007. At first, the proposal was that both block A and block C should be acquired, but at some stage during the second part of 2007 block C dropped out of the picture and HSAL's interest was thereafter limited to block A.

[11] Mr David Parratt is a well-known surveyor who specialises in light cases. On 28 September 2007, he wrote a report on "Right of Light Problems relating to Building A". The report was written on the instructions of GSD Ltd, which was acting as agent for CHRL, but I think that it must have come fairly quickly to the attention of HSAL. Mr Parratt has in fact acted as an expert witness for the claimant in this action. His conclusions in his report of 28 September 2007 were: (i) the redevelopment of block A would not cause any actionable interference with the access of light to the Old Post Office, a side elevation of which is on Infirmary Street, but the owners of the building should be formally notified of the proposals, so that if they had a different view, they could put it forward; (ii) he gave the same advice with regard to block B, Cloth Hall Court; (iii) the Yorkshire Penny Bank building would suffer an actionable loss of light and the owner should be approached with a view to agreeing either that there would be no claim or to a calculation of the losses.

[12] On 1 October 2007, Mr Martin Dalley, of GSD Ltd, wrote to the defendant:

> Block A Cloth Hall Court, Infirmary Street, Leeds
>
> Additional two Storeys and Refurbishment Work
>
> I act on behalf of Cloth Hall Realty, the owners of the above building and I understand you own the Yorkshire Bank Chambers opposite. Having received planning permission for the extension and refurbishment of the building, my clients are currently finalising tender costs and are hoping to start work in the near future. In preparation for start of the re-development, my clients have asked that I write to you to consult over any potential rights of light issues.
>
> I enclose a copy of the planning documentation for which we have received planning permission, which shows you artistic images of the re-development together with floor plans, sections and elevations. I believe this should give you an understanding of what is proposed by Cloth Hall Realty Limited.
>
> There may not necessarily be any effect on light levels, but my clients are keen to raise this matter at an early stage and wish to deal with any claim in a reasonable manner.
>
> If you feel that the proposed alterations may cause a significant loss of daylight, please could you let us have drawings which show floor plans and (if you have them) sections and the elevation which faces Cloth Hall Court. In due course it may also be necessary for my client's right to light surveyor to enter the premises to take some measurements and I would be grateful if you could confirm whether this would be possible.
>
> If there are any matters you wish to discuss please contact me.

A [13] The defendant passed this letter to Mr Raymond Rowan, a commercial property consultant who had acted for the defendant for some years.

[14] On 10 December 2007, the board of the claimant approved a proposal to purchase block A from CHRL for £18.75m. The price, originally agreed on a subject-to-contract basis, had been £350,000 more than this sum, but had been negotiated downwards in order to allow for rights of light claims.

[15] Contracts were exchanged between CHRL and the claimant on 11 December 2007. The purchase was completed on 20 December 2007.

B [16] Between exchange of contracts and completion on 13 December 2007, Mr Rowan replied to Mr Dalley's letter of 1 October 2007. I should say that there had been some informal contact between the parties' representatives between those dates. Mr Rowan said:

> We have concerns, maybe significant concerns, regarding the potential loss of light at Yorkshire Bank Chambers and it may well be my client will be appointing a right of light specialist to investigate and act on his behalf. However, before taking this matter any further we require confirmation your clients will be responsible for all the fees and costs of the specialists and for my own fees and costs, whatever the outcome.

C [17] Mr Dalley replied on 25 January 2008. He informed Mr Rowan of the change of ownership of block A and he told Mr Rowan that both he (Mr Dalley) and Mr Parratt were now acting for the claimant. Mr Dalley concluded his letter:

> Highcross' lawyers, Shoosmiths, are shortly to issue formal letters regarding the right to light matter to all to all the adjoining affected properties to confirm the change of ownership and David Parratt's assessment of the situation regarding loss of light to the Penny Bank will be included within this correspondence. In the meantime if there are matters you wish to discuss, please contact me.

[18] During January 2008, officers of the claimant prepared for the board a paper seeking board approval to carry out the redevelopment of D block A. The overall project cost was put at £9.78m, of which £200,000 was "contingency sums for rights of light issues". Board approval was duly given.

[19] The main contractor engaged by the claimant was ISG Regions Ltd, and the first letter of intent was issued to that company in respect of preparatory works on 31 January 2008. ISG Regions Ltd went into occupation of the site and commenced the initial works on 18 February 2008. By March 2008, there was scaffolding and sheeting around block A and, in particular, along the part of block A that faces the left-hand elevation of the defendant's building.

E [20] On 3 March 2008, Mr Dalley wrote to Mr Rowan again concerning the likely loss of light:

> My client is keen to achieve an amicable settlement over this matter. To this end he has agreed, at this stage, to give a maximum undertaking of £4,000 plus VAT to cover your fees and costs and any specialist fees and costs your client incurs. All professional fees will be considered during the course of the negotiations over any potential damages claim which will take place on a without prejudice basis.

[21] At this point, I should record that there have been without-prejudice discussions between the parties. I know no more F than this. I do not have the dates on which the discussions began and ended, but it seems that they extended over quite a lengthy period. I do not know what was on offer in these discussions nor on what rock they foundered, nor whether either party was conducting the negotiations in an unreasonable manner.

[22] On 28 April 2008, the work of removing the mansard roofs and the plant rooms and lift motor rooms on the fifth floor of block A was commenced.

[23] On the same day, 28 April 2008, Mr Rowan wrote to Mr Dalley informing him that the defendant had instructed Drivers Jonas to act in respect of the light issue. The person responsible for the matter within Drivers Jonas was Ms Elizabeth de Burgh Sidley who is, like Mr Parratt, a specialist lighting surveyor.

[24] On 6 June 2008, Mr Dalley sent to Ms de Burgh Sidley a full set of architectural drawings of the redevelopment of block A. G

[25] Later in June 2008, one of Ms de Burgh Sidley's colleagues visited Leeds in order to consider the orientation of the buldings and the likely effect of the redevelopment on the defendant's right of light. On 23 June 2008, Ms de Burgh Sidley wrote to Mr Dalley:

> My initial view, seeing the drawings, is that there will be a serious infringement of our clients' rights of light.
>
> For this reason, please would you advise your clients not to undertake any work along the elevation of Infirmary Street which will take further light away from our clients' building. I would ask for your response to this letter by return confirming that no works will be undertaken? Please confirm that your clients H will meet all our clients' costs and fees in this matter.

[26] On the same day, 23 June 2008, the foundation works for the extension to block A were commenced. Steel erection to the extension started on 21 July 2008 and was complete from the basement to fifth-floor level by 25 August 2008.

[27] Up to August 2008, ISG Regions Ltd had been working on the basis of successive letters of intent. The building contract was signed on 14 August 2008. The contract price was £8,642,725 and the date for completion was 12 June 2009.

[28] Before looking at any further correspondence, it will be convenient to summarise in a single paragraph the progress of building J works over the following few months. The steel-frame erection to form the fifth, sixth and seventh floors started on the Infirmary Street wing of the building, the long arm of the L, on 22 September 2008. The steel-frame erection to form these floors started on the Toronto Street wing, the short arm of the L facing the side elevation of the bank building, on 13 October 2008. The steel frame was complete on both wings by 17 October 2008. All steel work to these upper floors was completed by 13 December 2008. The metal decks forming the floors on the sixth and seventh levels were completed, and the concrete to those floors was poured, by 22 December 2008.

[29] On 24 or 25 September 2008 (the precise date is not clear), K Mr Parratt wrote to Ms de Burgh Sidley with "an open offer of compensation to your client for the reduction in daylight to his building arising from the upward extension of Block A". The offer was formulated in the "book value" form conventional in light cases. I do not propose to set out the calculations put forward in the letter. It is sufficient to say that the offer was of £20,667, together with reasonable professional fees. The offer was expressly made open for acceptance for a period of 14 days.

[30] On 20 November 2008, Pinsent Masons came on the scene as the solicitor acting for the defendant. Its letter of that date to the claimant required the claimant, by 4pm on 28 November 2008, to do L the following:

> 1 expressly acknowledge that the Proposed Development will cause a substantial interference with our client's right that is both actionable and injunctable;
>
> 2 undertake not to carry out any part of the Proposed Development that will interfere with our client's right to light enjoyed at the Property.

The letter went on to warn the claimant that, absent this confirmation, "our client has instructed us to issue proceedings in the High Court for an injunction", which might be either an interim prohibitory injunction, preventing the continuation of works until a full hearing, or a mandatory M injunction, requiring the demolition of the offending works.

The letter also sought pre-action disclosure of all documentation relevant to the loss of light and an undertaking to pay the defendant's legal costs "in the first instance" of £5,000 plus VAT.

[31] Shoosmiths replied on behalf of the claimant on 28 November 2008. It accepted that its client's development had caused an actionable loss of light to the defendant's building, but it queried whether the defendant would obtain an injunction. In this connection, Shoosmiths referred in particular to: (i) "your client's inaction in the knowledge of our client's plans for over a year"; and (ii) "the vast discrepancy between the likely sum of damages your client might be awarded when compared with the value of the development to our client and the losses

which would accrue to our client". Shoosmiths went on to state that the claimant would proceed with the development, and it rejected the requests for pre-action disclosure and a costs undertaking.

[32] Further letters were written between solicitors on 3 and 5 December 2008. Pinsent Masons threatened an application for pre-action disclosure if disclosure were not given voluntarily by 8 December. Disclosure was not offered by Shoosmiths, the threat was not carried out and, at any rate so far as the open correspondence is concerned, silence descended as between the solicitors until 5 February 2009. On that date, Pinsent Masons wrote to say that it "w[as] instructed to proceed as outlined in earlier correspondence". Shoosmiths replied on 16 February 2009, confirming that it was instructed to accept service. Once again, the threat of proceedings was not carried into effect and silence descended, this time for rather longer than previously.

[33] The building works were completed on 10 July 2009, which was less than a month behind schedule, and at some stage block A was renamed Toronto Square. The final account agreed with ISG Regions Ltd was £8,884,971, which was not greatly in excess of the contract sum. The total cost of the works, including such additional matters as professional fees, additional electrical costs and carpeting, but excluding interest on borrowed money, was £9,775,759: see the final account statement dated October 2009, in vol 3, at p373, of the trial bundle. The total cost of the project to the claimant, including the cost of acquisition and finance charges, came to £35,814,161: see the CBRE appraisal summary dated 16 June, in vol 2, at p69P, of the trial bundle.

[34] The correspondence between solicitors was reactivated by Shoosmiths on 24 July 2009. In its letter of that date to Pinsent Masons, Shoosmiths taxed Pinsent Masons with delay, in that it had threatened to issue proceedings, had not done so and had been content to keep the threat of an injunction hanging over the claimant's head for eight months. Shoosmiths expressed the opinion that the defendant had, by his delay, lost the chance of obtaining any remedy, and it called on Pinsent Masons to issue proceedings by 7 August 2009, failing which Shoosmiths would do so. In the event, it was indeed Shoosmiths that caused the claim form to be issued on 19 August 2009.

[35] Finally, I must deal with the letting of the sixth and seventh floors of Toronto Square.

[36] The seventh floor was let to Zolfo Cooper, chartered accountants, from 1 August 2009 for a term of 10 years, with a break clause at the end of five years. I presume that there is to be a rent review after five years. The initial rent is of £27 psf on an agreed area of 7,050 sq ft (that is, £190,350 pa), but there are two years' rent-free to be taken as half-rent for each of the first, second, sixth and seventh years of the term. In addition there was a capital allowance of £18 psm (that is, £1.67 psf) to the lessee for carpeting.

[37] Maples & Calder, which is a firm of solicitors based in the Cayman Islands, agreed in principle to take a lease of part of the sixth floor on broadly similar terms, with an initial rent of £24 psf on an area of 5,560 sq ft (giving £133,440 pa). The intending lessee, however, withdrew from the proposed letting in December 2009, and its agent stated that it was doing so because of the dispute with which I am now dealing.

Expert evidence

Preliminary

[38] There are reports from seven expert witnesses. No application was made for permission to call experts to give oral evidence and counsel agreed that it would be impracticable to have such evidence within the confines of a three-day trial. This has not given rise to too much difficulty with regard to the specialist lighting surveyors, who are in agreement with each other as to the amount of light that has been lost, although they are not agreed as to the valuation of the loss. There are, however, real problems with some of the other experts, for the simple reason that significant differences between them, whether on general approach or on figures, have not been explored in cross-examination. These problems have not been wholly eliminated by the production during the trial of several supplementary reports, written with a view to commenting on the main reports of other experts.

Specialist lighting surveyors

[39] The lighting surveyors who provided reports were Mr Parratt, to whom I have already referred, for the claimant, and Mr Neil Lovell-Kennedy for the defendant.

[40] With regard to measurement of loss of light, the general principles are agreed. A point in a room is regarded as being adequately lit if, at a height of 2ft 9in (or 850mm) above floor level, 1-500th or 0.2% of the dome of the sky is visible. A room is regarded as being adequately lit for commercial purposes if 50% of the room is well lit. According to the textbook to which I referred earlier in this judgment: (i) if more than 50% of a room used for commercial purposes remains adequately lit after work carried out by an adjoining owner, there is no actionable loss; (ii) if work undertaken by an adjoining owner reduces that area to less than 50%, there is potentially actionable damage; (iii) if a room is already poorly lit to 25% or less, any reduction in light would almost inevitably be regarded as a serious loss and a plain case for an injunction (the available light is regarded as invaluable because there is so little of it.): see Bickford-Smith and Francis, *op cit*, in paras 12-42 and 12-45.

[41] The area that has suffered actionable loss of light is expressed in square metres or square feet "EFZ". EFZ stands for "equivalent first zone". This is, in effect, a process of weighting the more and less important parts of a room. A room is divided into four zones: the front zone, the first zone, the second zone and the makeweight zone, each comprising 25% of the total area. The front zone is that which is nearest to, and the makeweight zone the furthest from, the window or windows. Loss of sky visibility in the front zone is the most serious, and the area of loss is multiplied by 1.5; first zone loss is taken at par; and second and makeweight zone losses are reduced by factors of 0.5 and 0.25 respectively.

[42] Mr Parratt's calculations have been agreed by Mr Lovell-Kennedy, and I have summarised them on a room-by-room basis in the table that follows. Areas are given in square metres. Under the heading "Before", the first figure given is the area that was adequately lit before the erection of the sixth and seventh floors of Toronto Square, and the second figure is the percentage of the entire room that that area represents. Under the heading "After", there are equivalent figures for the present position. The column headed "Loss" shows the loss of adequately lit areas in raw figures, with the loss expressed as an EFZ figure in the final column.

No	*Room*	*Area*	*Before*	*After*	*Loss*	*EFZ*
1	Board room (ground floor)	20.9	13.9 (66%)	9.0 (23%)	4.9	3.20
2	Stair hall (ground floor)	31.2	10.1 (32%)	8.6 (27%)	1.5	1.50
3A	Toilet (ground floor)	2.9	2.9 (100%)	2.8 (96%)	0.1	0.05
	3B ditto	1.7	1.6 (94%)	1.5 (88%)	0.1	0.05
4	Open area	314.0	78.5 (25%)	76.3 (24%)	3.2	3.20
5	Bar annex/ Cloakroom (gf)	7.6 5.5	(72%) 4.1	(54%)	1.4	0.70
6	Office (mezzanine)	20.9	5.5 (26%)	5.0 (24%)	0.5	0.25
7	Small office (mezzanine)	9.9	1.5 (15%)	0.7 (7%)	0.8	0.80
8A	Toilet	2.9	2.9 (100%)	2.8 (96%)	0.1	0.03
8B	Toilet	1.6	1.6 (100%)	1.5 (94%)	0.1	0.03
9	Stairwell	11.4	3.6 (32%)	2.2 (19%)	1.4	0.04
10	Workshop	11.6	5.0 (43%)	3.1 (27%)	1.9	1.90
11	Upper Bar	306.0	72.7 (24%)	72.2 (24%)	0.5	0.75
12	Large Office (first floor)	51.5	35.5 (69%)	32.8 (64%)	2.7	1.35
13A	Toilet	11.0	5.4 (49%)	2.0 (18%)	3.4	0.85
13B	do	1.7	1.6 (94%)	1.4 (62%)	0.2	0.05
14	Stairwell	11.4	7.6 (67%)	4.6 (40%)	3.0	0.75
15	Office (first floor)	19.2	7.5 (39%)	3.0 (16%)	4.5	2.70

16A	do	7.3	3.5 (48%)	1.9 (26%)	1.6	1.60
16B	do	14.7	5.6 (38%)	3.2 (28%)	2.4	2.65
17	Office (2nd floor)	59.1	51.2 (86%)	51.2 (86%)	0.0	0.0
18A	Toilet	3.2	3.0 (94%)	2.8 (87%)	0.2	0.05
18B	do	1.6	1.5 (94%)	1.4 (87%)	0.1	0.03
19	Stairwell	11.4			3.0	0.75
20	Office (2nd floor)	18.8	16.3 (87%)	12.4 (66%)	3.9	1.95
21	do	21.8	16.3 (71%)	12.4 (53%)	3.9	1.95
24	Attic storeroom	12.1	3.6 (30%)	3.0 (25%)	0.6	0.60
25	Office	20.1	17.0 (85%)	17.0 (85%)	0.0	0.0

I have not included in the table rooms 22 and 23: 22 is a storeroom of $32.3m^2$, which has suffered no loss of light, and 23 is an unmeasured void area.

On these agreed calculation, the total loss of light expressed as EFZ is $27.78m^2$.

[43] When it comes to assessing the value of the light that has been lost, the experts part company.

[44] Mr Parratt has been advised that the rental value of the entire bank building as office space in October 2008 would have been around £102 psm, which would mean that loss of light should be calculated at £25.50 psm of EFZ loss: save that, for special reasons connected with room 24, the small area of loss there should be taken at only £20 psm. This gives an annual conventional book value loss thus:

$27.18m^2$ at £25.50	693.09
$0.60m^2$ at 20.00	12.00
	£705.09

On the assumption that the loss will be continued, and on the basis of advice that a yield of 10% would be appropriate, standard valuation tables give a multiplier of 10 to convert the annual loss into a loss in perpetuity of £7,051.

[45] Mr Lovell-Kennedy takes, on the basis of information provided by the defendant, a rental of £215.28 psm in October 2008. As Mr Parratt had done, he takes one-fourth of this figure as "light rental value", that is, £53.82 psm that, applied to 27.87m2 EFZ (a slightly higher figure than Mr Parratt's 27.78), gives an annual book value loss of £1,500. To capitalise this into a loss in perpetuity, Mr Lovell-Kennedy takes a yield of 6.5%, which gives a multiplier of 15.385, and a loss of £23,077.50. To this, he applies the so-called *Carr-Saunders* uplift, using a multiplier of 3.5, and arrives at a compensation figure of £80,771.25: see *Carr-Saunders* v *Dick McNeil Associates Ltd* [1986] 1 WLR 922*.

[46] There are therefore, on the face of their reports, three points at which these experts differ: (i) the rental value to be taken as the starting point; (ii) the appropriate yield to produce the multiplier for a loss in perpetuity; and (iii) the *Carr-Saunders* uplift. Mr Francis accepted that the application of such an uplift, in order to compensate the dominant owner for loss of amenity as well as financial loss, is conventional among specialist lighting surveyors, and it may be that Mr Parratt has simply left unsaid something that he assumed. Mr Francis said, however, that the multiplier usually taken is 2.66, which would bring Mr Parratt's figure up to £18,755, which is not far short of the open offer that had been made in September 2008: see [29].

Valuation

[47] The first report to consider under this heading is that of Mr Kevin Bramley, the chartered surveyor who acted as the claimant's valuation expert. I do not find it necessary to do more than describe his general approach and the resulting figures. Mr Bramley takes as his valuation date 13 October 2008. This was of course after the onset of the major banking crisis, the effects of which are still with us. What Mr Bramley does is to value Toronto Square as an investment. On one side of the sheet, he calculates the capitalised value of the anticipated rental income. On the other side, he places the outlay on the acquisition of the property, construction costs, professional fees and finance charges. He performs this exercise on two different assumptions, namely that: (i) at the valuation date, the claimant decided to proceed with the development as planned; (2) at that date, an injunction had been served and the claimant decided, rather than abandoning the development, to carry it through with modifications that would ensure that there would be no actionable interference with the defendant's right of light. Clearly, (ii) will produce a less favourable result than (i) because a reduced lettable area will pull down the rental income and alterations put in hand at a late stage of the construction will increase the outlay. In fact, each of the exercises carried out by Mr Bramley produced a loss on the investment: on (i), the "as built" scheme, a loss of £8,720,330; and on (ii), the modified scheme, a loss of £10,752,270.

[48] Next, there is the report of Mr Bruce Allan, who is also a chartered surveyor and who was instructed on behalf of the defendant. He takes as his starting point the sale of block C, Cloth Hall Court in October 2007. He takes the price paid for block C as equivalent to £231 psf. If one assumes that the same figure is applicable to block A at the time the claimant acquired block A, of the price of £18.75m, £14.679m will have been attributable to the building as it then existed and the balance of £3.021m for the development potential. The development potential was, however, attributable both to the extension and to the sixth and seventh floors and, again on a square footage basis, £1.865m can be regarded as having been paid for the top floors in isolation.

[49] Then, as a separate exercise, Mr Allan carried out an analysis of anticipated revenue and outlay. His methodology is not dissimilar from Mr Bramley's, but there are significant differences between them. In particular, Mr Allan's analysis is restricted to the sixth and seventh floors, and he uses a lower yield, 5.75% compared with Mr Bramley's 7.25%. The result would have been: (i) an anticipated development profit for the two upper floors of £2,398,740 in December 2007; (ii) with a change in yield to 6.5% in October 2008, an eroded profit figure of £1.445m.

[50] Mr Allan then concluded that, with an anticipated profit of (in round terms) £2.4m from the sixth and seventh floors in January 2008, he would have expected the parties to negotiate for a release of the defendant's rights within a range from 33-50% of that profit, namely a bracket of £800,000 to £1.2m.

[51] These reports were exchanged less than a month before trial. There seems to have been no meeting of the valuers, and there is certainly no joint statement. The reports are unfortunately ships that pass each other in the night, each ignorant (if I may pursue the analogy) of the cargo carried by the other. However, three supplemental reports, or notes, have been provided to me during the trial.

[52] First, there is a supplemental report from Mr Bramley, dealing with the bank building. His main points are these: (i) he has not conducted an inspection of the building; (ii) the valuation is on a vacant possession basis and on specified assumptions as to floor areas, rental value and yield; (iii) on this basis, he values the property at £4-4.5m plus or -10%; (iv) the loss of light would have only a limited effect on value; and (v) the redevelopment of Cloth Hall Court has enhanced the attractiveness of the bank building in a number of respects.

[53] Next, Mr Allan provided a note of observations on Mr Bramley's supplemental report. His main points are: (i) the failure to inspect the building; (ii) the building is let or occupied by the defendant and his trading company; (iii) the floor areas may well be wrong and the rents being received are much higher than the values assumed by Mr Bramley; (iv) the difference between Mr Bramley's lower figure less 10%, £3.6m and his higher figure plus 10%, £4.95m, is 37%, which is outside a normal tolerance range; and (v) he regards Mr Bramley's view on the limited effect of loss of light on value as "very difficult" and "subjective."

[54] Finally, Mr Allan produced a further report, which takes as its starting point an appraisal that was prepared in December 2007 for the meeting at which the board of the claimant was to consider whether to proceed with the purchase of block A. This appraisal set against

* Editor's note: Also reported at [1986] 2 EGLR 181

a capitalised value of anticipated rental income of £41,783,822 an anticipated outlay (once again covering acquisition and construction costs, professional fees and finance charges) of £34,875,075. This would give a profit of £6,908,747. Mr Allan then reran the appraisal, taking out 4,498 sq ft, that is the area by which, on Mr Bramley's evidence, the sixth and seventh floors would have to be reduced in order to avoid infringing the defendant's right to light. On this basis, he arrived at the following figures: capitalised rental, £39,682,163; outlay, £34,176,724; profit, £5,505,439. Accordingly, the difference between profit for the whole and profit excluding the cutback area is £1.408m, which Mr Allan regards as "the figure that ought to have been used in negotiations with the defendant in early 2008".

Architect

[55] Mr Leppington was instructed on behalf of the claimant. There is no corresponding expert witness on the defendant's side. Mr Leppington's instructions were to prepare scheme drawings for an alternative development at Toronto Square that would not cause an actionable interference with the right of light at the bank building. He calls this alternative "option B." Mr Leppington summarised the principal design changes that would be involved as follows: (i) the reinstatement of the fifth floor mansard roof and dormer windows; (ii) a reduction of the floor area at sixth-floor level; (iii) a reduction of the floor area at seventh-floor level; and (iv) reconfiguration of the tower at the corner of Infirmary Street and Toronto Street. Mr Leppington provides a table showing the loss of gross and net internal space as between Toronto Square "as built" and the redesigned building. I have simplified his calculations for the purposes of the following table by omitting the gross areas (which do not affect rental values) and the figures for the fifth floor (where, although there is some loss of gross internal area, there is a loss of only 3 sq ft of net internal area).

Floor	*As built*		*Option B*	
	Net area m²	*Net area sq ft*	*Net area m²*	*Net area sq ft*
Sixth	893	9,612	670	7,212
Seventh	652	7,015	461	4,962
Total	1,545	16,627	1,131	12,174
Loss			414	4,453

It will be observed that the area calculated as lost through cut-back 4,453 sq ft is slightly less than that taken by other witnesses, 4,498 sq ft, but the difference is not of any significance.

Quantity surveyors

[56] Quantity surveyors were instructed by each side, Mr John Pillinger, for the claimant, and Ms Kay Williams, for the defendant. They produced a joint report that demonstrated that they had reached agreement on one matter, namely the additional cost to the claimant if, in October 2008, it had revised its plans and had carried out the remainder of the construction work in a manner that would not have caused interference with the defendant's right to light. They agreed that the net additional cost (which is a figure derived from savings set against extra expenditure) would have been £1,010,494.

[57] One figure that appeared to me to be central to the main issue that I have to decide is the current cost of cutting back the sixth and seventh floors so that they do not interfere with the right of light. Ms Williams had provided a figure of £1,115,500, but as I pointed out to counsel at an early stage of the trial, there was nothing on the subject either in Mr Pillinger's report or in the joint statement. Mr Pillinger then produced a supplementary report and, most unfortunately, his conclusion is very different from that of Ms Williams, and as I have already indicated, it has not been possible to test their respective opinions through cross-examination. Mr Pillinger costs the work at £2,455,327, which is more than double Ms Williams' estimate. Of the difference between the figures, just over a half is accounted for by Mr Pillinger's regarding Ms Williams' sums as too low and just under a half by his inclusion of sums for matters for which Ms Williams has made no allowance. These additional matters are: (i) work required to existing beams, columns, floors and walls to accommodate changes to the sixth and seventh floors, £50,000; (ii) new escape stairs from the ground to the fifth floor, £150,000; (iii) reconfiguring toilets from the ground to the sixth floor, £280,000; and (iv) craneage and associated lifting costs, £75,000.

First issue: Injunction or damages
Law

[58] The law is clear, although how a judge is to apply the established principles in a given situation is often problematical. The present is a difficult case, as both counsel recognised. Mr Francis began his closing submissions by acknowledging that the burden of dissuading me from granting a mandatory injunction rested on him: the subtext appeared to me to be that although he had reached the summit, he had only just managed to complete the climb. Mr Bickford-Smith said in the course of his final speech that the court was faced with "a close call".

[59] The leading case is *Shelfer* v *City of London Electric Lighting Co Ltd (No 1)* [1895] 1 Ch 287, which was decided by the Court of Appeal in December 1894. That case has remained good law to this day. In *Regan* v *Paul Properties DPF No 1 Ltd* [2006] EWCA Civ 1391; [2007] Ch 135* (counsel in the case were Mr Francis and Mr Bickford-Smith), Mummery LJ (with whom Tuckey and Wilson LJJ agreed) said:

> 35 *Shelfer* is the best known case. It is a decision of the Court of Appeal. It has never been overruled and it is binding on this court. The cause of action was nuisance, as in this case, though in the form of noise and vibration rather than interference with a right of light.
>
> 36 *Shelfer* has, for over a century, been the leading case on the power of the court to award damages instead of an injunction. It is authority for the following propositions which I derive from the judgments of Lord Halsbury and Lindley and AL Smith LJJ. (1) A claimant is prima facie entitled to an injunction against a person committing a wrongful act, such as continuing nuisance, which invades the claimant's legal right. (2) The wrongdoer is not entitled to ask the court to sanction his wrongdoing by purchasing the claimant's rights on payment of damages assessed by the court. (3) The court has jurisdiction to award damages instead of an injunction, even in cases of a continuing nuisance; but the jurisdiction does not mean that the court is "a tribunal for legalising wrongful acts" by a defendant, who is able and willing to pay damages: *per* Lindley LJ, at pp315 and 316. (4) The judicial discretion to award damages in lieu should pay attention to well settled principles and should not be exercised to deprive a claimant of his prima facie right "except under very special circumstances": *per* Lindley LJ, at pp315 and 316. (5) Although it is not possible to specify all the circumstances relevant to the exercise of the discretion or to lay down rules for its exercise, the judgments indicated that it was relevant to consider the following factors: whether the injury to the claimant's legal rights was small; whether the injury could be adequately compensated by a small money payment; whether it be oppressive to the defendant to grant an injunction; whether the claimant had shown that he only wanted money; whether the conduct of the claimant rendered it unjust to give him more than pecuniary relief; and whether there were any other circumstances which justified the refusal of an injunction: see AL Smith LJ, at pp322 and 323, and Lindley LJ, at p317.
>
> 37 In my judgment, none of the above propositions has been overruled by later decisions of any higher court or of this court.

[60] I was taken by counsel through several of the modern cases in which the courts have had to consider whether the dominant owner should have an injunction or be left to an award in damages. The journey was a helpful one, but I do not propose to retrace it here. I have in mind what was said by Millett LJ in *Jaggard* v *Sawyer* [1995] 1 WLR 269†, at p288A-B (adopted by Nourse LJ in *Gafford* v *Graham* (1999) 77 P&CR 73‡, at p84):

> Reported cases are merely illustrations of circumstances in which particular judges have exercised their discretion, in some cases by granting an injunction, and in others by awarding damages instead. Since they are all cases on the exercise of a discretion, none of them is a binding authority on how the discretion should be exercised. The most that any of them can demonstrate is that in similar circumstances it would not be wrong to exercise the discretion in the same way. But it does not follow that it would be wrong to exercise it differently.

---

* Editor's note: Also reported at [2006] 3 EGLR 94
† Editor's note: Also reported at [1995] 1 EGLR 146
‡ Editor's note: Also reported at [1999] 3 EGLR 75

A [61] There is, however, one other modern case to which I should refer specifically because it re-emphasises an important point of principle. In a famous passage in his judgment in *Shelfer*, AL Smith LJ said, at pp322-323:

> In my opinion, it may be stated as a good working rule that
> (1) If the injury to the plaintiff's legal rights is small,
> (2) And is one which is capable of being estimated in money,
> (3) And is one which can be adequately compensated by a small money payment,
> (4) And the case is one in which it would be oppressive to the defendant to grant an injunction
> B then damages in substitution for an injunction may be given.

The modern case to which I wish to refer is *Jacklin* v *Chief Constable of West Yorkshire* [2007] EWCA Civ 181, in [48], in which Lloyd LJ (with whom Buxton and Rix LJJ agreed) stressed that, in order to avoid the grant of an injunction, a defendant must show that all four of AL Smith LJ's criteria have been satisfied:

> True it is that the *Shelfer* principles are only a working rule, although a long hallowed and reliable working rule, but it is clear that the four elements in that are cumulative and that it is necessary for a defendant to satisfy the first three, but that it is by no means sufficient for it to do so. There has to be
> C some additional factor, characterised in *Shelfer* and in *Jaggard* v *Sawyer* as oppression, to justify withholding the injunctive remedy, which is the claimant's prima facie right as ancillary to his property rights.

[62] I propose now to deal separately with each of the four *Shelfer* considerations. As will appear from what I have to say in a moment, all but one of them raises a difficult question.

(1) Small injury

[63] The first question is whether the injury to the defendant's legal rights is to be characterised as small.

[64] If one returns to Mr Parratt's calculations that I summarised
D in the table in [42], one finds that the total loss of adequately lit space expressed in EFZ is 27.78m2. However, it is necessary to make a detailed examination of the table.

[65] First of all, one finds two areas that were adequately lit before the works to block A and are no longer so. Area 1, the board room on the ground floor, which has suffered a serious deprivation of light on all three possible measures: percentage reduction, from 66% to 23%; actual area lost, 4.9m2; EFZ, 3.20m2. It is clear from the photographs that I have seen that the board room is the "star room", or one of the star rooms, of the refurbished bank building. Area 14 is a stairwell, not a working area, but I do not think that the loss here should be
E downgraded on that account: well-lit routes of passage enhance the feel of any building.

[66] There are then areas that were not adequately lit prior to the works but that have lost more than a minimal amount of light as a result. If one deals just in percentages, there is a loss of 5% or more in each of spaces 2, 7, 9 (a significant loss), 10 (another substantial loss), and 15, 16A and 16B (the last three are offices, each of which has suffered a serious loss on any measure).

[67] Finally, there are notable losses, although ones that would not be actionable on their own because they leave the rooms concerned with more than 50% of space adequately lit, in areas 5, 20 and 21.

F [68] Mr Francis relied heavily on the disproportion between the quantum of light deprivation and the total area of the bank building. Mr Bramley has taken the bank building to contain 40,000 sq ft. The EFZ loss expressed in Imperial measures (the papers are bedevilled by failure to use a single system throughout) is 300 sq ft. That is no more than 0.75% of the internal area of the building. Even if Mr Bramley (on whose figure doubt has been cast by Mr Allan) was a long way out and the net lettable area building extended to no more than 32,000 sq ft (which is the defendant's own figure), the EFZ loss would still be less than 1%.

[69] The other comparison made by Mr Francis is between the conventional book value damages for loss of light and the value of the bank building. Taking the damages at the upper end of the range that emerges from the lighting surveyors' evidence (see [44] and [45]) and G
an admittedly broad-brush stab at the value of the building (see [51] and [52]), the figures are £80,000 and £4m respectively. The former is only 2% of the latter.

[70] Mr Francis also reminded me that I am dealing with a commercial property. He contrasted the present case with that of *Regan*, where the loss of light was in the living room of the claimant's maisonette.

[71] It seems to me that the injury done to the defendant's rights in this case lies close to the margin of what is and what is not small for the purposes of the first of the *Shelfer* criteria. In such a situation, it is no more than realistic to recognise that the decision carries a subjective
element and is perhaps to quite a large extent dependent on impression H
as much as on analysis. I accept that comparisons of the kind put forward by Mr Francis are considerations that I have to take into account. I accept also that good light is relatively more important in a person's home than it is in industrial or commercial premises. However, these matters do not deflect me from the view that the injury done to the defendant's rights exceeds what can properly be defined as small. In my judgment, (i) the character of the defendant's building, (ii) the commitment that the defendant has demonstrated to the restoration of the property, in which he has invested heavily and in which he operates several businesses; and (iii) the extent to which the claimant's works
have reduced the flow of light to the building as summarised in [63] to J
[67] together establish that what has occurred constitutes real damage of a kind for which the defendant should not be expected to content himself with a money payment. In other words, the injury is not small.

[72] If the conclusion just stated is correct, one does not have to proceed any further because as soon as the claimant falls at just one of the *Shelfer* hurdles it has failed to dislodge the defendant from his prima facie right to an injunction. It is, however, possible that my view of the nature of the injury might be held to be wrong: if it were, an appellate court would wish to have my conclusions on the other *Shelfer* questions.

K

(2) Injury capable of being estimated in money

[73] Notwithstanding what is said in Mr Bickford-Smith's closing written submissions, I take the view that this criterion is satisfied. What the defendant has lost is not something so intangible as not to be susceptible of calculation in terms of money.

(3) Compensation by a small money payment

[74] There are two ways in which, in theory at least, the defendant could be compensated in money for the loss of light to his building, namely by an award of damages: (i) for nuisance at common law; or
(ii) in lieu of an injunction under what is now section 50 of the Senior L
Courts Act 1981, which is in a line of descent from section 2 of the Chancery (Amendment) Act 1858, commonly called "Lord Cairns's Act". An award under (i) would ordinarily follow the book value or conventional method. An award under (ii) would nowadays inevitably follow the "release fee" or "buy out" path, which, so far as the modern law goes, was first mapped out by Brightman J in *Wrotham Park Estate Co Ltd* v *Parkside Homes Ltd* [1974] 1 WLR 798*.

[75] Mr Francis said in the course of his submissions that, in answering the third of the *Shelfer* questions, the court should look only at the damages that would be awarded at common law. One has, he
said, to assume that the court is not looking at the price of a release M
from the grant of an injunction because one has not yet reached the stage of deciding whether to grant the injunction. There is an apparent logic in this, but I do not accept that Mr Francis is correct, and that for two reasons. First, it is artificial when one is looking at the question of remedies generally to exclude from consideration one remedy that may become available to the court. Second, when the Court of Appeal was considering whether the damages would be small in *Regan*, in [71] and [72], it did so expressly, and separately, in respect of diminution in value damages and release fee damages.

---

* Editor's note: Also reported at (1973) 229 EG 617

[76] For reasons that will be found later in this judgment, I would have assessed *Wrotham Park* damages in the sum of £225,000. This is not, in my judgment, a sum that can be regarded as small within the third *Shelfer* question, even if one looks at it, not in absolute terms, but in respect of the value of either of the properties involved or to the cost of the works that payment of the sum would avoid.

(4) Oppression

[77] Mr Francis assembled a catalogue of reasons for why, in his submission, the grant of an injunction would be oppressive in all the circumstances of the case. The reasons can be grouped under two headings: those connected with the effect of an injunction on the claimant and those connected with the conduct of the defendant.

[78] Most of the points made concerning the effect of an injunction on the claimant cannot be gainsaid. (1) The carrying out of the necessary works would be expensive. The best I can say about the cost is that one has a range from the evidence of the quantity surveyors, and if the range at least is accurate the cost of the works would fall, in round terms, between £1.1m and £2.5m. (2) There would be a consequential loss of prime office space, which is of value both to the claimant as an investment and to the commercial community more generally. (3) There would be general disruption within Toronto Square while the work was being carried out. (4) There would be particular disruption to the tenants of the seventh floor, which would have to be "decanted" to other premises, at least for the duration of the work. (5) A final point under this heading was that the claimant might encounter difficulty in obtaining planning permission for a rebuild. I do not regard this point as realistic. Mr Leppington stated in his report that "the overall architectural style [of a reconstruction] remains sympathetic to the existing development and its surroundings".

[79] Mr Francis was critical of the conduct of the defendant in two respects. (1) He failed to commence legal proceedings: it was, unusually if not uniquely, the infringing party that had to inject some certainty into the situation by commencing the present action. Mr Francis accepted that the defendant could not be heavily criticised for his failure to apply for an interim injunction, given the possible consequences of the cross-undertaking in damages. The defendant could, however, have demonstrated activity by giving instructions for the issue of a claim form and by seeking a speedy trial. (2) The defendant was dilatory in correspondence and in responding timeously to communications from the claimant. There were lengthy periods during which, at any rate so far as the open correspondence goes, nothing was heard from him or his representatives. When the silence was broken on the defendant's side, there were threats of litigation that were not carried into effect, inappropriate suggestions (such as the demand for pre-action disclosure) and attempts to impose unreasonable time limits. Mr Francis made good this criticism by a detailed analysis of the correspondence; the material on which he relied has been set out in an earlier section of this judgment: see, generally, [12] to [34].

[80] The points made by Mr Francis all have force, although I think that, in view of the total cost of the venture to the claimant, which exceeded £35m, too much cannot be made of the cost of eliminating the infringement of the defendant's right of light. None the less, there is a rock on which, in my judgment, Mr Francis's case, both on hardship to his own client and on the conduct of the defendant, comes to grief. I refer to the nature of the infringement of the defendant's rights by the claimant. (1) The infringement was not a trivial one. (2) The infringement was not inadvertent. It was, rather, committed in the knowledge that what was being done was actionable. (3) The infringement was committed with a view to profit. (4) The claimant was not driven by necessity, but could very easily, if somewhat less profitably, have built sixth and seventh floors of reduced dimensions. It would be wholly wrong for the court effectively to sanction what has been done by compelling the defendant to take monetary compensation that he does not want.

Second issue: Quantum

[81] I now come to what is, in view of conclusions reached earlier, yet a further theoretical question. This is the amount of damages that I should have awarded to the defendant if I had refused an injunction.

[82] Under the principle that was established by Brightman J in *Wrotham Park*, the court assesses damages in accordance with the figure that the party entitled to a right (in that case, the right to enforce a restrictive covenant) could reasonably have demanded for relaxing that right. In *Tamares (Vincent Square) Ltd* v *Fairpoint Properties (Vincent Square) Ltd* [2007] EWHC 212 (Ch)*, in [22], Mr Gabriel Moss QC, sitting as a deputy High Court judge, deduced from previous cases the principles that the court should apply when assessing *Wrotham Park* damages in a right of light case. I have found Mr Moss's statement most helpful, and I gratefully adopt it.

> (1) The overall principle is that the Court must attempt to find out what would be a "fair" result of a hypothetical negotiation between the parties;
>
> (2) The context, including the nature and seriousness of the breach, must be kept in mind;
>
> (3) The right to prevent a development (or part) gives the owner of the right a significant bargaining position;
>
> (4) The owner of the right with such a bargaining position will normally be expected to receive some part of the likely profit from the development (or relevant part);
>
> (5) If there is no evidence of the likely size of the profit, the Court can do its best by awarding a suitable multiple of the damages for loss of amenity;
>
> (6) If there is evidence of the likely size of the profit, the Court should normally award a sum which takes into account a fair percentage of the profit;
>
> (7) The size of the award should not in any event be so large that the development (or relevant part) would not have taken place had such a sum been payable;
>
> (8) After arriving at a figure which takes into account all the above and any other relevant factors, the court needs to consider whether the "deal feels right."

[83] There is in this case a controversy as to the date of the hypothetical negotiation that the court has to examine. Mr Francis argued for October 2008, when the claimant's interference with the defendant's right became permanent in the sense that it was set in the form of a building structure: any previous interference from scaffolding or sheeting should be disregarded as being temporary. Mr Bickford-Smith argued for the early weeks of 2008, before any works on the ground had been started and around the time when the first letter of intent was issued to ISG Regions Ltd.

[84] There is not, so far as I can see, any hard and fast rule that ties the court to a particular date as a point of principle: see *AMEC Developments Ltd* v *Jury's Hotel Management (UK) Ltd* (2001) 82 P&CR 22†, in paras 31 and 32‡. The court should, in my judgment, although dealing with a hypothetical situation: (i) plant its feet in the real world; and (ii) assume that both parties would act reasonably. In the circumstances of this case, a reasonable developer in the situation of the claimant would have wanted to conclude negotiations by March 2008, so as to be able to reach a decision one way or another as to the additions that were to be made to block A whether to build it as planned (if terms were concluded with the defendant) or with the sixth and seventh floors attenuated so as to accommodate the defendant's right of light or in some third, non-offending manner. Reasonable people want to know at an early date where they stand and a reasonable developer does not risk its money on works that it may be ordered to pull down.

[85] Counsel have come up with widely differing figures for *Wrotham Park* damages in this case. Mr Francis's primary submission is that there should be no such damages. If there are any, they should be a fair percentage, which he puts at 5%, of the value of the net benefit to the claimant created by the offending parts of the sixth and

* Editor's note: Reported at [2007] 1 EGLR 26

† Editor's note: Reported at [2001] 1 EGLR 813

‡ Judge's note: I do not read the decision in *Pell Frischmann Engineering Ltd* v *Bow Valley Iran Ltd* [2009] UKPC 45 as requiring a judge inflexibly to adopt the date of the breach

seventh floors. Mr Bickford-Smith's final bid, at the end of his closing submissions, was for the round sum of £500,000.

[86] I regret to say that I have not found the opinions of Mr Bramley or, so far as his first report goes, Mr Allan of any real assistance. Mr Bramley (in [47]) starts from what I regard as the wrong date. He postulates a capital loss for the claimant's venture that leads to the conclusion that the defendant's potential for obstructing the development as planned was worth little or nothing in negotiating terms: that conclusion simply does not correspond with reality. Mr Allan's first exercise (see [48]) rests on an assumption, which is untested, that block C can be used as a base from which to value the top floors of block A. His second exercise (see [49] and [50]) uses a method similar to Mr Bramley's, but there is nothing in evidence to persuade that the yield figure taken by either is more accurate than that used by the other, and so there is no starting point for an intelligent comparison of their respective conclusions.

[87] The one solid piece of evidence that I do have is the estimate of profit that was prepared for the meeting of the claimant's board in December 2007. This is dealt with in Mr Allan's further report: see [54]. The anticipated profit from the entire development was £6.908m. The claimant would have had this figure in mind at the time of the hypothetical negotiations, and would also have had in mind that the profit would inevitably have been less had it been forced to reduce the areas of the sixth and seventh floors. Mr Allan's calculation of a profit differential of £1.408m looks credible but it does not, in my judgment, support the conclusion that reasonable negotiators would have arrived at a figure anywhere close to the £500,000 for which Mr Bickford-Smith contended. Further, I am reluctant to use as a base a calculation that was produced at a late stage of the proceedings and one that, like so many other opinions that have been put forward in this case, has not been tested by cross-examination. Even if the figure is accurate, an award of more than one-third of the profit differential appears to me to be extravagant. In *Wrotham Park*, Brightman J said, at p815, that, on the facts before him, the court should "act with great moderation". The apparent reluctance of the defendant to issue legal proceedings is, I think, a factor that in this case should have a similar influence on the level of the award.

[88] The following are the matters that appear to me to be relevant in arriving at the figure on which the parties would have settled at the end of the hypothetical negotiation. (1) The claimant had obtained a reduction of £350,000 in the purchase price to take account of the lights issue, so to that extent was resting on a cushion that could be regarded as expendable. It would, however, have wanted to retain at least some of that cushion. (2) The claimant had budgeted for £200,000 as the price of settling all lighting issues, and in the event the defendant had turned out to be the only dominant owner with a claim. (3) In view of the price reduction, the claimant could be expected to go beyond the budgeted figure in order to obtain a settlement of the issue before starting work. (4) The uplift on £200,000 should be modest because the defendant's seeming reluctance to commence proceedings would have made it unlikely that he would have pushed unduly hard in negotiations.

[89] Taking these matters into account, I have arrived at a figure of £225,000. The evidence in the case is not such as would enable any judge to produce a scientifically justifiable figure. This makes all the more important the last part of the process, which is to stand back and ask myself whether the sum that I have mentioned "feels right" or, put another way, whether it looks either extravagant or mean. To my mind, £225,000 survives this final test.

Disposal

[90] Counsel will no doubt want to make submissions at a convenient date as to the terms of the injunction that is to follow this judgment and as to costs.

Claim dismissed; counterclaim allowed.

LAND REGISTRATION

Silkstone and another v Tatnall and another

CHANCERY DIVISION

[2010] EWHC 1627 (Ch)

2 July 2010

FLOYD J

Estates Gazette 2 October 2010

[2010] 39 EG 110

Land registration — Practice and procedure — Reference to adjudicator — Land Registration Act 2002 — Unilateral notice of right of way against title to land — Objection to landowner's application to cancel notice — Dispute referred to adjudicator — Objector seeking to withdraw objection — Adjudicator refusing request — Section 73(7) of 2002 Act — Whether objector entitled to withdraw objection — Whether adjudicator having discretion to permit withdrawal

The appellants and the first respondent owned adjoining residential properties. In February 2008, the appellants lodged a unilateral notice on the first respondent's registered title in respect of a right of way on foot that they claimed to have over the first respondent's garden. The first respondent applied to cancel the notice; the appellants submitted an objection to the cancellation. The matter was referred to a deputy adjudicator in June 2008, pursuant to section 73(7) of the Land Registration Act 2002.

Witness statements were exchanged and a hearing date was fixed for June 2009. In May, the appellants wrote to the adjudicator, indicating that they were finding it difficult to obtain certain evidence and requesting permission to withdraw from the proceedings to enable them to bring court proceedings at a later date. The adjudicator refused permission to withdraw on that basis at such a late stage and held a hearing on the merits in the appellants' absence. He decided that no case could be made for a right of way, directing that the unilateral notice should be cancelled but that a mutual right of access for maintenance and repair did exist and should be noted on the title to both properties.

The appellants appealed, arguing that: (i) the adjudicator could not refuse to accept their withdrawal and had lacked jurisdiction to rule on the merits; or, in the alternative, (ii) if he did have the power so to refuse, he had erred in his exercise of that power. The Chief Land Registrar joined the proceedings as the second respondent.

Held: **The appeal was dismissed. An objector cannot withdraw an objection unilaterally once a dispute has been referred to the adjudicator. Section 73(7) contemplates that an objection to an application to the registrar may be disposed of by agreement, including by unilateral withdrawal, prior to such a reference as a purely administrative procedure. By contrast, once reference is made to the adjudicator, then, in the absence of full agreement between the parties, the adjudicator will determine the mechanism for disposing of the objection. Such a reference provides a mechanism for determining the underlying rights, so as to dispose of applications in which the objection is not obviously groundless and the parties have not agreed the correct state of the register. The objection cannot be withdrawn by giving notice to the registrar. The adjudicator does not have the power to accept or reject a withdrawal of the original objection; his sole function is to determine the proceedings before him. Thus, any purported notice of withdrawal to the adjudicator will be ineffective. The proceedings before him had therefore continued notwithstanding the appellants' purported withdrawal. The first respondent had sought a substantive remedy and had been entitled to obtain a decision concerning his rights in that regard. The only way for the appellants to extricate themselves from the reference was by settling with the first respondent or conceding his entitlement to the relief that he sought. Since they had done neither, the case had continued to be referred to the adjudicator for his decision.**

No cases are referred to in this report.

This was the hearing of an appeal by the appellants, Leonard Silkstone and Gary Silkstone, from a decision of the adjudicator to HM Land Registry determining a reference made under section 73(7) of the Land Registration Act 2002 on the application of the first respondent, Simon Tatnall, with the Chief Land Registrar joined as the second respondent, notwithstanding the purported withdrawal of the appellants' objection to the application.

Bruce Monnington (instructed by Bates Wells & Braithwaite, of Sudbury) appeared for the appellants; the first respondent did not appear and was not represented; Timothy Morshead (instructed by the Treasury Solicitor) represented the second respondent.

Giving judgment, **FLOYD J** said:

[1] This is an appeal from the adjudicator to HM Land Registry. The appeal is from a decision of Deputy Adjudicator Mark dated 15 June 2009, and is brought with his limited permission granted by a separate decision on 31 July 2009. The appeal is limited to the question of whether the adjudicator can refuse permission to a party to withdraw from proceedings before the adjudicator at any time and the powers of the adjudicator in respect of the withdrawal. By an order dated 17 June 2010, I granted the Chief Land Registrar's application to be joined as a second respondent, since the registrar wanted to make submissions on matters affecting practice in the Land Registry. Mr Bruce Monnington argued the case for the appellants. Mr Timothy Morshead made the submissions of the Chief Land Registrar. The first respondent, Mr Simon Tatnall, has not participated in the appeal. It follows that I have not had the benefit of submissions in support of the deputy adjudicator's decision.

[2] The dispute concerns a right of way. The appellants, the Silkstones, own 2 Croft Cottages, Belchamp Water, Sudbury. The first respondent, Mr Tatnall, owns no 3, which is next door. The Silkstones claimed to have a right of way on foot through a gate in the rear of the boundary fence between no 2 and no 3, through the garden of no 3 and then through a covered passage to the street at the front.

[3] The Silkstones lodged a unilateral notice on the registered title of no 3 on 7 February 2008 in respect of this right of way. Mr Tatnall applied to cancel the unilateral notice, and the registrar gave notice of this application to the Silkstones, who in turn objected to the cancellation. A dispute having arisen, the matter was referred to the adjudicator on 18 June 2008, under section 73(7) of the Land Registration Act 2002 (the 2002 Act). Mr Tatnall was made the applicant in the proceedings before the adjudicator since he was the party applying for the unilateral

notice to be cancelled. The Silkstones were made respondents to the application. Throughout the proceedings, the case for the Silkstones was conducted by their father, Mr Geoffrey Silkstone, as their agent.

[4] At an intermediate stage in the proceedings, the Silkstones wrote to the adjudicator requesting disclosure relating to an alternative way of supporting the claim to the right of way, based on section 62 of the Law of Property Act 1925, as opposed to prescription. The application was dismissed on the ground that no case under section 62 had been pleaded by the Silkstones. Witness statements were exchanged, and the hearing was set down for Tuesday 2 and Wednesday 3 June 2009, with a site view to take place on the afternoon of Monday 1 June.

[5] On 28 May 2009, Mr Geoffrey Silkstone wrote to the adjudicator explaining that he was in difficulty obtaining certain evidence. He said:

> On behalf of the [Silkstones] I therefore request permission from the Adjudicator to withdraw from the proceedings on the 2nd and 3rd June (with site visit on 1st June) to the effect that the disputed application referred to the Adjudicator by the Land Registry should be given effect to leaving the way clear for the [Silkstones] to take out Court proceedings at a later date if they so choose.

[6] The adjudicator responded on the following day in the following terms:

> The Adjudicator is not at present prepared simply to permit you to withdraw your claim to a right of way on the basis that you can subsequently resurrect this claim in court. This dispute has been proceeding for almost a year and it appears to the Adjudicator that it could well be unfair to the applicant to permit you to withdraw on this basis just before the final hearing.

[7] The adjudicator indicated that the Silkstones could renew their application at the start of the hearing or apply for an adjournment. Also on 29 May 2009, in response to the adjudicator's communication, Mr Silkstone wrote to the adjudicator saying that if the adjudicator was not prepared to permit a withdrawal on the basis indicated in the letter of 28 May, he requested an adjournment of six weeks.

[8] Mr Silkstone sent two further faxes on 1 June 2009. In the first, which followed the site inspection, he confirmed that the Silkstones wanted to "withdraw their case due to lack of evidence". In the second, he wrote to confirm that he had sent two e-mails to Mr Tatnall "confirming the withdrawal". Finally, on 2 June 2009, Mr Silkstone wrote to confirm and clarify the content of its two faxes of 1 June "by stating that the [Silkstones] withdraw their objection to the Applicant's application".

[9] The deputy adjudicator nevertheless held a hearing on 2 June 2009, attended by Mr Tatnall in person, and issued a decision on the merits on 15 June 2009. By his decision, he held, in [28]:

> A purported notice of withdrawal, whether given to the Land Registry or to the adjudicator, will be ineffective until the adjudicator has ruled on the terms on which the notice will be given effect to and the terms have been accepted by the party seeking to withdraw.

[10] The deputy adjudicator then went on to exercise the discretion that he considered he held. He decided that he should not give permission to the Silkstones to withdraw their objection without a decision on the merits. He said, in [30]:

> I concluded that I had power to refuse permission to withdraw and that I would have that power even if, without my knowledge, the [Silkstones] had communicated that withdrawal to the Land Registry. I also concluded that in the present case it would be wrong to permit a withdrawal and that the litigation between the parties needed to be brought to a conclusion.

[11] As to the substance of the dispute, the deputy adjudicator came to the conclusion that:

> it was plain that no case could be made out under either section 62 of the Law Property Act 1925 or under the rule in *Wheeldon* v *Burrows*. There was also no admissible evidence to support a claim to a prescriptive easement.

[12] However, the deputy adjudicator did find that there was a right of access for maintenance and repair for the owners of nos 2 and 3 over the property of the other and directed that this should be noted on the register in respect of both titles. In the result, he directed the Chief Land Registrar to give effect to the application of Mr Tatnall to cancel the unilateral notice, but to note the rights of entry for the purpose of maintenance and repair on the titles of both properties.

[13] The Silkstones submit that the deputy adjudicator had no power to refuse to accept their withdrawal from the proceedings and had no jurisdiction to deliver a decision on the merits. If they are wrong about that, they submit that he exercised his discretion refusing them permission to withdraw on a flawed basis, and should in any event have said nothing about section 62, which was not a pleaded basis for the existence of the right of way. There is no attack, apart from this, on the substance of the decision.

Statutory framework

[14] The relevant statutory provisions creating the system for entry of unilateral notices, their removal and cancellation are in Part 4 of the 2002 Act, headed "Notices and Restrictions". I set out below the relevant parts of sections 34 to 36:

> **34 Entry on application**
>
> (1) A person who claims to be entitled to the benefit of an interest affecting a registered estate or charge may… apply to the registrar for the entry in the register of a notice in respect of the interest.
>
> (2) Subject to rules, an application under this section may be for
>
> (a) an agreed notice, or
>
> (b) a unilateral notice.
>
> **35 Unilateral notices**
>
> (1) If the registrar enters a notice in the register in pursuance of an application under section 34(2)(b) ("a unilateral notice"), he must give notice of the entry to
>
> (a) the proprietor of the registered estate or charge to which it relates, and
>
> (b) such other persons as rules may provide.
>
> (2) A unilateral notice must
>
> (a) indicate that it is such a notice, and
>
> (b) identify who is the beneficiary of the notice.
>
> (3) The person shown in the register as the beneficiary of a unilateral notice, or such other person as rules may provide, may apply to the registrar for the removal of the notice from the register.
>
> **36 Cancellation of unilateral notices**
>
> (1) A person may apply to the registrar for the cancellation of a unilateral notice if he is
>
> (a) the registered proprietor of the estate or charge to which the notice relates, or
>
> (b) a person entitled to be registered as the proprietor of that estate or charge.
>
> (2) Where an application is made under subsection (1), the registrar must give the beneficiary of the notice notice of the application and of the effect of subsection (3).
>
> (3) If the beneficiary of the notice does not exercise his right to object to the application before the end of such period as rules may provide, the registrar must cancel the notice.
>
> (4) In this section
>
> "beneficiary", in relation to a unilateral notice, means the person shown in the register as the beneficiary of the notice, or such other person as rules may provide;
>
> "unilateral notice" means a notice entered in the register in pursuance of an application under section 34(2)(b).

[15] So the system provides for the entry of unilateral notices by a "beneficiary" here, the Silkstones. Notice is given of the entry to the proprietor of the registered estate Mr Tatnall. The beneficiary of the notice may himself apply to the registrar for *removal* of his unilateral notice: section 35(3). That did not happen here. The proprietor of the registered estate may apply to the registrar for *cancellation* of the unilateral notice. That is what Mr Tatnall did here. Notice of the application to cancel must be given to the beneficiary of the notice, and if the beneficiary of the notice does not exercise his right to object the registrar must cancel the notice. All this is administrative and straightforward.

[16] Objections to applications are dealt with in Part 6 of the 2002 Act. Section 73 provides as follows:

A **73 Objections**

(1) Subject to subsections (2) and (3), anyone may object to an application to the registrar.

…

(3) In the case of an application under section 36, only the person shown in the register as the beneficiary of the notice to which the application relates, or such other person as rules may provide, may object.

(4) The right to object under this section is subject to rules.

(5) Where an objection is made under this section, the registrar

(a) must give notice of the objection to the applicant, and

(b) may not determine the application until the objection has been disposed of.

B (6) Subsection (5) does not apply if the objection is one which the registrar is satisfied is groundless.

(7) If it is not possible to dispose by agreement of an objection to which subsection (5) applies, the registrar must refer the matter to the adjudicator.

(8) Rules may make provision about references under subsection (7).

[17] The registrar may therefore throw out groundless objections: but anything that passes this very low threshold must be referred to the adjudicator. By section 108(1), the adjudicator has the following among other functions:

(a) determining matters referred to him under section 73(7),

C [18] There is, I was told, nothing relevant in the Land Registration Rules 2003 referred to in section 73(4). The Adjudicator to Her Majesty's Land Registry (Practice and Procedure) Rules 2003 set out the procedure to be followed in an adjudication, including disclosure of documents and the service of witness statements. Rule 5(3) of these rules provides that:

Following receipt by the adjudicator of a reference, the adjudicator must

…

(b) serve on the parties notice in writing of

(v) which party will be the applicant for the purposes of the proceedings and which party or parties will be the respondent.

D [19] In the present case, the adjudicator determined that Mr Tatnall should be the applicant and the Silkstones the respondents. This reflected the positions of the parties before the Land Registry where Mr Tatnall was applying to cancel the unilateral notice. He might equally have reversed the descriptors, since it was the Silkstones who were asserting the right of way.

[20] The rules also contain a rule similar to CPR 1 concerned with the overriding objective:

3 The overriding objective

E (1) The overriding objective of these Rules is to enable the adjudicator to deal with matters justly.

(2) Dealing with a matter justly includes, so far as is practicable

(a) ensuring that the parties are on an equal footing;

(b) saving expense;

(c) dealing with the matter in ways that are proportionate

(i) to the value of the land or other interests involved;

(ii) to the importance of the matter;

(iii) to the complexity of the issues in the matter; and

(iv) to the financial position of each party; and

(d) ensuring that the matter is dealt with expeditiously and fairly.

F (3) The adjudicator must seek to give effect to the overriding objective when he

(a) exercises any power given to him by these Rules; or

(b) interprets these Rules.

(4) The parties are required to help the adjudicator to further the overriding objective.

[21] Rule 55 gives the adjudicator power to impose a sanction on a party that has defaulted in complying with a direction. The adjudicator may direct the registrar to allow or cancel the original application, in which case the imposition of this sanction is deemed to count as the substantive decision. Apart from this sanction procedure, if the matter proceeds to a substantive decision without default, the adjudicator may make an order including requirements that the registrar give effect to the original application in whole or in part as though the objection had not been made or cancel the original application in whole or in part: G see r 41.

[22] To complete the picture, the 2002 Act contains provisions (section 110) to enable the adjudicator to direct a party to commence court proceedings to determine the matter. In these circumstances, the adjudicator will adjourn the proceedings before him once he has received notice that court proceedings have been commenced: see rr 8 and 9. Rule 11 provides a discretionary power to adjourn when proceedings are commenced otherwise than following a direction under section 110.

Appellants' submissions

H [23] Mr Monnington submitted that:

(i) the effect of the correspondence that I have recited was that, by 2 June at the very latest, the Silkstones had made it clear to the adjudicator that the objection to Mr Tatnall's application was withdrawn and that the cancellation of the unilateral notice was consented to;

(ii) the adjudicator had no power under the Act or rules to prevent the Silkstones from withdrawing in this way;

(iii) if, contrary to the last submission, the adjudicator had the power to refuse permission to withdraw, the adjudicator exercised his discretion wrongly in not permitting withdrawal;

J (iv) on such a withdrawal, the adjudicator no longer has jurisdiction to decide the matter referred on the merits; and

(v) in such circumstances, the adjudicator ought simply to have reported the withdrawal to the Land Registry and declined to deal with any further matter, apart from costs.

Discussion

[24] On the first of Mr Monnington's submissions, I think it is clear that the Silkstones were, in their letter of 28 May, seeking to withdraw their objection to Mr Tatnall's application to cancel the unilateral notice, and therefore no longer resisting an order for its cancellation. The statement that this would "leave the way clear for the Respondents to take out Court proceedings at a later date" was not seeking to K impose any condition on the withdrawal. It was merely a contention as to the effect of the withdrawal. Even if that is wrong, by 2 June at the latest it was clear that the Silkstones were purporting to withdraw their objection unequivocally and recognising that the unilateral notice would be cancelled.

[25] As to the remainder of Mr Monnington's submissions, there is an important distinction between: (a) a party that states that it does not want to take any further part in the proceedings in respect of the matter referred to the adjudicator; and (b) a party purporting to be able to withdraw the underlying objection that gives rise to the reference and consents to the cancellation of the notice. In the former case, given L that the party "withdrawing" is the respondent resisting an order for cancellation, one would not expect the withdrawal of participation to affect the jurisdiction of the adjudicator to go on and decide the matter. This is no more than an illustration of the general principle that any defending party that fails to appear or take part in any other type of legal proceeding opens itself to the possibility that, at the trial, its opponent will successfully establish the case against it in its absence. There is nothing in the Act or rules that prevents a party from taking this stance in proceedings before the adjudicator, and in consequence nothing that prevents the adjudicator from coming to a substantive decision on the merits in the absence of the objecting party. If it were otherwise, there M would be no means of disposing of the objection and cancelling the unilateral notice.

[26] The position may be different if a party can withdraw the objection that gave rise to the reference. Unlike the unilateral notice, where section 37(3) has a specific provision providing for an application for removal of the notice by the party that applied for it, there is no express provision governing the withdrawal of objections. The Act, in section 73(7), contemplates, at least before a reference to the adjudicator is made, that an objection may be disposed of by agreement, without the need for any application to withdraw the objection. The objection is after all the only obstacle that stands between the application to cancel the notice and its cancellation. The words "dispose by agreement"

must, as I interpret them, also encompass the case where the objection is unilaterally withdrawn. It would be inconvenient if the effect of the section were that a reference still had to be made to the adjudicator if the objection were unilaterally withdrawn. So, at least up to the point where the matter is referred to the adjudicator, the objection may be disposed of by agreement, including by unilateral withdrawal. The withdrawal of the objection in this case would have meant that the register was cleared of the notice asserting the existence of the right of way. The former beneficiary of the notice would not normally be prevented from asserting the claim to that right in separate proceedings.

[27] The purpose of the referral to the adjudicator is to provide a mechanism for disposal of applications that are objected to and in which the objection cannot be disposed of by agreement in this broad sense. Referral to the adjudicator involves a recognition that the objection is not obviously groundless, and that the parties have not achieved agreement as to the correct state of the register. The adjudicator is given the task of determining, by examination of evidence, those applications where there is a dispute as to the underlying rights.

[28] I think that the heart of the question that I have to decide is whether one should treat the matter referred to the adjudicator as a procedure whose purpose it is to establish the existence or otherwise of the right of way or other right in dispute, or whether the matter is better viewed as a subsidiary proceeding whose sole purpose is to provide a mechanism to dispose of applications to which there is a subsisting administrative objection. If the former, I would be inclined to hold that although the objector should be free to withdraw participation, this would not prevent the adjudicator from deciding the matter. If the latter, I think that it is clear that the adjudicator should not continue to deal with the substance of the matter after the administrative objection is withdrawn.

[29] I think that the reference to the adjudicator is better viewed as a proceeding whose purpose it is to determine the underlying right, quite unlike the administrative procedure in the Land Registry. That is illustrated by the fact that one way in which the adjudicator may carry out his function is by ordering a party to commence court proceedings, which would also have the effect of determining underlying rights. Proceedings before the adjudicator are triggered precisely because it is necessary to determine those rights in order to dispose of the objection. The procedure laid down by the rules is plainly one directed at determining those underlying rights.

[30] It follows, in my judgment, that there is no reason why administrative steps, such as unilateral withdrawal of an objection, which would be possible before the registrar, should be available as of right in proceedings before the adjudicator, which are of a different character. As the deputy adjudicator pointed out, once an application and its objection are transferred to the adjudicator, the matter is no longer in the hands of the registrar and is instead in the hands of the adjudicator. The adjudicator has the function, under section 108(1), of deciding the matter referred to him. While the matter is in his hands, the objector cannot apply to the registrar to withdraw the objection because the objection is no longer in the hands of the registrar. The only way of disposing of the objection is by means of the proceedings before the adjudicator.

[31] Section 73(5)(b) prevents the registrar from determining the application until the objection is disposed of. Both counsel submitted that an objection could be disposed of in more ways than those specifically contemplated by the 2002 Act and could be disposed of in various ways even after a reference to the adjudicator. Mr Morshead pointed to the fact that an objection might be concluded by a court decision other than in proceedings directed by the adjudicator. Section 73(5)(b) therefore was no obstacle to the registrar or the adjudicator accepting a withdrawal of the objection after a reference to the adjudicator.

[32] I am not persuaded by these submissions. I agree with the deputy adjudicator when he said that a purported notice of withdrawal of the objection given to the registrar while the matter is referred to the adjudicator will be ineffective. It would be wholly unsatisfactory if the same matter could be processed at the same time in two different fora. So, it is a complete answer to an application to the registry to withdraw an objection that the matter is pending before the adjudicator. On the other hand, absent full agreement between the parties disposing of the entire proceeding before the adjudicator, once proceedings before the adjudicator are on foot, the mechanism for disposing of the objection is via the determination of the adjudicator.

[33] I therefore also agree with the deputy adjudicator that a purported notice of withdrawal of the objection given to the adjudicator will be ineffective. There is no provision in the Act or rules for the adjudicator to receive or act on a withdrawal of the original objection. The objection is filed with the registrar, and the adjudicator does not take over the function of the Land Registry in this regard even though the registrar cannot himself act on such a withdrawal until the reference is disposed of.

[34] I part company with the deputy adjudicator only to the extent that he implies that a withdrawal of the underlying objection would be given effect to when "the Adjudicator has ruled on the terms on which the notice will be given effect to and the terms have been accepted by the party seeking to withdraw". I think the proper analysis is that the adjudicator does not have the power to accept or reject a withdrawal of the original objection. That simply cannot be done when the matter is before the adjudicator. The adjudicator's sole function is to determine the proceedings before him. If the parties are agreed as to the terms on which those proceedings are to be determined, he can give effect to that agreement by giving appropriate directions to the registrar. However, if the parties are not so agreed, there is no need or basis for him to give a ruling on any terms.

[35] On the other hand, as indicated earlier in this judgment, if a party wants to take no further part in the proceedings before the adjudicator, it is free to do so, and there is nothing in the Act or rules that requires it to seek the permission of the adjudicator. Where this occurs, the adjudicator may proceed with the reference and reach such conclusions as are justified by the evidence. Alternatively, there may come a point where the non-participating party has failed to comply with directions and the power to sanction under r 55 comes into play. However, none of this involves the implication of a power to accept or refuse the withdrawal of an objection. Withdrawal from the proceedings is a quite different matter from withdrawing the underlying objection, which is something that cannot be achieved while the matter is referred to the adjudicator.

[36] Accordingly, I would hold in the present case that the adjudicator's jurisdiction to determine the matter referred to him continued notwithstanding the purported withdrawal of the Silkstones. Mr Tatnall was seeking a substantive remedy, namely cancellation of the unilateral notice, and was entitled to demonstrate that he was entitled to it by obtaining a decision from the adjudicator to that effect. The Silkstones had not withdrawn the underlying objection by withdrawing it from the registrar and could not do so. They had to extricate themselves from the reference before the adjudicator. To do that they would either have to settle with Mr Tatnall or concede the relief he was seeking and his entitlement to it. They did not do so.

[37] Mr Monnington relied on section 35(3) as background to the question of the construction of the Act. He said that since the Silkstones could withdraw the unilateral notice under this provision (although they had not done so), they should have a corresponding right to withdraw the objection. I think that there are a number of answers to this. First, I am not persuaded that while the matter of the application to cancel and the objection are before the adjudicator, the registrar has the power to permit the withdrawal of the notice. Second, the Silkstones did not have a right to withdraw the unilateral notice: they had a right to apply to withdraw it. If such an application had been made, Mr Tatnall could have objected on the grounds that the cancellation of the notice was before the adjudicator and that it would be unjust to allow unilateral withdrawal. That matter too would have to be referred to the adjudicator. So, I do no think that section 35(3) compels a different result.

[38] Mr Monnington submitted that this approach to the matter ignored the fact that the proceedings as a whole were instigated by the Silkstones' assertion of the right of way, and that to view them as responding to a claim was merely a consequence of the rather

convoluted procedure in the Land Registry and the adjudicator's decision to nominate Mr Tatnall as applicant under r 5(b)(v). He submitted that the position was analogous to a party that sought a declaration as to the existence of a right of way in ordinary court proceedings, and discontinued those proceedings. The court would not then go on to reach a conclusion on the merits.

[39] I am not persuaded that the fact that it might have been possible for the Silkstones to withdraw, had the procedural arrangements been different, is a reason for allowing them to do so when they are aligned as they were. I am far from convinced that even if the Silkstones had been applicants before the adjudicator, their withdrawal from participation in the reference would have deprived the adjudicator of jurisdiction to decide the matter. The matter, namely the unresolved application before the registrar, would remain referred to him for decision.

[40] My conclusions are supported by the fact that if it were the case that the adjudicator was deprived of jurisdiction by the withdrawal of the objection, some basis would have to be found for a power to award costs and make the parts of his direction concerned with maintenance and repair. So far as costs are concerned, Mr Monnington suggested that a residual power to award the costs of the matter would continue, notwithstanding the loss of jurisdiction since the costs would have been incurred while the matter was properly pending before him. I do not think that this is a satisfactory answer, given that some costs could continue to be incurred after the withdrawal, for example costs at a hearing on the appropriate order for costs.

[41] I think that there is further support in a point made by the deputy adjudicator that if the withdrawal of the objection deprived the adjudicator of jurisdiction, it would have this effect even if exercised after the hearing when the party withdrawing has already been able to form a view as to the likely result. I am not attracted by a procedure that gives one party the election to abort the proceedings in this way.

[42] If I am right thus far, there is no question of an exercise of discretion in permitting the Silkstones to withdraw. I will nevertheless consider whether there is any basis for interfering with the exercise of the adjudicator's discretion on the basis that I am wrong. Mr Monnington's attack on the deputy adjudicator's exercise of his discretion was principally on the basis that he was wrong to take into account the fact that the Silkstones had at one point reserved the right to bring fresh proceedings. The right time, so he submitted, to take that matter into account would be when, if at all, the Silkstones sought to commence fresh proceedings. I reject that submission. It was material to the exercise of the deputy adjudicator's discretion, if he did have one, that one possible consequence of allowing the reference to terminate without a determination of the substantive rights would be that the dispute would resurface at a later time.

[43] The law has never favoured those who are willing to wound but afraid to strike. The Silkstones instigated the procedure before the Land Registry, provoked the reference by their objection and pursued it right up to the eve of a hearing at which they recognised that they would have inadequate evidence to establish their claimed right. There is no merit in their suggestion that they should be entitled as of right to deprive the adjudicator of jurisdiction at the last minute so as to preserve a right to prolong the litigation by relitigating it in a more expensive forum.

[44] The final point is the assertion that the deputy adjudicator should not have mentioned section 62, given that it had not been relied on by the Silkstones, and that they had been refused disclosure on this basis. I do not think that there is anything in this point. The decision of the deputy adjudicator was that the evidence did not establish the existence of the right of way. The Silkstones could have, but did not, take the opportunity to support their case based on section 62. There is no reason why the deputy adjudicator should not have recorded the absence of evidence on this basis, as well as on the existence of a right of way by prescription.

[45] It follows that I have reached the same conclusion as the deputy adjudicator, albeit by a slightly different route. I therefore propose to make an order dismissing the appeal.

Appeal dismissed.

Franks and another *v* Bedward and another

CHANCERY DIVISION

[2010] EWHC 1650 (Ch)

12 July 2010

BRIGGS J

Estates Gazette 16 October 2010

[2010] 41 EG 128

Land registration — Procedure — Cancellation order — Land Registration Rules 2003 — Application to register title to land cancelled by registrar on order of adjudicator and removed from day list — Adjudicator not granting stay of execution — Cancellation order overturned by High Court on appeal — Whether possible to restore application in day list with original priority date — Whether applicants for registration having to submit new application

In April 2005, the applicants applied to the Land Registry to register title by adverse possession to a strip of land bordering their property. The application was entered on the day list as being for the transfer of part of the land comprised in the title to the respondents' adjoining property. The respondents objected and the application was transferred to an adjudicator under section 73(7) of the Land Registration Act 2002.

In June 2006, the adjudicator ordered the registrar to cancel the registration application on the ground of procedural defaults by the applicants. No application was made or granted for a stay of execution of that order under r 45(1) of The Adjudicator to Her Majesty's Land Registry (Practice and Procedure) Rules 2003. The registrar deleted the record of the application on the day list a few days later. The High Court granted permission to appeal against the cancellation order and, in June 2007, allowed that appeal.

The registrar was notified only in December 2008 that the cancellation order had been set aside. Meanwhile, in October 2007 and June 2008, adverse interests, namely two charges in favour of mortgagees, were registered against the respondents' title. The registrar decided that the applicants' registration application could not be revived and re-entered on the day list with the same priority date of April 2005. The applicants therefore had to make a fresh application.

They applied to the court to reinstate their registration application with priority from April 2005. The Chief Land Registrar opposed the claim, arguing that the day list was a "real time" record, based on the actual time and date the entries were made on or cancelled from it, such that it was impossible to reverse the historical fact of the cancellation. The Attorney General appointed an advocate to the court to make submissions in support of the applicants' position.

***Held*: The application was allowed. (1) Where the court grants an appeal against a cancellation order, it can order the restoration of the registration application and its re-entry on the day list with its original priority date without the applicant having to make a fresh application. To hold otherwise would run counter to the general ability of an appeal court to restore a struck-out claim with retrospective effect. The understandable desire of the registrar to preserve the integrity of the day list as a real-time record is not sufficient to require the court to adopt a fundamentally different approach to appeals from the adjudicator. The registrar is required to record every pending application for registration in the day list; the effect of setting aside a cancellation order in respect of such an application is that that application will once again be pending: see r 12(1) of the Land Registration Rules 2003. (2) Where the court is aware of an intermediate incumbrancer that would be prejudiced by the restoration of a cancelled registration application, an order for the restoration of the application, and its re-entry on the day list with its original priority, should normally be made only on terms that require the applicant to respect the rights of the intermediate**

incumbrancer. If, on the hearing of the appeal, the court does not have the requisite information, it can either adjourn the appeal or make an order for re-entry on the day list but with leave for an aggrieved third party to apply to vary or set aside that order. In the latter case, the court's order should contain a direction to the registrar to notify any third party that appears to him, from the register and the day list, to be capable of being adversely affected. If any intermediate incumbrancer would be unfairly prejudiced by an order to restore the application with its original priority, for example, where the prejudice to it could not be remedied by any form of conditional order, the court might require the applicant to make a fresh application, with a direction that steps taken in the original application be taken into account in the new application so as to avoid unnecessary work and costs. (3) In the instant case, where the court was ignorant of the precise nature of the two mortgagees' rights or whether the restoration of the application would cause them real rather than theoretical prejudice, the appropriate order was to direct re-entry of the registration application on the day list with its original entry date but to allow any aggrieved third party to apply for a variation or setting aside.

***Per curiam*: It was desirable that the rules and practice relating to stay a pending appeal should be reviewed to make a stay more readily available as a sensible precaution against the difficulties revealed by the instant proceedings. In the meantime, the adjudicator might change his practice to ensure, as far as possible, that the registrar did not have to act on an order for cancellation until the expiry of a period that was sufficient for the applicant to obtain both permission to appeal and a stay, with time built in for obtaining representations from other parties. A two-month period before implementation of a cancellation order, built into the order when made, would be adequate to enable the necessary processes to be undertaken.**

No cases are referred to in this report.

This was the hearing of an application by the applicants, Alexander Franks and Irene Franks, for an order directing the Chief Land Registrar to reinstate an application for registration of title to land on the day list maintained by him under the Land Registration Rules 2003, with its original date of priority, following the setting aside on appeal of a cancellation order in respect of that application.

The applicants and respondents appeared in person; Amanda Tipples (instructed by the Treasury Solicitor) appeared as advocate to the court by appointment of the Attorney General; Timothy Morshead (instructed by the Treasury Solicitor) represented the Chief Land Registrar.

Giving judgment, **Briggs J** said:

[1] This is an application by Mr Alexander and Mrs Irene Franks for directions sufficient to ensure the effective reinstatement of an application they made to the Land Registry on 15 April 2005 for registration of title to land that they claim by adverse possession. Their application was ordered to be cancelled by the adjudicator on the ground of procedural default on 6 June 2006, but that order was set aside by me on the Franks' appeal, on 20 June 2007. On being belatedly informed of my order allowing the appeal, by letter from the adjudicator dated 19 December 2008, the Chief Land Registrar (the registrar) declined to re-enter the Franks' application in the day list maintained by him under r 12 of the Land Registration Rules 2003 (the LRR), suggesting that it was impossible for him to do so and that the Franks should instead make a fresh application to the Land Registry for registration. After further prolonged delay, caused in part by the Franks having no legal representation, they applied, by an application notice on 4 January 2010, to have their 2005 application reinstated.

[2] The unusual and protracted procedural history of the Franks' application for registration has thrown up some real difficulties of interpretation of the Land Registration Act 2002 (the Act) and of the LRR, as well as The Adjudicator to Her Majesty's Land Registry (Practice and Procedure) Rules 2003 (the AR), and revealed serious difficulties in the practice and procedure of the Land Registry adjudicator system, in connection with litigants' rights of appeal from the adjudicator. In short, where an application for registration is ordered to be cancelled by the adjudicator (whether, as here, for procedural default or on the merits) and the applicant subsequently succeeds in an appeal, unless the applicant has in the meantime obtained a stay of the order for cancellation the full reinstatement of the application risks causing loss to persons obtaining interests in the land the subject matter of the application during the period between cancellation and reinstatement, for which the scheme of the Act and the rules provide no compensation. Conversely, if the successful appeal does not lead to a full reinstatement of the application, the applicant's apparent success on appeal may prove to be a pyrrhic victory, since the application will lose priority over all other applications for or registrations of interests in the subject land made since the application was first initiated. Again, the successful appellant would have no remedy for compensation for the loss of priority suffered by him between the date of his application for registration and the date on which his appeal succeeded.

[3] Furthermore, the present practice and procedure of the Land Registry adjudicators appears to be such that, for reasons that will appear, it is in practice extremely difficult even for a well-represented applicant to obtain a stay of an order for cancellation before it has been acted on by the registrar, with the consequences to which I have just referred. It became apparent to me when giving directions for the hearing of this application that these difficulties were of a sufficient gravity and complexity to require careful submissions, both for the Franks (who remain litigants in person) and on behalf of the registrar, as guardian of the public interest in the simplicity and reliability of the system of registration now enshrined in the Act and the rules made pursuant thereto. As a result, the court has been greatly assisted by Ms Amanda Tipples, appointed by the Attorney General as an advocate to the court, to advance arguments in favour of a full restoration of the Franks' registration application, and by Mr Timothy Morshead, who appeared for the registrar. The Franks appeared in person, as did Mrs Angela Bedward, one of the registered owners of the land subject to the Franks' claim. Entirely understandably, none of them contributed significantly to the argument, since all of them remain, as they have been for some time, litigants in person.

Procedural history

[4] Mr and Mrs Franks made their registration application, dated 15 April 2005, on the prescribed form AP1, seeking the registration of a strip of land along the boundary between their property, known as 50 Norfolk Avenue, Slough, Berkshire SL1 3AD, and land registered in favour of a Mr and Mrs Bedward under title no BK361792. It was received by Gloucester District Land Registry on 18 April 2005 and entered in the day list for that day as an application for a transfer of part of the land comprised in that title. The application was objected to by the Bedwards, with the result that, on 15 July 2005, it was transferred to the office of the adjudicator pursuant to section 73(7) of the Act.

[5] There then ensued an unsatisfactory procedural history that included an unless order by the adjudicator, made on 28 November 2005, for service by the Franks of their statement of case and a further unless order on 11 May 2006, to the effect that unless the Franks served a full copy of their statement of case on the adjudicator and the Bedwards by 5pm on 18 May 2006, the adjudicator would require the registrar to cancel their registration application.

[6] On 6 June 2006, the adjudicator made an order requiring the registrar to cancel the registration application. It was posted both to the registrar and to the parties and, on 14 June, it was acted on by the registrar, by deleting the record of the registration application in the day list. From that date, no one searching the Bedwards' title (including for that purpose the day list) would know that the Franks' registration application had ever been made.

[7] At that time, the Franks were still legally represented, but, by 7 September 2006, they had become litigants in person. On 13 October 2006, the three-year time limit for applying for adverse possession registration as a transitional overriding interest under section 97 of and Schedule 12 to the Act expired. It is probable (although

I make no finding one way or the other) that the Franks were not then in occupation of the disputed land because, in proceedings in Slough County Court between the same parties, they had been restrained from entering onto the disputed land by a consent order pursuant to which the Bedwards had also undertaken not to enter onto it, in each case pending the outcome of the dispute as to ownership.

[8] Coincidentally, the Franks began to take steps on their own to challenge the cancellation of the registration application, initially by correspondence with the adjudicator, by e-mails sent on 19 and 24 October 2006, that the adjudicator decided to treat as an application for permission to appeal, which he purported to grant, albeit without a stay of his cancellation order, on 27 October 2006. He did so without obtaining representations from the Bedwards, contrary to AR45(4), which required him to do so, and although the time for appealing had expired before the arrival of the Franks' e-mails he did not purport to extend time.

[9] Permission to appeal out of time was in due course given by Henderson J on the Franks' paper application with detailed reasons on 26 March 2007, and the appeal came on for hearing before me on 20 June 2007, at which the Bedwards were represented by counsel and the Franks appeared in person.

[10] I allowed the appeal, taking the view that the cancellation of the Franks' registration application was not a proportionate response to the procedural muddle that had preceded the adjudicator's order, and I substituted for the adjudicator's order a further unless order in the following terms:

> Unless the Applicants/Appellants do attend the offices of the Respondents' solicitors Irwin Mitchell, at Riverside West, 1 Millsands, Sheffield S3 8NH on a date to be agreed during the month of August 2007 to identify the document or documents said to constitute their statement of case, for which purpose the Respondents' solicitors shall provide to the Applicants/Appellants a list of available dates, the Adjudicator will require the Chief Land Registrar to cancel the original application dated 18th April 2005 to register title based on adverse possession.

[11] It is now apparent that, as at 20 June 2007, no one had acquired or sought to apply to register an interest in the disputed land, either during the period following the cancellation of the Franks' registration application on 14 June 2006 or indeed at any time since the entry of that application on the day list on 18 April 2005. The Franks' registration application could therefore have been re-entered on the day list, with priority from April 2005, without causing any risk of harm to anyone relying on the register (or day list) in the meantime.

[12] No transcript has been obtained of the hearing on 20 June 2007 or of my *extempore* judgment given at its conclusion, and I cannot now recall whether I inquired (in particular of Mr Michael Bowmer, counsel for the Bedwards) whether a reinstatement of the Franks' application would prejudice any third-party rights. It is common ground that both I and the parties in court on 20 June 2007 assumed that the effect of my allowing the appeal, and making the order that I have recited, was that the Franks' application would be reinstated for all purposes as though it had never been ordered to be cancelled.

[13] It took until 31 July 2008 before a minute of the order that I had made on 20 June was sealed, after preparation by Mr Bowmer (to whom I had given that task, the Franks being in person). In the meantime, owing to a (possibly well-intentioned) failure by the Bedwards' solicitor to provide dates for the Franks pursuant to the June 2007 order, it had become necessary for me to vary that order by substituting November for August as the period for compliance. Compliance was in fact achieved by the Franks, with the result that the specified sanction for non-compliance never came into effect.

[14] For reasons that I have been unable to ascertain, it took a further year before the registrar was notified of the setting aside of the adjudicator's order for cancellation. Paragraph 23.8B(4) of CPR Practice Direction 52 provides that an appellant must serve a copy of the appeal court's decision on the registrar as soon as reasonably practicable and in any event within 14 days. In fact, notification occurred, as I have already described, during correspondence between the adjudicator and the registrar in December 2008. Nothing had therefore been done to record the reinstatement of the Franks' registration application in the day list, and in the meantime two adverse interests had been registered against title no BK361792 (including the disputed land). The first was a legal charge dated 29 October 2007 in favour of Southern Pacific Mortgage Ltd. The second was a general equitable charge dated 6 June 2008 in favour of Welcome Financial Services Ltd. It is not known whether either of the chargees made a priority search before entering into the transactions giving rise to those charges or, therefore, whether the chargees relied (potentially to their detriment) on the absence of any reference to the Franks' registration application in the day list.

[15] The registrar was notified of my June 2007 order in a letter from the adjudicator dated 19 December 2008, which concluded by inviting his view as to whether:

> the cancelled application can be in any way revived and re-entered on the Day List with the same original priority date (assumed to be 18th April 2005) so that the Adjudicator could make a further substantive order on it or whether you consider Mr and Mrs Franks will need to make a fresh application.

The response, by letter from the Gloucester office of the Land Registry dated 9 January 2009, was that:

> It is not possible to revive an old application once it has been cancelled. It appears that the applicant should make another application to Land Registry if the matter is to be revived.

[16] The Franks were understandably reluctant to have to start all over again, having won their appeal in June 2007. They sought my assistance, initially by telephone to my clerk and then by letter dated 17 February 2009. Recognising that a potentially important and wide-ranging point of principle might be at stake, I directed on 19 February 2009 that the Franks should elect between seeking an order that the registrar reinstate their registration application or making a fresh application. They chose the first of those two options on 17 April 2009, and after further delays caused largely by their lack of legal qualifications or experience, I was able to give directions for this hearing on 30 March 2010, enabling the registrar to be represented and having received confirmation that an advocate to the court would be appointed. The registrar had in the meantime set out in substantial detail his reasons for objecting to the reinstatement of the registration application with priority from April 2005. The essence of his detailed and carefully reasoned objection was that since the day list is a "real time" record, based precisely on the actual time and date on which entries are made in it or cancelled from it, it would be destructive of the integrity of the day list to include any entry capable of having retrospective effect, having regard to the important function that the day list performs in the context of registered conveyancing. I shall have more to say about the registrar's grounds of objection in due course.

Statutory scheme

[17] The essential purpose of the scheme recreated by the Act in 2002 was to provide a system of state-guaranteed registered title. Subject to limited exceptions, the register is intended to provide a comprehensive and accurate reflection of the state of title to registered land at any given time, and in the exceptional circumstances where it does not, for example where rectification is required, persons adversely affected are entitled to compensation: see, for example, *Megarry & Wade: The Law of Real Property* (7th ed), at pp130132, and para 1.5 of the Law Commission's report *Land Registration for the Twenty-First Century: A Conveyancing Revolution* (2001) Law Com No 271.

[18] The registrar is required to maintain a register of title that, as is apparent from LRR 2(2), includes an individual register for each registered estate vested in a proprietor. By section 68(1)(d) of the Act and Part 2 of the LRR (rr 10 to 12), the registrar must also keep an index or indices. They include an index map, an index of proprietors' names and the day list. His duties in respect of the day list are set out in LRR 12, of which the following part is material:

(1) The registrar must keep a record (known as the day list) showing the date and time at which every pending application under the Act or these rules was made and of every application for an official search with priority under rule 147.

The Franks' registration application is a type of pending application that the registrar is therefore required to record in the day list.

[19] By LRR 15, detailed provision is made as to the time at which an application for registration is to be taken as having been made. Subrule (1) provides as follows:

(1) An application received on a business day is to be taken as made at the earlier of

(a) the time of the day that notice of it is entered in the day list, or

(b) (i) midnight marking the end of the day it was received if the application was received before 12 noon, or

(ii) midnight marking the end of the next business day after the day it was received if the application was received at or after 12 noon.

Similar provision is made in respect of applications received otherwise than on a business day and, under subrule (3), an application is received when it is delivered to the designated proper office of the Land Registry, subject to irrelevant exceptions.

[20] LRR 20 provides that, subject to an irrelevant exception:

(1) Any entry in, removal of an entry from or alteration of the register pursuant to an application under the Act or these rules has effect from the time of the making of the application.

[21] Pausing there, if the Franks' registration application, recorded in the day list as having been made on 18 April 2005 had simply proceeded to adjudication and been successful, the registration of the Franks as owners of the disputed land would have taken effect, pursuant to LRR 20, from that date, with priority over every subsequent registered entry or application for registration made thereafter. Furthermore, any person considering acquiring an interest in the disputed land after that date would, on a priority search, have been notified of the Franks' pending application and could not have complained of the postponement of any interest thereafter acquired.

[22] Section 72 of the Act and LRR 147 to 149 make similar provision in respect of priority protection. In summary, an application for an entry in the register is protected if it is one to which a priority period relates and is made before the end of that period. By section 72(2), where an application for an entry in the register is protected, any entry made in the register during the priority period relating to that application is postponed to any entry made in pursuance of it. In order to protect persons dealing with the registered land from inadvertent loss of priority to those with priority protection, LRR 12(2) provides that:

The entry of notice of an application for an official search with priority must remain on the day list until the priority period conferred by the entry has ceased to have effect.

Part 3G of Schedule 6 to the LRR provides that an applicant for an official search is to be given notice of the entry of any relevant pending applications affecting the registered title entered on the day list.

[23] Mr Morshead submitted, and I accept, that the day list therefore plays an essential part in registered conveyancing by protecting persons contemplating acquiring interests in the registered land from doing so in ignorance of applications for registration that, under the provisions that I have summarised, would take effect in priority to that person's contemplated interest.

[24] I must briefly mention Mr Morshead's submission, which I also accept, that it is by no means clear that the provisions for indemnity in Schedule 8 to the Act would assist a person who had, during the period between the cancellation of an application and its restoration on appeal, searched the register (and day list) and then acquired an interest in the relevant registered land in ignorance of the temporarily cancelled application. Generally speaking, compensation is available where there has been either rectification (which itself usually depends on some mistake) or some other relevant mistake. The removal from the day list of a cancelled application is not a mistake merely because it might be restored on appeal. Furthermore, although the re-entry of an application restored on appeal might have something of the quality of rectification (loosely speaking), it would be rectification of an index rather than rectification of the register. For reasons to which I shall come, I do not consider that the day list is, within the meaning of the Act and of the LRR, either a register or part of a register.

[25] I turn to the provisions conferring powers on the adjudicator. For present purposes, I am concerned with the AR that were made in 2003 prior to their amendment in 2008, although the amendments are not in any respect material. Section 73(7) of the Act requires any objection to an application for registration that cannot be disposed of by agreement to be referred to the adjudicator. His power to determine matters referred under section 73(7) is conferred by section 108(1)(a). Section 109(2) provides for rules to be made to regulate the practice and procedure to be followed with respect to proceedings before the adjudicator. Section 111 provides for an appeal to the High Court by a person aggrieved by a decision of the adjudicator. That engages CPR 52.10(1) and (2)(a), by which the appeal court has all the powers of the lower court, including a power to affirm, set aside or vary any order or judgment made or given by the lower court. By section 112, an order of the adjudicator is to be enforceable as an order of the court.

[26] AR 39 provides that:

(1) Where there is a hearing, the substantive decision of the adjudicator may be given orally at the end of the hearing or reserved.

(2) A substantive decision of the adjudicator, whether made at a hearing or without a hearing, must be recorded in a substantive order.

(3) The adjudicator may not vary or set aside a substantive decision.

By AR 40:

(1) A substantive order must

…

(e) state any steps that must be taken to give effect to that substantive decision;…

By AR 40(2), the substantive order must be served by the adjudicator on every party to the proceedings, and where it requires the registrar to take action, the registrar.

[27] AR 55 confers a power on the adjudicator to impose sanctions for a failure to comply with a direction given by him either on the application of any other party or of his own motion. By subrule (2), a sanction may include requiring the registrar to cancel the original application in whole or in part. By subrule (4), a sanction of that kind is to be treated as a substantive decision on that matter. The surprising result is that where the adjudicator orders the registrar to cancel an application as a sanction for procedural default, AR 39(3) prevents him thereafter from varying or setting aside that decision.

[28] AR 45 gives the adjudicator a very limited power to stay the implementation of a decision of his, pending the outcome of any appeal. AR 45(1) provides that:

Where a party is granted permission to appeal, the adjudicator may, of his own motion or on the application of a party, stay the implementation of the whole or part of his decision pending the outcome of the appeal.

By subrule (2), a party wishing to apply for a stay is required to make its application to the adjudicator at the same time it applies to the adjudicator for permission to appeal. By subrule (4), the adjudicator must, before reaching a decision whether to grant permission to appeal a decision or to stay implementation of a decision, allow the parties the opportunity to make representations or objections.

[29] Mr Morshead submitted that the requirement in AR 3(3)(b) to interpret the adjudicator's rules in accordance with the overriding objective justified interpreting AR 45 as permitting the adjudicator both to grant a stay pending the hearing of an application to him (or to the High Court) for permission to appeal and to grant a stay where, after refusing permission to appeal himself, the unsuccessful party sought permission to appeal to the High Court. Although I agree that a wider power to grant a stay of the type to which Mr Morshead referred would probably serve the overriding objective more usefully than the power to

stay conferred by AR 45, I do not consider that the clear limitations in that rule can be ignored by a process of interpretation. Read as a whole, the rule unmistakably limits the adjudicator's power to grant a stay to the occasion on which he decides to grant permission to appeal. He can neither do so in advance of that decision or *a fortiori* if he decides not to give permission to appeal. Furthermore, he cannot grant a stay, even of his own motion, unless he has first allowed the parties an opportunity to make representations or objections. AR 45 is in those respects in sharp contrast with the generality of the court's power to grant a stay in CPR 3.1(2)(f). In my view, the contrast must have been deliberate.

[30] It appears that the present practice of the adjudicator when making a decision that includes a direction to the registrar to cancel an application is to communicate his decision more or less simultaneously to the registrar and to the parties and not to set a deferred date on which the registrar is to cancel the application, designed to give the unsuccessful applicant time both to seek permission to appeal and a stay. I say that this appears to be the present practice of the adjudicator both because of the facts of the present case, which I have described, and because counsel before me did not suggest that this case revealed anything atypical. It is fair comment that the adjudicator has not been invited to appear on this application (although he is aware of it) and that it may be that there is no such uniform practice. I shall none the less proceed on the assumption, but without a finding, that the practice is as it appears to be.

[31] The result of such a practice is that the unsuccessful applicant for registration, even if competently represented, may in practice find it impossible to take effective steps for the obtaining of a stay before the registrar has cancelled his application. If he applies to the adjudicator, he cannot expect a decision until time has been given to his opponent to object, and he will obtain no stay if the adjudicator refuses permission to appeal, even though the applicant may subsequently obtain permission to appeal from the High Court and in due course succeed on his appeal.

[32] If he applies to the High Court for a stay, I consider that the High Court judge's powers under the CPR are sufficient to enable him to grant a stay pending the hearing of an application (to the adjudicator or to the High Court) for permission to appeal. I was reminded by Mr Morshead that in this regard the High Court is exercising statutory appellate powers rather than any inherent jurisdiction and that, pursuant to CPR 52.10, those powers may in certain respects be limited to the powers available to the adjudicator. None the less, CPR 52.7 plainly assumes that the appeal court has the power to stay the order appealed against (as does Practice Direction 52, in para 23.8B) and the general power to stay in CPR 3 is expressed to be (like all other powers in r 3) in addition to any power conferred by any enactment. It is, however, hardly satisfactory that the aggrieved recipient of a cancellation order can obtain a stay in time only by an emergency application to the High Court.

[33] I turn to the question of whether the High Court can or should restore an application for registration on the hearing of an appeal from an order by the adjudicator that it be cancelled in circumstances where the order for cancellation has not been stayed and has been complied with by the registrar before the hearing of the appeal. The registrar's written case in response to the Franks' present application is that the cancellation of the registration application is "an irreversible historical fact which cannot now be altered it has taken place, with the application being removed from the Daylist on 14th June 2006 at 14.29.08". Anticipating a response from the Franks (which duly appeared in Ms Tipples' skeleton argument) that his power to alter the register under paras 2 and 5 of Schedule 4 to the Act might be sufficient, the registrar submitted that those powers related only to the register and deliberately not to the day list because of its vital status as a real time record. Paragraph 2 of Schedule 4 empowers the court to direct an alteration of the register either to correct a mistake or to bring it up to date. Paragraph 5 empowers the registrar to do the same, of his own motion, for the same purposes.

[34] The difficulty with Ms Tipples' otherwise attractive submission is (as she eventually acknowledged) the difficulty of treating the day list as or as part of the register. In my judgment, the day list is not the register or part of the register. It is clearly and distinctly identified in the LRR as an index, in a section separate from those dealing with the register itself. Furthermore, and conclusively in my judgment, LRR 12(3) shows that the distinction between the day list and the register was deliberate. It provides that:

> Where the registrar proposes to alter the register without having received an application he must enter his proposal on the day list and, when so entered, the proposal will have the same effect for the purposes of rules 15 and 20 as if it were an application to the registrar made at the date and time of its entry.

I consider that this subrule clearly shows that the day list is an index that contains information about alterations that may be made to the register so as to give appropriate warning to interested parties. This is not the register itself, nor part of it.

[35] For his part, Mr Morshead adopted a rather more pragmatic, less doctrinaire approach than the registrar had done in his written case. He began by pointing out, correctly, that in the real world it is frequently the case that the appeal court cannot restore the *status quo ante* when allowing an appeal from an order of the court or tribunal below that has not been the subject of a stay. He gave the obvious example of a judgment for the payment of a debt or damages, where the judgment creditor is paid before appeal and is unable to repay when the appeal is allowed. He submitted that the registrar has no alternative but to cancel the entry of an application for registration if ordered to do so by the adjudicator. He acknowledged that if the application for registration was restored by the court, its restoration ought to be recorded in some way in the day list. In my judgment, that concession is correct, if for no other reason than because LRR 12(1) requires the registrar to record in the day list every pending application for registration, and the effect of the setting aside of an order for the cancellation of a pending application is that, at least, it becomes a pending application again. His sticking point was that the new entry in the day list should not appear to suggest, contrary to the truth, that the application had been pending during the period between its cancellation and the order made for its reinstatement on appeal. Both he and Ms Tipples appeared for a time to be in agreement that the court could at least require the application for registration to be re-entered in the day list as an application made at the moment of its re-entry.

[36] In his helpful oral submissions, Mr Morshead eventually submitted that, at least in theory, the appeal court might make any one of the following orders by analogy with the adjudicator's power under AR 40(1)(e):

(a) An order that the registrar re-enter the registration application on the day list, with its original date of entry, but with liberty to any aggrieved third party with an interest in the relevant land to apply to have the order varied or set aside.

(b) Allow the appeal but treat the steps made in the cancelled application as steps in a new application for registration to be made by the appellant and waive any duplicate fees.

(c) Adjourn the appeal for enquiries as to the state of the register and of the day list so as, if possible, to avoid the need for liberty to apply under (a).

(d) Direct the re-entry of the registration application, but with a postponed date of entry.

[37] As a matter of discretion, Mr Morshead submitted that save in the most exceptional case, the court should ordinarily choose between (b) and (d) because the applicant for registration had sufficient alternative avenues for the protection of his priority not to need the assistance of the appeal court by directing a re-entry of his application with retrospective effect.

[38] An order in the form of Mr Morshead's alternative (b) or (d) would of course cause the applicant for registration to lose priority for the entire period between the date of his original application and the date of the entry (or re-entry) in the day list following the order made on appeal. He would be vulnerable, not merely to applications for and entries on the register made during the period following the cancellation of his registration application but applications already made before cancellation. Mr Morshead submitted that it would

A be unlikely that any adverse interest would be made the subject of registration or an application for registration for as long as the day list showed the appellant's own application (before cancellation), but with respect I disagree. There would, for example, be nothing surprising in a mortgage of a large registered title notwithstanding a pending application for a transfer of a very small part of it on the basis for a claim for adverse possession. The mortgagee would probably be unconcerned by a minor loss of subject matter. Similarly, a purchaser from the registered owner might be prepared (for an appropriate discount) to take the risk of the adverse possession application succeeding.

[39] The alternative remedies that Mr Morshead submitted were B available to an applicant for registration were as follows. First, he suggested that an applicant claiming adverse possession could apply for a unilateral notice to protect his interest. Ms Tipples did not suggest otherwise, but it seems to me bizarre to think that the statutory scheme required an applicant for registration not merely to lodge an application in the appropriate form but also to lodge a belt and braces unilateral notice in case that application was cancelled and then restored on appeal. By contrast, a unilateral notice entered only at the time of the hearing before the adjudicator, although it might provide protection during the period of temporary cancellation, would not provide any protection for interests registered or applied for during the period C from the original application for registration and the (*ex hypothesi*) much later date of the unilateral notice. Accordingly, although an ingenious suggestion, it seems to me to have been neither intended by the legislation nor to be practicable in the real world as a complete form of protection. It would plainly be inappropriate in the case of a cancellation order made on the merits.

[40] Mr Morshead's second alternative was that the claimant for adverse possession should protect his right by continued occupation of the land in question so as to confer on his rights the status of an overriding interest. Again, that might often be so, but it is not every case in which an applicant is able to insist on remaining in occupation. In D the present case, both the Franks and the Bedwards agreed, pursuant to a consent order, that neither should occupy the land in dispute, sensibly turning it into a sort of demilitarised zone. In any event, the problem thrown up by the present case is by no means limited to applications for registration based on adverse possession.

[41] The third suggested solution is of course an application made in good time for a stay of the cancellation order. This would, if obtained in time, be a complete solution but for the reasons that I have given, the present powers of the adjudicator to grant a stay, and the practice, apparently adopted, of the simultaneous communication of decisions requiring cancellation to the registrar and to the parties does not lend E itself to the obtaining of a stay in time.

[42] Finally, Mr Morshead suggested that the applicant for registration could always protect himself from a continuing loss of priority by the making of a fresh application, for example on learning of his failure before the adjudicator and pending an appeal. Again, this solution would give the applicant protection against loss of priority during the cancellation period but no protection against the loss of priority between the date of his original application and the date of the belt and braces new application. Furthermore, it would on the face of it render any appeal almost nugatory, save in the limited sense that he might, if otherwise successful, obtain an order that steps taken in the F first application be treated as steps in the second and the setting aside of any adverse order for costs. Finally, in the case of a cancellation order made on the merits, it would be tantamount to an abuse of process.

[43] In my judgment, the alternatives available to the court as a matter of jurisdiction when disposed to allow an appeal against an order for the cancellation of an application for registration are, broadly, those proposed by Mr Morshead, with the exception that I doubt whether his alternative (d) works, notwithstanding Ms Tipples' support for it. The difficulty with an order that an application for registration be restored with a postponed entry date is that it appears to run counter to the provisions of LRR 15 and 20. LRR 20 provides that an application for registration, if it succeeds, is to take effect from the time at which it was made and LRR 15 provides that an application is to be taken as G having been made on the earlier of the day (or the day after the day) on which it was actually received and the date of its entry in the day list. A restored application would, in my judgment, still have been made when it had originally been made even if re-entered in the day list at a much later date (namely after a successful appeal). Accordingly, Mr Morshead's alternative (d) would not protect bona fide intermediate incumbrancers who searched the register and the day list during the period of temporary cancellation and found nothing.

[44] Before looking at the remaining alternatives, I record that I have not been persuaded that I should adopt the doctrinaire approach favoured by the registrar, namely that an application once cancelled is H inherently incapable of being restored or re-entered on the day list on appeal. Although I appreciate the force of his objection, based on the preservation of the integrity of the day list as a real time record not admitting any retrospectivity, a conclusion that the court could never restore a cancelled application, so as to require it to be re-entered in the day list with its original priority, would seem to me to be capable of causing injustice to successful appellants without any corresponding benefit to anyone else. For example, in the present case, no third party would have been prejudiced by a re-entry in the day list of the Franks' application, with its original priority, after their successful appeal in June 2007 and before the registration of the first adverse incumbrance J some four months later. An order in the form of Mr Morshead's type (a) would simply not have led to any application to set aside by any interested party. However, it would have preserved for the Franks the ability to treat their rights as a transitional overriding interest. Those rights would be irretrievably lost by the time of their successful appeal if their application could not be restored with effect from a date prior to October 2006, when the three-year period for applications of that type expired.

[45] The view that an application for registration once cancelled pursuant to an order of the adjudicator cannot be restored on appeal runs counter to the general ability of an appeal court to restore a struck-out K claim with retrospective effect so as, for example, to save it from having become statute-barred in the meantime, and I am not persuaded that the understandable desire of the registrar to preserve the integrity of the day list as a real time record is sufficient to require the court to adopt a fundamentally different approach in respect of appeals from the adjudicator.

[46] None the less, I readily accept that where an intermediate incumbrancer would be prejudiced by the restoration of a cancelled application, this would be the most powerful reason why, as a matter of discretion, the court would not direct its restoration by way of appeal, L all the more so if compensation to the intermediate incumbrancer under the statutory scheme was, as it appears to be, unavailable. It follows that if, on appeal, the court is aware that there exists such an intermediate incumbrancer, it would not ordinarily direct the restoration of the application and its re-entry in the day list with its original priority, otherwise than on terms that required the successful appellant fully to respect the intermediate incumbrancer's rights. If, on the hearing of an appeal, the court did not have the requisite information available, there is no reason why the appeal could not be adjourned (Mr Morshead's alternative (c)) or, alternatively, why the court could not make an order for re-entry on the day list, giving liberty to any aggrieved third party to apply to have it varied or set aside (Mr Morshead's alternative (a)). In M that regard, I do not consider that the High Court would be incapacitated by AR39(3): see CPR 3.1(7). Any such order would need to include a direction to the registrar to notify any third party appearing to him (from the register and the day list) to be capable of being adversely affected.

[47] In a case in which there were intermediate incumbrancers that would be unfairly prejudiced by the making of an order for restoration of the application with its original priority, the court may have no alternative than to require the applicant to make a fresh application and to direct that steps taken in the cancelled application be treated as steps in the new application, so as to save the incurring of unnecessary work and cost. Such a case would arise if, for example, the prejudice that would otherwise be caused to an intermediate incumbrancer could not

A be remedied by any form of conditional order or where the appellant was unwilling to comply with the terms of any proposed conditions.

[48] I do not pretend that any of the alternatives that I have outlined above represent a simple, easy, neat or necessarily cost-effective solution to the problems thrown up by a successful appeal against a cancellation order not protected by a stay. The obvious solution to these difficulties is for the rules and practice relating to stay pending appeal to be reviewed so as to make a stay more readily available as the sensible precaution against the difficulties revealed by these proceedings. In that context, I bear in mind that the present jurisdiction of the adjudicators may be about to be replaced by that of the First Tier Tribunal, with B an appeal to the Upper Tribunal rather than to the High Court. In that context, it seems to me to be likely that a replacement of the AR and the relevant provisions of the Act will have to be undertaken in any event, and that the difficulties thrown up by the present case could be addressed as part of that process.

[49] Even in advance of that process, I can see no reason why steps cannot be taken by the adjudicator, by way of a change to the present apparent practice, that ensure as far as possible that an order for cancellation of an application does not have to be acted on by the registrar before the expiry of an appropriate time during which the applicant may obtain both permission to appeal and a stay. That C period would need to be specified in the order for cancellation and to be sufficiently long to enable the prospective appellant to apply for permission and a stay from the adjudicator (with time built in to permit representations or objections from other parties) and, if unsuccessful, to obtain permission and a stay from the High Court. I bear in mind in that context that AR 39(3) does not permit the adjudicator to vary his cancellation order by extending the time before which it is to take effect if he refuses permission to appeal so as to enable the prospective appellant to try again in the High Court. Doing the best I can, it seems to me that nothing less than a two-month period before the implementation of a cancellation order (built into the order itself D when first made) would be sufficient to enable the processes to be undertaken in a way that made a successful appeal, unprotected by a stay, wholly exceptional.

[50] Difficulties of this kind are rare in ordinary court proceedings because orders are generally made in court at a time when (often having received a draft judgment) the aggrieved party can immediately seek a stay in the presence of its opponent. They are much more likely to arise where a decision is published by letter, accompanied by an order that gives it immediate practical effect.

Present case

[51] My order in June 2007 allowing the Franks' appeal did not E expressly spell out any steps that the registrar was required to take in consequence. In that respect, it probably ran contrary to AR 40(1)(e) (to which my attention was not drawn at the time) as applied to the High Court by CPR 52.10(1). None the less, it provided by necessary implication for the restoration of the Franks' registration application F and was meaningless on any other basis. In my judgment, the registrar would, once belatedly apprised of it in December 2008, have been better advised to seek directions as to its implementation, bearing in mind his obligation to record pending applications for registration in the day list. None the less, the present hearing is in substance, even if not in form, capable of being dealt with as though he had.

[52] This is a case in which it would neither be just to the Franks to require them to make a new application (even with a provision that steps taken in the original application stand as steps in the new application) nor necessarily just to the two intermediate incumbrancers simply to make an order for the re-entry of the registration application G in the day list with priority from its original entry date of 18 April 2005. The first alternative would forever deprive the Franks of the availability of the right to treat their adverse possession as a transitional overriding interest. The second alternative may unfairly prejudice the two chargees.

[53] The court knows nothing about the precise nature of the rights of the two chargees, still less whether restoration of the Franks' application would cause real rather than merely theoretical prejudice. The adverse possession claim is for a modest part of the Bedwards' property, and it may well be that the value of the chargees' security rights may be more likely to be increased by a final determination of this long-running boundary dispute than decreased by the abstraction of some small part H of the property from the land comprised in their security.

[54] In those circumstances, the choice lies between making an order for the re-entry in the day list of the Franks' registration application with its original entry date, with liberty to the two chargees to apply to vary or set it aside, and a yet further adjournment of this long-running appeal to ascertain, if possible, the chargees' views.

[55] I prefer the first of those alternatives. It may well be that the chargees will, on enquiry, have no objection at all. Alternatively, if they do, the Franks may be able to provide satisfactory undertakings that preserve the chargees' rights. Either of those outcomes would save the costs and delay of a further hearing. I

[56] I am satisfied that it would be wrong to require the Franks to make a fresh application now unless compelled to that conclusion as the only way of avoiding irreparable prejudice to the chargees. That would deprive the Franks of the benefit of a transitional overriding interest through no fault of their own. That would merely provide a windfall benefit to the Bedwards, as the consequence of the adjudicator's order that, as I concluded on appeal, should not have been made.

[57] I will hear submissions on an appropriate form of order. It will need to include liberty to apply, not only to the two chargees but also to any third party aggrieved by the order, and for the registrar to be directed to notify any person appearing to be potentially prejudiced as appears from the state of the register and the day list on the day he J makes the re-entry.

Application allowed.

Somerfield Stores Ltd *v* Spring (Sutton Coldfield) Ltd (in administration) (No 2)

CHANCERY DIVISION: BIRMINGHAM DISTRICT REGISTRY

[2010] EWHC 2084 (Ch)

4 August 2010

HH JUDGE DAVID COOKE, sitting as a deputy High Court judge

Estates Gazette 27 November 2010

[2010] 47 EG 142

Landlord and tenant — Landlord and Tenant Act 1954 — New tenancy — Lease of business premises — Application for new tenancy — Landlord opposing application on ground of intention to redevelop within section 30(1)(f) of 1954 Act — Tenant applying for summary judgment dismissing ground of opposition — Whether landlord required to show requisite intention at date of anticipated trial or date of summary judgment hearing

The appellant held three leases of a supermarket and adjoining land for terms expiring in March 2008. In 2006, the respondent acquired the freehold with a view to redevelopment. In 2007, the appellant served notices on the respondent, under section 26 of the Landlord and Tenant Act 1954, requesting new tenancies. By its counternotices, the respondent indicated that it would oppose the grant of new tenancies under section 30(1)(*f*) of the 1954 Act, on the basis that, on the termination of the existing tenancies, it intended to demolish, reconstruct or carry out substantial works of construction to the demised premises.

The appellant applied to the court for the grant of the new tenancies and an order was made for a trial of the respondent's ground of opposition as a preliminary issue. In February 2009, the respondent went into administration. The application had not then proceeded to trial, but the appellant obtained the court's permission to continue the proceedings pursuant to para 43(6) of Schedule B1 to the Insolvency Act 1986: see [2009] EWHC 2384 (Ch); [2009] 3 EGLR 57; [2009] 48 EG 104. A directions timetable was set with an anticipated trial date of between March and May 2010. The appellant applied for summary judgment rather than wait for the full trial; that application was heard and dismissed in October 2009.

The appellant appealed. The central issue was the date at which the respondent's intention fell to be determined for the purposes of a summary judgment application in respect of a claim for new tenancies under the 1954 Act. The appellant contended that it fell to be determined by reference to the date of the summary judgment hearing, while the respondent argued that it was referable to the anticipated trial date.

***Held*: The appeal was dismissed. The date of the hearing at which the necessary intention must be shown to exist is always the date of the substantive trial of the landlord's ground of objection. The statute envisages a hearing at which evidence is tested and facts found for the purpose of a final determination of the landlord's ground of opposition. Although a preliminary issue hearing is a hearing of that nature, a summary judgment hearing is not, since no determination is made of disputed facts and such a hearing will result in the substantive issue coming to an end only if the decision goes one way, namely if the tenant makes a successful application. A summary judgment hearing determines whether a party has a real prospect of establishing its cause of action or defence at a future trial date, and the court can have regard to evidence that may realistically be expected to emerge between the date of the summary judgment hearing and the trial. The "real prospect" test applied on such an application is forward-looking, and the substantive nature of the cause of action or defence is not affected by the fact that the matter is considered at an earlier date than that of the trial. Accordingly, where summary judgment is sought on the ground of opposition under section 30(1)(*f*), the question is whether the landlord has a real prospect of forming, and proving that it has formed, the necessary intention by the anticipated future trial date, not the date of the summary judgment hearing. In so far as it is necessary to show a reasonable prospect of being able to commence work by reference to a particular date, that date also must be determined by reference to the anticipated trial date.**

The following cases are referred to in this report.

Betty's Cafés Ltd v *Phillips Furnishing Stores Ltd (No 1)* [1959] AC 20; [1958] 2 WLR 513; [1958] 1 All ER 607; (1958) 171 EG 319, HL
Cunliffe v *Goodman* [1950] 2 KB 237; [1950] 1 All ER 720, CA
Dutch Oven Ltd v *Egham Estate & Investment Co Ltd* [1968] 1 WLR 1483; [1968] 3 All ER 100; (1968) 19 P&CR 815, Ch
Somerfield Stores Ltd v *Spring (Sutton Coldfield) Ltd* [2009] EWHC 2384 (Ch); [2009] 3 EGLR 57; [2009] 48 EG 104
Tanfern Ltd v *Cameron-MacDonald* [2000] 1 WLR 1311; [2000] 2 All ER 801

This was an appeal by the appellant, Somerfield Stores Ltd, from a decision of District Judge Sheldrake, sitting in the Birmingham District Registry, dismissing an application for summary judgment on a claim against the respondent, Spring (Sutton Coldfield Ltd), a company in administration, for a new tenancy under the Landlord and Tenant Act 1954.

Mark Wonnacott (instructed by Hill Dickinson LLP, of Liverpool) appeared for the appellant; Nicholas Dowding QC and Greville Healey (instructed by Eversheds LLP) represented the respondent.

Giving judgment, **HH JUDGE DAVID COOKE** said:

[1] This appeal raises a question in respect of the renewal of business tenancies under the Landlord and Tenant Act 1954 (the 1954 Act) on which there is no direct authority at the level of the High Court or above. I am told that it has arisen in various other cases in the county court, some of which are awaiting this decision. The point is this: where a landlord opposes the renewal of a tenancy on the ground that it is its intention to redevelop the property, it must show that the requisite intention exists at a time that is determined by reference to the date of the trial of that ground of opposition. If the tenant does not wait for a trial but applies for summary judgment, is it entitled to that judgment if it can show that the landlord has no real prospect of establishing the requisite intention at a date determined by reference to the date of the summary judgment hearing, rather than the date of a prospective trial that may be many months into the future? If so, such an application would be a very valuable tactical weapon for a tenant because it would

A greatly reduce the time available to a landlord to formulate and put in place its proposals for the redevelopment.

[2] The appellant in this case is the tenant under three leases of a supermarket and adjoining land in Sutton Coldfield. The leases expired in March 2008. The defendant is the landlord, having acquired the freehold in September 2006 with the intention of redeveloping the site. In 2007, the tenant served on the landlord notices, under section 26 of the 1954 Act, requesting new tenancies, which were met by counternotices stating that the landlord would oppose the grant of the new tenancies on the ground set out in section 30(1)(*f*), namely:

> B that on the termination of the current tenancy the landlord intends to demolish or reconstruct the premises comprised in the holding or a substantial part of those premises or to carry out substantial work of construction on the holding or part thereof and that he could not reasonably do so without obtaining possession of the holding.

[3] The tenant then made an application to court for the grant of the new tenancies it sought and, on 11 February 2008, an order was made for trial of the ground of opposition as a preliminary issue. Directions were given with a view to a two-day trial in June or July 2008. It appears that the parties did not adhere to the directions timetable, preferring instead to negotiate. On 27 February 2009, C the landlord went into administration. The tenant promptly applied for permission to continue the proceedings pursuant to para 43(6) of Schedule B1 to the Insolvency Act 1986, which permission was granted by HH Judge Charles Purle QC on 12 June 2009. A further directions timetable was set, this time with a view to a trial between March and May 2010, but rather than wait for that trial the tenant made an application for summary judgment dismissing the ground of opposition, by notice dated 22 July 2009. That application came before District Judge Sheldrake on 27 October 2009, when he dismissed it with costs. In his judgment, in [9], Judge Sheldrake held that the tenant was required to establish clearly that the landlord would not be able to make out the ground of opposition either at the date of the summary judgment D hearing or within a reasonable time thereafter, and that the tenant had failed to satisfy that test. In view of the novelty of the point, the district judge himself gave permission to appeal.

[4] There was no dispute between counsel as to the framework provided by the 1954 Act in these cases. As to the quality of the intention the landlord must show, Mr Mark Wonnacott referred me to the following extract from *Woodfall: Landlord and Tenant*, no doubt with a nod to Mr Nicholas Dowding QC as one of the editors of that work:

> 22.106 It is not sufficient for the landlord merely to assert that he "intends" E since he may change his mind once he gets possession. An intention connotes that the landlord does more than merely contemplate; it connotes a state of affairs that he decides, so far as in him lies, to bring about, and which, in point of possibility he has a reasonable prospect of being able to bring about, by his own act of volition; the landlord does not "intend" if he has too many hurdles to overcome or too little control of events. The intention must be genuine and not colourable; it must be firm and settled, not likely to be changed. It must have moved out of the zone of contemplation the sphere of the tentative, the provisional and the exploratory and have moved into the valley of decision.
>
> Thus the landlord's intention is composed of two main ingredients; a fixed and settled desire to do that which he says he intends to do and a reasonable prospect of being able to bring about the desired result.

F I should say that Mr Dowding is not responsible for the slightly perplexing mixture of zones, spheres and valleys in that extract; the phrase is taken from the judgment of Asquith LJ in *Cunliffe* v *Goodman* [1950] 2 KB 237.

[5] Ground (*f*) requires that the necessary intention is one to redevelop "on the termination of the current tenancy". Section 64(1) of the 1954 Act has the effect that where an application has been made to the court for the grant of a new tenancy, the termination of the existing tenancy is postponed until "the expiration of the period of three months beginning with the date on which the application is finally disposed of". By section 64(2), an application is treated as having been finally disposed of on "the earliest date by which the proceedings on the application (including any proceedings on or in consequence G of an appeal) have been determined and any time for appealing or further appealing has expired…". That date was referred to as "the section 64 date".

[6] Mr Dowding submitted, in my view correctly, that at a trial of the landlord's ground of objection, the court would have to consider matters on the footing that the section 64 date would be three months and 21 days (the period allowed to appeal) after the conclusion of the trial; that is, the trial judge would have to assume that the landlord's objection would succeed and that there would be no appeal. The landlord would have to show at trial both that it had by then formed the necessary firm subjective intention and also that it had a reasonable prospect of being H able to commence the implementation of that intention on or at least within a short period after the section 64 date, as determined on the basis of those assumptions.

[7] The proposition that the landlord's subjective intention must have been formed by the date of the hearing of the landlord's opposition, and not at any earlier date (such as the date on which it served its counternotice), was established by the decision of the House of Lords in *Betty's Cafés Ltd* v *Phillips Furnishing Stores Ltd* [1959] AC 20*. Viscount Simmonds held that the relevant date for proof of the required intention was the date of "the hearing", saying, at p35:

> J At the hearing [the landlord] will oppose and prove his avowed intention. This seems to me, with all deference to those who take a different view, to be the plain English of section 26(6) and section 30(1)(*f*). I have already pointed out that it appears to accord also with the general purpose of the Act. It harmonizes also with the language of section 31(1) which contemplates the landlord satisfying the court upon any of the grounds upon which he is entitled to oppose the application.

Lord Somervell agreed with Viscount Simmonds. Lord Morton expressed a similar conclusion, at p41:

> Section 26(6) provides that the landlord's notice of opposition "shall state on which of the grounds mentioned in section thirty of this Act the landlord *will oppose* the application." The words "will oppose" must surely refer to some K date *after* the delivery of the notice in which the landlord states the ground on which he "will oppose" the tenant's application. And, in my view, that date can only be the date when the opposition of the landlord is heard by the High Court or the county court, as the case may be.

Lord Denning also agreed in the result, saying, at p51:

> Provided, however, that the notice is a good and honest notice when it is given, then it is clear to my mind that the ground stated therein must be established to exist at the time of the hearing… To succeed [the landlord] must satisfy the trial judge that, at the time when the court comes to make its order, he is *then* willing to provide alternative accommodation, or *then* intends to L reconstruct, or as the case may be…
>
> In short, it comes to this: the landlord must *honestly and truthfully* state his ground in his notice and he must establish it as *existing* at the time of the hearing.

Lord Keith dissented.

[8] The question of what constituted the relevant hearing was considered by Megarry J in *Dutch Oven Ltd* v *Egham Estate & Investment Co Ltd* [1968] 1 WLR 1483, when he approved an application that the landlord's ground of opposition be tried as a preliminary issue. Only if the landlord failed to establish its ground M would the court need to go on to address matters relating to the terms of any new tenancy. He held that the relevant hearing for the purpose of establishing the landlord's intention was the hearing to determine the preliminary issue and that if it failed on that occasion, so that a further hearing became necessary to determine the terms of a new tenancy, the landlord could not have a second bite of the cherry and seek to establish that its intention had been made good by the date of the later hearing.

[9] The question in this case could not have arisen prior to the introduction of the CPR in 1999 because until then it was not possible to apply for summary judgment in these cases. Now that it is,

* Editor's note: Also reported at (1958) 171 EG 319

Mr Wonnacott submitted that the hearing of such an application must be regarded as being similar to the hearing of a preliminary issue. In his argument, he used the expression "final hearing", although that was not the expression used by any of their lordships in *Betty's Cafés*, and submitted that although the determination of a summary judgment application is not normally regarded as a final hearing (because it does not necessarily determine the matters at issue whichever way the application is decided, see *Tanfern Ltd* v *Cameron-Macdonald* [2000] 1 WLR 1311), it *could* be the final hearing in the sense that it would in fact dispose of the landlord's grant of objection if it were the tenant's application and if it were successful. If the tenant was unsuccessful, of course, there would have to be a further hearing, at which the landlord's objection would be considered and decided on.

[10] In my judgment, this proposition runs counter to the essential nature of the summary judgment jurisdiction, which is to determine whether a party has a real prospect of establishing its cause of action (or defence as the case may be) at a future trial date. In most cases, of course, the facts relied on will have occurred at some previous date, so the issue at the summary judgment hearing in respect of those facts will be whether there is a real prospect that the evidence available at trial will be sufficient to establish that the alleged facts had already occurred. The "real prospect" test is a forward-looking one (as the word "prospect" implies) and means that the court is entitled to have regard to evidence that may realistically be expected to emerge between the date of the summary judgment hearing and the trial itself. However, crucially, the substantive nature of the cause of action (or defence) is not affected by the fact that the matter is considered at a date earlier than the trial itself.

[11] No doubt the jurisdiction in issue in the present case is an unusual one, in as much as the fact to be proved (that is, the formation of the necessary intention) is one that is required to exist at the trial, and not one that is required to be shown to have existed at some date in the past. The court can ask itself on a summary judgment application whether, looking forward, the landlord has a real prospect of forming and proving that it has formed the requisite intention at an anticipated trial date. It would, however, fundamentally change the substantive issue to be determined if the court were required to ask itself not whether the relevant intention is reasonably likely to be formed by the date of a future trial, but whether it already exists.

[12] Indeed, it seems to me that pursuing Mr Wonnacott's proposition leads to a number of absurdities arising from its essential circularity of reasoning. The circularity is that the summary judgment hearing is only "final", in the sense that it is the last hearing, if (on a tenant's application) the landlord has no real prospect of establishing the existence of the necessary intention. The date at which it must show a real prospect of establishing that intention is the date of the summary judgment hearing only if it is first assumed that the tenant's application will succeed. Thus, the requirement on the landlord to establish a "real prospect" at the summary judgment hearing arises only if it is first assumed that it will fail to do so.

[13] If the question is, as Mr Wonnacott submitted, to be considered by reference to the date of the summary judgment hearing, what can be meant by the expression "real prospect of succeeding" in CPR 24.2? Either the court would be asked to determine whether the evidence available at the date of the summary judgment hearing was sufficient to show that the necessary intention already existed or it would be required to ask itself whether, at some future trial date, the evidence by then produced would be likely to be sufficient to establish the existence of the necessary intention at the date of the earlier summary judgment hearing.

[14] If the hearing proceeded along the former lines, the court would not be considering a "prospect" at all but making a finding of fact as to whether or not the requisite intention existed. Evidence at a summary judgment hearing is normally in written form and considered without cross-examination, which would not be a suitable method of making determinations of contested fact.

[15] If the hearing proceeded on the latter basis, assuming the landlord got over the "real prospect" hurdle, there would have to be a further hearing at which the landlord's ground of objection would still be in issue and at which witnesses would be called and cross-examined, and the question would then arise as to whether the evidence was required to show that the necessary intention existed at the date of the summary judgment hearing or the later hearing. In this respect, the position would be different from that considered in *Dutch Oven* because the trial of the landlord's ground of objection as a preliminary issue, as considered in that case, would mean that, barring any appeal, there could be no further hearing at which that ground was still in issue.

[16] At any such further hearing, if the court were considering whether the necessary intention was shown to have existed at the previous summary judgment hearing, that would mean that the substantive issue for determination by the court had changed merely by virtue of the fact that an application (necessarily unsuccessful) had been made for summary judgment. If the court were instead to consider whether the evidence then showed that the necessary intention existed at the date of the later hearing, that would mean that the substantive question being tried was no longer that on which the landlord had been required to demonstrate a real prospect of success at the summary judgment hearing.

[17] In my judgment, much the preferable view is that the date of the hearing at which the necessary intention must be shown to exist is always the date of the substantive trial of the landlord's ground of objection. This accords with the passages from the judgment in *Betty's Cafés* that I have set out above; it seems to me plain that in all those extracts what is envisaged is a hearing at which evidence is tested and facts found for the purpose of a final determination one way or the other of the landlord's ground of opposition. It is consistent with the decision in *Dutch Oven* because the hearing at which the landlord's ground of objection is determined as a preliminary issue is a hearing of that nature. A summary judgment hearing, on the other hand, is not; no determination is made of any facts in dispute, and it can result in the substantive issue coming to an end only if the decision goes one way; for example, if it is the tenant's application, if that application is successful.

[18] It follows that at any summary judgment application, the question to be considered is whether, looking forward to the anticipated date of trial, the landlord can show a real prospect of being able to establish the necessary intention at that future date. In so far as it is necessary to show a reasonable prospect of being able to commence work by reference to a particular date, that date would also have to be determined by reference to the anticipated date of trial.

[19] Mr Dowding pointed out a further logical difficulty that would arise if the tenant's submission were correct. If the court at a summary judgment hearing were considering (on a tenant's application) whether the landlord held the necessary subjective intention at the date of that hearing, logically it would have to be an intention then existing to commence work three months and 21 days after the date of that hearing. However, the landlord would never be in a position to have possession and to commence work by that date because even if the landlord were successful at the summary judgment hearing, that would not result in the immediate dismissal of the tenant's application for a new tenancy. Instead, it would go on to a later hearing, at which the landlord's ground of objection would still have to be finally determined. It would be absurd to consider at the summary judgment hearing whether the landlord then had an intention, but only to commence work at a much later date after the inevitable further hearing.

[20] Finally, I should say that it seems to me that the interpretation the tenant contends for would introduce an unnecessary opportunity for the adoption of artificial tactics in an area of litigation that is, in my experience, already regarded by commercial parties as a costly tactical adjunct to their negotiations. It may be said that the present regime in which the landlord has all the time up to the final hearing to formulate and put in place its plans is somewhat favourable to the landlord, but that is what is determined by the decision of the House of Lords in *Betty's Cafés* to have been the intention of parliament. If it were open to the tenant to bring forward the date at which the intention must be formed by the simple expedient of putting on a summary judgment

A application, I have no doubt that the application would be made in virtually every case. This would result either in a transformation of the present balance to one that was strongly in favour of the tenant (which no doubt some may regard as desirable, but would be a wholly unintended consequence of the procedural changes introduced by the CPR) or it would lead to increased costs, with the parties fighting such applications, only in many cases to have to conduct essentially the same fight all over again at the final hearing. In my view, if the tenant has concerns that the landlord is unreasonably dragging things out in the hope of being able to put together a workable development proposal by B
the date of the final trial, the remedy is to proceed expeditiously with the directions timetable fixed by the court, and make any applications necessary to hold the landlord to it so that the trial comes on in good time.

[21] The result is that I would dismiss the appeal and uphold the decision of the district judge, although for reasons that I have expressed differently from those that he gave.

Appeal dismissed.

Roadside Group Ltd *v* Zara Commercial Ltd

CHANCERY DIVISION

[2010] EWHC 1950 (Ch)

30 July 2010

KITCHIN J

Estates Gazette 20 November 2010

[2010] 46 EG 118

Landlord and tenant — Commercial lease — User covenant — Construction — Law of Property Act 1925 — Underlease containing covenant not to use demised premises for parking motor vehicles for sale on any forecourt — Subtenant of underlessee displaying parked cars for sale around perimeter of premises — Whether underlessee thereby in breach of covenant — Whether covenant making underlessee liable for acts of subtenant — Whether underlessee to be deemed to covenant for itself and third parties deriving title under it by reason of section 79 of 1925 Act

The appellant held an underlease from the respondent landlord of premises that comprised a petrol station, a car showroom, a service garage and hardstanding. The underlease contained various user covenants, including a covenant "not to use the demised premises or any part thereof… for the parking of motor vehicles for sale on any forecourt". The appellant sublet part of the premises, comprising the car showroom and service garage, to a subtenant. Despite the covenants, the subtenant parked cars for sale around the perimeter of the site.

In proceedings between the parties, the respondent claimed that it was entitled to forfeiture of the lease and possession of the demised premises on the ground that the subtenant's activities had placed the appellant in breach of the parking covenant. The appellant sought a declaration that the covenant had not been breached or, alternatively, relief from forfeiture. It contended that, under the covenant, it was liable only for its own acts and not those of its subtenant because a covenant not to do something was not breached where that thing was done by a third party, rather than by the covenantor. It further relied on the difference between the language of the parking covenant and that of the other covenants, which prohibited the tenant both from doing certain acts and from allowing them to be done. The county court judge held that the appellant's construction was contrary to commercial common sense and found the appellant to be in breach of covenant, although he refused forfeiture in the circumstances of the case.

The appellant appealed. The respondent sought to support the judge's conclusion on the additional ground that, under section 79 of the Law of Property Act 1925, a covenant was deemed to be made by the covenantor not only for itself but also for those deriving title under it unless a contrary intention was expressed.

Held: **The appeal was allowed. (1) Although covenants must be construed in their own context and having regard to their ascertainable purpose and objective, there is a well-established distinction between a covenant expressed in the active voice, such as "not to use", and one in the passive voice, such as "shall not be used". All the user covenants in the underlease were active and in none of them did the appellant purport to covenant on behalf of anyone but itself. Moreover, the fact that the parking covenant was in a different form from the other user covenants, which contained a covenant not to do or permit others to do something, suggested that the parking covenant was intended to be narrower than those other covenants. An interpretation that made the appellant liable for the actions of its subtenant would instead be broader and that was unlikely to reflect the parties' intentions. Confining the covenant to the actions of the appellant was not contrary to commercial sense because any subtenant or stranger to the land would be bound by the parking covenant in equity so as to entitle the respondent to take injunctive proceedings against it. (2) Section 79 did not extend the scope of the parking covenant so as to make the appellant responsible for the acts of its subtenant. To read the words of section 79 into that covenant would substantially alter its scope and would be inconsistent with the purport of the underlease. The parties had expressed the necessary "contrary intention", thereby displacing the effects of the section. Accordingly, the appellant was not liable for the actions of the subtenant in parking cars for sale and the judge's decision was reversed on that ground. (3) However, the judge had correctly concluded that the subtenant's placing of the cars amounted to parking on the "forecourt" within the meaning of the covenant and his judgment on that point was upheld.**

The following cases are referred to in this report.

Berton v *Alliance Economic Investment Co Ltd* [1922] 1 KB 742, CA

Earl of Sefton v *Tophams Ltd (No 2)* [1967] 1 AC 50; [1966] 2 WLR 814; [1966] 1 All ER 1039, HL

Morrells of Oxford Ltd v *Oxford United Football Club Ltd* [2001] Ch 459; [2001] 2 WLR 128; (2001) 81 P&CR 13; [2001] 1 EGLR 76; [2001] 04 EG 147, CA

This was an appeal by the appellant, Roadside Group Ltd, from a decision of HH Judge Peter Langan QC, sitting in Newcastle upon Tyne County Court, in favour of the respondent, Zara Commercial Ltd, that the appellant was in breach of a user covenant in a lease of commercial premises by reason of the actions of its subtenant.

Jonathan Small QC (instructed by Pickworths, of St Albans) appeared for the appellant; Jonathan Gaunt QC (instructed by Sintons LLP, of Newcastle upon Tyne) represented the respondent.

Giving judgment, **KITCHIN J** said:

Introduction

[1] This is an appeal from a decision and order of HH Judge Peter Langan QC dated 2 July 2009 concerning the proper interpretation of a user covenant.

[2] The appellant (Roadside) is the tenant of premises comprising a petrol station, a car showroom (with two flats over it), a service garage and hardstanding on the west side of the Broadway on the Darras Hall Estate, Ponteland, Northumberland, pursuant to an underlease dated 8 July 1963.

[3] The respondent (Zara) is Roadside's landlord, holding under an intermediate lease and a headlease. The demised premises under the headlease include land adjacent to and to the south of Roadside's demised premises under the underlease.

[4] Roadside has in turn granted a sublease of the car showroom and the service garage (but not the petrol station and hardstanding) to a company called Triple Eight Motor Co Ltd (Triple Eight) that for some

A time has parked cars for sale in certain areas around the perimeter of the site.

[5] In this action, Zara contended that, as a result of the activities of Triple Eight, Roadside was in breach of a covenant in the underlease not to use the demised premises "for the parking of motor vehicles for sale on any forecourt" (the parking user covenant) and that it was entitled to forfeiture of the underlease and possession of the demised premises. Roadside sought a declaration that there had been no breach, alternatively relief from forfeiture.

[6] At trial, Zara was successful on the issue of the true construction of the underlease and so Roadside was held to be in breach. However,
B Zara failed in its bid to forfeit the underlease, the judge holding that the notice under section 146 of the Law of Property Act 1925 (the 1925 Act) was invalid and, in any event, that this was a clear case for relief from forfeiture.

[7] Triple Eight nevertheless wants to continue to park cars for sale just as it has in the past. Accordingly, with the leave of the judge, Roadside appeals on two points, namely whether:

(i) on the true construction of the parking user covenant, it is liable for the activities of its subtenant; and

(ii) the judge was right to conclude that certain areas on which Triple Eight has been parking cars for sale are part of the forecourt within the
C meaning of the parking user covenant.

Terms of the underlease

[8] The parking user covenant is contained in clause 2(xii)(b) of the underlease. This and certain other relevant covenants are set out below:

> 2. The company HEREBY COVENANTS with the Lessors as follows:
>
> ...
>
> (xi) Not to alter or permit to be altered any part of the buildings at any time on the demised premises or the fences or walls thereof without the plans and specifications thereof having been previously approved of by the Lessors and Superior Lessors
>
> D (xii)(a) Not without the written consent of the Lessors and the Superior Lessors to use the demised premises or any part thereof or the buildings thereon nor to permit the same to be used for the display of advertisements other than the name and type of business carried on and type of oil and petroleum products sold from the demised premises provided that the Lessors and Superior Lessors shall not unreasonably withhold their written consent to the display thereon of the advertisements of motor cars and motor accessories and service of motor vehicles of any kind
>
> (b) Not to use the demised premises or any part thereof for the sale of motor vehicles by auction or for the parking of motor vehicles for sale on any forecourt or for the erection or maintenance thereon of temporary building huts caravans or moveable dwellinghouse or for the sale of ale beer wine spirits or
> E other intoxicating liquors or for a club where intoxicating liquors are permitted to be consumed or distributed and further not to use the said piece of land otherwise than for a petrol and oil filling station service garage with showroom and residential accommodation therefor
>
> (xiii) Not to erect or permit to be erected any outside lighting except by attachment to a building and further not without the written consent of the Lessors to erect or permit to be erected any form of neon or tubular display or advertising lighting on any part of the demised premises or any building for the time being erected thereon
>
> (xiv) Not to do or permit to be done on the demised premises or any part thereof anything which may be or be likely to lead to be a nuisance damage grievance or annoyance to the owners or tenants of any of the adjoining
> F property or to the neighbourhood.

[9] The underlease was granted out of the intermediate lease dated 1 July 1963 that demised the same premises. The intermediate lease contains what is, in substance, the parking user covenant at clause 2(xii).

[10] The intermediate lease was in turn granted out of the headlease dated 2 February 1960. As I have mentioned, this demised larger premises that wholly encompass the premises the subject of the intermediate lease. It again contains what is, in substance, the parking user covenant at clause 2(xv), which applies to the entirety of the demised premises.

Demised premises

[11] A plan of the premises demised by the underlease is attached G
to the order of the judge and is reproduced in the appendix to this judgment [not reproduced here]. The premises occupy a quadrilateral site fronting and to the west of the Broadway, which runs in a generally north-south direction.

[12] In the middle of the site is the petrol station with pumps, canopy and kiosk from which Roadside carries on business. To the south of the site is a cluster of buildings that are sublet to Triple Eight and comprise, to the front, the showroom, and, to the rear, the service garage.

[13] For some time Triple Eight has, with the consent of Roadside, parked cars for sale in the shaded areas coloured red along the northern and western perimeters of the site and in the shaded area coloured blue H
along the southern perimeter of the site.

[14] The judge concluded that the areas coloured red, but not the area coloured blue, formed part of the forecourt of the premises within the meaning of the parking user covenant, and it is this finding that forms the second aspect of the appeal.

Liability for the subtenant

[15] Before the judge, Roadside relied on the proposition stated in *Woodfall: Landlord and Tenant* that a covenant not to do something is generally not broken if the prohibited thing is not done by the covenantor but by a third party, and also on the contrast in language J
between the user covenant of clause 2(xii)(b) and the other covenants contained in clause 2(xi), (xii)(a), (xiii) and (xiv), each of which is a covenant not to do various acts or permit them to be done.

[16] Zara countered that Roadside's construction made a nonsense of the parking user covenant. It was clear that the original landlord regarded as objectionable the parking of cars for sale on any forecourt and that the original lessee must have known this. Yet, on Roadside's construction, the original lessee could, a day or two after the grant of the lease, have avoided the covenant by simply subletting the premises without imposing a corresponding negative covenant on the subtenant.

[17] The judge expressed his conclusion in concise terms in [27] of K
his judgment:

> [27] The choice between these two approaches is almost a matter of impression, and it is not one which I can pretend to make with great confidence. In my judgment, Ms Temple's submission is to be preferred. The proposition taken from *Woodfall* is, no doubt correct, but it is a general proposition only and it must yield to particular considerations in an individual case. The real question here is whether the contrast in language on which Mr Bruce relies is so stark as to drive me to the conclusion that clause 2(xii)(b) should be given the restricted and literal construction for which he contends. In my judgment, the linguistic argument is not so powerful that it overwhelms the commercial sense which is the basis of Ms Temple's submission. I have put the matter quite shortly, but it L
> does not appear to call for further elaboration which would, in truth, be no more than repetition in other words of what I have said so far.

[18] On this appeal, Mr Jonathan Small QC, who has appeared for Roadside, contended that the judge fell into error in failing to accept the submissions advanced on behalf of Roadside. Mr Jonathan Gaunt QC, who has appeared for Zara, contended that judge was right for the reasons that he gave and, further, that his interpretation is confirmed and supported by section 79 of the 1925 Act, which reads, so far as relevant:

> (1) A covenant relating to any land of a covenantor... shall, unless a contrary M
> intention is expressed, be deemed to be made by the covenantor on behalf of himself, his successors in title and the persons deriving title under him or them, and, subject as aforesaid, shall have effect as if such successors and other persons were expressed.
>
> ...
>
> (2) For the purposes of this section in connexion with covenants restrictive of the user of land "successors in title" shall be deemed to include the owners and occupiers for the time being of such land.

[19] Covenants in leases are to be construed in their own context and having regard to their ascertainable purpose and objective. Nevertheless, the distinction between a covenant expressed in the active voice, such as "not to use", and a covenant expressed in the passive voice, such

as "shall not be used", is well established. As Atkin LJ said in *Berton* v *Alliance Economic Investment Co Ltd* [1922] 1 KB 742, at p759:

> It is clear that a person under a covenant not to use premises in a particular way cannot commit a breach of the covenant except by his own act or that of his agent. The same is true of a covenant not to permit. The user in the one case and the permission in the other must be something which can be predicated of the defendant or the defendant's agent. It is not sufficient to show that the premises have been used in a way which would constitute a breach of the covenant; it must further be shown that the user is by the defendant or his agent, or that it is permitted by the defendant or his agent.

[20] The crucial part of this passage was cited with approval by Lord Guest more than 40 years later in *Earl of Sefton* v *Tophams Ltd (No 2)* [1967] 1 AC 50, at p68B:

> In *Berton* v *Alliance Economic Investment Co* Atkin LJ said that a person under a covenant not to use premises in a particular way cannot commit a breach of the covenant except by his own act or that of his agent.

[21] In the case of the underlease in issue on this appeal, it is, I think, striking that all the covenants contained in clause 2(xi), (xii), (xiii) and (xiv) are cast in the active voice. In none of these covenants has Roadside purported to covenant on behalf of anyone except itself.

[22] Moreover, there is, as the judge recognised, a clear distinction between the words of the parking user covenant contained in subclause (a) of clause 2(xii) and the covenants contained in clause 2(xi), (xiii), (xiv) and subclause (b) of clause 2(xii). Each of these other clauses and subclause contains a covenant by Roadside not to do something or permit others to do the same. By contrast, this formulation is lacking from the parking user covenant. This would suggest that the parking user covenant was intended to be *narrower* than the other covenants. Yet, if the interpretation contended for by Zara is correct, the parking user covenant is a good deal *broader*. Indeed, it becomes, in effect, a covenant expressed in the passive voice, meaning "shall not be used". Subject to the points to which I must now turn, it seems to me to be most unlikely that this interpretation reflects the intention of the parties.

[23] The judge appears to have seen the force of these points but nevertheless held that this interpretation had to yield to the commercial sense that underpinned the interpretation contended for by Zara. However, as Mr Small submitted, and I accept, this ignores the fact that any subtenant, or indeed any stranger to the land, will be bound by the parking user covenant in equity, thus entitling Zara to take injunctive proceedings against the subtenant. It seems to me that this point therefore falls away.

[24] Finally, I must deal with section 79. Mr Gaunt submitted that the effect of the section is that the parking user covenant is deemed to have been made by Roadside on behalf of itself, its successors in title (which expression is deemed to include any occupiers for the time being of the premises) and on behalf of its undertenants. It follows, he continues, that Roadside is responsible for the acts of Triple Eight.

[25] Section 79 was considered by the Court of Appeal in *Morrells of Oxford Ltd* v *Oxford United Football Club Ltd* [2001] Ch 459*. In [35] of his judgment, Robert Walker LJ expressed a "tentative view" that section 79, where it applies and subject to any contrary intention, extends the number of persons whose acts and omissions are within the reach of the covenant in the sense of making equitable remedies available, provided that the other conditions for equity's intervention are satisfied; further, where the covenant is expressed in the active voice, that its normal effect is to turn "A covenants with X that A will not build" into "A (on behalf of himself and B) covenants with X that A (or, as the circumstances may require, B) will not build".

[26] Mr Small accepted, at least for the purposes of this appeal, that section 79, if applicable, would extend the scope of the parking user covenant to make Roadside responsible for the acts of Triple Eight. However, he continued, and as is apparent from the words of the section, it applies subject to the expression of any contrary intention. The approach to be adopted to this limitation was explained by Robert Walker LJ in *Morrells*, in [24] of his judgment:

> Ultimately there is a single question to be answered, that is whether the words provisionally read in by the statute can, in the commercial context of the particular case and having regard to the words of the particular document, usefully supplement those words, or whether any extended meaning read in by those means would on a fair reading be "inconsistent with the purport of the instrument"…

[27] In my judgment, the covenants in clause 2(xi), (xii), (xiii) and (xiv) have been defined with some care and the distinction between the covenants contained in clause 2(xi), (xiii), (xiv) and subclause (a) of clause 2(xii), on the one hand, and the parking user covenant contained in subclause (b) of clause 2(xii), on the other, is significant. For the reasons that I have given, I believe that it reflects the intention of the parties to limit the scope of the parking user covenant. To read the words of section 79 into the parking user covenant would be substantially to alter its scope and be inconsistent with the purport of the underlease.

[28] Finally, Mr Gaunt submitted that the order and declaration made by the judge is still correct, it being accepted and asserted by Roadside that Triple Eight is bound by the parking user covenant in equity. I am unable to accept this submission. Triple Eight is not party to these proceedings and I believe that it would not therefore be right for the declaration to stand.

Forecourt

[29] The second aspect of the appeal concerns the scope of the term "forecourt" in clause 2(xii). At trial, Zara contended that it comprises the whole of the demised premises with the exception of that part that is occupied by buildings; alternatively, so much of the land as lies to the north and west of the showroom. Roadside contended that it was limited to the area in front of the petrol station; alternatively, the area in front of Triple Eight's sales office; alternatively, that part of the demised premises that lies between these buildings and the Broadway.

[30] Various dictionaries were referred to before the judge, including the *Shorter Oxford English Dictionary*, which defines "forecourt" as:

> 1. An enclosed space in front of a building, an outer court. (b) spec The part of a filling station where petrol is supplied.

However, although dictionaries have their place, what is a forecourt in any given case must depend on all the circumstances, as the judge fully appreciated.

[31] The judge concluded that the term "forecourt" in clause 2(xii) includes the areas that lie to the north and west of the demised premises and that are shaded red on the plan annexed to this judgment. However, he held that it does not include the area that lies to the south of the demised premises and that is shaded blue on the plan. He reasoned as follows in [33] to [35] of his judgment:

> [33] Mr Bruce submitted that "forecourt" in clause 2(xii)(b) is limited to e area in front of the petrol pumps; (2) alternatively, the area in front of Triple Eight's sales office; (3) in the further alternative, only such part of the demised premises as lies between the buildings and the highway. Ms Temple submitted that "forecourt" is (4) the whole of the demised land save that part which is occupied by buildings; (5) alternatively, so much of the land as lies to the north and west of the showroom.
>
> [34] There are real difficulties with several of these formulations. As to (1), "forecourt" cannot be restricted to the apron, because that simply is not an area where anyone would park cars for sale. I can see no rational basis for (2). As to (4), if the prohibition had been intended to be so sweeping, the lease would have said so: the forecourt must, in my view, be something less than the whole of the open land.
>
> [35] (3) and (5) are more attractive propositions. Looking, as I am bound to do, at the particular circumstances of this case, I have come to the conclusion that (3) cannot be supported and that the right construction of the word, if it does not tally precisely with (5), comes very close to it. I referred earlier in this judgment to what I take to be the object of clause 2(xii)(b): that the site should not be, nor indeed appear to be, a used car lot. If, as I think must be the case, some limitation has to be put on the area regarded as forecourt, the criterion should be that forecourt must be something which can be seen from Broadway. On that basis, I conclude that the areas shown shaded on the plan at page 906,

* Editor's note: Also reported at [2001] 1 EGLR 76

are, as regards the parts running around the northern end and the western side of the demised premises, to be regarded as forecourt; but, as regards the smaller part at the southern end, not to be so regarded.

[32] On this appeal, Mr Small contended that the judge fell into error and failed to have adequate regard to the relevant context, which is that the demised premises contain a petrol station and car showroom that front the public highway. The showroom is of modest size, but there are areas between these buildings and the Broadway where one could park cars for sale. There are also areas to the rear and to the sides of these buildings where one could park cars for sale.

[33] In this context, Mr Small continued, there is ample scope to give the term "forecourt" its normal meaning, which is to say the open space in front of a building or structure. Further, the land let under the underlease was let for purposes including that of a showroom. It is usual for showroom proprietors to park cars for sale outside the showroom and that must be particularly so where, as here, the showroom is of modest size. Hence, the more natural expectation of the parties would be that the showroom operator would park cars for sale at the sides and rear of the premises since it was forbidden to park cars for sale at the front.

[34] Mr Small also submitted that there is a further contextual pointer in the same direction to which the judge did not refer at all. As I have explained, the underlease is carved out of superior leases that contain similar restrictions but that, in the case of the headlease, demised premises with a long frontage to the Broadway, extending some way to the south of the showroom. The submission continues that since the head lessor had a large piece of land that fronted the Broadway, the natural understanding is that it did not want cars parked for sale along that frontage. On the other hand, it is equally clear that the demised premises under the underlease were always to be used as a petrol station and showroom and, in the circumstances, there is no reason to suppose that the headlessor had any problem with the parking of cars for sale to the sides and rear of those buildings.

[35] Attractively though these submissions were presented, I am unable to accept them. First and importantly, to the front of the petrol station and showroom there are large grassed areas that the judge described as the "apron". These apart, there is room between the buildings and the grass for only a single vehicle to pass. Hence, there is no part of the premises demised by the underlease that lies between the buildings and the Broadway that could sensibly be described as a "forecourt" and on which cars for sale could be parked.

[36] Second, as a matter of English, the term "forecourt" is broad enough to encompass an outer court, not necessarily located at the front of a building or at a road frontage.

[37] Third, the judge sought to ascertain the meaning of the term "forecourt" in the context of the underlease by asking himself what the parties must have intended the parking user covenant to restrict and what they must have understood its object to be. In my judgment, he concluded, correctly, that the parties intended that the demised premises should not be nor appear to be a used-car lot. Accordingly, he reasoned, the forecourt must be something that can be seen from the Broadway. He therefore concluded the forecourt encompasses the areas to the north and west of the demised premises shaded red on the plan annexed to this judgment [not reproduced here], but not the area shaded blue because that is not visible. Accordingly, the parking of cars for sale is prohibited along the north and west perimeters of the demised premises but is permitted along the southern perimeter at the rear of the showroom.

[38] Finally, although I accept that the premises demised in the headlease have a long frontage and it is a natural understanding that the lessor did not want cars parked for sale along that frontage, I also consider that the parties must have been concerned particularly with that part of the premises ultimately demised by the underlease that were always intended to comprise a petrol station, a service garage and showroom. The reasoning of the judge is therefore as apposite to the headlease as it is to the underlease.

Conclusion

[39] It follows:

(i) Roadside is not liable for the activity of its subtenant. On this point, the appeal therefore succeeds.

(ii) The judge was right to conclude that the areas shaded red on the plan annexed to this judgment and on which Triple Eight has been parking cars for sale are part of the forecourt within the meaning of the parking user covenant. On this point, the appeal therefore fails.

[40] I will hear argument as to the appropriate form of order if it cannot be agreed.

Appeal allowed.

Austin *v* Southwark London Borough Council

SUPREME COURT

[2010] UKSC 28

LORD HOPE, DEPUTY PRESIDENT, LORD WALKER, LADY HALE, LORD BROWN AND LORD KERR

23 June 2010

Estates Gazette 4 September 2010

[2010] 35 EG 94

Landlord and tenant — Housing Act 1985 — Succession — Possession order — Revival of tenancy — Tenant under secure tenancy failing to comply with terms of suspended possession order — Tenant in occupation until death — Tenant's brother applying to postpone possession date in order so as to revive tenancy with view to exercising succession rights under section 89 of 1985 Act — Whether tenancy continuing until execution of possession order — Whether possible to vary terms of possession order under section 85 after death of former tenant

The appellant's brother was a secure tenant of the respondent council. In 1986, the respondents brought possession proceedings against him on the ground of rent arrears. This led, in 1987, to a conditional suspended possession order, not to be enforced for so long as he complied with the terms governing the payment of arrears. He failed to meet those terms and the order became enforceable. He none the less remained in the premises, paying rent together with contributions towards the arrears until he died in 2005.In 2006, the respondents served a notice to quit on the appellant, who was then living in the flat; they subsequently brought possession proceedings against him. Those proceedings were stayed pending the outcome of the appellant's application, under CPR 19.8, to be joined to the 1986 possession proceedings as a representative of his brother's estate. He sought to apply in that capacity, under section 85 of the Housing Act 1985, to postpone the date for possession under the suspended possession order so as to revive the secure tenancy retrospectively; he claimed that if the tenancy were revived, he was a person entitled to succeed to it under section 89.The appellant's application was dismissed in the county court, and appeals to the High Court and Court of Appeal were also dismissed: see [2008] EWHC 499 (QB), [2009] EWCA Civ 66; [2009] 2 EGLR 41; [2009] 25 EG 138. It was held that the right to revive a secure tenancy was personal to the former tenant and did not survive his death.The appellant appealed to the Supreme Court. He argued that a secure tenancy did not end on the breach of a conditional suspended possession order, but continued until that order was executed; since that had not occurred, his brother's tenancy subsisted at the time of his death and the appellant could succeed to it. In the alternative, he submitted that the right to revive the tenancy survived the death of the former tenant and passed to his estate.

***Held*: The appeal was allowed. (1) Established case law indicates that where a suspended possession order is made against a secure tenant under section 82 of the 1985 Act, the tenancy determines where the terms of the order are breached. Although the wording of section 82(2) could be read as contemplating that the tenancy ended only on the date on which possession would actually be relinquished under a warrant for possession that was duly executed or acted on, that is not the view that has prevailed in the established case law. It would not now be appropriate to reverse the law previously regarded as settled. The approach taken by the Housing and Regeneration Act 2008, which has prospective effect only, is to avoid the creation of further tolerated trespassers and, where it restores tenancy status to existing tolerated trespassers, to do so not by reviving former tenancies retrospectively but by creating new tenancies. A reversal of the previously established case law on the effect of possession orders would undermine that carefully crafted system and raise the problems that it had been designed to avoid, including the possibility of damages claims against social landlords that had assumed that those who failed to comply with the conditions of a suspended possession order had no right to enforce repairing covenants. (2) The fact that a former secure tenant has died does not deprive the court of jurisdiction to exercise the power conferred on it by section 85(2) to postpone the date of possession under a possession order. It was irrelevant in that connection whether the right to revive the tenancy was a right capable of being inherited at common law. Section 85(2) provides for the court's powers to be exercisable at any time before the execution of the possession order and makes no mention of the possibility that the former tenant might have died in the meantime. The view that the court retains its section 85(2) jurisdiction in such circumstances is supported by other provisions in that part of the Act: see sections 87 and 90. It was therefore open to the appellant, who sought to represent the estate of the person who had been served with a claim for possession, to apply, under CPR 19.8, for the date for possession to be postponed.**

An order was made appointing the appellant to represent the deceased's estate under CPR 19.8(1)(b), and his section 85(2) application was remitted to the county court for determination.

The following cases are referred to in this report.

Austin v *Southwark London Borough Council* [2008] EWHC 499 (QB), [2009] EWCA Civ 66; [2009] 2 EGLR 41; [2009] 25 EG 138; [2010] HLR 1

Brent London Borough Council v *Knightley* (1997) 29 HLR 857, CA

Bristol City Council v *Hassan*; *Bristol City Council* v *Glastonbury* [2006] EWCA Civ 656; [2006] 1 WLR 2582; [2006] 4 All ER 420; [2006] HLR 31

Burrows v *Brent London Borough Council* [1996] 1 WLR 1448; [1996] 4 All ER 577; [1997] 1 EGLR 32; [1997] 11 EG 150; (1997) 29 HLR 167, HL

Fitzleet Estates Ltd v *Cherry (Inspector of Taxes)* [1977] 1 WLR 1345; [1977] 3 All ER 996

Greenwich London Borough Council v *Regan* (1996) 72 P&CR 507; (1996) 28 HLR 469

Harlow District Council v *Hall* [2006] EWCA Civ 156; [2006] 1 WLR 2116; [2006] 2 P&CR 16; [2006] HLR 27

Hesperides Hotels v *Muftizade*; *sub nom Hesperides Hotels* v *Aegean Turkish Holidays* [1979] AC 508; [1978] 3 WLR 378; [1978] 2 All ER 1168, HL

Horton v *Sadler* [2006] UKHL 27; [2007] 1 AC 307; [2006] 2 WLR 1346; [2006] 3 All ER 1177

Jones v *Secretary of State for Social Services*, *Hudson* v *Secretary of State for Social Services*; *sub nom R* v *National Insurance Commissioner, ex parte Lloyd-Jones*; *R* v *National Insurance Commissioner ex parte Jones*; *R* v *National Insurance Commissioner, ex parte Hudson* [1972] AC 944; [1972] 2 WLR 210; [1972] 1 All ER 145, HL

A *Knuller (Publishing, Printing & Promotions) Ltd* v *DPP*, *Hall* v *DPP*; *Stansill* v *DPP*; *Keen* v *DPP*; *sub nom R* v *Knuller (Publishing, Printing & Promotions) Ltd* [1973] AC 435; [1972] 3 WLR 143; [1972] 2 All ER 898, HL

Miliangos v *George Frank (Textiles) Ltd (No 1)* [1976] AC 443; [1975] 3 WLR 758; [1975] 3 All ER 801; [1976] 1 Lloyd's Rep 201, HL

R v *Howe* [1987] 1 AC 417

R v *Kansal (Yash Pal) (Change of Law)*; *sub nom R* v *Kansal (Yash Pal) (No 2)* [2001] UKHL 62; [2002] 2 AC 69; [2001] 3 WLR 1562; [2002] 1 All ER 257

R v *Secretary of State for the Home Department, ex parte Khawaja* [1984] AC 74; [1983] 2 WLR 321; [1982] 1 WLR 948; [1983] 1 All ER 765, HL

B *Rees* v *Darlington Memorial Hospital NHS Trust* [2003] UKHL 52; [2004] 1 AC 309; [2003] 3 WLR 1091; [2003] 4 All ER 987

Street v *Mountford* [1985] AC 809; [1985] 2 WLR 877; [1985] 2 All ER 289; (1985) P&CR 258; [1985] 1 EGLR 128; (1985) 274 EG 821, HL

Thompson v *Elmbridge Borough Council* [1987] 1 WLR 1425; (1987) 86 LGR 245; (1987) 19 HLR 526, CA

Vestey v *Inland Revenue Commissioners (No 1)*; *Baddeley* v *Inland Revenue Commissioners*; *Payne* v *Inland Revenue Commissioners (No 1)* [1980] AC 1148; [1979] 3 WLR 915; [1979] 3 All ER 976, HL

White v *Knowsley Housing Trust*; *Islington London Borough Council* v *Honeygan-Green*; *Porter* v *Shepherds Bush Housing Association*; *sub nom Knowsley Housing Trust* v *White* [2008] UKHL 70; [2009] 1 AC C 636; [2009] 2 WLR 78; [2009] 2 All ER 829; [2009] 1 P&CR 22; [2009] 1 EGLR 131; [2009] HLR 17

This was an appeal by the appellant, Barry Austin, from a decision of the Court of Appeal, dismissing his appeal from a decision of Flaux J, sitting in the Queen's Bench Division, upholding a decision of HH Judge Welchman, sitting in Lambeth County Court, refusing an application, under CPR 19.8, to join the appellant as a party to possession proceedings as the representative of his deceased brother's estate so as to permit the making of an application to vary the terms of a possession order granted to the respondents, Southwark London Borough Council, under section 85 of the Housing Act 1985.

D Jan Luba QC and Desmond Rutledge (instructed by Anthony Gold Solicitors) appeared for the appellant; Richard Drabble QC and Shaw Kelly (instructed by the legal department of Southwark London Borough Council) represented the respondents.

Giving the first opinion, **Lord Hope** (with whom Lord Brown and Lord Kerr agree) said:

[1] The appellant is the brother of the late Mr Alan Austin (the deceased), who was the tenant of a dwelling-house at 52 Michael Faraday House, Thurlow Street, London. Southwark London Borough Council (the council) granted the deceased a tenancy of the premises E on 12 July 1983. It was a secure tenancy under the Housing Act 1980. The provisions of that Act were consolidated in the Housing Act 1985 (the 1985 Act). In June 1986, the council brought a claim against the deceased in Lambeth County Court for possession of the premises, relying on the fact that he was in arrears of rent. An order for possession was made against him on 4 February 1987. It was a conditional suspended possession order, issued in the form then current that was form N28. It provided that it was not to be enforced so long as the deceased paid the arrears of rent, amounting to £3,312.98, by 4 March 1987. The deceased failed to comply with the terms of the order so, on 4 March 1987, it became enforceable. However, he remained in the F premises, paying rent plus amounts towards the arrears, until his death 18 years later, on 8 February 2005.

[2] Prior to his death, the deceased had been suffering from a chronic illness that proved to be terminal. The appellant maintains that he moved in to live with his brother in October 2003 and became the deceased's full-time carer. The council do not accept the appellant's claim that he resided in the premises prior to the deceased's death. This has not yet been established as a fact. The deceased died intestate. No grant of probate or administration has been made, and he has no personal representative. On 11 September 2006, the council served a notice to quit on the appellant and, in January 2007, they issued proceedings against him in Lambeth County Court for possession of the premises. Those proceedings are stayed, pending the outcome of this appeal.

[3] On 29 March 2007, the appellant lodged an application in G Lambeth County Court seeking to be appointed to represent the deceased's estate in the possession proceedings that had been brought in 1986, pursuant to CPR 19.8. If so appointed, his intention is to apply to the county court, under section 85(2)(*b*) of the 1985 Act, for the postponement of the date for possession in the order that was issued against the deceased in 1987. If that application is successful, it would have the effect of reviving the deceased's secure tenancy. The appellant would then seek to show that he was resident in the premises throughout the period of 12 months that ended with the deceased's death. If that is found to have been the case, the deceased's secure tenancy will vest in the appellant by virtue of section 87 of the 1985 Act. H

[4] On 5 September 2007, the appellant's application to be appointed to represent the deceased's estate in the original possession proceedings was dismissed by HH Judge Welchman. On 29 January 2008, Flaux J dismissed the appellant's appeal against its dismissal: see [2008] EWHC 499 (QB). On 16 February 2009, the Court of Appeal (Pill, Arden and Longmore LJJ) dismissed the appellant's appeal against the order of Flaux J: see [2009] EWCA Civ 66*. He now appeals to this court.

Issues

[5] It has until now been assumed that a secure tenancy ends at the moment the tenant is in breach of the terms of a conditional suspended J possession order. In his annotations to the 1985 Act in *Current Law Statutes* the annotator, Mr Andrew Arden, seems to have entertained no doubt on the point. He said of section 82(2) that when the tenancy is brought to an end by the court under that subsection it ends on the date the court specifies, not at any later date when the tenant is actually evicted. In *Thompson* v *Elmbridge Borough Council* [1987] 1 WLR 1425, at pp1430H-1431A, Russell LJ described the effect of the subsection in this way:

> In my judgment, once the defendant in proceedings of this kind where there is a suspended order for possession, ceases to comply with the conditions of the order, namely, "the punctual payment of the current rent and arrears," and there K is a breach of the terms of the order, the tenancy, whatever it may be, from that moment comes to an end.

[6] In *Burrows* v *Brent London Borough Council* [1996] 1 WLR 1448†, at p1457, Lord Jauncey of Tullichettle, referring to the decision in *Thompson*, said that if the court makes an order but postpones the date of possession, the tenancy will not be terminated under section 82(2) until any condition imposed under subsection (3) has been breached by the tenant. He added these words, at p1457D-E:

> However, the court's power to make an order postponing the date of possession is not restricted to exercise on the first application for an order for possession L but may be exercised on the application of either party at any time prior to execution of that order and even after the secure tenancy has ended by reason of section 82(2). This is made clear by the words in section 85(2) "or at any time before the execution of the order:" see also *Greenwich London Borough Council* v *Regan*, 28 HLR 469, 476, *per* Millett LJ. In such an event the secure tenancy is reinstated or revived subject to any conditions imposed under subsection (3).

[7] Encouraged by observations to the contrary effect by Lord Neuberger of Abbotsbury in *White* v *Knowsley Housing Trust* [2008] UKHL 70; [2009] 1 AC 636‡, in [91], the appellant now seeks to challenge this assumption. He submits that a secure tenancy does M not end on the breach of a conditional suspended possession order but endures until the order for possession is executed. That is the first issue in this appeal. On the assumption that a secure tenancy ends on the breach of the terms of a conditional suspended possession order, a former tenant who continues in occupation assumes an unusual status. It has come to be known as that of a "tolerated trespasser". Commenting on this expression in *White*, in [3] to [4], Lord Walker of Gestingthorpe described the phrase as rather unfortunate, but he concluded that it was

* Editor's note: Reported at [2009] 2 EGLR 41

† Editor's note: Also reported at [1997] 1 EGLR 32

‡ Editor's note: Also reported at [2009] 1 EGLR 131

too firmly embedded to be dislodged. If the appellant is right on the first issue, however, the deceased was still a tenant when he died. The description of him and so many other secure tenants who are in the same position as he was as "tolerated trespassers" will be consigned to history.

[8] The appellant has an alternative argument, should he not succeed on the first issue. Until now, it has been assumed that the right of a former secure tenant to apply for a postponement of the date for possession to enable him to remedy the default, and thus revive a secure tenancy under section 85(2) of the 1985 Act according to the principles described by Millett LJ in *Greenwich London Borough Council* v *Regan* (1996) 28 HLR 469, at p476, did not survive the death of the former secure tenant. In *Brent London Borough Council* v *Knightley* (1997) 29 HLR 857, at p862, Aldous LJ said that the right to apply for a postponement of an order for possession was not an interest in land capable of being inherited. It was held that the daughter of a deceased tolerated trespasser who had resided with her could not apply under that subsection for the revival of the tenancy. The appellant submits that that case was wrongly decided. So, the second issue is whether the former tenant□s statutory right to apply to postpone the date for possession, and thus revive the secure tenancy, survives his death and passes to the estate of the deceased former tenant.

[9] In support of the position that he seeks to advance on the second issue, the appellant submits that the statutory right to apply to the court under section 85(2) is a "possession", the enjoyment of which is protected within the meaning of Article 1 of Protocol 1 of the European Convention on Human Rights. If so, he submits that a construction of section 85(2) that holds that the right determines on death would be contrary to that provision and should not be adopted. Those are the third and fourth issues. The fifth issue is whether, if the statutory right to apply under section 85(2) endures beyond the death of a former tenant, the deceased was a person who had an interest in a "claim" for the purposes of CPR 19.8, and, if so, what claim. However, the council accept that if the claimed right did exist at the date of the deceased's death, the appellant would be entitled to apply to the court under that rule.

Statutory provisions

[10] It is not necessary to do more to describe the general background than to refer to Lord Neuberger's summary of the law governing residential security of tenure in *White*, in [30] to [47]. There were three appeals in that case. They raised issues concerning the effect of suspended possession orders on the status and rights of secure tenants under the 1985 Act and assured tenants under the Housing Act 1988 (the 1988 Act). It was held that an assured tenancy subject to a suspended possession order ended only when possession was delivered up, but that reconsideration of the approach that had been adopted to secure tenancies under the 1985 Act was inappropriate because it derived from long-standing authority, had been applied in numerous cases and would be resolved by section 299 of and Schedule 11 to the Housing and Regeneration Act 2008 (the 2008 Act), which had amended and clarified the law. I shall come back to examine that part of the decision in *White* later. Since we are concerned in this case only with secure tenancies, I shall concentrate on the relevant provisions of the 1985 Act as originally enacted and on the amendments that were introduced by the 2008 Act. It should be noted that the relevant provisions of the 2008 Act were for the most part prospective only. They took effect from the commencement date of Schedule 11, which was 20 May 2009: the Housing and Regeneration Act 2008 (Commencement No 5) Order 2009 (SI 2009/1261).

[11] Section 79(1) of the 1985 Act provides that a tenancy under which a dwelling-house is let as a separate dwelling is a secure tenancy at any time when the conditions described in sections 80 and 81 as the landlord condition and the tenant condition are satisfied. Section 80(1) provides that the landlord condition is that the interest of the landlord is vested in one or more of the authorities or bodies listed in that subsection, which include a local authority. Section 81 provides that the tenancy condition is satisfied if the tenant is an individual and occupies the dwelling-house as his only or principal home.

[12] Section 82 deals with security of tenure. As originally enacted, subsections (1) to (3) of this section were in these terms:

(1) A secure tenancy which is either —
(*a*) a weekly or other periodic tenancy, or
(*b*) a tenancy for a term certain but subject to termination by the landlord,
cannot be brought to an end by the landlord except by obtaining an order of the court for the possession of the dwelling-house or an order under subsection (3).

(2) Where the landlord obtains an order for the possession of the dwelling-house, the tenancy ends on the date on which the tenant is to give up possession in pursuance of the order.

(3) Where a secure tenancy is a tenancy for a term certain but with a provision for re-entry or forfeiture, the court shall not order possession of the dwelling-house in pursuance of that provision, but in a case where the court would have made such an order it shall instead make an order terminating the tenancy on a date specified in the order and section 86 (periodic tenancy arising on termination of a fixed term) shall apply.

[13] As amended by para 2 of Part 1 of Schedule 11 to the 2008 Act with effect from 20 May 2009, the opening subsections of section 82 now provide:

(1) A secure tenancy which is either —
(a) a weekly or other periodic tenancy, or
(b) a tenancy for a term certain but subject to termination by the landlord,
cannot be brought to an end by the landlord except as mentioned in subsection 1(A).

(1A) The tenancy may be brought to an end by the landlord —
(a) obtaining —
(i) an order of the court for the possession of the dwelling-house, and
(ii) the execution of the order,
(b) obtaining an order under subsection (3), or
(c) obtaining a demotion order under section 82A.

(2) In the case mentioned in subsection (1A)(a), the tenancy ends when the order is executed.

[14] Section 83 provides that the court shall not entertain proceedings for the possession of a dwelling-house let under a secure tenancy unless the landlord has served on the tenant a notice complying with the provisions of that section. Section 84 deals with the grounds on which an order for possession may be made. In cases where the tenant is in arrears of rent, they require the landlord to satisfy the court that it is reasonable to make such an order before the court will order possession. In such a case, the court is given very wide and flexible powers by section 85. Subsections (1), (2) and (3) of this section, as originally enacted, were in these terms:

(1) Where proceedings are brought for possession of a dwelling-house let under a secure tenancy on any of the grounds set out in… ([being] cases in which the court must be satisfied that it is reasonable to make a possession order), the court may adjourn the proceedings for such period or periods as it thinks fit.

(2) On the making of an order for possession of such a dwelling-house on any of those grounds, or at any any time before the execution of the order, the court may —
(*a*) stay or suspend the execution of the order, or
(*b*) postpone the date of possession,for such period or periods as the court thinks fit.

(3) On such an adjournment, stay, suspension or postponement the court —
(*a*) shall impose conditions with respect to the payment by the tenant of arrears of rent (if any) and rent or payments in respect of occupation after the termination of the tenancy (mesne profits), unless it considers that to do so would cause exceptional hardship to the tenant or would otherwise be unreasonable, and
(*b*) may impose such other conditions as it thinks fit.

By para 3(2) of Part 1 of Schedule 11 to the 2008 Act, the reference to mesne profits in section 85(3)(*a*) was omitted with effect from 20 May 2009.

First issue

[15] The question to which this issue is directed concerns the effect of section 82(2) of the 1985 Act. Does the secure tenancy end when the

tenant is in breach of the conditions of the suspended possession order? Or does it end only when possession is delivered up when the possession order is executed? As I said earlier, it was for a long time assumed that the law was as stated in *Thompson*. In that case, a secure tenant had fallen into arrears of rent. The local authority obtained an order for possession in the same form as that which was set out in the order that the council had obtained against the deceased in this case. It stated that it was not to be enforced for 28 days in any event and for so long thereafter as the tenant paid the arrears of rent by stated instalments in addition to the rent. She failed to comply with the terms of the order and the question was whether, and if so when, the secure tenancy had come to an end. Russell LJ rejected an analogy that was sought to be drawn with the position under the Rent Acts. Basing his decision simply on the terms of section 82(2), he said that where there was a breach of the terms of the order the tenancy came to an end from the moment of the breach: see [5] above, where the words that he used are set out.

[16] In *Burrows,* a secure tenant had fallen into arrears with the rent and the council obtained a final order for possession of the premises. Shortly before the order was due to take effect, the council entered into an agreement with the tenant that she would not be evicted provided that she complied with certain conditions, which she failed to do. The question was whether a new tenancy had been created by this agreement. Reference was made to the Court of Appeal's decision in *Thompson*, but the soundness of that decision was not questioned by either side. Lord Browne-Wilkinson referred to it at p1453, when he was summarising the local authority's argument that the old tenancy had come to an end when the tenant failed to comply with the conditions imposed by the possession order. Lord Jauncey, with whom Lord Griffiths and Lord Steyn agreed, at p1457, went further. He summarised the law in a way that indicated that, in his opinion, *Thompson*, which by then had been followed by the Court of Appeal in *Greenwich*, was rightly decided.

[17] There, the matter rested until it came under the scrutiny of the House of Lords more than a decade later in *White*. Lord Neuberger observed in that case, in [73], that a number of cases in the Court of Appeal had proceeded on the assumption that the law was as stated in *Thompson*: see, for example, *Harlow District Council* v *Hall* [2006] EWCA Civ 156; [2006] 1 WLR 2116. He also said that there would also have been tens of thousands of cases in the county courts, many negotiations, much legal advice and many actions that had also proceeded on that assumption. In [91], however, having added that he strongly suspected that, if the point had not been determined in *Burrows*, he would have reached the same conclusion with regard to section 85 of the 1985 Act as that which he had reached with regard to section 9 of the 1988 Act in respect of assured tenancies, he said:

> There is a powerful case for saying that "the date on which the tenant is to give up possession in pursuance of the order" in section 82(2) of the 1985 Act can, and therefore should, mean the date specified in a warrant of possession which is duly executed (or acted on by the tenant). Furthermore section 121 of the 1985 Act [circumstances in which the right to buy cannot be exercised] appears to me to be arguably inconsistent with the decisions in *Thompson* [1987] 1 WLR 1425 and in *Hall* [2006] 1 WLR 2116, in that it appears expressly to assume that a tenant who "is", as well as a tenant who "will be", obliged to give up possession pursuant to a court order, would remain entitled to pursue the right to buy, and only a person who is a secure tenant can have that right.

[18] Despite these reservations, he refrained from moving the House to hold that *Thompson* and *Harlow* were wrongly decided so far as secure tenancies were concerned. He set out his reasons for doing so in [92] and [93]. One was the response that was received from counsel in that case, all highly experienced in the field of social housing law. Mr Jan Luba QC said that it was simply too late to take that course, as *Thompson* had been assumed to be right, and had been acted on, in many tens of thousands of cases over the past 20 years or so. The other was that, by section 299 of and Schedule 11 to the 2008 Act, parliament had amended and clarified the law so that secure and assured tenancies would end only when the order for possession was executed by means of provisions that were largely prospective in their effect. Agreeing with these submissions, Lord Neuberger said that it would be wrong for the House effectively to go back on its previous approval of *Thompson* in *Burrows*, when there was in place amending legislation having the same effect as such a reversal, in which parliament had decided to amend the law only prospectively.

[19] Mr Luba's position for the appellant in this case, as he frankly accepted, was the reverse of that which he adopted in *White*. He submitted nevertheless that this court should now take the step from which the House held back in that case and hold that *Thompson* was wrongly decided. There are, then, two questions that must be addressed. The first is whether, as Lord Neuberger indicated, section 82(2) can be read as meaning that, notwithstanding that the tenant is in breach of the conditions in the possession order, the tenancy continues until the date specified in a warrant for possession that is duly executed or acted on. The second is whether, if it can be so read, it should now be held that this is indeed its meaning and *Thompson* should be overruled.

[20] As to the first question, it is a remarkable fact that a conclusion on the meaning of section 82(2) that, admittedly with the benefit of hindsight, is so obviously unsatisfactory and conceptually confusing should have been reached with so little reasoning. The Court of Appeal's decision on the point in *Thompson* was expressed by Russell LJ in a single sentence, at pp1430-1431. It amounted to little more than an assertion. No attempt was made to see whether the meaning that he attributed to the subsection was consistent with provisions that were to be found elsewhere in the 1985 Act. As for the references to *Thompson* in *Burrows*, I think that Lord Neuberger was perhaps being a little generous when he said in *White*, in [72], that Lord Browne-Wilkinson and Lord Jauncey expressly stated that *Thompson* was rightly decided. Lord Browne-Wilkinson referred to that case at p1453, when summarising the submissions that were before the court in *Greenwich* and accepting counsel's analysis of that case. Lord Jauncey went further in endorsing what was said in *Thompson*, but I do not detect in his treatment of it any attempt on his part to reach a view of his own as to whether section 82(2) had to be read in the way in which Russell LJ had indicated. I would not attach much weight to the uncritical way in which the decision in *Thompson* was treated by the House of Lords in *Burrows*.

[21] The wording of section 82(2) needs, therefore, to be examined more closely in its context. There are, of course, other ways in which a tenancy may come to an end. However, since the section as a whole is concerned with security of tenure, it deals with the steps that must be taken by the landlord. The landlord must first obtain an order for possession of the dwelling-house. In the case of a secure tenancy for a term certain with a provision for re-entry or forfeiture, the court is required to make an order terminating the tenancy on a date specified in the order, which — unless the court orders that both termination and possession are to take effect on the same date — will be followed by a periodic tenancy: see section 82(3). Section 82(2) adopts a different approach. It does not say, as it could have done, that the date specified in the order is to be the date on which the tenancy terminates. It refers instead to the date on which the tenant "is to give up possession in pursuance of the order". That phrase can, I think, be read as indicating that the date on which the tenancy is to terminate is to be found in the possession order itself. That is how Russell LJ read it in *Thompson*, at pp1430-1431. However, the words "is to give up possession" can also be read, as Lord Neuberger said in *White*, in [91], as contemplating the date on which possession will actually be given up under a warrant for possession that is duly executed or acted on.

[22] I think that the context tends to favour Lord Neuberger's indication that the tenancy ends only when the order for possession is executed. The fact that the court is given such wide powers by section 85, including the power to discharge or rescind the possession order if the conditions are complied with, suggests that it was envisaged that the tenancy could still be in existence during the period when the court can exercise this control. Then, there are the indications in section 121, to which Lord Neuberger referred in [91], that the section was drafted on the assumption that a tenant who is obliged to give up possession pursuant to a court order is nevertheless still entitled to

A exercise the right to buy, which is a right that, as section 118 makes clear, only a secure tenant can have. The reference in section 85(3) to payments after the termination of the tenancy and mesne profits might seem to indicate the contrary. However, as Lord Neuberger pointed out in [87] and [88], this is best seen as an example of torrential drafting because the same wording appears in section 100(3) of the Rent Act 1977, where it cannot have that effect. Apart from this point, I do not find anything in the context that supports the interpretation that was given to section 82(2) in *Thompson.*

[23] The conclusion that I would draw is that there is much to be said for Lord Neuberger's interpretation. Perhaps the strongest point B in its favour is that reading section 82(2) in the way he has suggested avoids what he described in Knowsley, in [80], as the anomalous and potentially retrospectively reversible status of tolerated trespassers. The conceptual problems that this gives rise to do not seem to have been anticipated by the judges who guided the law in a different direction in *Thompson* and *Burrows.* A fair and practical reading would, as Lord Neuberger suggests, eliminate these difficulties. However, it seems to me that the contrary view is not unarguable.

[24] The question, then, is whether this court should now hold that the interpretation of section 82(2) that Lord Neuberger has suggested is indeed what this subsection means and that *Thompson* should be C overruled. As is of course very well known, the House of Lords issued a practice statement on 26 July 1966 that stated that it would still treat former decisions of the House as normally binding, but that it would depart from a previous decision when it appeared right to do so: see Practice Statement (Judicial Precedent) [1966] 1 WLR 1234. Its application was considered and applied from time to time by the appellate committee during the 40 years or so that were to elapse until 1 October 2009, when the appellate jurisdiction was transferred from the House of Lords to this court: see, for example, *R* v *National Insurance Commissioner, ex parte Hudson* [1972] AC 944, at p966, *per* Lord Reid, *R* v *Knuller (Publishing, Printing & Promotions) Ltd* [1973] D AC 435, at p455, *per* Lord Reid, *Miliangos* v *George Frank (Textiles) Ltd (No 1)* [1976] AC 443, *Fitzleet Estates Ltd* v *Cherry (Inspector of Taxes)* [1977] 1 WLR 1345, at p1349 *per* Lord Wilberforce, *Hesperides Hotels* v *Aegean Turkish Holidays* [1979] AC 508, *Vestey* v *Inland Revenue Commissioners (No 1)* [1980] AC 1148, *R* v *Secretary of State for the Home Department, ex parte Khawaja* [1984] AC 74, *R* v *Howe* [1987] AC 417, *R* v *Kansal (Yash Pal) (No 2)* [2001] UKHL 62; [2002] 2 AC 69, *Rees* v *Darlington Memorial Hospital NHS Trust* [2003] UKHL 52; [2004] 1 AC 309 and *Horton* v *Sadler* [2006] UKHL 27; [2007] 1 AC 307, in [29], *per* Lord Bingham of Cornhill.

[25] The Supreme Court has not thought it necessary to reissue E the practice statement as a fresh statement of practice in the court's own name. This is because it has as much effect in this court as it did before the appellate committee in the House of Lords. It was part of the established jurisprudence relating to the conduct of appeals in the House of Lords that was transferred to this court by section 40 of the Constitutional Reform Act 2005. So, the question that we must consider is not whether the court has the power to depart from the previous decisions of the House of Lords that have been referred to, but whether, in the circumstances of this case, it would be right for it to do so.

[26] In *Knuller*, at p455C-E, Lord Reid made the following observations about the practice statement that I think are particularly F in point in this case:

> I have said more than once in recent cases that our change of practice in no longer regarding previous decisions of this House as absolutely binding does not mean that whenever we think that a previous decision was wrong we should reverse it. In the general interest of certainty in the law we must be sure that there is some very good reason before we so act. ... I think that however wrong or anomalous the decision may be it must stand and apply to cases reasonably analogous unless or until it is altered by Parliament.

In *National Insurance Commissioner*, at p966, on the other hand, he said that it might be appropriate to do so if to adhere to the previous decision would produce serious anomalies or other results that were plainly unsatisfactory. In *Rees*, in [31], Lord Steyn said that, without trying to be exhaustive, a fundamental change in circumstances such G as was before the House in *Miliangos*, or experience showing that a decision of the House results in unforeseen serious injustice, might permit such a departure. In *Horton*, in [29], Lord Bingham said that the power had been exercised rarely and sparingly, but that too rigid an adherence to precedent might lead to injustice in a particular case and unduly restrict the development of the law. The House, he said, will depart from a previous decision where it appears right to do so.

[27] Two previous decisions of the House are before us in this case: *Burrows* and *White*. Of these, the one that gives rise to most concern is *White*. In [92] of his speech in that case, Lord Neuberger addressed the question directly as to whether, given that the point at issue was H not actually debated in *Burrows*, *Thompson* and *Harlow* were rightly decided. His answer, as he explained in [93], was that it should not reconsider the view expressed by the House in *Burrows* that *Thompson* was rightly decided so far as secure tenancies were concerned. It is true that this was the approach that had been contended for by counsel. However, there was much more in it than that. The fact that the decision in *Thompson* was of such long-standing and had been acted on in so many cases was a powerful factor in his assessment. So, too, was the way the need for an amendment of the law had been dealt with by parliament.

[28] I am not persuaded that we should depart from the decision that J the House took in *White* that the view expressed concerning *Thompson* in *Burrows* should not be reconsidered and departed from. It is true that we have had the benefit of a more complete argument on the point than was before the House in *White*. However, the fact remains that the law was regarded as having been settled by *Thompson* and the effects of reversing that decision now are incalculable. As Lord Neuberger said, it has been assumed to be right and has been acted on in many tens of thousands of cases. The area of greatest concern is the effect that a retrospective reversal would have on social landlords that for so long have assumed that those who had failed to comply with the conditions in a suspended possession order were no longer tenants with K a right to enforce the implementation of repairing covenants. Although we have not seen any direct evidence on the point, it is a reasonable assumption that the consequences of reviving these covenants and the opportunity that this would give for claiming damages for breach of these obligations was one of the factors that led to the decision that the law should be amended by the 2008 Act only prospectively.

[29] In August 2007, Communities and Local Government issued a consultation paper on tolerated trespassers. It set out four options for changes to the legislation relating to secure and assured tenancies. One was to do nothing. The second was to deal only with tenants who were subject to future possession orders. The third was to restore tenancy L status to all existing tolerated trespassers. The fourth was to restore this status to tolerated trespassers who had complied with the terms of the possession order. In its summary of responses to the consultation in April 2008, the department noted that the overwhelming majority of those who responded were in favour of legislation to prevent the creation of future tolerated trespassers. It also noted that although there was strong support for amending the legislation to restore the tenancies of all existing tolerated trespassers, there were dissenting voices. Those who opposed this option suggested that it could be seen to reward tenants who had repeatedly failed to meet their obligations and that it would remove potential leverage against difficult tenants. They M drew attention, too, to the need to protect landlords from liability for actions taken in accordance with the law at the time. The government's position was that the opportunity should be taken to prevent the creation of future tolerated trespassers and restore tenancy status to all existing tolerated trespassers. In an impact assessment issued in November 2008, it was indicated that, following consultation, the options had been narrowed down to two: do nothing and restore tenancy status to all tolerated trespassers.

[30] The 2008 Act received royal assent on 22 July 2008. Section 299 introduces Schedule 11. Part 1 of the Schedule amends the relevant legislation for the future. Part 2 restores tenancy status to existing tolerated trespassers. These provisions were brought into force by the

Housing and Regeneration Act 2008 (Commencement No 5) Order 2009 (SI 2009/1261). Significantly, however, what Part 2 of the Schedule does is to provide for the creation of replacement tenancies. As para 16 states, these are new tenancies. Paragraph 18 provides that the new tenancy is to have effect on the same terms and conditions that were applicable to the original tenancy immediately before it ended. Absent from the Schedule, however, is any provision for the revival or restoration of the pre-existing tenancy. This carefully crafted system avoids the problems to which the dissenting voices had drawn attention during the consultation process. For us now to declare that *Thompson* was no longer good law would undermine the system, with the result that those problems would, after all, become unavoidable. Such a result would contradict the will of parliament.

[31] Reverting to what Lord Reid said in *Knuller*, at p455, I think that far from there being some very good reasons for reversing the decision that the House took in *White*, the position in this case is the other way round. There are very good reasons for accepting that the law as declared in *Thompson*, however unsatisfactory it can now be seen to be, should not be disturbed. I would therefore reject the appellant's argument on the first issue.

Second issue

[32] This issue is directed to the effect of section 85(2) of the 1985 Act. The question is whether the former tenant's statutory right to apply to postpone the date for possession, and thus revive the secure tenancy, survives death and passes to the estate of the deceased former tenant. The answer to it depends on how the wording of this subsection should be construed in the context of the scheme of Part IV of the 1985 Act read as a whole.

[33] In the Court of Appeal, it was submitted that the short answer to this question was that the point was decided in *Knightley* and that that case was binding authority in that court for the proposition that the right to apply for the postponement of the order for possession under section 85(2) is not an interest in land capable of being inherited. So it could not survive the deceased's death. Having examined the decision, Arden LJ, with whom Pill LJ agreed, concluded that it bound the court to hold that the former tenant's right to apply under section 85(2) terminates on the tenant's death: see [42]. Longmore LJ said that even if the matter had not been expressly decided by *Knightley*, the correct view must be that the right is not exercisable by a deceased's personal representatives: see [54].

[34] The appellant in *Knightley* was the daughter of the tenant under a secure tenancy. On her mother's death, she claimed the right to succeed to the tenancy under section 89 of the 1985 Act. The council refused to accept her claim on the ground that since the mother had died, there was no tenancy to which the daughter could succeed because it had been terminated pursuant to a conditional possession order for non-payment of rent. In other words, the mother was no more than a tolerated trespasser. The judgment of the Court of Appeal was delivered by Aldous LJ. At p862, he noted that the law as to the effect of section 82(2) had recently been clarified by the decision of the House of Lords in *Burrows*. Given that, at her mother's death, there was no tenancy to which the daughter could succeed, everything depended on whether it was open to her to apply for an order, under section 85, to postpone the order for possession and revive the tenancy. He concluded that her submissions to that effect were untenable.

[35] Aldous LJ set out the reasoning on which that conclusion was based in the following paragraph, at p862:

> The right to apply for a postponement of an order for possession is not an interest in land capable of being inherited. Further, the right to apply under section 85 is a right given to the tenant and in subsection (5) to the tenant's spouse or former spouse. Section 87 also gives a right to apply to a person who is qualified to succeed as a tenant under a secure tenancy. That section only applies where there is a tenancy in existence. That was not the case here. To be a tolerated trespasser of the kind contemplated in Burrows, the person must be a trespasser tolerated by the law. The appellant was not such a person. In my view, there is no right given to a person in Miss Knightley's position to apply to revive a tenancy and no tenancy existed at the time when her mother died.

[36] Endorsing that decision in the Court of Appeal, Longmore LJ introduced his remarks with this comment, in [53]:

> Aldous LJ, with whom the other members of the court agreed, said in terms that the right was "incapable of being transmitted". That conclusion applies to transmission by will or on intestacy just as much as any other transmission eg by assignment between living persons.

What Aldous LJ actually said, in the passage that I quoted in [35], was that the right was not "capable of being inherited". There is an important distinction between these two phrases. Had Aldous LJ appreciated that the question that is really at issue here is whether the right is capable of being "transmitted", not whether it is capable of being "inherited", he might perhaps have arrived at a different answer. The right to apply to the court for the exercise of the powers that are given to it by section 85(2) is a right conferred by the statute. So, the answer to the question of whether the right can be exercised after the tenant has died is to be found by construing the statute. It does not depend on whether it is thought to be a right that is capable of being inherited at common law.

[37] Addressing himself, in [54], to the terms of the statute, Longmore LJ said that parliament plainly intended that the person who would otherwise be entitled to a secure tenancy if it had been revived should not be able to revive it in his or her own name. To hold that the deceased's estate can apply to revive it for the purpose of enabling that person to obtain a secure tenancy would be to circumvent, if not flout, that intention by means of a legal device. I have to confess that I do not follow this reasoning. It seems to confuse the provisions on succession on the death of the tenant under a secure tenancy with the situation that section 85 is dealing with, which is the exercise of powers by the court on the making of a possession order. Of course, one thing may lead to another. However, the sections must be taken in the order in which they appear, and they must be taken separately.

[38] The first thing that strikes one on reading the words of the subsection is that the powers that it refers to are said to be exercisable "at any time before the execution of the order". The possibility that the tenant may have died in the meantime is not mentioned. If it had been the intention that the powers should not be exercisable on the tenant's death, it would have been easy to say so. Indeed, given the width of the phrase that is actually used, one would have expected words to that effect to have been inserted. Cases could arise, for example, where the tenant has died before the possession order has taken effect to end the tenancy. There would seem to be no reason why the deceased's personal representative should not be able to seek the exercise of the power to postpone giving effect to the possession order, for example to enable the deceased's affairs to be put in order and any licensee or subtenant to be rehoused. Another example of a case where one would expect the personal representative to be able to apply would be where the deceased tenant, having made good a previous default, has applied for the date for possession to be postponed but dies the day before his application is to be heard. The wording of the subsection does not compel a reading that would deny the jurisdiction of the court to exercise its powers in such circumstances.

[39] There are other indications in this Part of the Act that support this approach. When it contemplates what is to happen on death, it says so. Sections 87 to 90, which deal with succession on the tenant's death, constitute the prime example. It is worth noting, too, that section 90, which deals with fixed-term tenancies, contemplates that the tenancy may continue after the secure tenant dies and vest in someone else in the circumstances referred to in subsection (3). One would expect the powers under section 85(2) to be exercisable after the tenant's death in such circumstances, and there is nothing in its wording that suggests the contrary. Also, the rights that were given to the tenant's spouse or former spouse who is in occupation when proceedings for possession are brought by section 85(5) are not said to come to an end when the secure tenant dies. That subsection was repealed by section 299 of and Schedule 11, paras 1 and 3, to the 2008 Act. However, it was there when section 85 was enacted, and it indicates the width and variety of the

A circumstances in which the powers under section 85(2) were intended to be available.

[40] For these reasons, I would hold that *Knightley* was wrongly decided and that it should be overruled. In my opinion, the fact that the former secure tenant has died does not deprive the court of its jurisdiction to exercise the power conferred on it by section 85(2)(*b*) of the 1985 Act to postpone the date of possession under a possession order. It follows that it is open to the appellant, who seeks to represent the estate of the person who was served with a claim for possession, to apply under CPR 19.8 for the date for possession to be postponed. I have to say that I regard this solution to his case to be preferable to
B the solution for which Mr Luba contended under the first issue. It is directed precisely to the situation that arises where a former tenant who has become a tolerated trespasser has died. Above all, it preserves the discretion of the court under section 85(2) to do what is just in all the circumstances. This is a protection for the landlord that would be entirely absent if the first solution were to be adopted.

Conclusion

[41] In view of the conclusion that I have reached in the appellant's favour on the second issue, I do not need to say anything about the remaining issues. Mr Richard Drabble QC, for the council, did not
C suggest that if the second issue were to be answered in the appellant's favour, the deceased was not a person who had an interest in a "claim" for the purposes of CPR 19.8 that, in the events that have happened, could be invoked by his estate. The claim that the appellant can invoke for this purpose is the claim for possession that was issued against the deceased in 1986.

[42] I would allow the appeal. I would order that the appellant be appointed to represent the estate of the deceased under CPR 19.8(1)(b) and remit his application under section 85(2) of the 1985 Act for postponement of the 1987 possession order to Lambeth County Court for determination.

D

Giving the second opinion, **Lord Walker** said:

[43] I am in full agreement with the reasoning and conclusions in Lord Hope's judgment. I have nothing to add except to express my admiration for the concurring judgment of Lady Hale, who has written the definitive obituary of the "tolerated trespasser". Indeed, her trenchant analysis clearly demonstrates that this unfortunate zombie-like creature achieved a sort of half-life only through a series of judicial decisions in which courts failed or did not need to face up to the theoretical and practical contradictions inherent in the notion. However, in common with all the members of the court, I agree that parliament is best fitted to
E give the tolerated trespasser his *quietus*, as it has by the 2008 Act.

Giving the third opinion, **Lady Hale** said:

[44] I agree that this appeal should be allowed for the reasons given by Lord Hope, but I wish to add a few words on the issue of "tolerated trespassers". In my view, had it not been for parliament's intervention, it would have been the duty of this court to set the matter right. There is no reason to believe that parliament intended that such an anomalous status should arise as a result of the provisions of the 1980 and 1985 Acts. There is little reason to believe that the full implications of its decision were apparent to the Court of Appeal when it decided
F *Thompson* v *Elmbridge Borough Council* [1987] 1 WLR 1425. That decision was assumed to be correct by the House of Lords in *Burrows* v *Brent London Borough Council* [1996] 1 WLR 1448, but it suited both parties in that case for them to do so. Moreover, the issue did not strictly arise in *White* v *Knowsley Housing Trust* [2008] UKHL 70; [2009] 1 AC 636, which was concerned with a different statutory regime. Thus, there is no House of Lords case that has addressed the issue full on and reached a reasoned conclusion on it. If there had been, it would have had to address all the conceptual and practical problems that have arisen since *Thompson*. These were forcefully spelt out by Lord Neuberger in *White*, but it is worth reiterating them here because, in *White*, the parties did not want the House to address the matter, whereas in this case the court has expressly been asked to do so.

[45] A tolerated trespasser is an oxymoron. A trespasser is someone G
who should not be there. However, tolerated trespassers were allowed to be there. Indeed, in some cases, the local authority had no right to evict them. The Court of Appeal decided in *Harlow District Council* v *Hall* [2006] EWCA Civ 156; [2006] 1 WLR 2116 that if the order fixed a date for possession but postponed its enforcement on terms, the tenancy came to an end on the date fixed, even if the trespasser complied with the terms. In other cases, the local authority had expressly agreed that the trespasser could stay. The House of Lords decided in *Burrows* that even a written agreement not to evict the trespasser if she complied with certain terms did not create a new tenancy. Even without such an agreement, the local authority were often quite uninterested in H
enforcing the order. They may not have realised that the order had been breached; they may have realised that the order had been breached but also that this was not the trespasser's fault but the result of the way the housing benefit system worked; they may have obtained the order without any intention of actually evicting the trespasser, but in order to obtain a money judgment and encourage more punctual payment of what both still regarded as rent; and they may not have wanted to have to rehouse a trespasser, who was by definition homeless, if she was in priority need. These were not people whom the local authority were reluctant to have there and were waiting for the machinery of eviction to take its course. These were people whom the authority wanted to J
have there, provided that they could be persuaded to pay most, if not all, of their rent.

[46] In normal circumstances, this would give rise to some sort of right to be there, whether a licence or (more probably, given *Street* v *Mountford* [1985] AC 809*) a tenancy, and if the landlord and tenant conditions required by section 79 were satisfied, this would be a secure tenancy. However, the House of Lords was persuaded in *Burrows* to hold that a new tenancy would not arise, save in special circumstances. The practical reason for this was that local authorities did not want to have to go back to court for a new possession order if the new agreement was breached. The chain of reasoning that persuaded the K
House relied mainly on the decision in *Greenwich London Borough Council* v *Regan* (1996) 28 HLR 469, that the old tenancy could be revived by a successful application under section 85(2) at any time before the possession order was actually executed. No one argued that the same sensible policy result could have been reached by overruling *Thompson*.

[47] So, we had a situation in which people became trespassers in their own homes, whether they or their landlords knew that this was so and irrespective of whether the landlords were content for them to stay. During the time that they were trespassers, neither the landlord nor the tenant could enforce the covenants under the tenancy agreement, L
although the tenant might be able to sue the landlord for nuisance. The statutory scheme for determining the rent did not apply. The trespasser could not exercise the right to buy even if he was now fully paid up. His spouse, partner or member of his family living there with him could not succeed. Technically, they were all homeless. Yet all these consequences could be retrospectively reversed by a successful application under section 85(2). The tenancy miraculously sprang back into life and it was as though the trespasser had been a tenant all along.

[48] Whether the court always realised that this would be the effect of its order may be doubted. The standard form of possession order granted in the county courts has changed over the years. The pre-1993 M
form N28, which was used in this case, did not specify a date on which possession was to be given up. It merely adjudged that the landlord "do recover against the defendant possession of the land mentioned". It ordered that the judgment for possession should not be enforced for 28 days in any event "and for so long thereafter as the defendant punctually pays to the plaintiff or his agent the arrears of rent, mesne profits and costs by 4th March 1987". This was in fact 28 days after the date of the order, which was 4 February 1987. The order was also most unusual in ordering the defendant to pay off the entire arrears plus costs

* Editor's note: Also reported at [1985] 1 EGLR 128

A within that time rather than by the more usual instalments. This does not induce confidence that either the plaintiff landlord or the court had addressed their minds to the exact consequences if this were not done. Was it an order for possession forthwith, postponed for 28 days and suspended on terms? Or was it an order for possession on 4 March? Or was it an order for possession at some indeterminate future date? The standard form changed in 1993 and again in 2001. Paragraph 1 of the 2001 order required that "the defendant give the claimant possession of [...] on or before [...]". If this form was used, the Court of Appeal held in *Harlow* that the tenant became a trespasser on that date, even if she faithfully complied with the terms for postponing enforcement. B However, the standard form did not have to be used and, in *Bristol City Council* v *Hassan* [2006] EWCA Civ 656; [2006] 1 WLR 2582, the Court of Appeal approved an order providing that "the date on which the defendant is to give up possession of the premises to the claimant is postponed to a date to be fixed by the court on an application by the claimant". It is a fair assumption that there are many old possession orders around that had an effect that the court making them would have avoided if it had known how to do so. The acquisition of trespasser "status" was accidental, not intentional. It was also very common. Strict compliance with the terms of suspension would, in Lord Neuberger's view, be rare.

C [49] All this nonsense could have been avoided if a different construction had been put on section 82(2) of the 1985 Act. The entire edifice was built on the *extempore* judgment of a two-judge Court of Appeal in *Thompson*. Section 82(1), so far as is material, provides that a secure tenancy "cannot be brought to an end by the landlord except by obtaining an order of the court for the possession of the dwelling house…". This does not affect the ways in which the tenancy may be brought to an end by the tenant or by agreement between the landlord and the tenant. It does mean that the landlord cannot end the tenancy without obtaining a possession order. Section 82(2) then provided that "where the landlord obtains an order for the possession of the D dwelling-house, the tenancy ends on the date on which the tenant is to give up possession in pursuance of the order".

[50] Clearly, the construction put on section 82(2) in *Thompson* and in subsequent cases is a tenable one. The subsection did say "is to give" rather than actually "gives" up possession. Equally clearly, as Lord Neuberger demonstrated in *White*, it is not the only tenable construction. The order made in this case did not specify a date on which the tenant had to give up possession, so why choose an unspecified and indeterminate date on which the tenant puts himself in a position whereby he may be required to give up possession in pursuance of the order? Even if the order does specify a date or an event on which possession is in theory to E be given up, that is never the end of the story. Unless the tenant leaves voluntarily, the landlord will have to get and have executed a warrant for possession. The tenant is not obliged to leave until a warrant has been obtained and cannot be forced to do so until the date specified in the warrant. So an even more tenable interpretation is that it refers to the date specified in a warrant of execution. However, that too is not the end of the story because the landlord may obtain a warrant and never execute it. That, in fact, is what happened in *Thompson*; the possession order was obtained on 31 January 1985; its terms were breached at the latest by 5 September 1985; and a warrant was obtained on 8 January 1986; the proceedings by the tenant's husband then F ensued. It is therefore difficult to say that the tenant "is to" give up the property until he actually does so, whether of his own accord or with the encouragement of the bailiffs.

[51] This construction is reinforced by section 85(2), which allows the court to stay or suspend execution or postpone the date of possession "at any time before the execution of the order". If the tenancy continues until then, there is no need for it to be resurrected G with retrospective effect. This construction is also, as Lord Neuberger observed in *White*, arguably more consistent with section 121. This provides that the right to buy cannot be exercised by a "tenant" who "is obliged to give up possession of the dwelling-house in pursuance of an order of the court", thus assuming that even if a person is currently obliged to give up possession he is nevertheless still a tenant until he actually does so.

[52] Arguably inconsistent with this construction was section 85(5), which gave the current or former spouse or civil partner of a tenant the same rights in respect of adjournment, stay, suspension or postponement of possession proceedings under section 85 as he or she would have had H his or her "home rights" not been affected by the termination of the tenancy. This may be explained as an example of "torrential drafting" in which the same provision is inserted into different statutory schemes irrespective of its applicability.

[53] Thus, there is indeed a powerful case for construing the date referred to in section 82(2) as "the date specified in a warrant of possession which is duly executed (or acted on by the tenant)", as Lord Neuberger put it in *White*, in [91]. When the linguistic case is put together with the conceptual and practical problems that arise from any other construction, the case becomes overwhelming. Legislation designed to protect residential tenants should be clear, simple and J consistent in its effects, not dubious, complex and arbitrary. It is scarcely surprising that the government's view, when consulting on what became Schedule 11 to the 2008 Act, was that the result of *Thompson* had been "unintended".

[54] Were it not for that Act, I would consider it "right" for this court to sort out the matter. The decisions in *Thompson* and the cases that proceeded on the unquestioned basis that *Thompson* was correct were not merely wrongly decided. They set the law on a course that was wrong in principle and wrong in practice. They produced a position with which no one was happy — neither the landlords nor the tenants — as is shown by the response to the government's consultations. Even K if some local authority landlords might have welcomed not being under a contractual obligation to repair properties for which the occupier was not paying the full rent, they would also have acknowledged that it could not be right for them to be able to charge the equivalent of the full rent that was calculated on the basis that they did have an obligation to repair. In such circumstances, it would ordinarily be our duty to recognise that the law had always been what we hold it to be.

[55] Does the 2008 Act make a difference? I am persuaded that, in this case, it does. It has abolished the problem for tenancies granted after it came into force. It has given those formerly considered "tolerated trespassers" a new tenancy that is in most respects the same as the L tenancy they would otherwise still have had. In other respects, of which repairing covenants are likely to be the most important, the court has the discretion to tailor a just solution. The only gap that counsel have identified is the gap exemplified by this case — where the tolerated trespasser has died before an application under section 85(2) has been made or determined. However, this case solves that problem.

[56] Parliament has therefore recently devised a considered and carefully balanced solution to the problem. We would be obliged to respect the will of parliament if it had devised a wholly new scheme or amended a scheme that we thought had been properly interpreted by the courts. I am persuaded that we should also do so in this case, even M if we believe that the premise that led it to devise the new scheme was wrong. In agreement with the other members of the court, therefore, I would reluctantly dismiss the appeal on the first issue but happily allow it on the second.

Appeal allowed.

A

Vision Enterprises Ltd (t/a Universal Estates) *v* Tiensia; Honeysuckle Properties *v* Fletcher and others

COURT OF APPEAL

[2010] EWCA Civ 1224

11 November 2010

B THORPE, SEDLEY AND RIMER LJJ

Estates Gazette 11 December 2010

[2010] 49 EG 80

Landlord and tenant — Housing Act 2004 — Assured shorthold tenancy — Tenancy deposit scheme — Dwelling — Landlords failing to protect tenants' deposits within 14 days of receipt by complying with initial requirements of authorised tenancy deposit scheme or to notify tenants of such compliance as required by section 213 of 2004 Act — Whether landlords liable to sanctions prescribed by C ***section 214 for failure to comply with those requirements — Whether compliance within 14-day period necessary to avert sanctions — Whether compliance by date of hearing sufficient***

In each of the two joined appeals, a landlord received a deposit on granting an assured shorthold tenancy of a dwelling. Both landlords failed, within 14 days, to protect that deposit by complying with the "initial requirements" of an authorised scheme and giving the prescribed information to the tenant concerning such compliance, contrary to section 213(3) and (6) of the Housing Act 2004 respectively.

In proceedings brought by the landlords relating to unpaid rent, the tenants counterclaimed, under section 214 of the 2004 Act, for a sum D representing three times the deposit, on the grounds of the landlord's failure to comply with the initial requirements of an authorised tenancy deposit scheme or to provide them with the prescribed information pursuant to the notification requirement. In each appeal, the landlord took steps before the hearing to register the tenants' deposit with an authorised scheme and to produce a certificate confirming this. However, the tenants contended that the section 214 sanctions had been triggered by the landlord's failure to fulfil the statutory requirements within the 14-day period laid down by section 213 and that it could not avoid those sanctions by late compliance.

In the first appeal, the tenant obtained summary judgment on E its counterclaim, but the landlord's appeal was allowed; the tenant appealed. In the second appeal, the tenants' counterclaim was allowed at first instance; the landlord appealed. On the first appeal, a further issue arose as to the effect of a time limit imposed by the particular deposit scheme; that chosen by the respondent landlord contained a requirement in similar terms to those in section 213, requiring it, within 14 days of receiving a deposit, to protect that deposit within the scheme.

Held: **(Sedley LJ dissenting): The first appeal was dismissed; the second appeal was allowed. (1) The precondition of a tenant's application to the court under section 214 is not a failure by the landlord to meet the 14-day time limit for complying with the** F **"initial requirements" or informing the tenant thereof, but the failure to comply with those obligations at all. The inquiry directed by the statutory language is whether the initial requirements and notification obligations have been performed, and not whether they were performed within a particular period that has passed: see section 214(1)(a) and 214(2)(a). Accordingly, where a landlord is late in complying with its section 213 obligations but does so before the tenant brings section 214 proceedings, the tenant will have no cause of action under that section. The landlord's compliance with its dual obligations under section 213 will furnish it with a complete defence to any section 214 claim. If the tenant has already brought such a claim, the landlord will have a defence if it complies with the requirements by the date of the hearing of that claim. The relevant date for determining whether the initial requirements are satisfied is** G **the date of the hearing of the tenant's section 214 claim, not the date on which it issues its claim or counterclaim: see section 214(2)(a) and (b). That interpretation is consistent with the purpose of the legislation, which seeks to achieve the due protection of deposits paid by tenants, not to punish landlords that might be late in securing such protection. That construction also avoids encouraging the ambushing of landlords by tenants who have grounds for believing that the landlord has not complied with its section 213 obligations. (2) The "initial requirements" referred to in section 213(3) and (4) of the 2004 Act do not include a requirement imposed by a particular scheme as to the time within which the landlord must** H **secure the deposit. Since section 213(3) imposes a time limit for compliance with the initial requirements, it is necessarily implicit that a time limit imposed by a particular scheme is not one of those requirements. Otherwise, uncertainty and confusion could arise as to the relationship between the section 213(3) time limit and the scheme time limit.**

The following cases are referred to in this report.

Draycott v *Hannells Letting Ltd (t/a Hannells Letting Agents)* [2010] EWHC 217 (QB); [2010] 3 All ER 411; [2010] HLR 27

UK Housing Alliance (North West) Ltd v *Francis* [2010] EWCA Civ 117; J [2010] 3 All ER 519; [2010] 2 EGLR 81; [2010] 18 EG 100; [2010] HLR 28

These were appeals by Christelle Tiensia, the appellant tenant in the first appeal, from a decision of HH Judge Ellis, sitting in Croydon County Court, allowing an appeal by the respondent landlord, Vision Enterprises Ltd (t/a Universal Estates), against a summary judgment by Deputy District Judge Clarke dismissing a claim for possession and rent arrears and allowing the appellant's counterclaim under section 214 of the Housing Act 2004; and by Honeysuckle Properties, the appellant landlord in the second appeal, from a decision of District Judge Stonier, sitting in Manchester County Court, dismissing a claim against the respondent tenants, James Fletcher, Frank McGrory and K Matthew Whitworth, for unpaid rent on the ground that, in each appeal, the tenants had a valid counterclaim under section 214 of the 2004 Act.

James Bowen (instructed by Wandsworth & Merton Law Centre) appeared for the appellant in the first appeal; Matthew Hutchings (instructed by Brian McKenna & Co) appeared for the respondent; Jonathan Manning (instructed by Bury Walkers LLP, of Leeds) represented the appellant in the second appeal; the respondents appeared in person.

Giving the first judgment, **RIMER LJ** said:

L

Introduction

[1] These two appeals were heard together and raise the same issue, one of general importance concerning tenancy deposit schemes established under the Housing Act 2004 (the 2004 Act). Putting it generally, section 213 imposes various obligations on a landlord that receives a tenancy deposit, including to: (i) comply with the "initial requirements" of an authorised scheme in respect of the deposit (which means taking steps to protect the deposit); (ii) do so within 14 days of its receipt; (iii) give to the tenant prescribed information relating to such protection; and (iv) do so within the same 14-day period. Section 214 requires the court, in certain events, to impose sanctions on a landlord M that has been or is in breach of its section 213 obligations, one sanction being an order requiring the landlord to pay the tenant a sum equal to three times the amount of the deposit.

[2] The issue is as to the circumstances in which such sanctions must be imposed. Again, put generally, the tenants' argument is that if the landlord fails to perform its section 213 obligations within the 14-day period, it will automatically be beyond redemption and will have no answer to a claim to impose the section 214 sanctions on it. The landlord's argument is that section 214 deliberately avoids any focus on the 14-day requirements but instead focuses exclusively on whether by the time of the hearing of the tenant's application for the imposition of the sanctions the landlord is still in breach of its obligations (i)

and (iii) summarised in [1] above. If it is, the court must impose the sanctions. If it is not, it cannot do so.

[3] In one of the appeals, *Vision Enterprises Ltd (t/a Universal Estates)* v *Tiensia*, Universal, the respondent landlord, brought a claim against Ms Christelle Tiensia, the appellant, for possession and arrears of rent in Croydon County Court, a claim that was met by her section 214 counterclaim. Her counterclaim succeeded before Deputy District Judge Clarke on 23 February 2009, who (*inter alia*) ordered Universal to pay to Ms Tiensia's solicitor £7,200 (a sum equal to three times her deposit), but Universal's appeal was allowed by HH Judge Ellis on 17 June 2009. Lloyd LJ gave Ms Tiensia permission to appeal to this court on the basis that it was arguable that Judge Ellis was wrong and that the point was of sufficient general importance to justify a second appeal. Ms Tiensia was represented before us by Mr James Bowen; and Universal by Mr Matthew Hutchings. Both also appeared before Judge Ellis, although neither before Judge Clarke.

[4] In the other appeal, *Honeysuckle Properties* v *Fletcher*, Honeysuckle (the trading name of Ms Melissa Moore, the appellant landlord) brought a claim in Manchester County Court against its three tenants for unpaid rent, a claim that was met by the tenants' section 214 counterclaim. The claim and counterclaim were heard by District Judge Stonier on 30 January 2009, who found in favour of the tenants on their counterclaim and ordered Honeysuckle to pay them £3,408, an amount including a sum equal to three times that of their deposit. HH Judge Holman gave Honeysuckle permission to appeal and transferred the appeal to this court, bearing in mind the importance of the issue that it raised. Mr Jonathan Manning represented Honeysuckle before us and the respondent tenants (Mr James Fletcher, Mr Frank McGrory and Mr Matthew Whitworth) appeared in person, Mr McGrory being their spokesman.

Legislation

[5] Chapter 4 in Part 6 of the 2004 Act is headed "Tenancy Deposit Schemes". It came into force on 6 April 2007. It set up tenancy deposit schemes to safeguard tenancy deposits paid in connection with assured shorthold tenancies, provided for the imposition of sanctions for failure to comply with the requirements of such schemes and set up structures to facilitate the resolution of disputes in connection with such deposits. The schemes are either custodial ones, in which the deposit is paid by the landlord into a designated account and held by the scheme administrator until it falls to be paid back (either wholly or in part) to the landlord or tenant; or insurance-based, in which (in short) the landlord keeps the deposit but its return is protected by insurance cover maintained by the scheme administrator. In both cases before us, the deposit was protected by an insurance-based scheme.

[6] I will set out the material provisions of sections 212 to 215, although the arguments focused primarily on sections 213 and 214.

212. Tenancy deposit schemes

(1) The appropriate national authority must make arrangements for securing that one or more deposit schemes are available for the purpose of safeguarding tenancy deposits paid in connection with shorthold tenancies.

(2) For the purposes of this Chapter a "tenancy deposit scheme" is a scheme which

(a) is made for the purpose of safeguarding tenancy deposits paid in connection with shorthold tenancies and facilitating the resolution of disputes arising in connection with such deposits, and

(b) complies with the requirements of Schedule 10.

(3) Arrangements under subsection (1) must be arrangements made with any body or person under which the body or person ("the scheme administrator") undertakes to establish and maintain a tenancy deposit scheme of a description specified in the arrangements.

(4) The appropriate national authority may

(a) give financial assistance to the scheme administrator;

(b) make payments to the scheme administrator (otherwise than as financial assistance) in pursuance of arrangements under subsection (1).

(5) The appropriate national authority may, in such manner and on such terms as it thinks fit, guarantee the discharge of any financial obligation incurred by the scheme administrator in connection with the arrangements under subsection (1).

(6) Arrangements under subsection (1) must require the scheme administrator to give the appropriate national authority, in such manner and at such times as it may specify, such information and facilities for obtaining information as it may specify.

(7) The appropriate national authority may make regulations conferring or imposing

(a) on scheme administrators, or

(b) on scheme administrators of any description specified in the regulations,

such powers or duties in connection with arrangements under subsection (1) as are so specified. ….

Section 212(8) defines a "shorthold tenancy" as an assured shorthold tenancy within the meaning of Chapter 2 of Part I of the Housing Act 1988 (the 1988 Act); a "tenancy deposit" as any money intended to be held (by the landlord or otherwise) as security for: (a) the performance of any obligations of the tenant; or (b) the discharge of any liability of his arising under or in connection with the tenancy; and "custodial scheme" and "insurance scheme" as having the meaning given by para 1(2) and (3) of Schedule 10.

213. Requirements relating to tenancy deposits

(1) Any tenancy deposit paid to a person in connection with a shorthold tenancy must, as from the time when it is received, be dealt with in accordance with an authorised scheme.

(2) No person may require the payment of a tenancy deposit in connection with a shorthold tenancy which is not to be subject to the requirement in subsection (1).

(3) Where a landlord receives a tenancy deposit in connection with a shorthold tenancy, the initial requirements of an authorised scheme must be complied with by the landlord in relation to the deposit within the period of 14 days beginning with the date on which it is received.

(4) For the purposes of this section "the initial requirements" of an authorised scheme are such requirements imposed by the scheme as fall to be complied with by a landlord on receiving such a tenancy deposit.

(5) A landlord who has received such a tenancy deposit must give the tenant and any relevant person such information relating to

(a) the authorised scheme applying to the deposit,

(b) compliance by the landlord with the initial requirements of the scheme in relation to the deposit, and

(c) the operation of provisions of this Chapter in relation to the deposit,

as may be prescribed.

(6) The information required by subsection (5) must be given to the tenant and any relevant person

(a) in the prescribed form or in a form substantially to the same effect, and

(b) within the period of 14 days beginning with the date on which the deposit is received by the landlord. …

(9) The provisions of this section apply despite any agreement to the contrary.

(10) In this section

"prescribed" means prescribed by an order made by the appropriate national authority. …

214. Proceedings relating to tenancy deposits

(1) Where a tenancy deposit has been paid in connection with a shorthold tenancy, the tenant or any relevant person (as defined by section 213(10)) may make an application to a county court on the grounds

(a) that the initial requirements of an authorised scheme (see section 213(4)) have not, or section 213(6)(a) has not, been complied with in relation to the deposit; or

(b) that he has been notified by the landlord that a particular authorised scheme applies to the deposit but has been unable to obtain confirmation from the scheme administrator that the deposit is being held in accordance with the scheme

(2) Subsections (3) and (4) apply if on such an application the court

(a) is satisfied that those requirements have not, or section 213(6)(a) has not, been complied with in relation to the deposit, or

(b) is not satisfied that the deposit is being held in accordance with an authorised scheme, as the case may be.

(3) The court must, as it thinks fit, either

(a) order the person who appears to the court to be holding the deposit to repay it to the applicant, or

(b) order that person to pay the deposit into the designated account held by the scheme administrator under an authorised custodial scheme,

within the period of 14 days beginning with the date of the making of the order.

(4) The court must also order the landlord to pay to the applicant a sum of money equal to three times the amount of the deposit within the period of 14 days beginning with the date of the making of the order.

A (5) Where any deposit given in accordance with a shorthold tenancy could not be lawfully required as a result of section 213(7), the property in question is recoverable from the person holding it by the person by whom it was given as a deposit.

(6) In subsection (5) "deposit" has the meaning given by section 213(8).

215. Sanctions for non-compliance

(1) If a tenancy deposit has been paid in connection with a shorthold tenancy, no section 21 notice may be given in relation to the tenancy at a time when

(a) the deposit is not being held in accordance with an authorised scheme, or

(b) the initial requirements of such a scheme (see section 213(4)) have not been complied with in relation to the deposit.

B (2) If section 213(6) is not complied with in relation to a deposit given in connection with a shorthold tenancy, no section 21 notice may be given in relation to the tenancy until such time as section 213(6)(a) is complied with.

(3) If any deposit given in connection with a shorthold tenancy could not be lawfully required as a result of section 213(7), no section 21 notice may be given in relation to the tenancy until such time as the property in question is returned to the person by whom it was given as a deposit.

(4) In subsection (3) "deposit" has the meaning given by section 213(8).

(5) In this section a "section 21 notice" means a notice under section 21(1)(b) or (4)(a) of the Housing Act 1988 (recovery of possession on termination of shorthold tenancy).

C I need say no more than that a section 21 notice, as defined in section 215(5), is a notice the service of which is a precondition of the obtaining of a mandatory possession order on or after the coming to an end of an assured shorthold tenancy.

Facts of each case and the decisions below

A. Tiensia v *Vision Enterprises Ltd (t/a Universal Estates)*

[7] On 17 May 2008, the respondent, Universal, let a dwelling at 1 Rosedene Avenue, Croydon to the appellant, Ms Tiensia, under an assured shorthold tenancy within Part I of Chapter II of the 1988 Act. It was for a fixed term of six months starting on 19 May 2008 at a monthly rent of £1,200 payable on the 19th of each month, the first payment D falling due on 19 May. Ms Tiensia moved in on 19 May 2008 and paid a total deposit of £2,400 in four instalments between 24 April and 4 June 2008. Page 1 of the tenancy agreement included the following statement under the heading "Deposits":

> If the landlord takes a deposit, the landlord must, within 14 days from the date of payment, give the tenant and any person who has paid the deposit on the tenant's behalf, certain written information about the way the deposit is protected. See the Housing Act 2004 s213(5) and The Housing (Tenancy Deposits)(Prescribed Information) Order 2007 S1 2007/797. The landlord may not require a deposit which consists of property other than money.

E Although section 213(6)(a) requires the section 213(5) information to be given to the tenant "in the prescribed form or in a form substantially to the same effect," no form has been prescribed. The Housing (Tenancy Deposits)(Prescribed Information) Order 2007 simply prescribes what information is to be given.

[8] Ms Tiensia soon fell into arrears with her rent. This led to Universal's notice to her dated 20 June 2008, under section 8 of the 1988 Act, informing her that it intended to apply for an order for possession, a letter before claim and, on 5 August 2008, the issue of a possession claim in Croydon County Court based on and related to rent arrears then totalling £2,400 (reliance was placed on grounds 8, 10 and F 11 of Schedule 2 to the Housing Act).

[9] The claim came before the court on 19 September 2008, when it was adjourned on Ms Tiensia stating her wish both to defend and counterclaim, the latter being apparently intended to be based on a complaint concerning the shower in the property. Ms Tiensia was still £2,400 in arrears with her rent.

[10] It is common ground that Universal had not, by the end of September 2008, complied with its obligations under section 213(3) of the 2004 Act: it had neither complied with the "initial requirements" of an authorised scheme within the 14-day period nor at all. On 2 October 2008, however, it produced a deposit protection certificate confirming that Ms Tiensia's deposit of £2,400 was protected by an insurance-based deposit protection scheme operated by Tenancy Deposit Solutions Ltd (TDS), a scheme that quite apart from the 14-day G requirement imposed by section 213(3) itself required the landlord "within 14 days of receiving a deposit… [to] protect the deposit with the scheme…". The certificate inaccurately recorded the deposit as having been collected from Ms Tiensia on 19 May 2008 and described the period of protection as running from "2 October 2008 until the date the deposit is unprotected plus 90 days". It was signed at some point (which is in dispute) by both Universal and Ms Tiensia. There is, however, no dispute that, by 2 October 2008, Universal had, albeit late, complied with the "initial requirements" of an authorised scheme.

[11] As necessarily follows, it is also common ground that at no point before October 2008 had Universal provided Ms Tiensia with H the information required by section 213(5), let alone within the 14-day period prescribed by section 213(6)(b). Although I understand it to be accepted that Universal has now provided Ms Tiensia with that information, there is a dispute as to whether it had done so by the time of the hearing before HH Judge Clarke, to which I shall come. Universal claims that it did so on 2 October 2008, when, it says, Ms Tiensia signed the certificate and was provided with the leaflet about the TDS scheme (which "IMPORTANT INSTRUCTIONS" on the certificate say must be passed to the tenant). By contrast, I understand Ms Tiensia to assert that it was not until around 13 October 2008 that she saw, was asked to and did sign the certificate and also to put in J issue whether she had been provided with all the required information by the date of the Clarke hearing. We are not concerned to resolve that factual dispute.

[12] On 3 October 2008, Ms Tiensia served a defence and counterclaim, which she amended on 3 December 2008. It admitted arrears of rent then due of £2,400 but claimed to set off the damages she claimed for an alleged breach of covenant by Universal in respect of the shower and also the liquidated sums for which she counterclaimed. The counterclaim asserted that Universal had breached its tenancy deposit obligations under section 213. It asserted that Universal had failed to comply with the "initial requirements" of an authorised deposit scheme by 20 June 2008 and had not done so until 2 October 2008: K see section 213(5). (I am not clear why 20 rather than 18 June 14 days from the last deposit instalment was regarded as the relevant date but nothing turns on it). It asserted that Universal had not provided her with the information required by section 213(5) either within 14 days of the receipt of the deposit or at all. It counterclaimed for: (i) a repayment of the £2,400 deposit within 14 days (section 214(3)(a)); and (ii) a payment of £7,200, also within 14 days, being three times the amount of the deposit: see section 214(4).

[13] Ms Tiensia's summary judgment application for those two payments was heard by Judge Clarke on 23 February 2009. It succeeded. L Judge Clarke ordered Universal to pay: (i) £2,400 into the designated account of an authorised custodial scheme (section 214(3)(b)); and (ii) £7,200 to Ms Tiensia's solicitor: see section 214(4). Her judgment reflects that the argument before her apparently turned on whether, as a matter of the interpretation of section 213, it is possible for a landlord to comply with its requirements under that section after the expiry of the prescribed period of 14 days. Her conclusion was that it is not. The Act, she said, seemed to be in unambiguous terms requiring compliance by the landlord within 14 days and "contains no express power to waive, vary, or mitigate the consequence of that failure". Late performance by the landlord did not therefore provide any basis for a judicial discretion M to excuse him from the sanctions imposed by section 214.

[14] On 17 June 2009, Universal's application for permission to appeal against Judge Clarke's order (on notice to Ms Tiensia and with the appeal to follow if permission were given) came before HH Judge Ellis. He gave permission to appeal, allowed the appeal and set aside Judge Clarke's order. He was invited to treat the issue before him as a preliminary issue raising a question of law as to when and in what circumstances the mandatory section 214 sanctions bite.

[15] Judge Ellis, in an impressive *extempore* judgment, made polite criticisms of the drafting of the statutory provisions, whose interpretation he did not find easy, and said that he regarded the arguments as finely balanced. It was common ground that Universal

A had not met either the 14-day time limit imposed by section 213(3) for compliance with the "initial requirements" or the 14-day time limit imposed by section 213(6)(b) for providing the tenant with the information required by section 213(5). The question was whether either breach triggered the entitlement to the sanctions mandatorily imposed by section 214(3) and (4).

[16] Judge Ellis held that neither breach did so. He explained that section 214(1)(a) was specific in providing that the tenant's entitlement to apply to the court arose in a case in which "the initial requirements of an authorised scheme (see section 213(4)) have not, or section 213(6)(a) has not, been complied with in relation to the deposit;" B and section 214(2)(a) likewise specifically provided that the sanctions in section 214(3) and (4) could be imposed only in a case in which the court "is satisfied those requirements have not, or section 213(6)(a) has not, been complied with in relation to the deposit…". (The alternative provisions in sections 214(1)(b) and (2)(b) are not relevant to this case.) What the legislation had not done was to provide for a right of application to the court or the imposition of sanctions on the landlord, on proof that the landlord had failed to comply with the "initial requirements" within 14 days or had failed to perform the section 213(6)(a) obligation within 14 days. The focus of section 214 was not therefore on whether the landlord had failed to comply with C the 14-day obligations but on whether it was in continuing breach of its obligations to comply with the "initial requirements" and to provide the tenant with the required information as to such compliance. This interpretation was supported by the language of section 214(2)(a), which used the phrases "have not" and "has not", a tense showing that the focus of relevant interest was whether the landlord was in breach by the time of the hearing (or possibly by that of the issue of the tenant's application, although in the post-judgment discussion, Judge Ellis said that he regarded the hearing date as the relevant one). Had the intention been to punish the landlord for a failure to comply with his obligations within the 14-day period, then, said the judge, the phrases "were not" D and "was not" would more probably have been used.

[17] This construction was said to receive support from section 215(1)(b), as to which Judge Ellis made the point that it cannot have been the intention of section 215 to bar forever the service of a section 21 notice if the landlord had committed a 14-day breach; the more rational interpretation was that the landlord is simply so barred until it has remedied its section 213(4) default. Finally, Judge Ellis made the policy point that the tenant's construction would be potentially harsh on landlords: if for good reason they were unable to comply within the 14-day time limits, they would still be faced with the section 214(3) and (4) sanctions.

E [18] Universal has since recovered possession of the property from Ms Tiensia, although the money issues raised by the proceedings remain alive.

B. Honeysuckle Properties v *Fletcher*

[19] Honeysuckle (the appellant: it is the trading name of Ms Moore) granted an assured shorthold tenancy to Messrs Fletcher, McGrory and Whitworth (the respondents) on 19 October 2007 of flat 42, Thomas Telford Basin, Manchester. The term was six months commencing on 26 October 2007. The rent was £995 monthly, payable on the first F of the month. As required by the tenancy agreement, the tenants paid Honeysuckle a deposit of £1,100.

[20] In August 2008, Honeysuckle issued proceedings against the tenants for unpaid rent and other money. That led, on 23 September 2008, to their counterclaim, including a claim for three times the deposit under section 214(4) on the basis that Honeysuckle had neither complied with the "initial requirements" under section 213(3) nor (as followed) provided them with the section 213(5) information. Both assertions were correct as matters of fact. Ms Moore claims to have believed in fact mistakenly that she had protected the deposit in October 2007. Having discovered that she had not, she registered the deposit under the tenancy deposit scheme operated by TDS (the same one as in the other appeal) on 27 September 2008; and, on 29 September 2008, she despatched the deposit protection certificate, including the prescribed G information, to the tenants, who admit its receipt.

[21] The claim and counterclaim came before District Judge Stonier on 30 January 2009, by when the "initial requirements" and the giving of the required information had been complied with. Ms Moore and the tenants all appeared in person. Judge Stonier allowed the tenants' application for three times the deposit, or £3,300, under section 214(4), which she held Honeysuckle was bound to pay. Her reasoning was short, namely that:

> The Act clearly provides that if the provisions with regard to the tenancy deposit are not complied with the court must order a sum equivalent to the three times the deposit and that must be my order in respect of the tenants. There is no H way around that provision. It is mandatory and the legislation was designed to protect tenants in these circumstances.

She engaged in no close consideration of the language of the statutory provisions. She did not explain why, given the mandatory nature of section 214(3), she did not also make an order under that subsection.

Appeals

[22] We had arguments advancing the tenants' case from Mr Bowen for Ms Tiensia, the appellant tenant in the *Tiensia* appeal, and Mr McGrory for the respondent tenants in the *Honeysuckle* appeal. We had arguments advancing the landlords' case from Mr Hutchings J in the *Tiensia* appeal and Mr Manning in the *Honeysuckle* appeal. We reserved our judgments and, in the course of preparing them, considered it necessary to request the parties to provide further written submissions on one particular issue. I would express my gratitude for those further submissions. As a result of that further exercise, the promulgation of these judgments has taken longer than the court would otherwise have wanted.

[23] Mr Bowen pointed out that the effect of section 213(2) is that no deposit may be required by a landlord if it is not to be dealt with in accordance with an authorised scheme. Having taken such a deposit, the landlord has 14 days within which to comply with the K requirements of the scheme (section 213(3)) *and* to give the tenant the information about such compliance: see section 213(5) and (6). He submitted that section 214(1)(a) ought to be read as referring not just to whether "the initial requirements" have been complied with. The more natural sense of its language is, he said, that its reference to "the initial requirements" means "such requirements imposed by the scheme as fall to be complied with by a landlord on receiving such a tenancy deposit", which includes the section 213(3) 14-day obligation. He submitted that the omission of section 214(1)(a) to refer to section 213(3) (which alone refers to the 14-day period) should not deter the court from this expansive interpretation. L

[24] Judge Ellis was, he said, also wrong to be influenced by the policy consideration that a landlord that is a day late in complying with the 14-day time limits would or could face the harsh consequences of the section 214 sanctions. The merits will necessarily vary from case to case and the answer to the question before the court must turn on what the statute requires. The scheme of section 213 is plainly that compliance with it is required within 14 days and the interpretation favoured by Judge Ellis namely that a landlord can avoid the section 214 sanctions provided that he complies with the section 213(3) and (5) requirements by the time of the hearing (or, presumably, any adjourned hearing) effectively robs section 214 of all practical effect. That is M because the cost of compliance is relatively modest and no landlord will ordinarily decline to bear it and so incur disproportionate financial penalties when faced with a section 214 claim.

[25] A point to which Mr Bowen drew attention in his original written argument but did not, as I understood it, either emphasise or develop during his oral submissions was that to which I adverted in [10] above, namely that the TDS scheme *itself* required the landlord "within 14 days of receiving a deposit… [to] protect the deposit with the scheme…". Landlords (such as Universal) that used that scheme were therefore subject not just to the 14-day time limit for compliance imposed by section 213(3), they were also subject to the same time limit for compliance imposed by the scheme itself. The point on

which the court sought further assistance from the parties was whether compliance with such a time limit imposed by the scheme can *itself* be regarded as part of the "initial requirements" of the scheme. If so, the court invited submissions on whether the failure to comply with such a scheme time limit would amount to a failure to comply with the scheme's "initial requirements" for the purposes of that part of section 214(1)(a) referring to section 213(4). If the answer is yes, it may not help the landlords to demonstrate that, on the true interpretation of section 214, the legislation does not seek to penalise a landlord that fails to comply with the *statutory* time limit imposed by section 213(3). In short, the court invited assistance on the relationship between a *scheme* 14-day requirement and the *statutory* 14-day requirement imposed by section 213(3). I will call this argument "the scheme time-limit point".

[26] Mr Bowen's further submission on this point was that when, as here, the scheme itself imposes a 14-day time limit for compliance, compliance with such a requirement is itself part of the scheme's "initial requirements" within the meaning of section 213(4). The consequence of any failure to comply with such time limit is, therefore, that, on the expiry of that time limit, there will be no answer to the tenant's claim under section 214 that the section 213(4) "initial requirements" of the scheme have not been met, with the further consequence that the tenant will have an unchallengeable right to the mandatory relief prescribed by section 214(3). In particular, late compliance will provide the landlord with no defence to such a claim. It makes no difference, he submitted, whether the scheme time limit is, like the statutory time limit, a 14-day one or whether it is longer or shorter: compliance with it is what the landlord has agreed and such compliance is part of the scheme's "initial requirements". Mr Bowen submitted that, in so far as Tugendhat J decided in *Draycott* v *Hannells Letting Ltd (t/a Hannells Letting Agents)* [2010] EWHC 217 (QB); [2010] 3 All ER 411 that a scheme time limit such as that in the TDS scheme is not part of the scheme's "initial requirements", the decision was wrong.

[27] By way of an alternative argument to his two main arguments, Mr Bowen submitted, even if the failure to comply with the section 213 "initial requirements" within the 14-day period (whether that imposed by section 213(3) or the scheme itself) does *not* automatically trigger the mandatory section 214 sanctions, it cannot be the case that the landlord can then have until the *hearing* to comply with the "initial requirements" and so avoid the sanctions. A more rational interpretation of the legislation, which would give it real practical effect, is that the landlord's cut-off time for compliance with the "initial requirements" and notification obligations can be no later than the date of the issue of tenant's claim or counterclaim under section 214. If, therefore, at the date of the issue of either such claim the landlord has still not complied, subsequent compliance will not enable it to avoid the section 214 sanctions.

[28] Mr McGrory supported Mr Bowen's argument and, in the course of his succinct and clear representations, helpfully referred us to extracts from the TDS website. In response to the court's request for further submissions on the scheme time-limit point, he too emphasised that a landlord using a scheme such as the TDS scheme has to comply with its "initial requirements", not just within the time limit imposed by section 213(3) but also within the same time limit imposed by the scheme itself. He said that compliance with the latter was in the nature of a contractual obligation between the landlord and the scheme and that it follows that a failure to comply with a scheme time limit will be a failure to comply with its "initial requirements". The consequence will be that, on the expiry of that time limit, the landlord will have no answer to a section 214 claim for relief under section 214(3).

[29] For Universal, Mr Hutchings supported Judge Ellis's decision, essentially for the reasons the judge gave. Focusing first on the interpretation of sections 213 and 214, he emphasised that the right under section 214 to apply for the imposition of its sanctions was specifically linked by section 214(1)(a) to an omission either to comply with the "initial requirements" of a scheme or to provide the tenant with the required information as to such compliance. What, however, such right was apparently deliberately not linked to was an omission so to comply within the 14-day period prescribed by section 213(3). He said that this interpretation was consistent with the sense of section 215(1)(b). That, too, could not have intended that any non-compliance within 14 days would raise a permanent bar on a section 21 notice: its sense was plainly that such a bar would apply only during such time as there had been no compliance at all. He said that the mandatory obligations of section 214(3) repay the deposit or protect it by an order for its payment into a custodial scheme also made no sense (or at any rate the latter alternative did not) if, by the date of the hearing, the landlord had already protected the deposit in accordance with its statutory obligations. He emphasised the penal nature of the tenants' argument and submitted that there could be no justification for it. The purpose of the legislation is to achieve the due protection of the tenant's deposit and so long as it becomes duly protected if necessary, as the result of proceedings brought by the tenant the aim of the legislation will have been achieved. He recognised that a landlord that leaves it until the 11th hour to comply will be likely to be visited with a costs order in favour of the tenant. In support of his submission as to the purpose of the legislation, Mr Hutchings referred us to *UK Housing Alliance (North West) Ltd* v *Francis* [2010] EWCA Civ 117* in which (in a judgment with which Lord Neuberger of Abbotsbury MR and Smith LJ agreed), Longmore LJ said, in [7]:

> Chapter 4 of Part 6 of the 2004 Act was intended to deal (inter alia) with the notorious abuse of landlords requiring deposits from prospective tenants but not keeping the sums paid in any separate account or refusing to repay such sums at the end of the tenancy.

[30] Mr Hutchings also submitted that the landlord will have until the date of the hearing of the tenant's application within which to comply with its section 213 "initial requirements" and notification obligations so as to avoid the imposition of the section 214 sanctions. If, as Mr Bowen had argued in the alternative, the cut-off point is the date of the tenant's application (meaning, presumably, the issue of his application), the landlord will no longer have any incentive to achieve the protection for the deposit intended by the legislation: it will at that point have no choice but to submit to the hit imposed by the section 214 sanctions.

[31] In response to the court's request for submissions on the scheme time-limit point, Mr Hutchings pointed out (I consider correctly) that this point did not form part of Ms Tiensia's grounds of appeal. He also pointed out that the terms of the scheme had not been put before Judge Clarke and that no respondent's notice was before Judge Ellis contending that the particular terms of the scheme provided an alternative basis for summary judgment. He nevertheless dealt with the point on its merits. His submission was that a time limit imposed by the scheme itself is outside the definition of "the initial requirements" in section 213(4). The only time limit for compliance with such requirements is that imposed by section 213(3). He submitted that if he is correct in his submission that section 214 was deliberately drafted so as to exclude a liability for a sanction for an omission to comply with the statutory time limit, it is improbable that the legislation could have intended that a breach of a time limit imposed by the scheme could itself trigger a liability to such a sanction.

[32] Mr Hutchings developed this submission by pointing out that there is no statutory requirement that a tenancy deposit scheme must itself impose a 14-day or any time limit within which the deposit must be protected. By contrast, the section 213(3) time limit applies statutorily to all schemes. Moreover, the basis of the scheme time-limit argument is anyway artificial. A landlord has a choice between scheme providers and, until it has protected the deposit with a particular scheme, it is neither committed to any particular scheme nor to any particular scheme time limit. Universal received the deposit in June 2008 and protected it with the TDS scheme only in October 2008. On what basis, he asked, can it be said that, before it so protected the deposit, it was subject to a 14-day time limit imposed by the scheme? Mr Hutchings also pointed out that since TDS was content to accept the deposit late, there is an argument available to Universal that TDS thereby waived

* Editor's note: Reported at [2010] 2 EGLR 81

any need for it to comply with the 14-day period. If it did so waive such need, how can it be said that Universal was in breach of such time limit so as to trigger a liability to the section 214 sanctions? If, however, the scheme time-limit point leads to a result that a failure to comply with it puts the landlord in irredeemable breach, the landlord is not only faced with an unanswerable claim to have such sanctions imposed on it, it will be forever barred from serving a section 21 notice in respect of the tenancy: that is because the interpretation of section 215(1)(b) must be the same as the relevant part of section 214(1)(a). That, too, is said to be an improbable legislative intention.

[33] Mr Manning, for Honeysuckle, supported Mr Hutchings' submissions and also made clear his like submission that the cut-off point for compliance by the landlord is the date of the hearing, not a date immediately prior to the tenant's section 214 application. He said that the language of the legislation provided no support for any date prior to the hearing date. The principal purpose of section 214 is to ensure that deposits are protected, and provided that they are protected by the date of the hearing, why should the section 214 sanctions be imposed? He recognised that the tenant may, in the event of post-application compliance, recover less than its full costs, but submitted that that is not a reason for interpreting the legislation differently.

[34] As for the scheme time-limit point, Mr Manning submitted that the natural interpretation of section 213(3) and (4) is that the reference in it to "the initial requirements" of a scheme cannot include a reference to any time limit for securing the deposit imposed by the scheme itself, since the legislation itself specifies the time within which the deposit must be secured. The meaning of "the initial requirements" within those subsections is therefore necessarily confined to *what* you have to do in order to protect the deposit, and does not extend to *when* you have to do it. That interpretation was supported by the decision in *Draycott*, which Mr Manning submitted was correctly decided. In addition, if a scheme time limit is an "initial requirement", the result would be that a scheme could select a time limit different from the 14-day period specified in section 213(3), which would have the potential to produce odd consequences. If, for example, the scheme limit was seven days, section 213(3) would in theory still require compliance only within the longer period of 14 days; conversely, if the scheme limit was 28 days, section 213(3) would require compliance within the shorter period of 14 days. Either consequence would, submitted Mr Manning, be nonsensical. Mr Manning developed the potential for absurdity by further examples.

Discussion and conclusion

[35] In my judgment, Mr Hutchings and Mr Manning are correct in their submission that the natural interpretation of the phrase "the initial requirements" as used in section 213(3) and (4) is that it does not include any requirement imposed by a particular scheme as to the time within which the landlord must secure the deposit. The quoted phrase must mean the same in both subsections, and since section 213(3) itself imposes a time limit within which "the initial requirements" must be complied with, it is necessarily implicit in the subsection that it cannot be recognising that one such requirement is the (perhaps different) time limit for securing the deposit that may be imposed by a particular scheme. If it were otherwise, there would be the potential for unacceptable uncertainty and confusion as to the relationship between the section 213(3) time limit and the scheme time limit. The natural interpretation of section 213(3) and (4) is, in my judgment, that its reference to "the initial requirements" of a scheme is only be to those requirements for protecting a deposit other than any time limit for doing so that may be imposed by the scheme. Tugendhat J in *Draycott* (in [29]), was also of the view that a scheme time limit is not one of "the initial requirements" within the meaning of the definition in section 213(4). I respectfully agree with him. The real issue in dispute in this case turns on the arguments that were originally advanced to us. I turn to those arguments.

[36] In my judgment, Judge Ellis was correct to hold that the precondition of a tenant's application to the court under section 214 is not a failure by the landlord to comply with the "initial requirements" or the notification thereof to the tenant within the 14-day period specified in section 213. It is the failure to comply with either of those obligations at all. It follows, in my judgment, that if, therefore, the landlord is late in complying with its dual section 213 obligations but it nevertheless duly does so before any section 214 proceedings are brought by the tenant, the tenant will have no cause of action under section 214 and any claim that he brings under it will fall to be dismissed. I consider that both Judge Clarke and Judge Stonier, neither of whom applied any apparent consideration to the precise terms of the legislation they were concerned to apply, interpreted it wrongly. My fuller reasons are these.

[37] Section 214(1)(a) entitles the tenant to apply to the court in a case in which "the initial requirements of an authorised scheme (see section 213(4)) have not, or section 213(6)(a) has not, been complied with in relation to the deposit…". In the case of both alternatives, the focus is, apparently deliberately, not on whether there was compliance within the 14-day period but on whether there has been compliance at all. Had the *time* for compliance been in section 214's sights, the "initial requirements" reference would have been to section 213(3) rather than to 213(4); and the notification reference would have been either to section 213(6)(b), or else to section 213(6) as a whole, rather than simply to section 213(6)(a). Exactly the same points can be made in respect of the language of section 214(2)(a) in which the reference to "those requirements" is plainly a reference back to the first limb of section 214(1)(a) and in which there is a repeat reference to section 213(6)(a).

[38] Moreover, as Mr Hutchings rightly pointed out, the tense of the language in section 214(1)(a) and 214(2)(a) is consistent only with an inquiry as to whether the "initial requirements" and notification obligations have been performed at all, and not with whether they were performed within a particular period that is now past. That is the natural sense of "have not" and "has not" in both subparagraphs. There can plainly be no right to make any application to the court under section 214 *before* the 14-day period for compliance has expired, and so if the crucial trigger to the section 214 cause of action was the landlord's non-compliance within such 14-day period, the more natural choice of language for the draftsman to have used in section 214(1)(a) and 214(2)(a) would have been "were not" and "was not" respectively.

[39] I also agree with Mr Hutchings that this interpretation finds support in section 215. Once again, section 215(1)(b) focuses exclusively on section 213(4), not on section 213(3), and so avoids any focus on the 14-day period within which compliance is required. Its plain sense is that, so long as the initial requirements have not been complied with (including after the expiry of the 14-day period), there will be a bar on the service of a section 21 notice. It is impossible to interpret section 215(1)(b) as intending to impose a permanent such bar in the event of a failure to comply with the 14-day requirement. Like points can also be made in respect of section 215(2), which opens by referring to section 213(6) generally (thus including a reference to an omission to comply with the 14-day requirement imposed by section 213(6)(b)), but then goes out of its way to make clear that the giving of a section 21 notice is only barred for so long as section 213(6)(a) is not complied with.

[40] That interpretation of the legislation means that late, but nevertheless due compliance by the landlord with its dual obligations under section 213(3) and (6) will furnish it with a complete defence to any claim by the tenant under section 214. Such interpretation appears to me to be not only firmly supported by what I would regard as the carefully chosen statutory language, it is also a properly precise, or strict, one to apply to legislation such as section 214, which is manifestly penal in intent. Moreover, it is an interpretation that is consistent with the purpose of the legislation. That purpose is to achieve the due protection of deposits paid by tenants, ideally within the 14-day period but, if not, then later. It cannot be its purpose to punish landlords that may for example, for innocent reasons, be just a day late in securing such protection.

[41] The other question raised by the arguments is whether (as Judge Ellis held and Mr Hutchings and Mr Manning submitted) the landlord has until the *hearing* of the tenant's section 214 application to comply with its section 213 "initial requirements" and notification

A requirements. If it has until then to remedy any prior section 213 default, and does so, it must follow that the tenant's section 214 application will fail, although no one suggested that, in such a case, the tenant would not ordinarily be entitled to recover from the landlord the costs of his claim. The contrary case, advanced by Mr Bowen, was that the cut-off point for section 213 compliance is not the *hearing* date but the issue of the tenant's section 214 claim or counterclaim. The theory is, presumably, that at that point the tenant's cause of action under section 214 has accrued and cannot be defeated by the landlord's subsequent compliance with its section 213 obligations by the time of the hearing. It would, it is said, be unjust if the claim could be so B defeated and the tenant then left with at most a right to a probably less than full indemnity for his costs from the landlord.

[42] I agree with Judge Ellis, Mr Hutchings and Mr Manning that the date of the hearing is the relevant date and do not accept Mr Bowen's contrary argument. First, there is, in my view, nothing in the legislation that points to any date earlier than the date of the hearing date. On the contrary, the use of the present tense "is satisfied" in section 214(2)(a) appears to me to support the case for the hearing date; and, consistently with that (albeit in respect of the different breach with which it is concerned), I consider that section 214(2)(b) shows unambiguously that the relevant date for its own purposes *is* the hearing date. If that is the C relevant date for section 214(2)(b) purposes, it would be odd if it were not also the relevant one for section 214(2)(a) purposes.

[43] Second, the 2004 Act was enacted at a time when the culture of the conduct of civil litigation had become one under which in ordinary circumstances a claimant should endeavour to avoid the need for litigation by applying any applicable pre-action protocol or otherwise by writing a letter before claim. A tenant ought, therefore, to write such a letter before making a section 214 claim and so give the landlord the opportunity to remedy its shortcoming and avoid proceedings. The landlord would, I should imagine, ordinarily be ready, willing and anxious to do so (it may, like Honeysuckle, have mistakenly believed D that it had already duly protected the deposit) so as to bring to an immediate end the possibility of a section 214 claim and the visiting on it of a painful penalty. Although, however, the tenant *ought* to write a letter before claim, and his omission to do so may present him with a costs risk if he does not, if Mr Bowen's argument is right the prospect of a section 214(4) order in his favour may be sufficiently attractive to encourage him instead to ambush the landlord with an unheralded claim and run such costs risk as there may be a risk that he may regard as making the game well worth the candle. Of course if (as in both the cases under appeal) the claim is made by a counterclaim in the landlord's action, there will ordinarily be no occasion for the tenant to E write a letter before claim. I recognise that section 214 claims are, in practice, probably more commonly going to be made by counterclaim than by claim, but it appears to me helpful to consider the alternative position that I have. I note that *Draycott* was a tenant's claim, not a counterclaim.

[44] These considerations do, in my view, also tend away from an interpretation of the legislation to the effect that the cut-off date for compliance is the issue date of the tenant's claim or counterclaim. As I have said, the objective of the legislation is not the punishment of landlords but the achieving of proper protection of tenants' deposits. The legislation should not be interpreted in a sense that implicitly F encourages the ambushing of landlords by tenants who have grounds for believing that the landlords have not complied with their section 213 obligations. It should be interpreted in a way that avoids litigation. Litigation will or should be avoided if, following a letter before claim, the landlord promptly puts its house in order. If the landlord declines or fails to do so, then of course it is open to the tenant to pursue his section 214 claim. If the landlord later (before the hearing) repents and remedies its defaults, the claim will still fail, although the tenant will ordinarily recover his costs. He may not recover his full costs, but there is nothing unusual about a claimant not doing so. The tenant will bring his claim knowing of that risk.

[45] The only remaining point that I should deal with is Mr Bowen's point that the interpretation of the legislation that I favour will have the practical consequence of robbing section 214 of virtually all its G force. That is because it will be an unusual landlord that will not, faced with a section 214 claim, ensure that by the time of the hearing it has fulfilled its outstanding obligations under section 213, with the consequence that, in practice, section 214 will be likely to bite only in the most exceptional and unusual cases. I recognise all of that. Equally, however, it can also be said that in the overwhelming majority of cases, the net result will be that the legislation will have achieved its primary objective, that of the due protection of the tenant's deposit. What more can reasonably be asked of it?

[46] I comment that in *Draycott* the deposit was protected late and the section 213(5) information was necessarily also provided late. H The tenant's section 214 claim was commenced only subsequently. Tugendhat J held that, in those circumstances, the court could not be satisfied under section 214(2)(a) and so could not make an order under section 214(3) and (4). I respectfully agree with his decision. He did not, however, also have to decide, as we have had to, whether the landlord will have until the hearing to put its house so in order as to avoid the making of an order under those subsections.

[47] I would dismiss the appeal in *Vision* and allow the appeal in *Honeysuckle*.

Giving the second judgment, **SEDLEY LJ** said: J

[48] As Rimer LJ notes, this legislation was passed to deal with a widespread abuse in a situation of housing scarcity the retention by lessors of deposits, commonly of a month's rent, at the termination of the tenancy. It makes compulsory provision for safeguarding tenants' money and backs the scheme up by penalising defaulting landlords in triple damages. This much is plain.

[49] If the meaning of the statute is, however, that a landlord is not culpably in default if, months or maybe years after the expiry of the fortnight given by law for compliance, it eventually complies with the initial requirements, then the scheme is a dead letter. The extent to which it is drained of effect is emphasised by Rimer LJ's conclusion K (which I agree follows from his reading of the legislation) that the landlord's opportunity for penitence extends not only to the date of issue of the tenant's claim or (more probably) counterclaim but to the eve of judgment. If this is right, no tenant could ever sensibly be advised to sue or counterclaim for the penalty.

[50] This is one result that the court has to face up to. The other is said to be the imposition of a penalty by less than clear words, an outcome against which the courts always lean heavily. However, I do not think that we are in that situation here because neither side's reading of the statute is non-penal. They differ only, albeit crucially, as to when and in what circumstances the landlord becomes liable to the penalty. L The fact that, on the tenants' reading, the penalty falls on the inadvertent and the unlucky equally with the devious and the dishonest is morally questionable but perfectly explicable: parliament has decided that recovery is to be in the tenant's hands, that it is to depend on the simple question of compliance or non-compliance, and that strict liability for non-compliance will catch the devious and encourage the others. That is a matter for legislators, not for us.

[51] What is a matter for us is the converse: that to sanction compliance at any time up to the moment of judgment is to eviscerate the legislative scheme. Not only the innocent defaulter but the scheming one can evade the penalty with ease. The only lessor likely to be taken M to judgment is one that has appropriated the money and is no longer worth powder and shot (since otherwise it would have paid at the 11th hour into a scheme).

[52] There is, however, a further sanction for non-compliance. Section 215(1) prohibits the giving of notice to recover possession on termination of a shorthold tenancy where the deposit is not being held in accordance with an authorised scheme or the initial requirements of a scheme have not been complied with. Although the first of these conditions might be met by a late payment into the scheme, the second could not be, at least on the tenants' construction, once the initial fortnight had gone by. It would seem to follow that, if the tenants' argument is right, the defaulting lessor, whether culpable or inadvertent,

is debarred indefinitely from recovering possession. Even in a statutory scheme designed to concentrate minds on the protection of tenants' deposits, this appears a remarkable result, akin to a forfeiture. It is a result avoided by the construction of section 213 preferred by the majority of the court, but that in itself is not enough. One has to see what meaning section 213 is properly capable of bearing.

[53] The two critical subsections of section 213, (3) and (4), are set out in Rimer LJ's judgment. Section 213(4) defines the initial requirements as requirements imposed by an authorised scheme that fall to be complied with by the landlord on receiving a tenancy deposit. This, if I may say so with respect to the drafter, is thoroughly unhelpful. Until the deposit is handed over by the tenant, no obligation exists to pay it into any scheme: see section 213(1). On receipt of it the landlord has to select a scheme in which to secure the deposit. It is only at that point that the initial requirements, which will always be those of a particular scheme, can be known. So the subsection has to mean that the initial requirements of a scheme are those that become operative not on receiving the deposit but on entry into the scheme: for example (see Schedule 10) the payment of the full amount of the deposit into a custodial scheme account or the necessary undertakings on entry into an insurance scheme.

[54] If this is so, the provision of section 213(3), that the initial requirements are to be complied with within 14 days from the receipt of the deposit, is not the vacuous provision on which the landlords' counsel rely. It now has some purchase. The 14 days, I accept, cannot logically be derived from the scheme itself, since it is only on entry into it that its terms become operative. However, they can and, in my judgment, do constitute a statutory time limit for the obligatory entry into one authorised scheme or another. It would certainly have been clearer if section 213(3) had said "Where a landlord has received a deposit he must secure it in an authorised scheme within 14 days of receiving it". However, that, however elliptically expressed, is what section 213(3) means.

[55] I do not, with great respect, consider that the *indicia* in section 214 on which Rimer LJ relies for his contrary conclusion are significant. It may be, as he suggests, that the intended purpose would be more correctly expressed by the use in section 214(1)(a) of a different tense "the initial requirements were not complied with" to signify a past rather than an ongoing failure to comply. However, any lay person reading this text (and legislation like this is or ought to be written for lay people) would understand well enough what was intended: that a landlord that did not put a deposit into safekeeping within 14 days of receiving it was from then on at risk of having, among other things, to pay three times the deposit as a penalty. They would not read it as meaning that the 14 days could be safely ignored because the landlord could always escape a penalty by safeguarding the deposit if and when the tenant sued. The reason would not be that they saw and forgave the draftsman's grammatical solecism; it would be that "the initial requirements have not been complied with" is a comprehensible if not an ideal or strictly correct way of describing a past failure to do something. Indeed, the drafter might legitimately have replied that, once the 14 days have elapsed, there will be an irremediable failure that will therefore be ongoing at the date of any future claim, justifying the use of the perfect tense. Such an understanding would also give some effect to the word by which parliament has chosen to describe this and the associated requirements: "initial".

[56] Rimer LJ, in reading this provision, gives weight to the submission that the parenthetic reference in section 214(1)(a) is not to section 213(3) but to section 213(4) in other words to the definition of "initial requirements" rather than to the 14-day requirement itself. He takes the same approach to a similar parenthesis in section 215(1)(b). I regret to say that I cannot agree. The only purpose of the parenthesis seems to me to be to help readers to get their bearings by reminding them where the definition of "initial requirements" is to be found. The differential use of "initial requirements" in subsections (3) and (4) respectively is open to stylistic criticism, but nothing turns on it in these appeals. For reasons that I gave earlier, the distinction between the two subsections for the purposes of construing section 214(1)(a) and section 215(1)(b) is, in my respectful view, a distinction without a difference.

[57] The same applies, in my respectful opinion, to Rimer LJ's reasoning in respect of section 213(6). I do not consider that the provisions that refer back to the requirement in section 213(6) for prescribed notices to be given within 14 days of receipt of the deposit either add anything to or subtract anything from the interpretation of section 214(1)(a). The two elements of the scheme no doubt stand or fall together, but that is all.

Conclusions

[58] For the reasons that I have given, there is nothing legally (as contrasted with morally) objectionable in the strict liability to a penalty, and no way of avoiding a penal construction of the legislation. There is, however, something approaching overkill in the additional prohibition in section 215 on the recovery of possession. We are left with an intractable dilemma: to drain the legislative scheme of all effect by reducing the remedy for non-compliance to near-impotence, or to give what, in my judgment, was without doubt the intended meaning to the prescribed 14-day limit, with irreversible economic and proprietary consequences for landlords that fail, even if only through misfortune, to meet it.

[59] We have not been shown and have not asked for the parliamentary proceedings, but I find it impossible to believe that legislators thought that they were voting for the first of these outcomes. I find it almost as difficult to believe that they appreciated that the effect of what they were voting for was as drastic as it appears to be. In many cases turning on the meaning of a statute, the absurdity or oppressiveness of one reading can legitimately steer the court to a less apparent but more equitable interpretation. However, where, as here, it steers the court to an interpretation that is every bit as unacceptable, the canons of construction fall silent.

[60] Left alone, therefore, with this hostile text, it seems to me that the tenants' reading is the more proximate to the language and configuration of the statute. Accordingly for my part, and without much confidence that the enactment actually makes the provision that parliament meant to make, I would dismiss the appeal in *Honeysuckle* and allow it in *Vision*.

Giving the third judgment, **Thorpe LJ** said:

[61] I have had the advantage of reading the draft judgments of Rimer LJ and Sedley LJ on this particularly difficult point of construction. At the conclusion of the oral argument, I was persuaded that the language of the statute compelled the conclusion expressed by Rimer LJ. I regret that conclusion because, as Sedley LJ persuasively explains, it means that landlords are not disciplined as parliament must surely have intended. Nevertheless, I concur with the outcome proposed by my lord, Rimer LJ.

First appeal dismissed; second appeal allowed.

Ashdown Hove Ltd *v* Remstar Properties Ltd

LEASEHOLD VALUATION TRIBUNAL

28 June 2010

PROFESSOR JAMES DRISCOLL LLM LLB SOLICITOR (LAWYER CHAIR), MR NEIL CLEVERTON FRICS AND MR ROGER WILKEY FRICS

Estates Gazette 18 September 2010

[2010] 37 EG 138

Leasehold enfranchisement — Leasehold Reform, Housing and Urban Development Act 1993 — Collective enfranchisement — Purchase price — Deferment rate — Flats — Enfranchisement of block of flats in Hove — Whether uplift to generic 5%* Sportelli *rate for flats justified in light of greater obsolescence and management risks and lower capital growth rates than in prime central London area

The applicant company was the nominee purchaser for the purposes of a claim by qualifying tenants of flats in a block in Hove, East Sussex, to acquire the freehold of the building under the collective enfranchisement provisions of the Leasehold Reform, Housing and Urban Development Act 1993. The building comprised an L-shaped, brick-built block with a flat roof, containing 125 flats, all of which, save for a caretaker's flat, were let on long leases. A management company controlled by the leaseholders managed the building pursuant to the terms of the leases, although these provided for the respondent freeholder to take over the role should the management company fail to act properly or at all. The management company held a lease of the common parts.

The applicant applied to the leasehold valuation tribunal, under section 24 of the 1993 Act, for a determination of the purchase price. The sole issue in dispute was the deferment rate to be applied to the freehold vacant possession value. The applicant argued for a higher rate than the generic 5% rate for flats laid down in *Earl Cadogan* v *Sportelli* [2007] 1 EGLR 153, on the ground that the building differed in a number of respects from properties in the prime central London (PCL) area on which the *Sportelli* rate was based. The applicant contended for additions of: (i) 0.25% to reflect obsolescence and deterioration, on the basis that although the average values of the flats were lower than in the PCL, maintenance costs were not; (ii) 0.5% to reflect the difference in capital growth rates, on the basis that these were significantly lower outside the PCL; and (iii) 0.25% in respect of the special management problems of the building and the increased complexity of leasehold management following the changes made by Part 2 of the Commonhold and Leasehold Reform Act 2002.

***Decision*: A purchase price of £1,732,109 was determined, applying a deferment rate of 6%. (1) Notwithstanding the full repairing covenants in the leases, deterioration was more likely in the block than in the PCL, which was not reflected in the vacant possession values. The difference between the value of the flats in the block and those considered in *Sportelli* was significant, while building costs were comparable to those in London. It followed that it was likely to remain economically viable to repair high-value flats in the PCL area for much longer than would be the case for similar-sized flats in the application property. (2) An investor that was examining long-term growth would not be confident that the PCL rate of growth would be achieved in the Hove and Brighton area and would accordingly adjust its bid for the block downwards. (3) Although the existence of a leaseholder-controlled management company on its face absolved the freeholder from the pressures and risks of flat leasehold management, there was still a risk that the management company could fold, leaving the freeholder to assume the management responsibilities, as it was entitled to do under the leases. An investor would require an adjustment in respect of the application property to reflect the greater management responsibilities associated with flats compared with houses, to take account of the possibility that it would have to take over the management of the block should the management company collapse. A further relevant factor was the new stricter legal framework governing flat management and the market's awareness of its effects. (4) Additions should be made to the 5% generic rate to reflect the above factors, comprising: (i) 0.25% for obsolescence; (ii) 0.5% for the additional risk that growth would not be achieved; and (iii) 0.25% for the risks involved in managing the block. Accordingly, a deferment rate of 6% should be applied to the freehold vacant possession value for the building.**

The following cases are referred to in this report.

Daejan Investments Ltd v *Benson* [2009] UKUT 233 (LC); [2010] 2 P&CR 8

Earl Cadogan v *Sportelli*; *Earl Cadogan* v *27/29 Sloane Gardens Ltd*; *Earl Cadogan* v *Grandeden Property Management Ltd*; *Howard de Walden Estates Ltd* v *Maybury Court Freehold Co Ltd*; *Bircham & Co Nominees (No 2) Ltd* v *Clarke* [2008] UKHL 71; [2009] 2 WLR 12; [2009] 1 P&CR 17; [2009] 1 EGLR 153; [2009] RVR 42, [2007] EWCA Civ 1042; [2008] 1 WLR 2142; [2008] 2 All ER 220; [2008] 1 EGLR 137; [2007] RVR 314, LRA/50/2005; [2007] 1 EGLR 153; [2006] RVR 382, LT

Lethaby, Re; sub nom flats 1 and 2, 245 Glyn Road, London E5 0JP, Re [2010] UKUT 86 (LC)

7 Sudeley Street Brighton CHI/00ML/OCE/2008/0023 unreported 23 September 2008

Zuckerman v *Trustees of the Calthorpe Estate* [2009] UKUT 235 (LC); [2010] 1 EGLR 187

This was a determination on an application by the applicant, Ashdown Hove Ltd, to the Southern Leasehold Valuation Tribunal, for a determination of the price payable to the respondent freeholder, Remstar Properties Ltd, on a collective enfranchisement of a building under the Leasehold Reform, Housing and Urban Development Act 1993.

Anthony Radevsky (instructed by Osler Donegan Taylor Solicitors) appeared for the applicant; Wayne Clark (instructed by Howard Kennedy LLP) represented the respondent.

Giving the tribunal's decision, PROFESSOR JAMES DRISCOLL said: The price payable by the nominee purchaser for the acquisition of the freehold of the subject premises is the sum of £1,732,109. This is based on applying a deferment rate of 6%.

Introduction

[1] This is an application by the nominee purchaser under section 24 of the Leasehold Reform, Housing and Urban Development Act 1993 (the 1993 Act). It seeks a determination of the price payable for the acquisition of the freehold of the subject premises. The nominee

A purchaser is a company that was incorporated to act on behalf of the leaseholders who are participating in this claim. The respondent is the owner of the freehold and is the reversioner under the claim. Under the flat leases, there is a management company that, as the name might suggest, has the responsibility for the management of the property. The landlord has the right under the lease to take over management should the management company fail to act properly or at all. (There is also a headlease of common parts.)

[2] An initial notice dated 2 October 2007 was given under section 13 of the 1993 Act seeking to exercise the right to acquire the freehold under the provisions in Part I of the Act. The notice claimed the B freehold to the specified premises (that is, the building containing the flats) and certain land and buildings lying to the north of the specified premises. A price of £940,000 was proposed for the freehold to the specified premises and the sum of £950 for the additional property.

[3] In response, a counternotice dated 13 December 2007 was given under section 21 of the 1993 Act admitting that the participating leaseholders are entitled to exercise the right to enfranchise. However, the respondent made counter-proposals on price, by claiming £2,064,856 for the freehold to the specified premises and the sum of £272,500 for the freehold to the additional property. Since the parties could not agree on the prices payable, application was made to the C tribunal on 16 May 2008.

Application

[4] Provisional directions were given on 17 May 2008 and further directions on 11 January 2010. The latter followed a period during which this tribunal agreed to stay the hearing of the applications when it first appeared that the parties were likely to reach agreement on the price. When this proved unsuccessful, and the application for a hearing was renewed, these further directions were given. One hearing date was vacated at the request of the parties, and shortly before the adjourned hearing fixed for 16 March 2010 it had emerged that, unknown to the parties, there exists a lease of the common parts (which has not been D registered). The participating leaseholders considered an application in the county court to amend their section 13 initial notice to include the acquisition of this lease, but, in the event, they decided not do so. They are not seeking to acquire this lease. A revised statement of facts was signed by the valuers in April 2010.

Premises

[5] The subject premises were built in the early 1970s as a detached L-shaped block of flats on seven floors holding 125 flats, all but one of which is held on a long lease. The one exception is the flat that is used as a caretaker's flat.

E [6] There is a flat roof on top of the premises, which is a brick-built building. The windows to the flats are single-glazed. The accommodation ranges from one-bedroom units to three-bedroom larger penthouses. Some of the flats on the upper floors have roof terraces. In addition to the flats, there is a basement flat (no 29), which is a caretaker's flat, and a separate caretaker's store and workshop. There is also a launderette and a sauna suite. The building has two lifts.

[7] The accommodation consists of 23 one-bedroom flats, 69 two-bedroom flats, 24 three-bedroom flats and eight penthouse flats. All the flat leases are coterminus and expire by effluxion of time in 2094. At the valuation date (which the parties agree is 2 October 2007), F all the flat leases had 87 years unexpired. There is therefore no marriage value payable: see para 4[(2A)] of Schedule 6 to the 1993 Act.

[8] A company called Ashdown (Eaton Road) (Hove) Ltd holds a headlease, but, as indicated above, this is not registered against the freehold title and its existence was only recently discovered by those advising the parties. This headlease includes all the common parts, including the caretaker's store and workshop, the launderette and the sauna suite. It does not include, however, all the car parking and excludes some of the surface parking and the underground parking. This headlease is dated 1 June 1976 and is coterminus with the flat leases.

[9] Outside the building are areas used as a communal garden and a hard-surfaced car park with 40 parking spaces. There are also 17 other spaces to the north and the north-west of the main building. Further parking is to be found in an underground car park with an additional G 60 car-parking spaces. None of these car- parking spaces is demised under the flat leases. In some cases, the landlord leases them to the flat leaseholders; others are rented to traders.

Inspection

[10] The tribunal inspected the property prior to the hearing in the presence of representatives of the applicant and respondent (that is to say, the instructing solicitors, counsel and the valuers). Ashdown is a large residential development comprising 124 flats in a seven-storey block together with parking beneath part of the main building and additional parking on site. In addition, there is a caretaker's flat and one of the flats has been converted into a sauna and meeting room. The H block was probably built in the 1970s and occupies a central location within easy reach of shopping facilities and Hove railway station. The attention of the tribunal was drawn to various parts of the property that were inspected in the presence of the parties. In particular, an inspection was made of flats 5, 44 and 15, which are typical of the one-, two- and three-bedroom flats in the building. The internal standard of decorations, kitchen and bathroom fittings (and so on) varies in the three flats. The windows in each flat are the original single-glazed sliding aluminium type. Some have been fitted with secondary glazing. It is apparent that the windows are in poor order and beyond economic repair. This will be J a significant expense for the future management of the building.

[11] We went onto a south-facing sun terrace at roof level and were shown the problems that had arisen to this part of the block. Movement has occurred to a wall, which as a result is crumbling. We understand that a firm of structural engineers had been consulted and the recommended work will soon be put in hand. The tribunal was told that this is relatively minor problem and will not affect the main structure.

[12] Flat 30 has at some time been converted into a sauna, launderette and meeting room for the use of the residents. The sauna is no longer in use and the interior of this unit was tatty and in need of redecoration and upgrading. K

[13] The tribunal inspected the parking beneath part of the main building. There are signs of water penetration through the concrete ceiling and this is a continuing problem.

[14] On the north side of the block is open parking. It cannot be used for this purpose because the adjacent cricket ground has a right of way over the access road, and when a match is in progress media vans and equipment block the access. This has been the case for many years.

Hearing

[15] In their supplemental statement, the valuers were able to agree the figure for the capitalisation of the ground rents at £714,274; the long unimproved values of the leasehold flats (set out in the spreadsheets L accompanying this statement); the caretaker's flat (£160,000); and the car parking at £843,000. In addition, at the end of the headlease, the sauna suite and the north-side parking spaces will revert to the landlord. The valuers are divided on one key factor: the deferment rate, which according to the valuer for the nominee purchaser should be 6%, while the valuer advising the landlord proposes 5%.

[16] It follows that, in the absence of disputes over the terms of the transfer (or the payment of costs under section 33 of the 1993 Act), the sole issue for our determination is the deferment rate. Applying a deferment rate of 5% would produce a price of £1,956,089; applying a rate of 6% produces a price of £1,732,109. M

[17] Those advising the parties are fully aware of the developments in the law and, in particular, the passage of the *Earl Cadogan* v *Sportelli** litigation in which the Lands Tribunal (LT) (now called the Upper Tribunal (UT)) decided (among other matters) that generic rates should be used: 4.75% for house enfranchisement claims and 5% for flat new lease or collective enfranchisement claims (the additional fraction of 0.25% to reflect the greater potential risks caused by the problems of managing blocks of leasehold flats). The LT also decided that expert evidence from the financial markets was to be used instead of property

* Editor's note: Reported at [2007] 1 EGLR 153

market evidence because the latter is considered to have been (in a sense) "tainted" by the very statutory rights that allow leaseholders to demand as of right new or extended leases or the freehold. In the valuation formulae in the Leasehold Reform Act 1967 (for house leases) and in Schedules 6 and 13 to the 1993 Act, one has to disregard several factors, including the statutory rights of leaseholders. This approach is sometimes referred to as valuing in the "no-Act" world.

[18] The LT sanctioned the use of a different approach by adopting these generic rates based on the formula DR = risk-free rate (of 2.25%) *less* the real growth rate (2%) *plus* risk premium (4.5%) and an increased figure of 0.25% for the management risks for flats rather than houses. If this was not noteworthy in itself, that tribunal also stated that the generic rates should be used generally for all residential leasehold property regardless of its location. The properties that were considered by that tribunal are all to be found in very expensive areas of London, called "prime central London" (or the PCL). However, the LT emphasised that the generic rates should be used regardless of location.

[19] Appeals and cross-appeals followed, and the Court of Appeal ([2007] EWCA Civ 1042*) upheld the approach of the LT holding that the rejection of property market evidence in favour of a generic rate approach drawn from the financial markets was not an irrational decision. Nor was the rule or guidance from the LT, that leasehold valuation tribunals across England and Wales should apply the generic rates, irrational or wrong in law.

[20] There was no appeal against these decisions of the Court of Appeal (although there was an appeal to the House of Lords against the court's decision on an issue known as "hope value", which is not an issue in this case): [2008] UKHL 71†.

[21] There have been several LT decisions since the Court of Appeal's decision that upheld the use of the generic rates and, until recently, all such appeals urging the departure from the generic rates have failed. However, earlier this year, the LT allowed an appeal in *Zuckerman* v *Trustees of the Calthorpe Estates* [2009] UKUT 235 (LC)‡. It made three adjustments to the guideline rate of 5%: (i) 0.5% to reflect different growth rates; (ii) 0.25% for obsolescence; and (iii) 0.25% (in addition to the 0.25% sanctioned in *Sportelli*) to reflect management problems regarding flats, as opposed to houses. The case concerned a number of claims for new leases.

[22] A number of submissions were made in this application. These include written submissions by counsel representing the parties, an expert valuation prepared for the applicant nominee purchaser (undated) and an expert report dated 12 March 2010 prepared for the respondent landlord. The two valuers later signed a "supplemental statement of facts", signed and dated respectively on 1 and 6 April 2010.

[23] Mr Andrew Pridell FRICS advises the nominee purchaser. His company has completed more than 1,400 enfranchisement and lease extension cases. These have been mainly in the South East of England (including London) and as far north as Grimsby, east to Broadstairs and as far west as Cardiff. He carried out an inspection of the subject premises on 12 January 2005.

[24] He is critical of the *Sportelli* decision because it was based on properties in the PCL, which, he argues, is significantly different from the other parts of the country: see his report, in para 5.18. Before the decision, he says that he adopted a DR of 7% in many cases and this was upheld when challenged in the LVT. Now that the *Zuckerman* decision has been made, he is of the opinion that the three elements that led to a departure from the generic 5% rate apply equally to the Ashdown development.

[25] First, was the element based on obsolescence and deterioration? Mr Pridell argues that, with the exception of the three penthouse flats, the average value of the flats is around £220,000, which is quite different to properties in the PCL. As with many blocks of this style and construction, Ashdown has a number of maintenance problems, which include: flooding in the underground car park; the single-glazed windows that will need a major renewal programme in the foreseeable future; the energy-efficiency of the building is poor; and the hot-water system is communal and costly to maintain. Although there is a huge difference in the capital values of these flats and those in the PCL, this is not the case with the costs of repair and maintenance. Relying on his experience in building surveying, Mr Pridell says that building costs are virtually the same in the PCL or elsewhere in the South East. On this point, he concludes that "there is no material difference between Ashdown and Kelton Court and I have accordingly adjusted the risk premium upwards by 0.25%".

[26] Turning to the capital growth factor, he argues that values in the PCL have risen at dramatically higher rates than elsewhere in the country, including Hove and Brighton. He has not found evidence of long-term property value in the changes in the PCL and along the south coast. Instead, he has used a number of house price indices, including those produced by Nationwide and Halifax.

[27] With the agreement of Mr Rutledge (the valuer for the leaseholders in *Zuckerman*), he has adapted a graph based on various indices (including that produced by Knight Frank) by amending it to show growth in the outer South East region and by "superimposed sales evidence from Ashdown itself": see his report, at 5.67. With this information, he argues, a prudent investor would adjust his bid since he or she would be less confident of achieving the same growth rate outside the PCL. The same adjustment should be made here "because there is no reason to make a distinction between Kelton Court and Ashdown in this respect": see his report, at 5.71.

[28] As to the management issues, Mr Pridell describes how complex leasehold management has become since the changes made by Part 2 of the Commonhold and Leasehold Reform Act 2002. He gives examples of leasehold properties he has had experience with. In his opinion, the management of blocks of flats is a good deal more unpalatable than it used to be, and he concluded that this should be reflected in the discount rate because it presents an investor with a less attractive investment than was the case before the law was changed.

[29] He notes, however, that as here there is a management company (under tripartite leasing arrangements) the freeholder is somewhat removed from management. As against this point, he refers to clause 7 of the flat leases, which in terms states that if the management company is dissolved or otherwise unwilling or unable to manage, the freeholder is entitled to take over the management role. Mr Pridell then describes the many practical problems in managing Ashdown and gave many examples of serious problems that have emerged over the years. He points out that many of the leaseholders are older people on limited means and that finding suitable people to serve as directors has been difficult. He believes that the current uncertainty over the management arrangements warrants an additional 0.5% adjustment, but with the management arrangements in place an additional 0.25% should be added. He maintained this position during vigourous cross-examination.

[30] Mr Stewart Gray FRICS gave evidence on behalf of the respondent. Mr Gray is the principal in his firm and he has practised in the Hove and Brighton area for more than 20 years. He signed his report on 12 March 2010 and describes the property as a "very substantial building in a prime residential location in central Hove": see his report, in para 10.4. His position is that there should be no departure from the Sportelli rate. Mr Gray suggests that the property market had been buoyant over a sustained period, although just before the valuation date the market was affected by the crises that emerged in the banking sector in September 2007.

[31] Turning to the real growth rate, he is of the opinion that an investor in Hove and Brighton would be confident of achieving the 2% rate proposed by the LT in *Sportelli,* although he accepts that this may not be true of the investor in the West Midlands. In his view, the evidence of residential property sales since 1974 shows a higher rate of growth for properties in the outer South East than the national average, while those for the West Midlands are below that average. Other statistical evidence suggests that, in 2007, the growth rate in Brighton

* Editor's note: Reported at [2008] 1 EGLR 137

† Editor's note: Reported at [2009] 1 EGLR 153

‡ Editor's note: Also reported at [2010] 1 EGLR 187

A was running at 11% in comparison with Birmingham, where the rate was 1%.

[32] As to the flats in Ashdown, looking at the growth rate and allowing for inflation from the headline growth suggests a real growth rate of 2.46%. The LT in *Sportelli*, he argues, concluded that the growth rates in different parts of the country converge long-term, with the result that long-term growth comparisons between the PCL and other areas is largely irrelevant (a point that he suggest was not covered in *Zuckerman*: see his report, in para 11.1 b vii. In conclusion on these points, Mr Gray says that a long-term investor would be confident that, given that Ashdown is a prestigious block of substantial size and
B in a good-quality location, the growth rate would easily exceed the *Sportelli* average.

[33] Turning to the obsolescence and deterioration component of the risk premium, he argues that there is nothing in the design or the layout or construction that distinguishes Ashdown from thousands of other blocks of flats.

[34] On the risk factor for flats, he points to the existence of the management lease and states that the respondent has not had to be involved in management issues since the development was completed. He argues that it is not in the interests of the leaseholders, who are members of the management company, to allow it to fail. He
C therefore sees no reason to depart from the 0.25% adjustment for flats in *Sportelli*.

[35] During the adjournment for lunch, the decision of this tribunal in *7 Sudeley Street Brighton* CHI/00ML/OCE/2008/0023* in 2009 came to our attention. This was a missing landlord collective enfranchisement case, in which Mr Gray gave expert evidence on behalf of the claimant leaseholders. In that case, he argued for a 6% deferment rate. This was because in the *Sportelli* decision, he argued, it was reasonable to reject market evidence for the PCL. However, in Hove and Brighton, or elsewhere outside the PCL, he believes that the LT would have utilised market evidence to arrive at a higher deferment rate. He had looked at
D various recent sales investments in the locality and found that (ignoring hope value) the deferment rate is between 5.75% and 6.65%. He argued for a 6% deferment rate in that case and his submissions were accepted by the tribunal.

[36] Unfortunately, this decision and his evidence in it was not referred to in the expert evidence that he prepared for this application. In answer to questions on these omissions from his report, he claimed that he had not properly understood the *Sportelli* decisions when he gave evidence to this tribunal in 2009. Mr Gray was unable, in our view, to offer a convincing explanation for failing to mention the 2009 decision in his evidence to this tribunal. Further, he told us that
E he did not consider the *Sportelli* decision controversial. However, Mr Anthony Radevsky, for the applicant, showed him a printout from his firm's website in which he states the opposite.

[37] We were surprised and disappointed with his failure to disclose that he had given a quite different analysis of the effects of the *Sportelli* decision in a different application to this tribunal only last year. Had we not discovered this decision, we would not have known that Mr Gray had not only expressed a different opinion of the effects of *Sportelli* last year but we would also have been ignorant of his view expressed then that there is local market evidence that higher rates than 5% are being realised for freehold reversions.

F [38] This inevitably casts some doubt on the value of his evidence to the tribunal in this case. It is difficult to avoid the conclusion that in those areas where the experts differ Mr Pridell's evidence should be preferred.

[39] Counsel addressed the tribunal and referred to the revised skeleton argument they each prepared before the hearing. Mr Radevsky relied on the decision of the UT in *Zuckerman*. That decision concerned Kelton Court, part of an estate built in the 1970s in Edgaston in the West Midlands. Here, the UT made three adjustments to the guideline rate of 5%. First, there was an upwards adjustment of 0.5% to reflect the different growth rates between the PCL and the West Midlands.
G Mr Radevsky argued that there is evidence in this case for the same adjustment. Second, the UT made an upwards adjustment of 0.25% for obsolescence, and Mr Radevsky argued that given the similar profile of the building in this application to that in Kelton Court the same adjustment should be made. Third, he argued there should be an adjustment of a further 0.25% in this application, as there was in the Kelton Court matter, to reflect the special management problems with Ashdown house.

[40] Mr Wayne Clark, for the respondent, argued that there is no justification for the uplift in the risk premium. He did not accept that any adjustment need be made on the obsolescence and deterioration
H grounds. There is nothing unusual or extraordinary about Ashdown, he argued, that would not simply be reflected in the vacant possession value. On the flat management point, he argued that the risk of the current management arrangements collapsing is only theoretical. He did not consider that there is evidence that Ashdown suffers from far more onerous management problems than those considered in *Sportelli*. He firmly rejected the argument that there should be any adjustments to the generic rate.

[41] Mr Radevsky called Mr Nicholas Beck, a leaseholder in Ashdown since 1997 and chair of the board of the nominee purchaser company and chair also of Ashdown (Eaton Road) (Hove)
J Management Ltd, to give evidence. Mr Beck signed a witness statement dated 28 May 2010. He describes in detail the problems that he has experienced in the management of the block. Mr Beck says that part of the problem is that many leaseholders are elderly and there is a general apathy among their ranks, and it has proved difficult to persuade leaseholders to become directors. He told us that the running of the company is extremely fragile and that it could collapse at any time. A considerable sum needs to be spent on maintenance and repairs and he thinks that the management committee will have difficulty in collecting the necessary funds from the leaseholders.

Reasons for our decision

K [42] Having considered the evidence and the submissions, we have concluded that the deferment rate to be applied to the freehold vacant possession value for Ashdown House is 6%. Our conclusions are that to the 5% generic rate should be added: 0.5% for additional risk that growth will not be achieved; 0.25% for the obsolescence factor; and an additional 0.25% for the risks involved in the management of this block of leasehold flats.

[43] As noted above, the decision of the LT in the *Sportelli* cases, that the usual deferment rate to be applied in flat lease claims should be 5%, was upheld by the Court of Appeal. However, the court also stated, in [102], that:

L
> The issues within the PCL were fully examined in a fully contested dispute between directly interested parties. The same cannot be said in respect of other areas. The judgement that the same deferment rate should apply outside the PCL area was made, and could only be made, on the evidence then available. That must leave the way open to the possibility of further evidence being called by other parties in other cases directly concerned with different areas. The deferment rate adopted by the Tribunal will no doubt be the starting point; and their conclusions on the methodology, including the limitations of market evidence, are likely to remain valid. However, it is possible to envisage other evidence being called, for example, on issues relevent to the risk premium for residential property in different areas.

M [44] Until recently, the LT had consistently rejected appeals arguing the case for a different deferment rate. However, different deferment rates have recently been sanctioned by the LT in two appeals concerning flat claims: *Zuckerman,* on new lease claims, and *Re Lethaby* [2010] UKUT 86 (LC) on a collective enfranchisement claim.

[45] We will explain our reasons for departing from the generic 5% rate in this case by comparing the conclusions of the UT in *Zuckerman* and *Re Lethaby* with our assessment of the evidence in this application. In *Zuckerman*, the first case in which the UT sanctioned a departure from the generic rate in flat claims, it was said, in [77], "my conclusion in this appeal differs from the one reached by the Tribunal in *Culley,* namely that there was no justification for a departure from

* Editor's note: Unreported 23 September 2008

A the *Sportelli* starting point. In *Culley*, as in the present appeal, the basis of the Tribunal's decision was the expert evidence adduced by the parties".

[46] In *Zuckerman*, the UT heard evidence that the values of the properties in *Sportelli* were of a different magnitude to those in Kelton Court. In the instant case, the difference between the value of the flats and those considered in *Sportelli* is significant. Added to that is Mr Pridell's experience that building costs in the local area are comparable to those in London. It follows, in our view, that it is likely to remain economically viable to repair high-value flats in the PCL for much longer than will be the case for similar sized flats in Ashdown.
B Even though the flats are leased on full repairing covenants, it seems to this tribunal that there is a much greater risk of deterioration at Ashdown than there is in PCL, and this is not reflected in the vacant possession values. This leads us to the conclusion that a purchaser of the freehold reversion to Ashdown would expect an additional 0.25% in the risk premium to reflect this factor.

[47] Turning to the prospects for future growth, Mr Pridell produced evidence that, he argues, shows a much slower growth rate for properties in the local area. He adapted the methodology used in *Zuckerman* in appendix 7 of his report. As the UT noted in *Zuckerman*, there are limitations on the use of statistical evidence, but there, as in
C this case, the available evidence suggested a significantly lower rate than in the PCL. Conversely, in *Re Lethaby,* the leaseholders did not produce any evidence to compare growth rates in east London (where that property is located) and the PCL. Mr Gray did not produce any such evidence in comparison with the PCL. The evidence put forward leads us to conclude that an investor that is examining long-term growth would not be confident that the PCL rate of growth would be achieved in the Hove and Brighton area, and he would adjust his bid for Ashdown accordingly. Following the reasoning in *Zuckerman*, we conclude that the prospect of such a reduction in an investors bid should be assessed by a further increase in the risk premium by 0.5%.

D [48] This brings us to a third factor that can affect the deferment rate, namely the difference between the rates for houses and flats to reflect the greater management risks with the latter. In *Sportelli*, the UT concluded, in [95], that an adjustment to the deferment rate should be made to reflect the greater management problems associated with flats. The tribunal added that "although we do not consider it appropriate to differentiate between flats that are the subject of headleases and those which are not. … Even where flats are efficiently managed, service charge and repairs problems inevitably occur, and the management exercise in itself is, we feel, sufficiently more complex to warrant a generalised 0.25% additions for flats."

E [49] In *Zuckerman*, the UT took this a stage further and decided that the effect of the Service Charges (Consultation Requirements)(England) Regulations 2003 meant that the market would have been more aware of the challenges that face landlords under the regulations than was the case when *Sportelli* was decided. This time, the tribunal decided, in [74], that "in the eleven cases with which I am currently concerned, investors would have required an addition of 0.5% to reflect the greater management problems associate with flats than with houses".

[50] The tribunal added that had the headlease still been in existence, it would not have considered it appropriate to depart from the *Sportelli* uplift. It is difficult to reconcile the two positions taken by the LT on the
F effects of there being a headlease on this aspect of the risk factor. We have concluded that we should apply the principle adumbrated by the LT in *Sportelli* because its decisions on the deferment rate were upheld by the Court of Appeal. In other words, one should not differentiate between cases where the flats have headleases and those that do not.

[51] There is nevertheless a difficulty in a case such as this where, although there is no headlease between the flat leases and the freehold, there is a leaseholder-controlled management company that, on the face of it, absolves the freeholder from the pressures and risks of flat leasehold management.

[52] In *Re Lethaby*, the tribunal did not make an additional uplift in a case in which, with two flats, where the leaseholders shared the repair and upkeep and roof and foundations of the building: the landlord's responsibilities in that case are minimal. The risk to the landlord was G
therefore much less than it was in *Zuckerman,* where the landlord's responsibilities included the maintenance of the structure and common areas of the flats and a separate block of garages, private roads and amenity areas. We think that *Re Lethaby* is quite different from this case, a complex development with 125 flats.

[53] In this case, there is no headlease. However, the complicating factors are the leaseholder-owned management company and the lease of the common parts. Nevertheless, on balance we prefer the submissions and evidence (including Mr Beck's evidence) given by the applicant and find that there is a clear risk that the management company could fold leaving the freeholder to consider assuming the H
management responsibilities (as it is entitled to do under the leases).

[54] It is also the case that leasehold management responsibilities appear to be at least as complex (and probably greater) as they are in *Zuckerman.* We conclude that an investor would require an addition of 0.5% to reflect in this case the greater management responsibilities with flats than with houses. An investor would take account of the possibility that it would have to intervene should the management company collapse by exercising the right to take over the management of the block. This is a substantial block of flats with a history of difficult management issues that pose a real risk to an investor. J

[55] The risks that apply to flat management were illustrated by another recent decision of the LT in *Daejan Investments Ltd* v *Benson* [2009] UKUT 233 (LC)*, where the landlord's appeal against the refusal of a leasehold valuation tribunal to exercise its discretion to dispense with the consultation requirements in section 20 of the Landlord and Tenant Act 1985 was dismissed. Carnwath LJ, the senior president of the UT, stated, in [97]: "The potential effects draconian on one side and a windfall on the other are an intrinsic part of the legislative scheme." In that appeal, the landlord's charges were around £270,000. As a result of its failure to comply in full with the statutory consultation requirements, it could recover only £250 from each of the five leaseholders. This is an example of how the stricter legal framework governing leasehold K
flat management carries additional financial risks. As the LT concluded in *Zuckerman,* the market has become aware of the effects of the new legal framework.

[56] In summary, we conclude that the evidence produced by the applicant justifies a departure from the generic deferment rate by adding to that 5% rate: an additional 0.25% for obsolescence; an additional 0.5% because of the risk that the potential growth may not be achieved; and an additional 0.25% relating to the differences between managing houses and flats. This results in a 6% deferment rate in this particular case. We would add that this additional 1% reflects the fact that the evidence shows here that investors would require this by comparison to L
the risks of investing in dwellings in the PCL.

[57] There is also indirect evidence to support this higher deferment rate. This is the evidence that Mr Gray gave last year in *7 Sudeley Street Brighton* that there is local property market evidence to support deferment rates between 5.45% and 6.5%. Based on this expert evidence, the tribunal determined that a deferment rate of 6% should be applied. The valuation date in that case was 26 June 2007 (the valuation date is 2 October 2007 in this case). Mr Gray told the tribunal in that case that he believed that outside the PCL comparable sales of ground rent investments are the most appropriate evidence available: "He M
considers that if the *Sportelli* cases had been dealing with a provincial property in the Brighton area, the Lands Tribunal would have utilised market evidence to arrive at a higher deferment rate": see para 19.

Summary

[58] On the basis of the evidence adduced in this case, we have concluded that the applicant's case for a departure from the generic rates propounded in *Sportelli* has been made out. The deferment rate to be applied in this case is 6%. This results in a premium of £1,732,109 to be paid in this claim. Since the deferment rate is the only issue

* Editor's note: Reported at [2010] 2 P&CR 8

that divides the parties, we did not consider it necessary to append a valuation to this decision.

Purchase price of £1,732,109 determined, applying a deferment rate of 6%.

Hosebay Ltd *v* Day and another; Lexgorge Ltd *v* Howard de Walden Estates Ltd

COURT OF APPEAL

[2010] EWCA Civ 748

1 July 2010

LORD NEUBERGER MR AND LLOYD AND MOORE-BICK LJJ

Estates Gazette 25 September 2010

[2010] 38 EG 106

Leasehold enfranchisement — Leasehold Reform Act 1967 — Enfranchisement of house — Terraced properties originally built as single residences — Some properties furnished and let as bedsits for short-term holiday lets — One property long used entirely as office accommodation in breach of user covenant in lease requiring residential use of upper floors — Whether properties constituting houses for purpose of section 2(1) of 1967 Act — Whether leaseholders entitled to enfranchise

These were two joined appeals concerning the definition of a "house" in section 2(1) of the Leasehold Reform Act 1967, in connection with applications by the respondents, as long leaseholders of properties, to acquire the freehold of those properties from the appellant landlords by leasehold enfranchisement under the Act. Both appeals concerned large terraced properties that had originally been constructed and occupied as houses. The first appeal related to three properties in South Kensington, London SW7, the leases of which described the demised premises as a "messuage or dwellinghouse" and stipulated the permitted use as, in the case of two of the properties, residential flats retaining the external appearance of a private dwelling-house, and, in the case of the third, a single family residence or high-class furnished property for up to 20 occupiers. In fact, the individual rooms in the properties had been converted into self-catering units, each with bathroom and cooking facilities, and were let as short-term accommodation for tourists and other visitors to London.

The second appeal related to a property in Marylebone, London W1, the lease of which described the demised premises as a "messuage or residential or professional premises" and stipulated the permitted use as professional offices on the lower two floors and self-contained flats or maisonettes on the upper two, with associated storage in the basement. In fact, all four upper floors had long been used as professional offices.

Both respondents' leasehold enfranchisement claims were allowed in the county court. In the first appeal, the judge held that each of the properties was a "house" fulfilling both limbs of the section 2(1) definition, since it had been both originally designed and subsequently adapted for living in and could reasonably be called a house. In the second appeal, only the second limb of the definition was in issue; the judge rejected an argument by the appellant that, on an application of the judgment in *Prospect Estates Ltd* v *Grosvenor Estates Ltd* [2008] EWCA Civ 1281; [2009] 1 EGLR 47; [2009] 02 EG 86, the property could not reasonably be called a house in the light of its actual use as offices and the fact that that was the only permitted use of around half of the internal area. The appellants appealed.

***Held*: The appeals were dismissed. (1) In determining whether premises have been adapted for living in, it is necessary to look at the most recent works of adaptation and assess objectively whether they have resulted in the property being adapted for living in. The subjective intention of the person responsible for the works will rarely, if ever, be relevant since the issue is how the building has been adapted, not why. The use to which the building is put at the date of the tenant's notice is also of little relevance. (2) A building that had been designed or adapted for living in will be excluded from being a "house" under the second limb of the section 2(1) test only if the circumstances are such that nobody could reasonably call the building a house; that is a high hurdle for landlords to cross. The question is to be determined by reference to the external and internal character and appearance of the building. It is unlikely to be affected by whether the building is actually lived in at the relevant date. The *ratio* of *Prospect Estates*, excluding certain properties from the definition of a "house" on the grounds of their permitted use under the lease combined with their actual use, should be treated as limited to cases where residential use is either prohibited entirely, or is restricted to a very small part of the building, and the actual use accords with that. (3) In the first appeal, each of the three properties had been designed for living in and the effect of the most recent conversion works, if they were works of adaptation, had also been to adapt the properties for living in. Each of the three properties could reasonably be called a house by reason of its internal and external structure and appearance, the fact that it was fitted out for residential occupation, albeit multiple occupation, and the requirement in the lease for residential use. In the second appeal, the property was designed or adapted for living in since no work had been done to the upper two floors to adapt them away from their original design. It was also a house reasonably so called, notwithstanding the actual use of the entire property for office purposes, in the light of its external and internal character and appearance, at least on the upper two floors, the description of the property in the lease and, to the extent that it was relevant, the terms of the lease restricting the use of the upper two floors to residential.**

The following cases are referred to in this report.

Boss Holdings Ltd v *Grosvenor West End Properties* [2008] UKHL 5; [2008] 1 WLR 289; [2008] 2 All ER 759; [2008] 1 EGLR 51; [2008] 15 EG 174

Boss Holdings Ltd v *Grosvenor West End Properties Ltd*; *Mallett & Sons (Antiques) Ltd* v *Grosvenor West End Properties Ltd* [2006] EWCA Civ 594; [2006] 1 WLR 2848; [2006] 2 EGLR 43; [2006] 26 EG 156

Hosebay Ltd v *Day* [2009] PLSCS 318

Howard de Walden Estates Ltd v *Aggio*; *Earl Cadogan* v *26 Cadogan Square Ltd*; *sub nom Earl Cadogan* v *26 Cadogan Square Ltd* [2008] UKHL 44; [2009] 1 AC 39; [2008] 3 WLR 244; [2008] 4 All ER 382; [2008] 2 P&CR 19; [2008] 2 EGLR 57; [2008] 34 EG 94; [2008] RVR 236

Lake v *Bennett* [1970] 1 QB 663; [1970] 2 WLR 355; [1970] 1 All ER 457; (1969) 213 EG 633, CA

Lexgorge Ltd v *Howard de Walden Estates Ltd* [2010] PLSCS 54, CC

Prospect Estates Ltd v *Grosvenor Estates Ltd*; *sub nom Grosvenor Estates Ltd* v *Prospect Estates Ltd* [2008] EWCA Civ 1281; [2009] 1 WLR 1313; [2009] 2 P&CR 10; [2009] 1 EGLR 47; [2009] 02 EG 86

Tandon v *Trustees of Spurgeons Homes* [1982] AC 755; [1982] 2 WLR 735; [1982] 1 All ER 1086; (1982) 44 P&CR 307; [1982] 2 EGLR 73; 263 EG 349; 4 HLR 1, HL

This was an appeal by Hugo Day and Lady Hilary Day, the appellants in the first appeal, from a decision of HH Judge Hazel Marshall QC, sitting in Central London County Court, allowing a leasehold enfranchisement claim by the respondent, Hosebay Ltd, under the Leasehold Reform Act 1967; and by Howard de Walden Estates Ltd, the appellant in the second appeal, against a decision of HH Judge Marc Dight, also sitting in Central London County Court, allowing a similar claim by the respondent, Lexgorge Ltd.

Edwin Johnson QC (instructed by Pemberton Greenish) appeared for the appellants in the first appeal; Katharine Holland QC (instructed by Speechly Bircham LLP) appeared for the appellant in the second appeal; Anthony Radevsky (instructed by Boodle Hatfield & Wallace LLP) represented the respondents in both appeals.

Giving judgment, **LORD NEUBERGER MR** said:

A [1] These two appeals concern the meaning and effect of section 2(1) of the Leasehold Reform Act 1967 (the 1967 Act). In particular, they require one to focus on the question of what constitutes a "house" for the purpose of that section.

Statutory background

[2] Section 1 of the 1967 Act (section 1) enables certain tenants of houses under long leases to serve notice on their landlords to enfranchise that is, to acquire their freeholds from their landlords "on fair terms" (or to acquire extended leases). As originally enacted, its application was limited to relatively low value houses, let on low rents, to tenants who, at the date of their notice, resided there, and had resided B there for at least five of the preceding 10 years, as their only or main residence.

[3] Over the succeeding 35 years, the ambit of section 1 has been successively extended by statutory amendment, as is reflected by the rather bewildering appearance of section 1 in its current form, supplemented by sections 1A, 1AA and 1B. The value limitation was raised, and has now been removed altogether in the majority of cases. The low rent limitation was raised, and has now, in the majority of cases, been removed altogether. The residence requirement on the tenant was initially reduced, and has now, in the majority of cases, been removed altogether and replaced with a requirement of two years of C ownership of the lease.

[4] The consequence of this last amendment, effected by section 138 of the Commonhold and Leasehold Reform Act 2002 (the 2002 Act), is that, not only can an individual tenant enfranchise even if he or she has never lived in the house, but: (a) a corporate tenant can enfranchise (which it could not have done previously, as a company cannot reside); and (b) a tenant who holds a number, even a large number, of houses on long leases can simultaneously enfranchise all of them: see (albeit in respect of a statute relating to collective enfranchisement of leases of flats) *Howard de Walden Estates Ltd* v *Aggio* [2008] UKHL 44; [2009] 1 AC 39*. The disadvantageous consequences for landlords of D this amendment were apparently mitigated by a new subsection (1B) introduced into section 1 by the 2002 Act, which excluded a tenant who had the protection of Part II of the Landlord Tenant Act 1954 (the 1954 Act) from being able to enfranchise. However, as the facts of one of the instant cases show, that is in practice rarely of much value to landlords.

[5] Section 1, and sections 1A, 1AA and 1B, are thus concerned with identifying the purpose and ambit of the 1967 Act, and most of the ensuing sections effectively expand on some of the terms in those sections, or set out the procedural aspects.

[6] In particular for present purposes, section 2 is concerned with the definition of "house" (as only a tenant of a "house" can E enfranchise under the 1967 Act) and of "house and premises" (as the enfranchisement right extends to "premises", as defined, if let with the house). Unlike section 1, which has been significantly amended on eight occasions, section 2 is in precisely the same form as it was when first enacted.

[7] Section 2(1) of the 1967 Act (section 2(1)) provides as follows:

> "house" includes any building designed or adapted for living in and reasonably so called, notwithstanding that the building is not structurally detached, or was or is not solely designed or adapted for living in, or is divided horizontally into flats or maisonettes; and
>
> (*a*) where a building is divided horizontally, the flats or other units into F which it is so divided are not separate "houses", though the building as a whole may be; and
>
> (*b*) where a building is divided vertically the building as a whole is not a "house" though any of the units into which it is divided may be.

[8] Section 3 explains what is meant by a "long tenancy", and sections 4 and 4A are concerned with the meaning of "low rent", in the circumstances when it still applies. Section 5 has "General provisions as to claims to enfranchisement or extension", and it brings in Schedule 1, which contains the procedures for enfranchisement, and deals with the tenant's notice. Sections 6, 6A and 7 apply to cases where the tenancy is held on trust or the tenant has died. Section 8 contains the "Obligation G to enfranchise" on a landlord on which a notice is served, and section 9 deals with the "Purchase price and costs of enfranchisement…". It is unnecessary to refer to any of the other 32 sections or the other six schedules of the 1967 Act for present purposes.

* Editor's note: Also reported at [2008] 2 EGLR 57

Facts relating to the two appeals

[9] *Hosebay Ltd* v *Day* concerns three properties in a terrace of buildings, 29, 31, and 39 Rosary Gardens (29, 31, and 39 respectively), which were originally constructed, and first occupied, as large houses, in South Kensington, London SW7. Each of "the three properties", as I will refer to them, was let on a long lease, which described the demised premises as "the messuage or dwellinghouse". Twenty-nine H and 39 were each let by a lease granted in 1966 for some 55 years, and 31 by a lease granted in 1971 for some 60 years. The leases of 29 and 39 stipulated that the demised premises be used only "as 16 residential flatlets", with a resident caretaker, and that the external appearance of the demised premises be that of "a private dwellinghouse". The lease of 31 restricted its use to that of "a single family residence" or "a high class furnished property" for up to 20 occupiers, with a resident caretaker.

[10] Hosebay Ltd (a company owned by a Mr and Mrs Morris) acquired all three leases (the three leases) in 1996, and remains the tenant under each of the three leases. It served notices on its landlords, under J section 8 of the 1967 Act, to acquire the freeholds of the three properties. Its claim to be entitled to acquire the freeholds was challenged by the landlords, and the issue came before HH Judge Hazel Marshall QC*.

[11] The three properties are being used to provide what the judge described in the course of her full and clear judgment, in [8], as "short term accommodation for tourists and other visitors to London". The judge, who made a full inspection of 29, 31, and 39, said, in [9], that each of the three properties has been "fully adapted to provide individual rooms for letting out", with the exception of two rooms in 31, one of which is used for office and reception purposes, and the other of which is used for storage. The judge, in [10], thought that the most K apt description of the great majority of each of the other rooms was "room with self-catering facilities". Each such room has between one and four beds, a similar number of dining chairs, a table, a television, a telephone, limited storage and clothes-hanging space, and no easy chairs. Each such room also has, in a confined space, a sink, cooking facilities and cupboards that have cooking and eating equipment, and it also has a small "wet room" with shower, basin and WC. Fresh bed linen and room cleaning, but no other services, are provided to those staying in the rooms.

[12] There were two main issues before the judge. The first was whether nos 29, 31 and 39 were in each case a "house" within L section 2(1). The second issue was whether Hosebay was excluded from enfranchising each of the three properties by section 1(1B).

[13] On the first issue, the judge rejected the landlords' case that each of the three properties was not a "house"; that case was advanced on two grounds, namely that each of the three properties: (i) was not "designed or adapted for living in"; and (ii) was not a "house reasonably so called", in the light of the use to which it was currently put. With the judge's permission, the landlords appeal on both points.

[14] On the second issue, the judge concluded that Hosebay did not enjoy the protection of the 1954 Act, because, even though the three properties were being occupied for the purpose of a business (namely M provision of short-term serviced accommodation), Hosebay had lawfully sublet each of them to an associated company, Hindmill Ltd, and it was Hindmill that carried on the business. Hosebay therefore did not occupy the three properties, or any part of them, for the purpose of a business that it carried on, and, as the 1954 Act applies only to premises that are occupied by the tenant for the purpose of its business, it did not apply to the three leases. Although one might characterise the arrangement between Hosebay and Hindmill as rather artificial, the judge concluded that it was genuine, and therefore the landlords failed

* Editor's note: Reported at [2009] PLSCS 318

on the second main issue as well. The landlords (realistically in my view) do not appeal on that issue.

[15] *Lexgorge Ltd* v *Howard de Walden Estates Ltd* concerns 48 Queen Anne Street, in Marylebone, London W1 (the property). It was originally built, in around 1760, as a house (comprising five floors, including a basement), in a terrace of substantial houses, and for many years, it was occupied for that purpose. It was the subject of a lease that was granted in 1952 for a term of some 109 years. In the lease, the property is described as a "messuage or residential or professional premises", and its use is restricted (subject to landlords' consent) to "self-contained flats or maisonettes" on the upper two floors, professional offices on the first and ground floors, with the use of the basement being restricted to "storage… and lavatory… in connection with other parts of the demised premises".

[16] In 1978, the lease was acquired by Lexgorge Ltd, which more than 25 years later served notice on the landlord to acquire the freehold. The landlord objected on the basis that the property was not a house within section 2(1), and that issue came before HH Judge Marc Dight*.

[17] It appears likely that, from 1961 or thereabouts, the use of all four upper floors was as offices, until the landlords objected to that use of the upper two floors, and that use ceased shortly after Lexgorge served its notice seeking to enfranchise. Certainly, it seems clear that the office use of the upper two floors, which had started in around 1961 (and had never been consented to by the landlord), continued until after the notice was served, but, by the time the judge inspected the property, those floors were occupied as a maisonette. The ground and first floors are being, and have since 1978 been, used as solicitors' offices, one of the solicitors being a Mr Pearson, effectively the owner of Lexgorge; it would seem that the upper two floors have often been used for the same purpose, but have also been occupied by quantity surveyors.

[18] The only issue before the judge was whether the property was a "house… reasonably so called". Rejecting the landlords' argument in his well-articulated judgment, the judge held that the fact that all the property was being used for office purposes at the date of the notice, and the fact that, under the lease, around half the internal area of the property could not be used other than for office purposes did not prevent the property from being a "house reasonably so called". With the judge's permission, the landlords appeal against that decision.

Case law

[19] The issue of whether premises constitute a "house" has troubled the courts in a large number of different contexts. Particularly because it is a word with a relatively wide range of possible meanings, and because it is defined in unique terms in section 2(1), it is dangerous in the present context to try and find assistance in judicial dicta as to the meaning of the word in other contexts. All the more so as the effect of section 2(1) has been considered in four relevant previous authoritative cases, one of which was concerned with the expression "designed or adapted for living in", and three of which focussed on the words "house… reasonably so called". I now turn to consider those cases.

[20] The authority on the effect of the word "designed or adapted for living in" is the unanimous decision of the House of Lords in *Boss Holdings Ltd* v *Grosvenor West End Properties* [2008] UKHL 5; [2008] 1 WLR 289†.

[21] I said in [18] of an opinion with which the other members of the committee agreed:

> [T]he words "designed or adapted for living in", as a matter of ordinary English, require one first to consider the property as it was initially built: for what purpose was it originally designed? That is the natural meaning of the word "designed", which is a past participle. One then goes on to consider whether work has subsequently been done to the property so that the original "design" has been changed: has it been adapted for another purpose, and if so what purpose? When asking either question, one is ultimately concerned to decide whether the purpose for which the property has been designed or adapted, was "for living in".

In [18] and [19], I also made the point that the word "designed" is a past participle referring back to the date of the works of design, and, it is implicit in those paragraphs, and in any event it must follow as a matter of consistency of language and approach, that the word "adapted" refers back to the date of the works of adaptation.

[22] In [26], I raised the question of "whether a property would be a 'house' if it had been designed for living in, but had subsequently been adapted to another use", and said:

> As a matter of literal language, such a property would be a house, because "designed" and "adapted" appear to be alternative qualifying requirements. At least at first sight, such a conclusion seems surprising, so there is obvious attraction in implying a qualification that, if a property has been, and remains adapted for a purpose other than living in, the tenant cannot rely upon the fact that it was originally designed for living in. However, a term is not easily implied into a statute, and further reflection suggests that the literal meaning of the words is not as surprising as it may first appear, particularly bearing in mind the existence of the residence requirement in section 1(1) of the original Act. It is unnecessary to decide this point, and, particularly as it was only touched on in argument, I do not think we ought to do so.

[23] The first decision on the issue of "'house'… reasonably so called" is the unanimous decision of the Court of Appeal in *Lake* v *Bennett* [1970] 1 QB 663*, where, at at p670C-D, Lord Denning MR, said that "there have been many statutes which have used the word 'house' without defining it" and that "the courts have given it a wide interpretation". He also observed, at pp670H-671A, that, although he would not seek to define the effect of the words "reasonably so called", he did not think, for example, that "a tower block of flats would reasonably be called a 'house'". Salmon LJ explained, at p672C, that the words "reasonably so called" were necessary to exclude buildings such as "the Ritz Hotel… Rowton House, and a large purpose-built block of flats". He went on to make the point that the fact that a building could reasonably be called something other than a house would not mean that it could not also reasonably be called a house, a point also made by Cross LJ in his brief judgment at p673A-B.

[24] The reasoning and decision in *Lake* was approved by the majority of the House of Lords in *Tandon* v *Trustees of Spurgeons Homes* [1982] AC 755†.

[25] In that case, Lord Wilberforce, in the minority, said, at p760F-Gp, that "a mixed unit consisting in part of a shop and in part of a dwelling [was] not a house within the [1967] Act." Lord Fraser of Tullybelton, also in the minority, said at p762A-G, that "the original design" of the building, "the setting in which it is placed", "the proportion of residential use to the non-residential use" and the relative values of the parts used for those purposes were all relevant factors.

[26] Lord Roskill, who gave the majority's view, in an opinion with which Lord Bridge of Harwich and Lord Scarman agreed, said, at p765F-G:

> [I]f the premises might also be called something other than a "house"…, that fact alone [does] not prevent those premises from being a "house… reasonably so called."… [P]remises used for non-residential as well as for residential purposes [can] in law be a "house" within the definition, and… it depend[s] upon the character of the premises in question whether by reason of their mixed user they [fall] within or without the definition.

Later, Lord Roskill said, at p767B-C:

> as long as a building of mixed use can reasonably be called a house, it is within the statutory meaning of "house," even though it may also reasonably be called something else;… if the building is designed or adapted for living in, … only exceptional circumstances… would justify a judge in holding that it could not reasonably be called a house. They would have to be such that nobody could reasonably call the building a house.

[27] In the most recent decision on the issue, *Prospect Estates Ltd* v *Grosvenor Estates Ltd* [2008] EWCA Civ 1281; [2009] 1 WLR 1313‡,

---

* Editor's note: Reported at [2010] PLSCS 54

† Editor's note: Also reported at [2008] 1 EGLR 51

* Editor's note: Also reported at (1969) 213 EG 633

† Editor's note: Also reported at [1982] 2 EGLR 73

‡ Editor's note: Also reported at [2009] 1 EGLR 47

A the Court of Appeal unanimously reversed a first instance finding that premises could reasonably be called a house; the premises in question had been originally constructed as a house, but could be used under the lease only as to 11.5% for residential purposes, the remainder of the premises being limited to commercial use. Mummery LJ accepted that "a building designed for living in does not… cease to be a house reasonably so called, simply because a part of it is no longer used for living in": in [14]. However, in [18], he said that "It would, however, be wrong to take [Lord Roskill's] approach too far" and that it was "necessary to take account of all the relevant circumstances in order to decide whether a building is a house reasonably so called".

B [28] Mummery LJ then said, in [19], that the judge "paid insufficient attention to the peculiar, even exceptional…, circumstance of prescribed and predominant office use in compliance with the lease", which he described as "the overwhelming and decisive feature of this case". In the next paragraph of his judgment, he said that the judge should have given "due weight to the prescriptive terms of the lease, the actual uses of the building and the relative proportions of the mixed use at the relevant date", and that, had he done so, the judge would have held that the premises in question could not reasonably be called a house. Goldring LJ agreed, in [23], implying that the essential point was that a "building can[not] reasonably be called a house although no one C can lawfully live in virtually 90% of it." Smith LJ also agreed with Mummery LJ in a reasoned judgment.

Are 29, 31 and 39 Rosary Gardens each "designed or adapted for living in"?

[29] Judge Marshall QC concluded that each of the three properties was "designed or adapted for living in", on the ground that, although they had been "adapted" from the original design, very probably before the three leases were granted, the adaptation was "for living in". Mr Anthony Radevsky, for Hosebay, supported the conclusion on three grounds. First, given that the three properties were originally designed for living in, it is unnecessary to consider the issue of adaptation; D second, if that is wrong, the most recent works (the works) to the three properties were insufficient to have "adapted" them, so the original "design" stands; third, if that is wrong, the works adapted each of the three properties "for living in", that is, the judge's reason. (The better characterisation of this third argument may be that, since each of the three properties was "designed… for living in", that original design subsists as the works did not "adapt" the properties away from that design.) I shall take those three arguments in turn.

[30] I turn to Mr Radevsky's first argument. It is clear that each of the three properties was "designed… for living in", as each of them was constructed as a house for single occupation that is, for "living in" E on any view of that expression. Picking up on the point that was left open in *Boss Holdings*, in [26], the contention is that, even if the three properties were subsequently adapted away from residential use, that does not matter as they were "designed *or* adapted" for living in that is, that satisfaction of either alternative suffices even if the other alternative is not satisfied.

[31] The point has now been fully argued, and, although it is not necessary to decide it for reasons that will become apparent, it is right to record that I would reject it. It would seem somewhat illogical that one could "adapt into" residential use, but that one could not "adapt out of" such use. Further, although, as was accepted in *Boss Holdings*, F the literalist meaning of "designed or adapted" is that either alternative will do, that is not by any means what the words naturally convey: that is well evidenced by the fact that the tenant in that case (and the tenant in the case heard together with it in the Court of Appeal [2006] EWCA Civ 594; [2006] 1 WLR 2848*) did not raise the point. I must confess to having started, in *Boss Holdings*, in [26], what I now think is a hare by suggesting an over-literalist approach to the language used by the legislature.

[32] As to Mr Radevsky's second argument, we heard very little by way of submissions on the issue, and we were not taken to any evidence as to the precise nature of the works effected to enable the G three properties to be used as contemplated by the three leases. In those circumstances, particularly as it is unnecessary to do so (in the light of my conclusion on Mr Radevsky's third argument), I do not think that it would be right to interfere with the judge's conclusion on the point.

[33] I turn to the third argument. Assuming that the works carried out to convert nos 29, 31 and 39 were works that "adapted" each of the three properties, it seems to me that, despite Mr Edwin Johnson QC's argument for the landlords to the contrary, the works adapted the three properties for living in. My primary reason for that conclusion is that, in order to determine whether premises are adapted for living in, one looks at the most recent works of adaptation, and assesses objectively, H whether they resulted in the property being adapted for living in.

[34] I emphasise in that last sentence the words "works of adaptation" and "objectively". One looks at the effect of the works that altered the building, not furnishings or furniture; they are not works of adaptation, a view supported by the reasoning in *Boss Holdings*. It also seems to me that the subjective intention of the person responsible for the works, so far as the intended use is concerned, will rarely, if ever, be of any relevance: one is concerned with how the building was adapted, not why it was adapted. Thus, in *Boss Holdings*, in [17], when considering the expression "designed or adapted for living in", it was said that "for present purposes one is largely concerned with the J physical state of the property".

[35] The actual or intended use of the building may, however, sometimes have some relevance. For instance, if the building was intended to be, and was actually, used for living in immediately after the works were completed, it may help undermine the argument that it was not adapted for living in. Also of little if any relevance, in my view, is the use to which the building is actually put at the date of the tenant's notice. Again, that use could conceivably be relevant, but it appears to me to be even more unlikely to be of any significance, as it is a use that *ex hypothesi* occurs after, possibly many years after, the works of adaptation. K

[36] In this case, I consider that the effect of the most recent works of conversion to the three properties, if they were works of adaptation, adapted those properties for living in. Ignoring one or two rooms, each room in the three properties is a self-contained unit of accommodation, with its own basic small shower room/WC, and its own even smaller and more basic cooking facilities. As Moore-Bick LJ pointed out in argument, the rooms are entirely appropriate for letting to students on three-year degree courses, and, as Mr Johnson rightly accepted, if they had been, all the rooms, and therefore the three buildings, would have been used for living in. Even if, as Mr Johnson argued and I am prepared to assume without deciding, the current use of the three L properties is not for living in, that certainly does not mean that, viewed objectively, the three properties were not adapted for living in.

[37] As the judge indicated, the works of adaptation were in all probability carried out before the leases were granted: if the works had been carried out during the currency of the three leases, one would have expected the landlords to be able to produce, or at least to identify, a permission or contract to carry out works of alteration (either in the three leases or in a separate document) together with plans, yet no such documentary evidence was available. On that assumption, given the user covenants in the three leases, it seems to me pretty clear that, even if the subjective purpose for which the works of adaptation were carried M out, and the use to which the building was then put, is of relevance, that purpose was for "living in". Mr Johnson said that the intention of the person who carried out the works, and the actual use once the works had been completed, were matters of speculation. It is true that the intended and actual uses were not clearly established on the evidence, but, for the reasons that I have briefly given, on the balance of probabilities, they were the use described in the three leases.

Are 29, 31, and 39 Rosary Gardens each a "house… reasonably so called"?

[38] Each of the three properties was designed and constructed for use, and first used, as a residence in single family occupation (no

---

* Editor's note: Also reported at [2006] 2 EGLR 43

doubt with a number of domestic servants); externally, each of the three properties has the appearance of being a relatively large town house; internally, each of the three properties has been converted so that almost every room can be used as a self-contained unit for one or more individual, with cooking and toilet facilities. Ignoring for the moment the reasoning in *Prospect Estates*, I find it hard to see how the judge could be faulted for concluding that, even if each of the three properties might be called something else as well, they could each reasonably be called a house.

[39] As already mentioned, in *Tandon*, at p767C, Lord Roskill observed that the circumstances "would have to be such that nobody could reasonably call the building a house" before a building that had been "designed or adapted for living in" was excluded from being a "house" under section 2(1). That appears to present the landlords with a high hurdle to cross. It was suggested by Mr Johnson that the observation should no longer be regarded as sound, as it was made at a time when, *ex hypothesi*, any building to which it referred would have been occupied, at least in part, by the tenant as his only or main residence.

[40] Although I accept the premise, I do not agree with the conclusion. Lord Roskill's observation referred to the interrelationship of two provisions of section 2(1), a subsection that has remained unchanged since the 1967 Act was passed; further, the meaning of the provisions of the subsection cannot have changed as a result of the significant amendments to section 1: see *Boss Holdings*, in [23]. Further, given that a building must be "designed or adapted for living in" before it can be a "house", it seems to me that, as a matter of ordinary English, the question of whether it was a "house… reasonably so called" is unlikely to be affected by the fact that it is, or, as the case may be, it is not, actually lived in at the date the issue falls to be judged. All the more so in the light of my conclusion that a building originally designed for living in, but adapted for some other purpose, is not "designed or adapted for living in" (unless it is subsequently readapted for living in).

[41] As Mr Radevsky said, a further difficulty in the way of the argument that each of the three properties could not reasonably be called a "house" is that each of them was described in the relevant lease as a "messuage or dwellinghouse". As already mentioned, it seems more likely than not that the present layout and design of the three properties has not significantly changed since the grant of the three leases, 40 to 45 years ago. Given that "messuage" is a somewhat archaic legal term for a house, it adds little if anything to the essential point that it seems almost perverse to contend that a building cannot reasonably be called a "house" if the predecessors of the two parties actually described it as a "dwellinghouse", given that, in the intervening period, the meaning of those two words has not changed, and little work of significance to the internal or external structure or appearance of any of the three buildings appears to have been done.

[42] In his argument for the landlords that each of the three properties could not "reasonably be… called" a "house", Mr Johnson understandably placed considerable reliance on the reasoning in *Prospect Estates*. Even assuming in the landlords' favour that the effect of that case is that the actual use of the three properties can be taken into account as a major factor in the decision as to whether they can reasonably be called "house[s]", it appears to me that the judge's decision that they could reasonably be so called was right. As already mentioned, they were each designed and originally occupied as residences in single occupation, they each have the external appearance of being a relatively large London townhouse, they are each fitted out for residential occupation, albeit multiple occupation, as "bedsits", and they are each required under each of the three leases to be used for residential purposes. In this connection, it is worth mentioning that it seems clear that a five-storey building constructed as a house, but converted internally into five self-contained flats, would, at least absent any special factors, be a "house… reasonably so called", following the reasoning and decisions in the judgments in *Lake* and in the speech of Lord Roskill in *Tandon*.

[43] There can be no doubt that the external and internal appearance of the properties are highly relevant factors on this issue, and it is clear from *Prospect Estates* that, in so far as user is significant, the permitted use under the lease is a relevant factor. In those circumstances, even assuming that actual use is also relevant, I find it hard to see how it can be sensibly said that each of the three properties cannot "reasonably [be] called" a "house". To hold otherwise would involve concluding that the actual user, even where it involved people occupying virtually all the rooms in the building for relaxing, sleeping, cooking and washing, albeit on a short-term basis, trumped all the other factors to the extent of disabling the building from being able to be a "house… reasonably so called".

[44] I would therefore hold that, even assuming that *Prospect Estates* had the effect for which the landlords contend, the judge was right in concluding that each of 29, 31, and 39 Rosary Gardens was a "house… reasonably so called".

[45] It is appropriate to go a little further in considering *Prospect Estates*. Although it is, in principle, binding on us so far as its *ratio* is concerned, I must confess to some difficulty in identifying and agreeing with that *ratio*. Goldring LJ appeared to consider that the crucial point was the fact that only around 11% of the building concerned could lawfully be used for residential purposes under the lease in question. That is tolerably clear in its general effect: one concentrates on the permitted use, and, if it forbids residential use, or allows it only to a relatively exiguous extent, the building cannot be a "house… reasonably so called".

[46] Although the observation is tolerably clear, I am not sure that I would agree with it. It appears to place decisive weight on the user covenant in a lease, whereas it seems to me that the thrust of the judgments in *Lake* and the opinion of Lord Roskill in *Tandon* (as opposed to the dissenting opinion of Lord Fraser in that case) is that the question whether a building is a "house… reasonably so called" is to be determined essentially by reference to its external and internal physical character and appearance.

[47] The view that the question of whether a building is a "house… reasonably so called" depends, at least in the main, by reference to its physical appearance and character also seems to me to accord with the natural meaning of those words, the other provisions of section 2(1), and the analysis in *Boss Holdings*. One could, it seems to me, quite naturally describe a building built as a town house, which had subsequently been internally converted into offices, as a "house used as offices": hence it would "reasonably [be] called" a house, even though it was not used for residential purposes, and even if it was not permitted to be so used. If most people were asked whether a building could reasonably be called a house, I am not convinced it would occur to them to ask about the permitted use under any lease, or that they would be influenced if told what the permitted use was. Section 2(1) also refers to buildings that are "structurally detached", "divided horizontally" and "divided vertically" all concerned with physical state, not use. In *Boss Holdings*, in [17], the House of Lords said that, when deciding whether a building had been "designed or adapted for living in", "one is largely concerned with the physical state of the property".

[48] In his judgment in *Prospect Estates*, Mummery LJ (and likewise Smith LJ in hers) may have been saying the same thing as was being suggested in Goldring LJ's very brief judgment. Thus, in [19], Mummery LJ appears to consider that it was the combination of the very small extent to which residential use was permitted and enjoyed that prevented the building in that case from being "reasonably… called" a house. However, he also went on to suggest that a number of relevant factors should be taken into account. There is no indication as to how they are to be taken into account, and whether they are always to be taken into account for instance, whether the actual use should be taken into account only if it is expressly permitted under the lease. It is hard to accept that only 20% residential user would be inconsistent with a building being "reasonably… called" a house, when, on one view, only 25% of the property in question was being used for residential purposes in *Tandon*.

[49] For my part, I would suggest that the *ratio* of *Prospect Estates* should be treated as being limited to a case where residential use is either prohibited entirely, or restricted to a very small part of the building, and the actual use accords with that. The combination of the very limited permitted residential use and very limited actual residential use was, after all, described by Mummery LJ, in [19], as "the overwhelming and decisive feature of this case". On that basis, the decision is of even less assistance to the landlords' case than I have assumed so far.

Is 48 Queen Anne Street a "house... reasonably so called"?

[50] The centrally relevant facts are that the property was designed and built and initially used as a town house residence for a single family (again no doubt with staff quarters), that it maintains its physical appearance externally, that internally some works have been done so that the ground and first floors can be used as offices, that under the lease the two upper floors are restricted to residential use and the ground and first floors are restricted to office use, and that the property was used wholly as offices at the date of the notice, but the landlord had objected to that use in respect of the two upper floors.

[51] The property was, it is sensibly accepted by the landlord, "designed or adapted for living in", not least because, on any view, no work had been done to the upper two floors to undermine the point that those floors remained designed for living in and had not been adapted away from that design. The only ground for challenging the judge's conclusion that the property was a "house... reasonably so called" is the fact that the whole of it was used for office purposes, and none of it for residential purposes, at what is rightly accepted to be the centrally relevant day, namely the date of the notice.

[52] In my judgment, the fact that the property was used wholly for office purposes does not mean that it was, as the landlord, through Ms Katharine Holland QC, contended, not a "house... reasonably so called". The issue is not disposed of in the landlord's favour by the decision in *Prospect Estates*, because, as stated above, it seems to me that the reasoning in that case applies only where both the permitted use and the actual use of the building concerned exclude residential use or limit it to a very small proportion of the building.

[53] If the upper two floors of the property had been empty, I have little doubt but that the property could reasonably have been called a house, bearing in mind its external character and appearance (a classic town house in London's West End), its internal character and appearance at least on the upper two floors (which were, as I understand it, substantially as constructed), the description of the property in the lease as "messuage or residential or professional premises", and, to the extent that it is relevant, the terms of the lease (restricting the use of the upper two floors to residential). I find it hard to see why the fact that the upper two floors had been used (even for many years) as offices (in contravention of the terms of the lease) should wreak such a change that the property could no longer reasonably be called a house.

[54] Accordingly, I agree with the judge in this case as well. There were arguments as to whether: (i) when assessing the proportion of the property restricted to residential use, the judge should have taken into account the basement as non-residential; and (ii) the judge had rightly analysed the planning position. In my view, it is unnecessary to resolve those issues: even if the landlord is right, and: (i) the basement should be treated as non-residential; and (ii) office use of the upper two floors is lawful in planning terms, that would not alter my conclusion.

Conclusion

[55] For these reasons, I would dismiss these appeals; in each case, the tenant has the right to acquire the freehold of the properties at stake.

[56] Like Judge Marshall in *Hosebay*, I reach my conclusion with no particular enthusiasm. The 1967 Act was originally intended to assist residential tenants occupying their houses as their only or main residence to acquire their freeholds. Partly to extend its reach, and partly to defeat the device of company lettings, the legislature ditched the residence requirement, as a result of which the extension of the Act may well have gone further than the legislature intended or anticipated. It now applies to empty and substantially commercial buildings, even if nobody recently lived there or is even intending to live there (provided that they satisfy section 2(1)) as *Boss Holdings* shows, and it is open to tenants to enfranchise many properties at the same time, even if they do not live in any of those properties, as *Aggio* shows. If I am right on these appeals, it can extend to buildings exclusively used for business purposes.

[57] I rather doubt that the amendments made to section 1 in 2002 and, in particular, the removal of the residence requirement, were intended by the legislature to have this sort of effect. Significant amendments to statutes often provide good instances of the law of unintended consequences, and this may well be an example. However, the issue that we have to decide is not what we think the legislature would have said had it fully appreciated the consequences of the primary amendment it made to the 1967 Act (removal of the residence requirement), but what we think that the Act means in the light of that amendment.

[58] Accordingly, the two appeals must be dismissed.

Lloyd LJ said:

[59] I agree.

Moore Bick LJ said:

[60] I also agree.

Appeals dismissed.

Power *v* Stanton

CENTRAL LONDON COUNTY COURT

20 September 2010

HH Judge Marc Dight

Estates Gazette 23 October 2010

[2010] 42 EG 110

Leasehold enfranchisement — Leasehold Reform, Housing and Urban Development Act 1993 — Lease extension — Deceased landlord — Landlord leaving property to son in will and naming him as an executor — No application for grant of probate — Tenant of property seeking to exercise right to new lease under Chapter II of 1993 Act — Whether identity of landlord unknown in absence of grant of probate — Whether tenant entitled to new lease by vesting order under section 50 — Whether required to serve section 42 notice — Whether method of service under section 18(1) of Law of Property (Miscellaneous Provisions) Act 1994 available to tenant of deceased landlord for service of section 42 notice

The respondent was the long leaseholder of a residential property of which the deceased owned the freehold until her death in 1997. The appellant was the deceased's son and one of the executors and trustees under her will. The deceased had left her estate, including her freehold interest in the property, to the appellant for life and thereafter to her grandchildren. The appellant and his co-executors did not apply for a grant of probate of the will.

Twelve years later, the respondent sought to acquire a new long lease of the property pursuant to Chapter II of the Leasehold Reform, Housing and Urban Development Act 1993. Rather than serving a notice under section 42 of the Act, he applied to the court for the grant of the new lease by means of a vesting order under section 50, on the ground that the landlord could not be found or his identity ascertained within the meaning of section 50(1)(b). The claim form named as defendants the deceased's personal representatives and gave the appellant's correct address as the place for service of the proceedings.

The deputy district judge held that the appellant was not entitled to defend the proceedings because he had not been properly constituted as a personal representative of the deceased. He held that since probate had not been granted in respect of the will, the landlord of the property was unknown, section 50 applied and the respondent was entitled to a vesting order.

The appellant appealed. He contended that: (i) section 50 did not apply where an identified landlord was known to have died; and (ii) the requirements of section 5(1)(b) had not been met on the facts of the case. He submitted that the tenant of a deceased landlord could serve a section 42 notice by using the machinery in section 18(1) of the Law of Property (Miscellaneous Provisions) Act 1994 and in CPR 19.8 for service on "The Personal Representatives" of a named deceased at the deceased's last-known address. Moreover, in such circumstances, no lacuna arose that section 50 needed to fill.

***Held*: The appeal was allowed. (1) The death of a landlord does not mean that his or her successor in title cannot be identified or found within the meaning of section 50(1)(b). Title will devolve in one of the usual ways on the deceased landlord's personal representatives. So long as those personal representatives can be identified and found, the condition in section 50(1)(b) cannot be satisfied. In the case of property left under a will, an executor's title to real property derives from the will and not from the grant of probate; the deceased's property vests in him from the moment of the testator's death. A grant of probate is not required for the court to receive evidence of the existence of a will or its contents. The production to the court of the deceased's will was sufficient evidence of the appellant's title and should therefore have been accepted as such. On that evidence, the necessary conditions for the operation of section 50 were not made out.**

Moreover, the structure of section 50 is predicated on the tenant's inability to serve a notice under section 42. If it is possible to serve a section 42 notice, section 50 will not operate. Since the respondent had the executors' correct details, he could have served a section 42 notice on them in the usual way. Alternatively, he could have used the method of service contained in section 18(1) of the 1994 Act. The words of section 18 were apt to cover the case and the fact that it was a "saving" provision of general application did not alter its applicability or relevance to the circumstances. Accordingly, there was no lacuna in the 1993 Act, relating to the service of notices on personal representatives, that section 50 was intended to fill.

(2) Once proceedings were in contemplation, subject to the filing of notices and counternotices the appropriate defendants would have been the executors. However, in the instant case the appropriate course would have been to appoint the appellant to represent the estate of the deceased, on the application of the respondent, pursuant to CPR 19.8(2)(b).

The following cases are referred to in this report.

MA Holdings Ltd v *George Wimpey UK Ltd*; *sub nom R (on the application of George Wimpey UK Ltd)* v *Tewkesbury Borough Council*; *George Wimpey UK Ltd* v *Tewkesbury Borough Council* [2008] EWCA Civ 12; [2008] 1 WLR 1649; [2008] 3 All ER 859

Whitmore v *Lambert* [1955] 1 WLR 495; [1955] 2 All ER 147, CA

This was an appeal by the appellant, Anthony Stanton, as the son and named executor of the deceased, from a decision of Deputy District Judge Evans, sitting in Central London County Court, allowing a claim by the respondent, Thomas Power, for the grant of a new long lease by vesting order under section 50 of the Leasehold Reform, Housing and Urban Development Act 1993, and barring the appellant from defending that claim.

Samuel Laughton (instructed by Magrath LLP) appeared for the appellant; Lloyd Sefton-Smith (instructed by Bryan O'Connor & Co) represented the respondent.

Giving judgment, **HH Judge Marc Dight** said:

[1] This is an appeal from the decision of Deputy District Judge Evans made on 27 August 2009, whereby he held that the claimant was entitled to a new lease of the upper maisonette known as 21A Wotton Road, London NW2 (the flat), pursuant to Chapter II of the Leasehold Reform, Housing and Urban Development Act 1993 (the 1993 Act) and made various consequential orders to give effect to his decision. The learned deputy district judge held that the appellant, despite alleging to be an executor and main beneficiary of the estate of his mother, the late Ms Yetta Martha Stanton (the deceased), did not have any standing in the claim because there had been no grant of probate of her will, and he went on to make orders relying on section 50 of the 1993 Act (the court's power to make a vesting order where the landlord cannot be found) on the basis that, in effect, the landlord had not been found, the claim was undefended and the estate of the deceased unrepresented.

[2] The problem in this case has largely been caused by the fact that notwithstanding the death of the deceased more than 12 years ago, the appellant and his co-executors have failed to take a grant of probate to her estate.

[3] The appellant contends that the learned deputy district judge was wrong to make the order that he did because he wrongly held that section 50 of the 1993 Act applied, he failed to recognise the appellant's true status and failed to take account of section 18 of the Law of Property (Miscellaneous Provisions) Act 1994 (the 1994 Act) and r 19.8 of the Civil Procedure Rules (CPR), which, it was submitted, contained the appropriate provisions to be applied where a landlord from which a new lease is sought pursuant to the 1993 Act has died. The appellant concludes by saying that the learned judge should have dismissed the claim and held that the tenant should have served a notice pursuant to section 42 of the Act if he wanted to seek a new lease.

[4] The respondent contends that the learned deputy district judge reached the right conclusion for the reasons that he gave.

Factual background

[5] The deceased has at all material times been registered at HM Land Registry as the sole proprietor of the freehold interest in 21 Wotton Road, London NW2, under title no MX 208834. By a lease dated 19 March 1976 (the lease) made between Mr Harry MacLeod Burghes and Ms Phyllis MacCleod King (1) and Mr Thomas Simmonds and Ms Maureen Janis Reed (2), the flat was let for a term of 99 years from 25 December 1976 at an initial rent of £25 pa to the claimant's predecessors in title. Title to the lease is registered at HM Land Registry under number NGL 280313. The claimant/respondent has been registered as sole proprietor of the lease since 29 August 2002. The deceased is the successor in title to the grantors of the lease. It appears to be common ground that the respondent is a qualifying tenant within the meaning of the 1993 Act and is entitled to acquire a new lease of the flat.

[6] The deceased passed away on 29 December 1997. The death certificate dated 29 December 1997, a copy of which is attached to the claim form, identifies the appellant as the person who informed the registrar of his mother's death, described him as her son and stated his usual address.

[7] By her will dated 27 August 1997, the deceased appointed the appellant and a Dr Lawrence Shaw as her executors and trustees and, after a small number of pecuniary bequests, left her estate to the appellant for life and thereafter to her grandchildren. As I have already mentioned, no grant of probate has been taken in respect of the estate. It is accepted by the appellant that he had not produced a copy of the will to the respondent prior to the commencement of proceedings, although there had been correspondence between the parties.

[8] By his claim form issued in Barnet County Court, the respondent sought a vesting order pursuant to section 50 of the 1993 Act and an order that the premium for the new lease be determined by the leasehold valuation tribunal (LVT). In the claim form, it is said that: "The claimant has through his solicitors traced Anthony Trevor Stanton who has stated that he is entitled to apply for a grant of probate in relation to his mother's estate but that no grant has been taken out."

[9] The claim form names as the defendants the personal representatives of the deceased and gives the appellant's correct address as the place for service of the proceedings.

[10] Since the tenant took the view that his landlord could not be found, he had not initiated the usual process provided in Chapter II of the 1993 Act for the acquisition of a new lease of a flat by serving a notice under section 42 of the Act. Such a notice would have, *inter alia*, specified the premium that the tenant proposed paying and the

A terms that he proposed for the new lease. It would also have triggered an obligation on the part of the landlord to serve a counternotice if he did not accept that the tenant had a right to a new lease or disagreed with the premium or terms that had been proposed. Failing agreement, proceedings could have been taken. Instead, the tenant started these proceedings seeking a vesting order.

[11] The claim came on before the learned deputy district judge, who apparently heard from the solicitor for the respondent and from the appellant, who produced the will to him. He held, *inter alia*, as follows:

> B … I have suggested that I do not know how the claimants have brought this action on you because you were not properly constituted as personal representative. That is you at the moment do not have any standing in this action. That is where I started off. I have effectively taken the view that it is proper to proceed in the manner in which they have proceeded because your mother has passed away, unfortunately, therefore, the landlord is currently unknown and so section 50 applies and I am satisfied that it applies and I am satisfied that this order, therefore is the proper order. If you believe otherwise then you ought to do what you should have done some time ago and I am sure you have your own reasons. You should have made an application for a grant of probate. The moment you have got an occasion for a grant of probate you can take part in the litigation. You can apply to be joined as a defendant and you can
> C do whatever you want. Until then Mr Stanton, my view stands.

[12] Criticism is made of the fact that the transcript included in the appeal bundle is inadequate because it is plain from the extract quoted above that the learned deputy district judge had already expressed part of the reasons for his conclusions earlier in the course of argument and that a transcript of the entire hearing should therefore have been before me on appeal. It is undoubtedly true that the learned deputy district judge appears to have indicated his views on the merits of the appellant's stance prior to the passage that I have quoted, but, in my judgment, the transcript with which I have been provided contains sufficient of his reasoning for me to understand the true basis for his decision. His
D reasons must have been that because the landlord had passed away and there had been no grant of probate, the landlord could not be found or his identity could not be ascertained: *per* section 50(1)(b) see below.

[13] By his order, the learned deputy district judge effectively declared that the respondent was entitled to a new lease of the flat in the form that the respondent had proposed. His ancillary orders provided for the grant of the new lease to be made via a vesting order "of the Defendant's interest in the premises" (sic) and execution of the new lease by a district judge of the court on payment into court of a price to be determined by a surveyor to be selected by the president of the Lands Tribunal, after the deduction of the respondent's costs of the
E proceedings. In other words, the estate was effectively prevented from taking part in the process by which the terms and price for the new lease were to be arrived at.

Locus to appeal

[14] As a preliminary point, it was submitted that by virtue of the reasoning of the Court of Appeal in *George Wimpey UK Ltd* v *Tewkesbury Borough Council* [2008] EWCA Civ 12; [2008] 3 All ER 859, the definition of "appellant" in CPR 52.1(3)(d) is sufficiently wide to embrace a non-party in the position of the appellant that may be adversely affected by the proceedings. On behalf of the respondent,
F it was not strenuously argued otherwise. In any event, I would hold, in the absence of agreement, that because of his status as executor of the deceased's estate the appellant has locus by virtue of the provision of the CPR mentioned above, as explained by *George Wimpey*, to challenge the outcome of these proceedings in the court below by way of appeal.

A vesting order?

[15] A vesting order is defined by section 51(1) of the 1993 Act as:

> an order providing for the surrender of the tenant's lease of his flat and for the granting to him of a new lease of it on such terms as may be determined by a leasehold valuation tribunal to be appropriate with a view to the lease being granted to him in like manner…as if he had, at the date of his application, given notice under section 42 of his claim to exercise the right to acquire a new lease G
> of his flat.

[16] Section 50 of the 1993 Act, which is headed "Applications where landlord cannot be found", provides, in so far as material, as follows:

> (1) Where
>
> (a) a qualifying tenant of a flat desires to make a claim to exercise the right to acquire a new lease of his flat, but
>
> (b) the landlord cannot be found or his identity cannot be ascertained,
>
> the court may, on the application of the tenant, make a vesting order under this subsection. H
>
> (2)…
>
> (3) The court shall not make an order on any application under subsection (1)… unless it is satisfied
>
> (a) that on the date of the making of the application the tenant had the right to acquire a new lease of his flat; and
>
> (b) that on that date he would not have been precluded by any provision of this Chapter from giving a valid notice under section 42 with respect to his flat.
>
> (4) Before making any such order the court may require the tenant to take such further steps by way of advertisement or otherwise as the court thinks proper for the purpose of tracing the person in question; and if, after an application is made for a vesting order under subsection (1) and before any lease is executed in pursuance of the application, the landlord is traced, then no J
> further proceedings shall be taken with a view to a lease being so executed, but (subject to subsection (5))
>
> (a) the rights and obligations of all parties shall be determined as if the tenant had, at the date of the application, duly given notice under section 42 of his claim to exercise the right to acquire a new lease of his flat; and
>
> (b) the court may give such directions as the court thinks fit as to the steps to be taken for giving effect to those rights and obligations, including directions modifying or dispensing with any of the requirements of this Chapter or of regulations made under this Part.
>
> (5) An application for a vesting order under subsection (1) may be withdrawn at any time before execution of a lease under section 51(3) and, after it is withdrawn, subsection (4)(a) above shall not apply; but where any K
> step is taken (whether by the landlord or the tenant) for the purpose of giving effect to subsection (4)(a) in the case of any application, the application shall not afterwards be withdrawn except
>
> (a) with the consent of the landlord, or
>
> (b) by leave of the court,
>
> and the court shall not give leave unless it appears to the court just to do so by reason of matters coming to the knowledge of the tenant in consequence of the tracing of the landlord.

[17] The appellant submits, first, that section 50, on its proper construction, is not applicable to a situation where an identified landlord is known to have died. Second, it is submitted that the statutory L
criteria contained in subsection 50(1)(b) were not satisfied because identities and addresses of the personal representatives of the deceased were readily ascertainable from the terms of the documents and correspondence available to the respondent. Third, it is submitted that where a landlord has died, as opposed to disappeared, machinery is to be found in section 18(1) of the 1994 Act and r 19.8 of the CPR for the service of an initial notice under section 42 of the 1993 Act and for the constitution of proceedings prior to the grant of probate. Therefore, it is said, there is no lacuna in the 1993 Act relating to deceased landlords that has to be filled by the use of section 50.

[18] The respondent submits that sections 50 and 51 of the 1993 Act M
provide a complete and concurrent scheme for acquisition of a new lease, separate and distinct from that provided by section 42, and entirely apposite to a situation in which a landlord has died but no grant of probate has been obtained. Second, it is submitted that, prior to grant, the statutory criteria of section 50 are satisfied in this case and that, once satisfied, the court has jurisdiction to make a vesting order. As to the appellant's third submission, it is said that section 18 of the 1994 Act is permissive only; it is a saving provision of general application and merely plugs a gap where there may not be someone to accept notices and proceedings and would not fulfil the intended function of section 42, that is to identify and engage a party with which to negotiate over the terms of grant of a new lease.

[19] The respondent further submits that where 12 years after the death of the landlord probate has still not been sought, "a court would need fairly compelling evidence to conclude that the appropriate legal identity could now be ascertained and that there was not some other reason connected with IHT avoidance, fraud etc". That led to the further submission that the learned deputy district judge was not entitled to conclude that the appellant was the personal representative and was therefore correct to hold that the personal representative were unidentified and/or untraceable.

[20] Turning first, therefore, to the proper construction of section 50. The words that lie at the heart of this dispute are to be found in subsection 50(1)(b): "the landlord cannot be found or his identity cannot be ascertained". Unless that condition is satisfied, section 50 has no application. Prior to death, the landlord was both identified and her address known. The question is whether death renders both (or either) of those criteria uncertain. The section does not specifically refer to death. Further, as a matter of logic, it seems to me that it does not follow that because a landlord has died his successor in title as landlord cannot be identified or found. Title will devolve in one of the usual ways on the landlord's personal representatives. So long as those personal representatives can be identified and found, it seems to me that the condition in subsection 50(1)(b) cannot be satisfied.

[21] The respondent submits, however, that the situation is different where, as here, there has been no grant of probate because, until that point, the identity of the personal representatives is uncertain. So far as property left by will is concerned, it seems to me that such a submission is ill founded. An executor's title to real property derives from the will, not from the grant of probate, and the property of the deceased vests in him from the moment of the testator's death: see *Whitmore* v *Lambert* [1955] 2 All ER 147, *per* Sir Raymond Evershed MR, at p151 and following. In so far as the learned deputy district judge may, by necessary inference, have held otherwise, he was, in my judgment, wrong.

[22] The routes to the grant of a new lease contained in sections 50 and 42 are said by the appellant, in reinforcement of his primary submission, to be mutually exclusive. The respondent disagrees and suggests that they provide concurrent routes to the grant of a new lease. The terms of subsections 51(1) and 50(3)(b), (4) and (5) not only contemplate a situation in which a notice under section 42 has not been served on the landlord but could not be given, either because the landlord could not be identified or found. Moreover, subsection 50(4) provides, in effect, that where a landlord is traced part-way through the vesting order process, that process is to come to an end and the usual route for grant of a new lease followed. I agree with the appellant that the entire structure of section 50 is predicated on the inability of the tenant to serve a notice under section 42 and that therefore, in my judgment, it follows that if it is possible to serve a notice under section 42 there is no room for the operation of section 50.

[23] The question then arises as to whether, in the instant case, the statutory criteria of section 50(1)(b) are, properly construed, satisfied. Although I have great sympathy with the respondent's submissions on this point, it seems to me that the production of the will to the court was sufficient evidence of the appellant's title, in which respect I would again refer to the judgment of the Master of the Rolls in *Whitmore,* in which he held that there need not be a grant of probate before the court can receive evidence of the existence of a will or its contents. In those circumstances, the will should, in my judgment, have been accepted as evidence of the title of the appellant and his fellow executor, on which the learned deputy district judge should have come to the conclusion that, as a matter of fact, the necessary condition for the operation of section 50 was not made out.

[24] Turning next to the machinery said to have been operable in the circumstances. Section 18 of the 1994 Act provides, in so far as material, as follows:

> (1) A notice affecting land which would have been authorised or required to be served on a person but for his death shall be sufficiently served before a grant of representation has been filed if
>
> (a) it is addressed to "The Personal Representatives of" the deceased (naming him) and left at or sent by post to his last known place of residence or business in the United Kingdom, and
>
> (b) a copy of it, similarly addressed, is served on the Public Trustee.
>
> (2) …
>
> (3) The method of service provided for by this section is not available where provision is made
>
> (a) by or under any enactment, or
>
> (b) by an agreement in writing,
>
> requiring a different method of service, or expressly prohibiting the method of service provided for by this section, in the circumstances.

[25] The learned editors of *Hague on Leasehold Enfranchisement* (5th ed) apparently express their view on this issue without any comment, in para 34-17, as follows:

> Service on a dead landlord
>
> If the landlord is dead, a notice, eg section 13 or section 42 notice, is sufficiently served before a grant of representation has been filed if it is addressed to "The Personal Representatives of" the deceased (naming him) and left at or sent by post to his last known place of residence or business in the United Kingdom, and a copy of it, similarly addressed, is served on the Public Trustee: section 18(1) of the Law of Property (Miscellaneous Provisions) Act 1994.

[26] Taking the words of section 18 at face value, they are entirely apt in this case and set out a procedure that could have been followed by the tenant in order to serve a notice under section 42 of the 1993 Act. I have been taken to nothing in the 1993 Act itself that would have the effect, in accordance with subsection 18(3)(a), of prohibiting the method of service contained in section 18(1) from being used in this case. The fact that the provision is of general application or is "saving" in nature does not alter its applicability or relevance. In my judgment, even though the respondent appears to have had the correct details for the executors and could have served on them a notice under section 42 in the usual way, it was open to him also to use the machinery afforded by section 18 of the 1994 Act. It follows that I also agree with the submission that there is no lacuna in the 1993 Act relating to the service of notices on personal representatives that section 50 was intended to fill.

[27] Once proceedings were in contemplation, subject to the filing of notices and counternotices the appropriate defendants would have been the executors. However, in this case, it would plainly have been open to the deputy district judge to appoint the appellant to represent the estate on the application of the respondent pursuant to CPR 19.8(2)(b) and, in my judgment, the respondent should have so applied and the learned judge should have made an order appointing the appellant to represent the estate of the deceased.

Premium

[28] In para 5 of his grounds of appeal, the appellant asserts, and the respondent now accepts, that the learned deputy district judge erred in directing that the premium for the new lease be fixed by a surveyor selected by the president of the Lands Tribunal and the premium for the new lease (and the terms) should have been referred for determination by the LVT.

Conclusion

[29] For the reasons that I have given above, I would hold that section 50 of the 1993 Act has no application in this case, that the learned deputy district judge misdirected himself in reaching his conclusions, and that his order ought to be set aside. In its place, I would appoint the appellant to represent the estate of the deceased and dismiss the claim with costs.

Appeal allowed.

MORTGAGE

Anfield (UK) Ltd *v* Bank of Scotland plc and others

CHANCERY DIVISION

[2010] EWHC 2374 (Ch)

24 September 2010

PROUDMAN J

Estates Gazette 4 December 2010

[2010] 48 EG 86

Mortgage — Subrogation — Unjust enrichment — Land Registration Act 2002 — Bank advancing loan by way of remortgage — Registrable charge granted to bank but not registered — Further charge subsequently registered by another lender — Whether that lender unjustly enriched by bank's failure to register its own charge — Whether bank entitled to be subrogated to earlier charge securing loan repaid from bank's advance so as to obtain priority over subsequent charge

In 2006, the respondent received an executed charge over a registered property. This acted as security for a loan advanced to the owner by way of remortgage, to enable it to redeem an existing registered charge, made in 2000, in favour of a building society. The respondent failed to register its interest as a legal charge. Further charges were subsequently registered in favour of other lenders, including the appellant, whose equitable charge, obtained by way of an interim charging order made in a pending land action, was registered in April 2008. Meanwhile, the owner of the property was made bankrupt. In 2009, the respondent registered a unilateral notice in respect of its advance by remortgage.

In proceedings between the parties, the respondent claimed that it was entitled to be subrogated to the building society charge of 2000 so as to rank in priority to any subsequent charges, including the equitable charge held by the appellant. The county court accepted that contention, holding that the respondent was entitled to be registered as the proprietor of the building society charge to the extent of the moneys advanced to discharge it.

On an appeal from that decision, the central issue was whether the appellant had been unjustly enriched, so as to invoke the remedy of subrogation, in circumstances in which the respondent's failure to obtain the first legal charge that it had expected to receive was attributable solely to its failure to register the charge under the Land Registration Act 2002. The appellant submitted that the respondent had received all it bargained for in the way of a charge because registration was a matter solely for the lender, not for the borrower. Thus, in such circumstances, the appellant had not been unjustly enriched at the respondent's expense.

***Held*: The principle underlying subrogation is equitable in origin and is primarily aimed at preventing unjust enrichment. It may enable a lender whose advances have been used to discharge a secured debt owned to another lender to step into the shoes of that other lender with regard to the security and thereby gain priority over intermediate lenders also holding security over the same property. Although intermediate lenders are necessarily enriched by the discharge of a prior security, some unconscionable or unjust factor must also be identified for the principle to be engaged. In considering that matter, the court must look at the justice or otherwise of the intermediate lender's enrichment, judged by reference to the position as between the party seeking subrogation and the party enriched. The fact that the party seeking subrogation has itself been careless in failing to obtain the security it desired should not of itself deprive it of its right to resist an unjust enrichment gained at its expense. The availability of subrogation does not turn entirely on the existence of an unfulfilled mutual intention or agreement between the party seeking subrogation and the enriched party. The factor that renders the enrichment unjust is the non-fulfilment of the lender's expectation as to the security that it was to receive, which expectation formed the basis of its decision to advance funds. The question of whether the lender got what it bargained for should be considered in the wider sense of the envisaged transaction rather than by an inquiry, akin to the process of contractual construction, into the question of whether the precise terms of the bargain between the lender and borrower had been fulfilled. The appellant's enrichment had been unjust because the respondent had funded the repayment of the building society charge on the understanding that it would obtain a legal charge, whereas, owing to the failure to register under the 2002 Act, it had obtained only a subsequent equitable charge. It was irrelevant that the borrower had performed the terms of the bargain between the borrower and the lender. *Obiter* statements to the contrary in *Burston Finance Ltd* v *Speirway Ltd* [1974] 1 WLR 1648 were not followed; the approach in *Cheltenham & Gloucester plc* v *Appleyard* [2004] EWCA Civ 291; [2004] 13 EG 127 (CS) was preferred.**

The following cases are referred to in this report.

Banque Financière de la Cité SA v *Parc (Battersea) Ltd* [1999] 1 AC 221; [1998] 2 WLR 475; [1998] 1 All ER 737; [1998] EGCS 36, HL
Boscawen v *Bajwa*; *Abbey National plc* v *Boscawen* [1996] 1 WLR 328; [1995] 4 All ER 769; (1995) 70 P&CR 391, CA
Burston Finance Ltd v *Speirway Ltd* [1974] 1 WLR 1648; [1974] 3 All ER 735; 29 P&CR 34, Ch
Capital Finance Co Ltd v *Stokes* [1969] 1 Ch 261; [1968] 3 WLR 899; [1968] 3 All ER 625; 19 P&CR 791
Cheltenham & Gloucester plc v *Appleyard* [2004] EWCA Civ 291; [2004] 13 EG 127 (CS)
Congresbury Motors Ltd v *Anglo-Belge Finance Co Ltd* [1971] Ch 81
Cork & Youghal Railway Co, Re (1869) LR 4 Ch App 748
Eagle Star Insurance Co Ltd v *Karasiewicz* [2002] EWCA Civ 940
Filby v *Mortgage Express (No 2) Ltd* [2004] EWCA Civ 759
Orakpo v *Manson Investments Ltd* [1978] AC 95; [1977] 3 WLR 229; [1977] 3 All ER 1, HL

This was an appeal by the appellant, Anfield (UK) Ltd, from a decision of HH Judge Hazel Marshall QC, sitting in Central London County Court, upholding the respondent's entitlement to be subrogated to a legal charge over property so as to rank in priority to the appellant's security.

Richard Lander (instructed by Chandler Harris LLP) appeared for the appellant; Charlotte Eborall (instructed by Optima Legal Services Ltd) appeared for the first respondent; the second and third respondents did not appear and were not represented.

Giving judgment, **PROUDMAN J** said:

Issue on the appeal

[1] This is a case about the remedy of subrogation as it affects the priority of charges.

[2] It has been argued before me on behalf of the claimant and appellant, Anfield (UK) Ltd (Anfield), and Bank of Scotland plc (the bank), the first defendant. The case arises out of loans to the second defendant, Mr Siddiqui, a bankrupt, who is the registered proprietor of property (the property) in Stepney Green in London. Neither Mr Siddiqui nor his trustee in bankruptcy appeared at trial or on the appeal. There is another party to the action, London Scottish Finance Ltd (LSFL), joined after trial pursuant to an order of the trial judge on 3 March 2010. However LSFL (which is in administration) was not represented at and did not participate in the appeal.

[3] The following items appear on the charges register within the register of title in respect of the property, in the following order:

(1) A registered charge dated 30 June 2000 (the Halifax charge) to secure the moneys, including the further advances therein mentioned. The Halifax charge was registered on 29 September 2000. As is evident from earlier versions of the register, and is common ground, the Halifax charge was granted to Halifax Building Society. The bank is registered as the proprietor of the Halifax charge and the date of registration is 3 December 2007. Again, it is common ground that the bank advanced moneys to Mr Siddiqui to discharge the Halifax charge.

(2) A registered charge dated 28 June 2007, registered on 16 July 2007. LSFL is the proprietor of that charge and was registered as such on the same date.

(3) A unilateral notice in respect of a pending land action for a charging order registered on 4 October 2007, of which the beneficiary is Anfield, registered as such on the same date. On 3 April 2008, an equitable charge created by an interim charging order dated 25 March 2008 was registered.

(4) A unilateral notice in respect of an advance by remortgage (completed on 1 September 2006) was registered on 1 April 2009, of which the bank was registered as beneficiary on the same date.

[4] The bank lent Mr Siddiqui the money to redeem the Halifax charge and, in September 2006, received in return an executed charge (the 2006 charge) capable of being registered as a legal charge. However, the bank failed to register the 2006 charge. Anfield obtained and registered its charging order. Subsequently, the bank caused its unilateral notice to be entered in respect of the 2006 charge.

[5] The bank claimed to be entitled to invoke the remedy of subrogation to the Halifax charge. If it was so entitled, it would rank in priority to Anfield's equitable registered charge at (3) and LSFL's registered legal charge at (2).

[6] By order dated 28 January 2010, HH Judge Hazel Marshall QC, sitting at Central London County Court, held that the bank was entitled to be subrogated to and registered as proprietor of the Halifax charge to the extent of the moneys advanced to discharge it.

[7] Anfield appeals against that decision, relying on a similar reasoning to that of Walton J expressed in *Burston Finance* v *Speirway Ltd* [1974] 1 WLR 1648. Anfield's difficulty is that in *Cheltenham & Gloucester plc* v *Appleyard* [2004] EWCA Civ 291*, Neuberger LJ, giving the judgment of the full Court of Appeal, comprising also Lord Phillips of Worth Matravers MR and Kennedy LJ, expressed the view of the court that such reasoning was wrong in principle.

[8] Mr Richard Lander, on behalf of Anfield, submitted that: (a) the facts of *Cheltenham & Gloucester* are distinguishable from those of the present case; (b) the expression of opinion on the issue now arising was *obiter*; and (c) the Court of Appeal's opinion should not be followed.

Law

[9] A lender that has made advances that have been used to discharge a secured debt owed to another lender may be entitled to step into the shoes of the other lender as far as the security is concerned, thereby gaining priority over intermediate lenders also holding security over the same property.

[10] It is plain from the authorities cited and authoritatively analysed in *Cheltenham & Gloucester* (and particularly the decision of the House of Lords in *Banque Financière de la Cité SA* v *Parc (Battersea) Ltd* [1999] 1 AC 221) that the principle underlying such subrogation is equitable in origin and is now recognised as being primarily aimed at preventing unjust enrichment.

[11] Intermediate lenders are necessarily enriched by the discharge of a prior security. However, in order for the principle to be engaged it is necessary to identify some unconscionable or unjust factor. As May LJ (with whom the other members of the Court of Appeal agreed) said in *Filby* v *Mortgage Express (No 2) Ltd* [2004] EWCA Civ 759, in [62]:

> the remedy of equitable subrogation is a restitutionary remedy available to reverse what would otherwise be unjust enrichment of a defendant at the expense of the claimant. The defendant is enriched if his financial position is materially improved, usually as here where the defendant is relieved of a financial burden… The enrichment will be at the expense of the claimant if in reality it was the claimant's money which effected the improvement. Subject to special defences, questions of policy or exceptional circumstances affecting the balance of justice, the enrichment will be unjust if the claimant did not get the security he bargained for when he advanced the money which in reality effected the improvement, and if the defendant's financial improvement is properly seen as a windfall. The remedy does not extend to giving the claimant more than he bargained for. The remedy is not limited to cases where either or both the claimant and defendant intended that the money advanced should be used to effect the improvement. It is sufficient that it was in fact in reality so used. The remedy is flexible and adaptable to produce a just result. Within this framework, the remedy is discretionary in the sense that at each stage it is a matter of judgment whether on the facts the necessary elements are fulfilled.

[12] In *Cheltenham & Gloucester*, the Court of Appeal set out 13 established principles distilled from the authorities. Of particular relevance to the present case are seven, eight and 11, as follows:

> 38. Seventhly, a lender cannot claim subrogation if he obtains all the security which he bargained for, as in *Burston Finance* (applying *Capital Finance Co Limited* v *Stokes* [1969] 1 Ch 261) or where he has specifically bargained on the basis that he would receive no such security as in *Paul* v *Speirway Limited (in liquidation)* [1976] 1 WLR 220.
>
> 39. Eighthly, the fact that the lender's failure to obtain the security he bargained for was attributable to his negligence is irrelevant. It does not prevent him from claiming subrogation see *per* Lord Hoffmann at 235E-G in *Banque Financière*. The effect of that observation was probably impliedly to disapprove observations of Walton J in *Burston Finance* at 1657C and F. However, Walton J was concerned with a case where the lender obtained the security, but negligently failed to protect himself by registering it, whereas in *Banque Financière* the lender's negligence was in failing to check that he had obtained the security.
>
> …
>
> 42. Eleventhly, it is difficult, and may be impossible, for a lender who has obtained security to invoke subrogation where the security he has obtained gives him all the rights and remedies of security to which he claims to be subrogated (see *Burston Finance* at 1653D-E), or is a security in which the original security would naturally merge (see *Burston Finance* at 1653C and *per* Lord Diplock in *Orakpo* [v *Manson Investments Limited* [1978] AC 95] at 105B-C.

[13] The central issue in the present appeal is whether the enrichment of intermediate secured lenders (that is to say, Anfield and LSFL) is unjust in circumstances where the lender claiming subrogation (the bank) expects to receive a first legal charge over the property but does not do so solely because of its failure to register the charge under the Land Registration Act 2002 (the 2002 Act).

[14] Mr Lander submitted that the lender has all it bargained for in the matter of a charge since registration was a matter solely for the lender, not for the borrower. In those circumstances, he submitted that the fact that the failure to register may have been negligent is irrelevant.

[15] Two potentially conflicting views appear from the authorities as to the effect of non-registration. In *Burston Finance*, Walton J remarked that a lender that fails to obtain its desired security by reason of non-registration under the Land Registration Acts is not entitled to subrogation because it has obtained everything that it bargained for, namely a charge in registrable form.

* Editor's note: Reported at [2004] 13 EG 127 (CS)

A [16] By contrast, in *Cheltenham & Gloucester*, the Court of Appeal took the view that a lender that stipulates for a legal charge but that in fact obtains only an equitable charge because it fails to register under the Land Registration Acts does not obtain all that it bargained for. It bargained for a legal charge. Accordingly, subject to the fulfilment of all other relevant requirements, it will be entitled to be subrogated to an earlier legal charge that was discharged with his advance.

[17] The facts of both cases are distinguishable from those of the present case, but in both cases the court comments on the very situation that has now arisen.

B *Burston*

[18] In *Burston*, the chargee seeking subrogation had registered the charge at the Land Registry but had failed to register it under section 95 of the Companies Act 1948, with the effect that the charge was void as against the liquidator. The principal ground of Walton J's decision was that the plaintiff could not rely on its unpaid vendor's lien because it had been replaced by and merged with the legal charge. (As to which, see further *Orakpo**, at p105B-C.) In any event, relying on *Capital Finance Co Ltd* v *Stokes* [1969] 1 Ch 261, Walton J decided that, in the case of non-registration as a company charge, the charge was effective when made and therefore the chargee had obtained all that it bargained for. C The judge was bound by the decision in that case to reject the argument that what was bargained for was "a *valid* legal charge". He said, at p1657A:

> What the Court of Appeal there decided, it appears to me, was that it was quite impossible to equate a legal charge which only becomes void subsequently to its creation, and then only to a strictly limited extent, with the kind of charges in *Thurstan* v *Nottingham Permanent Benefit Building Society* [1902] 1 Ch 1 and *Congresbury Motors Ltd* v *Anglo-Belge Finance Co Ltd* [1971] Ch 81 which, for one reason or another, were wholly ineffective from their creation. See also *In Re Monolithic Building Co* [1915] 1 Ch 643 *per* Lord Cozens-Hardy MR, at p667 and Phillimore LJ, at pp667, 668.

D [19] On the one hand, therefore, there were cases in which a charge was invalid from its creation, as in *Thurstan,* where the loan was made to an infant without capacity, *Re Cork & Youghall Railway Co* (1869) LR 4 Ch App 748, where the borrowing was *ultra vires*, *Eagle Star Insurance Co Ltd* v *Karasiewicz* [2002] EWCA Civ 940, where the security did not bind a wife's beneficial interest, *Filby,* where the charge was tainted by forgery, or *Banque Financière,* where the party purporting to grant a postponement had no authority to do so. On the other side, there are the cases in which the charge became invalid subsequently and to a limited extent because of non-registration under the Companies Acts.

E [20] In considering what the parties bargained for, it could be argued, and was argued in *Burston*, that in the case of a company charge, the primary duty as to registration was on the borrower and thus that one of the matters bargained for was that the charge would be registered. That is not the case where the non-registration is under the Land Registration Acts. In both cases, however, the charge is not void from its inception. By contrast, in cases of non-registration, the lender's assumption as to validity becomes false only once the period for registration has elapsed because until that time it is within the chargee's power to obtain the security bargained for. Accordingly, Walton J went on to say, at p1657B:

> F Mr Price urged strongly that if the correct test was indeed… did the vendor obtain all that he bargained for?, then here he did not obtain it, because one of the things for which he bargained was that the defendants would fulfil their statutory duty of registration pursuant to section 96 of the Act of 1948. But this is to confuse substance with formalities. As regards the formality of registration under section 26 of the Land Registration Act 1925, nothing can at the end of the day turn upon this because it was duly effected. *But even had it not been, this is a matter wholly within the plaintiffs' own power, and therefore it could not be suggested that failure to register could have any conceivable effect on the plaintiffs obtaining "all that they bargained for."* As regards registration under section 95, once again, this is undeniably, as pointed out by Harman LJ, within the plaintiffs' own power, although the primary duty was cast on the defendants. G Nevertheless, even without such registration, the charge remained effective against the defendants and still so remains. …
>
> So all this demonstrates, to my mind, that there is no doubt that the plaintiffs got what they bargained for. They were merely lax in not taking steps to ensure that what they bargained for remained good against the world.

(Emphasis added.)

Cheltenham & Gloucester

[21] In *Cheltenham & Gloucester*, there was no failure to register in the sense considered by Walton J in *Burston.* The reason that the claimant did not obtain the legal charge for which it stipulated was H that an earlier incumbrancer, whose debt had been discharged with the claimant's advance, failed for reasons connected with its insolvency to consent to the registration of the claimant's charge. It was not within the power of the lender to obtain all it had bargained for.

[22] In *Cheltenham & Gloucester*, the appellants sought to rely on Walton J's *dicta* to say that the lender had obtained all that it had bargained for in that there was a valid and enforceable charge as between lender and borrowers. The Court of Appeal decided that the lender had not obtained all that it had bargained for; it had bargained for a legal charge, whereas non-registration under the Land Registration Acts gave it only an equitable charge with inferior remedies. The court treated J *Burston* and, indeed, *Capital Finance* as having been correctly decided on their facts, but expressed severe doubts (see below) concerning the correctness of Walton J's comments on the effect of non-registration under the Land Registration Acts. Neuberger LJ said, in [71]:

> … C&G did not obtain the security they bargained for. In the two earlier cases, the failure to register the charge under s95 did not affect the character of the charge obtained by the plaintiff in each case. Non registration under s95 did not prevent the charge in either case being a legal charge. It merely rendered it "void" as against a subsequent creditor or a liquidator. On the other hand, non-registration of the C&G mortgage at the Land Registry prevented it from being a legal charge: it was merely an equitable charge. Yet it was a legal charge for K which the parties bargained.

[23] Neuberger LJ went on to cite passages from *Capital Finance* and from *Boscawen* v *Bajwa* [1996] 1 WLR 328, at p339, in support of this analysis.

[24] Further, Neuberger LJ observed that in both *Burston* and *Capital Finance*, the court appears to have been influenced in refusing the remedy of subrogation by the fact that the lender had been negligent in failing to register its security. It is now settled after the decision of the House of Lords in *Banque Financière* that negligence or carelessness on the part of the lender is not a factor to be taken into account when L considering the claim for subrogation, "at least where the negligence relates to the obtaining of the security".

Am I bound by Cheltenham & Gloucester on the facts of this case?

[25] The first question for me is whether I am bound by the decision in *Cheltenham & Gloucester* on the facts of the present case. Although the rationale of the case would seem to apply to the present facts also, it seems that I am not strictly bound because the Court of Appeal expressly left the current issue open. Neuberger LJ said, in [75]:

> It may be (although we doubt it) that the lender's negligent failure to protect or perfect the security should be treated differently. M

[26] However, in the passage quoted above, the court has expressed the considered view that Walton J's *dictum* concerning the effect of non-registration under the Land Registration Acts was wrong. The view of the Court of Appeal must carry the strongest possible persuasive effect, and very good reason would have to be shown for departing from it.

[27] The judgment in *Cheltenham & Gloucester* contains two other specific criticisms of the *dictum*. First, in [39], Neuberger LJ states the court's view that the effect of Lord Hoffmann's observation at p235E-G in *Banque Financière* as to the irrelevance of the lender's negligence "was probably impliedly to disapprove observations of Walton J in *Burston Finance*, at 1657C and F". Second, he said, in [71]:

* Editor's note: *Orakpo* v *Manson Investments Ltd* [1978] AC 95

The fact that the effect of non-registration under the Land Registration act 1925 of the charge would prevent it from being a legal charge may have been overlooked by Walton J in *Burston Finance* at 1657C. If so, it is not particularly surprising, given that it was an *ex tempore* judgment and his observations on this aspect were purely *obiter*.

[28] Nevertheless, Mr Lander asks me to prefer the *obiter* reasoning of Walton J in *Burston* as to subsequent failure to perfect the lender's security to the *obiter* view expressed by the Court of Appeal in *Cheltenham & Gloucester*.

Negligence

[29] I see the force of Mr Lander's submission that negligence is irrelevant at the stage of perfecting a security if by that stage the lender can be said to have got all he had bargained for. In *Cheltenham & Gloucester*, Neuberger LJ refers to earlier courts being "influenced" by the negligence of the lender. He does not say negligence was the actual *ground* for denying subrogation.

[30] However, Neuberger LJ does rely on Walton J's reference to what he described as the plaintiffs' "lax" conduct. It does indeed appear to be Walton J's view that there is nothing unconscionable in a lender bearing the burden of its own mistake in circumstances where neither the borrower nor the defendant knew of the mistake or had any involvement in it. Such a view cannot survive the statement in *Banque Financière* that the lender's carelessness in obtaining the desired security does not by itself defeat a claim for subrogation. I accept Ms Charlotte Eborall's submission that to try and draw distinctions between carelessness at different stages of the process is perilously close to fashioning an exception to a principle that was expressed in general terms. The court is required, as the judge below found (see [77]), simply to look at the question of the justice of the enrichment of the defendant. You do not lose your right to resist an unjust enrichment gained at your expense simply because you have been careless in respect of the security.

[31] Of course there may potentially be harm to third-party lenders that can demonstrate reliance on the register. Mr Lander suggested that this is a basis for distinguishing between non-registration and other kinds of negligence or failures. However, I agree with the judge below, in [88], that the law is flexible enough to recognise that detriment to third parties might form a basis for withholding the remedy of subrogation on the basis that, in such circumstances, the enrichment of the defendant had not been unjust. I have in mind the kind of change of position defence available to a party that has acted in the reasonable but false belief that a particular state of affairs existed.

Banque Financière approach

[32] In my judgment, whether I am technically bound to apply it, the analysis in *Cheltenham & Gloucester* is to be preferred to that in *Burston* on the basis of the approach taken in *Banque Financière*. First, the focus has shifted from facts affecting the conscience of the mortgagor to the justice of the position as between the lender and the party enriched. Second, it is clear that the availability of subrogation does not turn entirely on the existence of an unfulfilled mutual intention or agreement between the party seeking subrogation and the enriched party: see at p227F, *per* Lord Steyn, at pp231E-234G, *per* Lord Hoffmann, with whom Lords Griffiths and Clyde agreed. Third, prominence is given to the intention of the lender.

[33] Applying these principles, the House of Lords held that the enrichment that would occur in the absence of subrogation would be unjust because the claimant had advanced funds on the mistaken assumption that it would obtain a postponement of the borrower's intra-group debts. This assumption was causative of the claimant's decision to advance money to the borrower.

[34] Thus, the factor that renders the enrichment unjust is the non-fulfilment of the lender's expectation as to the security, forming the basis of its decision to advance funds. This is to be contrasted with the *Burston* approach, which requires an inquiry, akin to the process of contractual construction, into the question of whether the precise terms of the bargain between the lender and borrower have been fulfilled.

[35] I agree with the judge below that in considering the question of whether the lender got what it bargained for, the bargain is to be interpreted in what she called, in [65] to [67] and [72], "the wider sense of 'the envisaged transaction'", analysing Lord Hoffmann's words "an essential part of the transaction under which it paid the money" in *Banque Financière*, at p235A.

[36] The enrichment of the claimant is unjust because the bank funded the repayment of the Halifax charge on the basis that it would obtain a legal charge. Because of the failure to register under the 2002 Act, it obtained only a subsequent equitable charge. It does not matter that the borrower has performed the terms of the bargain between the borrower and the lender.

[37] An argument could be advanced to the effect that the lender advances money on the faith only of a promise to supply a registrable charge, or, as Mr Lander put it, to supply documentation capable of being rendered a legal charge by the act of the lender itself. However, that argument potentially falls foul of Lord Hoffmann's warning (in *Banque Financière*, at p232A) against importing contractual requirements into the principles governing non-contractual forms of subrogation. The emphasis is, as the judge below said, in [64], on the overall position and the practical aspects of the enrichment of the party opposing subrogation.

Subverting the policy of the Land Registration Acts?

[38] Finally, I will deal briefly with Mr Lander's submission that to allow subrogation would be to subvert the policy of the 2002 Act. What this submission boiled down to was that the Act forms some sort of conclusive code, enabling those that lend to rely on the position on the face of the register. If that was correct, it would be difficult to claim subrogation in any case of registered land. It would, moreover, be difficult to reconcile with the clear view of the Court of Appeal as expressed in *Cheltenham & Gloucester*. In my judgment, the policy of the Act does not encroach on the principle of unjust enrichment. I agree with the judge below (for the reasons she gave in [85] to [88]) that any unfair consequences of subrogation can be dealt with by way of a defence founded on change of position.

[39] This is quite different from the cases concerning moneylenders where subrogation was denied on public policy grounds: see *Orakpo*, overruling *Congresbury Motors**.

Decision

[40] I therefore find that the appeal fails. In my judgment, HH Judge Marshall QC came to the right decision, that the bank is entitled to be subrogated to the Halifax charge, for the reasons that she gave.

Appeal dismissed.

---

* Editor's note: *Congresbury Motors Ltd* v *Anglo-Belge Finance Co Ltd* [1971] Ch 81

NEGLIGENCE

Nouri *v* Marvi and others

A

COURT OF APPEAL

[2010] EWCA Civ 1107

14 October 2010

RIX AND PATTEN LJJ AND SIR MARK WALLER

Estates Gazette 18 December 2010

[2010] 50 EG 64

B *Negligence — Solicitor — Limitation period — Limitation Act 1980 — Party purporting to be owner of property instructing solicitor to act on transfer — Transfer fraudulent — Actual owner alleging breach of duty of care by solicitor in failing to verify identity of client — Whether claim statute-barred under section 2 of 1980 Act — Whether cause of action accruing on completion of fraudulent transfer or on later registration of title by fraudulent transferee — Whether solicitor owing continuing duty of care after completion*

The appellant was the registered owner of a flat. In April 2001, his signature was forged on a fraudulent transfer of the flat to M, who obtained the transfer by purporting to be the appellant and instructing
C the respondent firm of solicitors to act on the sale. The appellant became aware that M had earlier obtained an unauthorised remortgage of the property but took no steps against him. M obtained registration of the property in his name in July 2001 and sold it to an unconnected third party in January 2003. The appellant's claim for rectification of the register was dismissed: see [2005] EWHC 2996 (Ch); [2006] 1 EGLR 71; [2006] 12 EG 196.

In July 2007, the appellant brought proceedings against various parties, including a claim against the respondent for negligence. He contended that, on receipt of the instructions from M in the appellant's name, it had assumed a responsibility to the appellant to handle the
D sale in a competent manner, including an obligation to confirm its instructions directly with the client and to satisfy itself of his identity by a face-to-face meeting and sight of his passport.

On a trial of a preliminary issue, the claim was held to be statute-barred for expiry of the six-year limitation period under section 2 of the Limitation Act 1980. The judge rejected the appellant's contention that the cause of action had not accrued until M obtained registration of title in July 2001; he held that the appellant had suffered actual damage on the completion of the sale to M. It had constituted an immediate blot on the appellant's title, producing an immediate financial detriment to him. He further rejected the appellant's alternative case that the respondent
E had been under a continuing duty of care, persisting after completion, to take steps to prevent the appellant from losing his registered title; that claim relied on a post-completion request by the respondent for sight of a passport, which had resulted in the production of a passport that was not the appellant's. The appellant appealed.

***Held*: The appeal was dismissed. (1) The loss suffered by the appellant on the completion of the transfer to M was actual loss and was not merely theoretical or contingent. Had a potential purchaser of the flat been told of the forged transfer in the period prior to registration of M's title, that would have reduced the price that he was willing to pay since the forged transfer, although of no legal effect, exposed any purchaser from the appellant to the risk of possible litigation. Accordingly, there had been a diminution in the** F **open market value of the flat following completion in April 2001. The appellant had therefore suffered actual damage, and his cause of action had accrued, at that date. It was irrelevant for the purposes of section 2 whether the appellant could realistically have discovered the fraud before M's registration of title in July 2001. Time starts to run under section 2 even if the claimant is unaware of the circumstances that have led to the cause of action; the existence of actual damage depends, not on the claimant's state of knowledge with regard to the breach of duty but on whether the breach has caused actual loss. Therefore, in determining whether the appellant could have maintained an action for damages against** G **the respondent following completion of the sale to M, it had to be assumed that: (i) the appellant had been aware of the forged transfer at that date; (ii) he was under a corresponding duty to alert a potential purchaser to the problem; and (iii) the open market value of the flat was accordingly the sum that a purchaser with knowledge of the forged transfer would pay. Although section 14A of the 1980 Act provides an alternative time limit of three years from the date of knowledge, that did not assist the appellant because he had had the requisite knowledge of the fraudulent sale to M by 2002 and the three-year time limit had therefore also expired. (2) There were no special facts to suggest that the** H **respondent had assumed a continuing duty to the appellant that survived the completion of the transaction for which it had been retained. It had been under no contractual duty to take specific steps in respect of the appellant's property or title. The receipt of the passport post-completion was indicative only of its general lack of care and diligence rather than any assumption of continuing responsibility towards the appellant. The respondent had, before the date on which M obtained registration of title, done all that was necessary to complete the sale and was not obliged to supervise the registration of that title or to carry out further investigations.** J **Moreover, even in the event of a continuing duty, the cause of action would still have accrued in April 2001, when loss was first suffered as a consequence of breach.**

The following cases are referred to in this report.

Axa Insurance Ltd (formerly Winterthur Swiss Insurance Co) v *Akther & Darby Solicitors* [2009] EWCA Civ 1166; [2010] 1 WLR 1662

Bell v *Peter Browne & Co* [1990] 2 QB 495; [1990] 3 WLR 510; [1990] 3 All ER 124, CA

Cartledge v *E Jopling & Sons Ltd* [1963] AC 758; [1963] 2 WLR 210; [1963] 1 All ER 341; [1963] 1 Lloyd's Rep 1, HL; [1962] 1 QB 189; [1961] 3 WLR 838; [1961] 3 All ER 482; [1961] 2 Lloyd's Rep 61, CA K

Forster v *Outred & Co* [1982] 1 WLR 86; [1982] 2 All ER 753, CA

Law Society v *Sephton & Co* [2006] UKHL 22; [2006] 2 AC 543; [2006] 2 WLR 1091; [2006] 3 All ER 401

Midland Bank Trust Co Ltd v *Hett, Stubbs & Kemp* [1979] Ch 384; [1978] 3 WLR 167; [1978] 3 All ER 571

Nouri v *Marvi* [2009] EWHC 2725 (Ch); [2010] PNLR 7

Pirelli General Cable Works Ltd v *Oscar Faber & Partners* [1983] 2 AC 1; [1983] 2 WLR 6; [1983] 1 All ER 65; [1983] 1 EGLR 135; (1982) 265 EG 979, HL

Wardley Australia Ltd v *State of Western Australia* (1992) 109 ALR 247; (1992) 175 CLR 514

This was an appeal by the appellant, Amir Nouri, from a decision of Norris J, sitting in the Chancery Division of the High Court, dismissing a claim against the respondent, Athi Kulisra Smith (a firm), for damages for professional negligence on the ground that the claim was statute-barred under section 2 of the Limitation Act 1980.

Philip Jones (instructed by Mackrell Turner Garrett) appeared for the appellant; Ben Lynch (instructed by Davies Arnold Cooper LLP) represented the respondent.

Giving judgment, **Patten LJ** said:

Introduction

[1] This appeal raises the familiar question of when a cause of action in tort has accrued so as to start time running for the purposes of the Limitation Act 1980 (the 1980 Act). Section 2 of that Act provides that:

> An action founded on tort shall not be brought after the expiration of six years from the date on which the cause of action accrued.

[2] A cause of action in tort (unlike one in contract) accrues only when the claimant suffers actual damage in the sense of measurable loss that can be quantified in money terms and made the subject of an award of damages at the relevant date. In *Forster* v *Outred & Co* [1982] 1 WLR 86, in the Court of Appeal, Stephenson LJ, at p94C-D, accepted the definition of actual damage put forward by Mr Stuart-Smith (as he then was) in these terms:

> What is meant by actual damage? Mr Stuart-Smith says that it is any detriment, liability or loss capable of assessment in money terms and it includes liabilities which may arise on a contingency, particularly a contingency over which the plaintiff has no control; things like loss of earning capacity, loss of a chance or bargain, loss of profit, losses incurred from onerous provisions or covenants in leases. They are all illustrations of a kind of loss which is meant by "actual" damage. It was also suggested in argument, and I would accept it, that "actual" is really used in contrast to "presumed" or "assumed." Whereas damage is presumed in trespass and libel, it is not presumed in negligence and has to be proved. There has to be some actual damage.

[3] The claimant, Mr Nouri, was from 1996 the registered proprietor of a leasehold flat at 8 Bron Court, Bronsbury Road, London, NW6 (the flat). He bought the flat when he came to England to study English and to enrol on a business studies course at Westminster University. It was purchased for £69,000 with the benefit of a mortgage from Barclays Bank for £39,000 and with £30,000 provided by his father.

[4] In 1998, the first defendant, Mr Amir Marvi, came to London and moved into the flat. He was a family friend. In July 1999, Mr Nouri left the UK but Mr Marvi remained living in the flat. By then, a further loan of £30,000 had been obtained from Barclays Bank on the security of the property. The arrangements with Mr Marvi were that he would pay sufficient moneys into the claimant's Barclays Bank account to meet the mortgage payments and the other outgoings on the flat. Mr Nouri also executed a power of attorney in favour of a Mr Musawi, a friend of his father's.

[5] After Mr Nouri left, the mortgage payments fell into arrears and, in October 1999, a notice of default was issued by the bank. These and other arrears were cleared in August 2000 when the Barclays Bank mortgage was redeemed and replaced by a new mortgage in favour of iGroup Mortgages. In earlier High Court proceedings before HH Judge Rich QC (which I will come to shortly), evidence was given that Mr Marvi had agreed to clear the indebtedness to Barclays using money provided by his family. However, the remortgage to iGroup was obtained by Mr Marvi in Mr Nouri's name whose signature on the mortgage deed was forged.

[6] In around December 2000, Mr Nouri and his father instructed Mr Musawi to obtain confirmation from Barclays Bank that its charge had been redeemed and he instructed a firm of solicitors (Nattu & Co) to do this. They also asked the bank for the return of Mr Nouri's life policy that had been charged to the bank as part of the security. By March 2001, the bank had told the solicitor that the charge had been redeemed but that the land certificate had been sent to another firm of solicitors. Mr Musawi then instructed Nattu & Co to obtain office copy entries of the registered title, which it did by 3 May 2001. These, of course, revealed that the flat was now charged to iGroup under a mortgage that Mr Nouri had not authorised.

[7] The mortgage to iGroup secured a loan from them in the sum of £105,000. After payment of the sums due to Barclays Bank, around £40,000 remained that Mr Marvi misappropriated. Not, however, content with this he then embarked on the steps that have led to this claim in negligence by Mr Nouri against the first defendants, Athi Kulisra Smith (AKS), a firm of solicitors.

[8] On 21 December 2000, holding himself out as Mr Nouri, Mr Marvi instructed AKS to act for him on a sale of the flat to Mr Marvi. Mr Marvi also instructed another firm of solicitors to act for him as the purchaser. On 2 April 2001, there was a simultaneous exchange of contracts and completion that resulted in AKS releasing to Mr Marvi's solicitor a transfer again containing Mr Nouri's forged signature. On 4 July 2001, Mr Marvi, using the transfer, obtained registration of the flat in his own name. The purchase price of the flat was stated to be £179,000. As part of the fraudulent transaction Mr Marvi was able to redeem the iGroup mortgage with a new loan from the Kensington Mortgage Co of £152,090. The balance necessary to complete the purchase was some £29,731.50 that was provided by a Dr Khan. On completion, a balance of £55,518.47 was due to Mr Nouri as vendor after the payment of expenses. Mr Marvi forged further letters of instruction that resulted in £30,731.50 of this sum being repaid to Dr Khan and the balance of £24,786.97 being returned to Mr Marvi. Subsequently, on 7 January 2003, the flat was sold on to an unconnected third party, a Mr Razaq, for the sum of £223,000. Mr Razaq was then registered as the proprietor.

[9] Despite becoming aware of the unauthorised mortgage to iGroup, Mr Nouri took no steps against Mr Marvi. This enabled the fictitious sale of the flat to Mr Marvi to be completed and his title registered and also allowed Mr Marvi to go on to sell the flat to Mr Razaq. It was not until April 2003 that Mr Nouri applied belatedly to the Land Registry for rectification of the register. They directed him to take proceedings against Mr Marvi, Mr Razaq and the Kensington Mortgage Co for rectification. These proceedings were concluded on 18 October 2005 when Judge Rich QC refused to order rectification largely on the basis that the registration of title in favour of Mr Marvi and subsequently Mr Razaq had occurred due to Mr Nouri's failure to take proper care of his own interests.

[10] There has been no appeal against that judgment. Instead, on 2 July 2007, Mr Nouri commenced these proceedings against Mr Marvi, the Land Registry and AKS. The claim against AKS is for negligence. The claim cannot be based in contract on the retainer because there was none. Instead, it is alleged that, on receipt of the instructions from Mr Marvi in Mr Nouri's name, AKS assumed a responsibility to the real Mr Nouri to handle the sale in a competent manner and knew that if it failed to do so Mr Nouri might suffer loss and damage. Its duty of care to him encompassed not only the conveyancing transaction itself but also an obligation to confirm their instructions directly with the client and to satisfy itself of his identity. This would include having a face-to-face meeting with him and verifying his identity from his passport. The case against AKS is that none of these steps was taken and that sight of the client's passport was not requested until 15 May 2001. When it was requested and produced, AKS failed to ensure that the passport was that of Mr Nouri.

[11] It is also alleged that the firm failed to take note of the advice contained in the Law Society's Green Card warning on property fraud. The warning signs generated by many of the unusual features of the transaction were simply ignored. As a result, AKS failed to warn Mr Nouri or the Land Registry, either before or after the sale, that this was or might be a fraudulent transaction and consequently allowed Mr Nouri to be removed as the registered proprietor and Mr Marvi to be registered in his place.

[12] It is common ground that the registration of Mr Marvi as the proprietor of the flat did cause actual damage to Mr Nouri. As a result of the provisions of section 69 of the Land Registration Act 1925, title was transferred to Mr Marvi by the act of registration. Mr Nouri therefore

A lost his title and would have had, at the very least, to commence rectification proceedings against Mr Marvi and his mortgagees in order to restore himself to the register. The six-year limitation period for the claim in negligence against AKS must therefore have begun to run by 4 July 2001 at the latest. It does not depend on the subsequent sale of the flat to Mr Razaq. However, AKS in its defence has taken a limitation point that time began to run earlier when the fictitious sale to Mr Marvi was completed on 2 April 2001. If this is correct, the limitation period expired on 2 April 2007 prior to the issue of the claim form in these proceedings on 2 July 2007.

[13] On 3 June 2009, Master Eyre ordered the trial of a preliminary
B issue on this limitation point. It was tried by Norris J on 7 October 2009: see [2009] EWHC 2725 (Ch)*. He found that the claim in negligence was statute-barred because Mr Nouri did suffer actual damage on completion of the sale and was not simply exposed to the contingent loss of his title that would depend on whether Mr Marvi applied for registration as proprietor based on the forged transfer released to his solicitor on completion.

[14] The judge expressed his reasoning as follows:

> 15. Although for present purposes I must and do focus upon the questions raised by forged transfers in registered conveyancing, the situation seems to me to raise the same sort of question as would be raised if a solicitor had negligently
> C released a genuine transfer before completion, or if a solicitor had negligently released a forged conveyance in unregistered conveyancing (where the vendor's title would never be defeated by the purchaser but might be defeated by a third party bona fide purchaser for value).
>
> 16. One tool that is sometimes used in circumstances of this sort is to apply what is called "the Bingham test", that proposed by Bingham LJ in *DW Moore* v *Ferrier* [1988] 1 WLR 267 at 282, namely that if, on the pleaded facts, an action for breach of contract were brought, would anything more than nominal damages be awarded? I do not think that this tool is of particular utility where the origin of the duty is not founded in any contract between Mr Nouri and the third defendants. Instead, I think I must postulate that on 1st May 2003 Mr Nouri laid the then facts before another solicitor and asked him to advise
> D what his position was. It seems to me unlikely that a solicitor would have said to Mr Nouri, "Relax. You are no worse off than if the third defendants had properly performed their duty. You will suffer no harm until the forged transfer is registered. Do nothing in the meantime. Wait and see if that happens. Come back if it does." I do not think that that is an adequate analysis of the situation in which Mr Nouri found himself.
>
> 17. It seems to me that until 2nd April 2001 Mr Nouri had an inviolable registered title. On 2nd April 2001 his solicitor (it must be assumed negligently) placed in the hands of a third party outside the control of Mr Nouri a document which could be used to defeat Mr Nouri's title. That constituted an immediate blot on Mr Nouri's title. It was not the position in which he ought to have found himself if the solicitor had properly performed the duties which the Particulars of Claim say should have been discharged. Mr Nouri would have been advised
> E that he was actually (not just potentially) worse off because the forged transfer was in circulation; the adviser would have gone on to say that it would be possible to prevent any consequences flowing from the circulation of the forged transfer and to restore the title to its former inviolable state, but that the removal of that detriment would cost money.
>
> 18. I would hold that, on the assumed facts and presumed law, the third defendants failed to preserve or protect an asset of Mr Nouri's which could and should have been protected by the proper performance of their duties in relation to a transaction in which they assumed a responsibility in relation to Mr Nouri's title. Putting into the hands of Mr Marvi a forged transfer constituted an immediate detriment.
>
> 19. In *Bell* v *Peter Browne*, to which I have already referred, a solicitor
> F failed to fulfil his obligations to his client in two respects. First, he failed to secure a formal document recording the bargain that had been made between his client husband and the wife, to whom the matrimonial home had been transferred. Secondly, he failed to protect by a caution the husband's interest under the agreement, given that title had been vested in the wife. The husband's interest could have been protected by a caution, and that caution could have been entered on the register at any time until the wife sold the house and the husband's interest was finally defeated. Nicholls LJ held that the second failure (to register the caution) occasioned immediate loss. In an oft-cited passage, which although considered extensively in subsequent authorities has never been criticised or doubted, Nicholls LJ said:
>
> "In considering whether damage was suffered in 1978 one can test the matter by G considering what would have happened if in, say, 1980 the plaintiff had learned of his solicitors' default and brought an action for damages. Of course, he would have taken steps to remedy the default. But he would have been entitled at least to recover from the defendants the cost incurred in going to other solicitors for advice on what should be done and for their assistance in lodging the appropriate caution. The cost would have been modest, but not negligible."
>
> In other words, he held that there was an immediate loss, even though there was a contingent greater loss (the defeasance of the husband's interest).
>
> 20. I would hold that the consequence of allowing into circulation a document which put at risk Mr Nouri's title was capable of remedy and that the costs of ameliorating that immediate detriment form a ready measure of the financial consequences of the solicitor's breach of duty. I would accordingly H hold that the cause of action in negligence was complete on 2nd April 2001 when the forged transfer was put into circulation. Mr Nouri suffered the detriment of having a blot upon his title, a detriment which is to be measured by the cost of removing the detriment.

[15] Mr Nouri now appeals to this court. There are two grounds of appeal. First, it is said that the release of the forged transfer on 2 April 2001 did not constitute an immediate loss to the claimant. Actual damage was not suffered until registration of Mr Marvi's title on 4 July 2001. Second, he contends that the duty of care owed by AKS was a continuing duty to take steps to prevent Mr Nouri from losing his registered title. This duty subsisted after completion and continued J at least until 4 July 2001. The judge was wrong to find (as he did) that no such continuing duty existed. AKS has filed a respondent's notice in which it seeks to uphold the judge's decision both for the reasons that he gave and on the additional ground that the value of the flat was materially diminished by the completion of the sale to Mr Marvi and the release of the forged transfer because any prospective purchaser, once aware of the existence of the transfer, would have known that there was a risk of his becoming embroiled in litigation to secure his title and was at risk of Mr Marvi being able to register himself as proprietor before the purchaser would be able to register his own title.

Actual damage K

[16] The starting point must be to identify the nature of the damage that Mr Nouri is alleged to have suffered as of the completion of the sale to Mr Marvi on 2 April 2001. The judge's analysis was that the transfer constituted a blot on Mr Nouri's title because it gave Mr Marvi the means to deprive Mr Nouri of registered ownership of the flat. Although steps could have been taken to prevent registration by Mr Marvi by obtaining an injunction against him and, in particular, by arranging for the registration of an inhibition against Mr Nouri's title, those steps would have cost money. There was therefore an immediate financial detriment to Mr Nouri. L

[17] Some support for this analysis can be found in the judgment of Nicholls LJ in *Bell* v *Peter Browne & Co* [1990] 2 QB 495, which was relied on by the judge. In that case, a husband agreed with his wife on divorce that a property registered in their joint names should be transferred into the wife's sole name and sold and that he should receive one-sixth of the net proceeds of sale. He arranged with his solicitor that his beneficial interest would be protected by a trust deed but no steps were taken in this respect. Instead, the solicitor allowed the house to be transferred to the wife who subsequently sold it and spent all the proceeds. The claim was dismissed on the ground that the limitation period began to run when the solicitor allowed the property M to be transferred to the wife without the protection of the trust deed and suitable entries being made on the register of title even though the omission could have been remedied by the registration of a caution at any time up to the date of sale. The husband had contended that no damage occurred until the sale took place.

[18] Nicholls LJ had to consider two alleged heads of negligence that bear a striking similarity to the claim against AKS in this case. There was the solicitor's failure to ensure that the agreement about the husband's one-sixth share of the proceeds of sale was properly recorded in a trust deed and there was its failure to protect that interest against the wife and any third party by the registration of a caution. However, in his view, actual loss was suffered by the husband as from the date of the

---

* Editor's note: Reported at [2010] PNLR 7

transfer to the wife. From then on the husband's ability (in the absence of a trust deed) to secure his share of the proceeds of sale depended on the wife's co-operation. She had title and could deny him his share. Second, although his rights could have been protected by a caution up to the date of any sale, the solicitor's file was closed on completion of the transfer so that the consequences of not having registered a caution were unlikely to be reviewed and would not have become apparent until after a sale. It should be noted in this regard that Nicholls LJ rejected the existence of any continuing duty in that case and I shall return to this when I consider the second ground of appeal.

[19] At p503, Nicholls LJ set out his answer to the question of whether these breaches of duty had caused any actual damage as of the date of completion rather than when the property was subsequently sold. He concluded that they had, see the passage cited by Norris J, in [19] of his judgment: in [14] above.

[20] The other members of the court came to the same conclusion but based their judgments on the diminution in value of the husband's interest in the property. At p510E-F, Beldam LJ said that:

> Due to the defendants' negligence, the plaintiff parted with his legal estate in the property conveyed to his wife in exchange for an equitable interest in the proceeds of sale. That equitable interest until secured by a charge or acknowledged by a deed of trust was clearly less valuable to the plaintiff. Unprotected against the interests of third parties by registration of a charge or of a caution, it was less valuable still. I consider therefore that the plaintiff's cause of action arose when he parted with his property or at the latest at the time when the careful solicitor would have affected registration either of a charge or of a caution. For those reasons I would dismiss this appeal.

[21] Mustill LJ, at p513C-D, put it this way:

> As to the claim in tort, I have little to add. The transaction caused the plaintiff to exchange his valid legal estate for an equitable interest in the proceeds of sale which was dependent on the goodwill and solvency of the wife unless and until protected by a formal declaration of trust and the lodging of a caution. The failure to see that these steps were taken promptly meant that the plaintiff was actually, and not just potentially, worse off than if the solicitors had performed their task competently. The sale in 1986 simply meant that the breach and its consequences were irremediable.

[22] Although expressed differently, I do not believe that the analysis of damage contained in the passage quoted from the judgment of Nicholls LJ is different in principle from that of the other two members of the court. In that case, the husband had not taken steps to secure his title to his share of the proceeds of sale by registration of a caution and had not therefore incurred the expense to which Nicholls LJ refers. The only loss that he can have suffered was a loss of value in his rights to the proceeds of sale. The costs of putting matters right is therefore simply a convenient way of calculating what the minimum amount of that loss would be, just as in a claim for damages for breach of a repairing covenant in a lease the cost of repairs is often taken to be the amount of the diminution in value of the reversion that the landlord has suffered. This is, I think, what Norris J had in mind in the passage in [21] of his judgment quoted above.

[23] The question therefore in this case is whether Mr Nouri suffered an equivalent diminution in the value of the flat owing to the completion of the fraudulent sale to Mr Marvi. Because Mr Nouri took no steps to prevent Mr Marvi from registering his title, he cannot claim for the cost of doing so. However, he will have had an actionable claim in negligence against AKS if post-completion the asset in his hands had become less valuable.

[24] Mr Philip Jones submitted that until the registration of Mr Marvi's title on 4 July 2001 any loss by Mr Nouri was purely theoretical and contingent. The forged transfer might never have been used and if Mr Nouri had wished to sell the flat in the meantime the intending purchaser would have acquired priority over Mr Marvi simply by making the customary search of the register. To support this argument he took us to the decision of the House of Lords in *Law Society* v *Sephton & Co* [2006] UKHL 22; [2006] 2 AC 543, where the House had to consider whether a purely contingent liability created by the negligence of solicitors caused actual damage at the date when the liability was incurred or only when the contingency was fulfilled and the loss actually materialised.

[25] Following the decision of the High Court of Australia in *Wardley Australia Ltd* v *State of Western Australia* (1992) 175 CLR 514, the House of Lords held that a purely contingent liability did not necessarily cause actual damage at the time when it was incurred. In order to argue for the earlier date, it was necessary to show some immediate measurable loss or diminution in the value of the claimant's assets that was attributable to the imposition of the liability. In *Law Society*, the potential liability on the Law Society's compensation fund caused by a solicitor's fraud was held not to have been actual damage so as to set time running for a claim in negligence. The effect of the solicitor's conduct was to expose the Law Society to a potential liability. It was not a case in which the solicitor's duty was to protect his client's property (as in this case) or to enable a client to acquire property through a transaction for which the solicitor was responsible.

[26] The speeches are, however, useful for the guidance that they give as to the basis of the decisions in cases such as *Bell*. In [22] of his speech, Lord Hoffmann said:

> Thus cases like *Bell* v *Peter Browne & Co* [1990] 2 QB 495 and *Knapp* v *Ecclesiastical Insurance Group plc* [1998] PNLR 172 are readily explicable as cases in which the damage was the difference between the plaintiff's position as it was and as it would have been if the defendant had performed his duty and in which it was possible to infer that the plaintiff's failure to get what he should have got from a bilateral transaction was quantifiable damage, even though further damage which might result from the flaw in the transaction was still contingent. The plaintiff had paid money, transferred property, incurred liabilities or suffered diminution in the value of an asset and in return obtained less than he should have got. But these authorities have no relevance to a case in which a purely contingent obligation has been incurred.

[27] Lord Walker, in [48], said:

> In all these cases the claimant has as a result of professional negligence suffered a diminution (sometimes immediately quantifiable, often not yet quantifiable) in the value of an existing asset of his, or has been disappointed (as against what he was entitled to expect) in an asset which he acquires, whether it is a house, a business arrangement, an insurance policy, or a claim for damages. Your Lordships have not, I think, been shown any case in which the imposition on a claimant of a purely personal and wholly contingent liability, unsecured by a charge on any of the claimant's assets, has been treated as actual loss. That would have been the position if the claimant in the *Forster* case [1982] 1 WLR 86 had given a personal covenant guaranteeing her son's debts (which she seems not to have done she paid them simply to prevent enforcement of the security on her farm) and if she had not given any security over any of her own assets.

[28] The application of the *ratio* in *Law Society* to the facts of particular cases of contingent liability has not been without its difficulties: see, for example, *Axa Insurance Ltd (formerly Winterthur Swiss Insurance Co)* v *Akther & Darby Solicitors* [2009] EWCA Civ 1166; [2010] 1 WLR 1662. However, further analysis of those difficulties is unnecessary for the disposal of this appeal. This is not a case concerning contingent liabilities. The only issue is whether Mr Nouri suffered loss in respect of the value of his property. If the flat diminished in value at the date of completion, actual damage was clearly sustained by Mr Nouri as at that date.

[29] There was no valuation evidence before the judge directed to this issue but Mr Jones accepted that had a potential purchaser been told of the forged transfer in the period prior to registration of Mr Marvi's title, this would undoubtedly have led to a diminution in the price that he would have been willing to pay. Although the contract and the transfer to Mr Marvi were of no legal effect, they did expose any purchaser to the risk of possible litigation even if that risk was more apparent than real. This is the point taken in the respondent's notice.

[30] On this basis, there will have been a diminution in the open market value of the flat following completion on 2 April 2001. Mr Jones contested this conclusion because it proceeds, he said, on a false assumption: that is, that a potential purchaser would have been made aware of the difficulties caused by the forged transfer. He submitted that there was no realistic possibility of Mr Nouri having discovered

A the fraud before registration on 4 July and therefore no question of his being under any obligation to disclose those matters to a potential purchaser. The facts were that he remained in ignorance of the fraud until much later.

[31] I do not accept this analysis. It is well established that a cause of action in tort can accrue for the purposes of the 1980 Act without the claimant being aware of it. The decisions of the House of Lords in *Cartledge* v *E Jopling & Sons Ltd* [1963] AC 758 and *Pirelli General Cable Works Ltd* v *Oscar Faber & Partners* [1983] 2 AC 1* led to limited changes to the Act to provide alternative time limits of three years running from the date of knowledge. These are now contained in
B sections 11 and 14A of the 1980 Act.

[32] In this case, Mr Nouri was not assisted by section 14A because he had the requisite knowledge of the fraudulent sale to Mr Marvi by 2002. However, the provision of these alternative time periods to address the problem of lack of knowledge has left intact the principle that time will run under section 2 even if the claimant is unaware of the circumstances that have led to it. The existence of actual damage for these purposes does not therefore depend on the claimant's state of knowledge in respect of the breach of duty or its consequences but on whether the breach has in fact caused actual loss.

C [33] For this purpose, the correct hypothesis must be to ask whether Mr Nouri could have maintained an action for damages against AKS following completion of the sale to Mr Marvi. That requires one to assume that he was aware of the breach and to assess the effect of that on his property or other assets as at that date. Consistently with this hypothesis, any diminution in the value of the flat attributable to the breach has to be determined on the basis that Mr Nouri was aware of the forged transfer and was under a corresponding duty to alert a purchaser to that problem. The flat was unsaleable without such disclosure. It must therefore follow that its open market value was what a purchaser would pay for it with knowledge of the forged transfer. There was therefore actual damage suffered by Mr Nouri as of 2 April 2001.

D

Continuing duty

[34] The second ground of appeal is that the judge was wrong to reject the claim based on a continuing duty. There are no particular facts pleaded to support this claim beyond the fact that AKS requested and obtained sight of the passport after completion. Mr Nouri is unaware of when the passport was received and does not allege that this occurred after 2 July 2001: that is, within the six-year limitation period. However, what is said is that the solicitor's duty to protect Mr Nouri's title subsisted until at least 4 July 2001.

E [35] The principal authority relied on by Mr Jones to support the claim that the duty was a continuing one is the decision of Oliver J in *Midland Bank Trust Co Ltd* v *Hett, Stubbs & Kemp* [1979] Ch 384. The solicitor's breach of duty consisted of a failure to register an option as an estate contract under the Land Charges Act 1925. The failure to register it enabled the guarantor of the option to convey the property to his wife free of the option. Oliver J held, at p438E, that there was no general or continuing duty on the part of the solicitors to consider the enforceability of the option but that on the facts of that case:

> The defendants here never treated themselves as functi officio in relation to the option. They kept the document on Geoffrey's behalf in their strongroom.
> F They opened a file relating to the matter. They were consulted about it at intervals over the next 6½ years. In my judgment the obligation to register which they assumed when they were first consulted continued to bind them. It was an obligation to protect the interest from third parties by registration and without their client's knowledge they failed to perform it until it ceased to be effectively capable of performance on August 17, 1967. It seems to me that it was then that the contract was broken once and for all.

[36] In Bell, the Court of Appeal treated the decision of Oliver J as very much dependent on its own facts. Mustill LJ, at pp512-513C, said:

> Certainly, a solicitor may have a continuing retainer from his client, and no doubt there are retainers which require the solicitor to be constantly on watch for new sources of potential danger, and to take immediate steps to nip them in the bud. I confess however that I cannot see the relationship between the present parties in any such light. The proposition entails that the defendants have two duties, one express and the other implied. The express duty would be to perform the task for which they were retained and paid, namely to put into effect in a legally appropriate manner the informal arrangement between the plaintiff and his wife. The second duty, implied and presumably gratuitous, and commencing immediately after the last moment when a careful solicitor would have taken the necessary steps to formalise and protect his client's interest in the future proceeds of sale, would be to exercise continuing vigilance to discover any mistake which they, themselves, might have made, and then to busy themselves in putting it right. Evidently this obligation would continue up to, but not beyond, the time when the mistake became irretrievable. I find it impossible to imply such a strange obligation from the mundane facts of the present case; and equally improbable to suppose that if it did exist the obligation would be broken at any time other than the time when the mistake should have been discovered and put right: namely, straightaway. To my mind the solicitors were employed to complete the transaction, and to complete it within the appropriate time. No more than that. Any further steps taken or not taken would relate to mitigating the consequences of a breach which had already occurred.
>
> As to the claim in tort, I have little to add. The transaction caused the plaintiff to exchange his valid legal estate for an equitable interest in the proceeds of sale which was dependent on the goodwill and solvency of the wife unless and until protected by a formal declaration of trust and the lodging of a caution. The failure to see that these steps were taken promptly meant that the plaintiff was actually, and not just potentially, worse off than if the solicitors had performed their task competently. The sale in 1986 simply meant that the breach and its consequences were irremediable.

[37] Beldam LJ, at p509A-F, also thought that the breach occurred when the property was conveyed without the necessary precautions being taken to secure the husband's interest:

> Whether in any particular case there has been a failure by a solicitor to exercise that degree of care and skill depends upon the circumstances of each case and on the finding of the court as to how the competent solicitor would discharge his duty in those circumstances. Where, as in the present case and in *Midland Bank Trustees Co Ltd* v *Hett, Stubbs & Kemp* [1979] Ch 384 and Bean v Wade, 2 TLR 157, the duty is to protect the interests of the client by giving notice or by registration of a charge or caution, I cannot believe that a competent solicitor would regard it as a satisfactory discharge of that duty to postpone action for months or years. Indeed there would be many cases when even postponement for a day would be regarded as negligent and a breach of duty. Further in a case in which it is necessary to take precautions to preserve a client's interest against the failure by another to fulfil an obligation whether by requiring the execution of a charge or a deed of trust, it seems to me clear that the exercise of reasonable care requires the transactions to be contemporaneous. I do not believe, therefore, that it is correct to say that in a case such as the present the obligation of the solicitor in contract was one which was capable of performance at any time up until the property was sold. If by chance he realised his mistake in time, he could take steps to prevent the consequences of breach of the obligation and to that extent repair the damage caused. I believe that the defendants were in breach of their duty when they permitted the plaintiff to execute a conveyance of his interest in the freehold property without securing by deed of trust or charge an interest in the proceeds of sale. They ought to have appreciated that the plaintiff had only executed at that stage in escrow and should have insisted on the performance by his wife of the condition to which she had agreed. Whilst it may not have been a breach of duty to fail to register the plaintiff's interest immediately or within a day or two on the facts of this particular case, nevertheless experience shows that such a duty should be discharged as soon as is practicable. Consequently, although the plaintiff only suffered nominal damage until the property was sold in July 1986, his cause of action arose when he conveyed his interest in the property to his former wife without proper precautions being taken to protect his future interest in the proceeds of sale.

[38] It seems to me that the same reasoning applies in this case. There are no special facts to suggest that the solicitor assumed a continuing duty to Mr Nouri that survived the completion of the transaction. This was not a case in which it was under a contractual duty to take specific steps in respect of Mr Nouri's property or title as in *Midland Bank*. Although it obtained the passport after completion, that was simply indicative of its general lack of care and diligence rather

G H J K L M

* Editor's note: Also reported at [1983] 1 EGLR 135

A than of any assumption of continuing responsibility towards Mr Nouri. Before 2 July 2001, it had done all that it considered necessary to complete the sale. It was not for it to supervise the registration of Mr Marvi's title or to carry out any further investigations once it had seen the passport.

[39] However, even if the duty owed by AKS was a continuing one of the kind alleged, it makes no difference. The cause of action in tort accrued when loss was first suffered as a consequence of the breach of duty. As explained earlier, the damage to the value of Mr Nouri's property was caused when completion took place. Although after 4 July 2001 it became necessary to seek rectification of the register in order to
B displace Mr Marvi as proprietor, which was obviously more expensive than merely registering an inhibition to prevent such registration taking place, this is simply an enhanced measure of damages that takes into account the full consequences of the breach and might be disputed on C
the ground that Mr Nouri had failed to mitigate. However, the loss remains the same. In either case, one is still dealing with a diminution in the value of the flat, which was Mr Nouri's only relevant asset. In these circumstances, the loss attributable to the breach of duty (continuing or not) would still have started time running on 2 April 2001.

[40] I would therefore dismiss this appeal.

Sir Mark Waller said:

[41] I agree.

Rix LJ said: D

[42] I also agree.

Appeal dismissed.

PROCEDURE

Gibbon v Manchester City Council; LG Blower Specialist Bricklayer Ltd *v* Reeves and another

COURT OF APPEAL

[2010] EWCA Civ 726

25 June 2010

SIR ANTHONY MAY P AND CARNWATH AND MOORE-BICK LJJ

Estates Gazette 11 September 2010

[2010] 36 EG 120

Civil practice and procedure — CPR 36 — Whether rejection of Part 36 offer to settle proceedings rendering offer incapable of acceptance at later date — Whether implied withdrawal of Part 36 offer possible — Whether later offer superseding earlier offer

The joined appeals concerned the CPR 36 provisions for making and accepting offers to settle claims. In the first appeal, the respondent council admitted liability to the appellant on a personal injury claim. The latter made a Part 36 offer to accept £2,500 in settlement of her claim, to which the respondents responded with a lower offer. The appellant rejected that offer and a subsequent one that equalled hers; she did not formally withdraw her offer. The respondents purported formally to accept her offer and sought a court declaration that they were entitled to do so. The appellant argued that her offer was no longer extant since: (i) the respondents' rejection of that offer had rendered it incapable of acceptance thereafter in accordance with general principles of law; and (ii) her indication that she was no longer willing to accept £2,500, in her refusal of the respondents' later offer in that sum, was an implied withdrawal of her offer. Those arguments were rejected by a district judge and by a judge on appeal. It was held that the appellant's offer had not been formally withdrawn and the respondents had been entitled to accept it; under CPR 36, it was for the appellant positively to withdraw an offer that was no longer available. The appellant appealed the costs award.

In the second appeal, the appellants made Part 36 offers to settle a claim by a building contractor for moneys due under an invoice, including a May 2007 offer to pay £8,023 inclusive of interest in full and final settlement of the claim, followed by two higher offers in August and November 2007. In January 2008, the appellants withdrew all offers save for that of May 2007. In February 2008, they purported to repeat their August offer, but inclusive of interest and costs. At trial, the district judge awarded the respondent £8,375, with interest and costs, and ordered the appellants to pay half of its costs from January 2008. The appellants appealed, contending that the order was too favourable to the respondent since the judgment was not materially more advantageous than the May 2007 offer. The judge held that this was irrelevant since all earlier offers had been superseded by the February 2008 offer and the respondent had won. The appellants appealed.

In both appeals, the central issue was whether CPR 36 embodied a self-contained code or was subject to the general law of offer and acceptance by not expressly providing otherwise.

***Held*: The appeals were dismissed. (1) Parties seeking to settle claims are not bound to use the CPR 36 mechanism, but they must follow its requirements if they want to take advantage of the consequences for costs and other matters that flow from a Part 36 offer. CPR 36 is drafted as a self-contained code, prescribing in detail the way in which an offer may be made and the consequences of accepting or rejecting it. Concepts of offer and acceptance underpin CPR 36, but it should not be understood as incorporating all the rules of law governing the formation of contracts. It was drafted to promote certainty and should be read and understood according to its terms without importing rules from the general law, save where clearly intended. (2) CPR 36 clearly states how a Part 36 offer may be made, varied and withdrawn. It may be accepted at any time unless the offeror has formally withdrawn it by serving notice on the offeree: see CPR 36.9(2). There is no provision for an offer to lapse or to become incapable of acceptance on being rejected by the offeree, as would occur at common law. To import the common law rule into CPR 36 would undermine the scheme. Likewise, there is no room for the concept of implied withdrawal of an offer. CPR 36 requires an express written notice making it clear to the offeree that the offer has been withdrawn: see CPR 36.3(7). In the first appeal, the letter relied on by the appellant, rejecting an offer by the respondents that equalled her own, was not a notice of withdrawal of her own offer of the kind required by the rules. The respondents were entitled to accept that offer. (3) The judge in the second appeal had erred in considering that the May 2007 offer had been superseded by that of February 2008. Whatever the intention of the later offer, it did not constitute a Part 36 offer because it did not comply with the requirements of CPR 36.2(2)(b) or (c). All offers save for that of May 2007 had been formally withdrawn. Thus, at the date of trial, the only offer for consideration was that of May 2007.**

***Per curiam*: The better view is that the offer is open for acceptance until the start of the trial or its withdrawal in the way prescribed by CPR 26.3(7) and it is not superseded by a subsequent offer. CPR 36 does not provide that only one offer may be available for acceptance at any one time or that a later offer should be treated as varying or revoking a previous offer. It would be inconsistent with the recognition of CPR 36 as a self-contained code to read such provisions into it.**

The following cases are referred to in this report.

Carver v *BAA plc* [2008] EWCA Civ 412; [2009] 1 WLR 113; [2008] 3 All ER 911

Multiplex Constructions (UK) Ltd v *Cleveland Bridge UK Ltd* [2008] EWHC 2280 (TCC)

Sampla v *Rushmoor Borough Council* [2008] EWHC 2616 (TCC)

This was the hearing of conjoined appeals by Susan Gibbon, the appellant in the first appeal, from a decision of HH Judge Holman, sitting in Manchester County Court, dismissing an appeal against a costs decision of District Judge Richmond in favour of the respondents, Manchester City Council; and by John and Anne Reeves, the appellants in the second appeal, from a decision of HH Judge Rubery, sitting in Shrewsbury County Court, dismissing an appeal against an order of District Judge Brown in favour of the respondent, LG Blower Specialist Bricklayer Ltd.

A Dr Mark Friston (instructed by Thompsons Solicitors, of Manchester) appeared for the appellant in the first appeal; Dr Timothy Hodgson (instructed by the legal department of Manchester City Council) appeared for the respondents in the first appeal; Thomas Plewman (instructed by Aaron & Partners LLP, of Manchester) appeared for the appellant in the second appeal; Frances Pigott (instructed by Andrews Law, of Bridgnorth) represented the respondents in the second appeal.

Giving the first judgment, **Moore-Bick LJ** said:

Issues

B [1] These appeals have been heard together because they raise similar issues concerning the interpretation and effect of CPR 36 of the Civil Procedure Rules. CPR 36 contains provisions that enable either party to the proceedings to make an offer to settle the dispute that has far-reaching consequences if the case goes to trial and the other party fails to obtain a judgment that is more advantageous to it than the offer. The central question raised on this appeal is whether CPR 36 embodies a self-contained code or is subject to the general law of offer and acceptance in so far as it fails expressly to provide otherwise.

Part 36

C [2] CPR 36 replaced the provisions of the Rules of the Supreme Court relating to payment into court by way of a formal offer of satisfaction of the claim. It also provided for the first time a means by which a claimant could offer to accept a sum of money less than the amount of its claim with protection in respect of costs comparable to that which had previously been available only to defendants. Its purpose is to encourage settlement and to enable those who make sensible offers to protect themselves against liability for the costs incurred in the continuation of proceedings to no ultimate advantage.

[3] The following are the most important provisions of CPR 36 as far as these appeals are concerned:

D **Scope of this Part**

(1) This Part contains rules about

(a) offers to settle; and

(b) the consequences where an offer to settle is made in accordance with this Part.

…

Form and content of a Part 36 offer

36.2

(1) An offer to settle which is made in accordance with this rule is called a Part 36 offer.

(2) A Part 36 offer must

(a) be in writing;

E (b) state on its face that it is intended to have the consequences of Section I of Part 36;

(c) specify a period of not less than 21 days within which the defendant will be liable for the claimant's costs in accordance with rule 36.10 if the offer is accepted;

(d) state whether it relates to the whole of the claim or to part of it or to an issue that arises in it and if so to which part or issue; and

(e) state whether it takes into account any counterclaim.

…

Part 36 offers general provisions

36.3

…

F (3) A Part 36 offer which offers to pay or offers to accept a sum of money will be treated as inclusive of all interest until

(a) the date on which the period stated under rule 36.2(2)(c) expires; or

(b) if rule 36.2(3) applies, a date 21 days after the date the offer was made.

…

(6) After expiry of the relevant period and provided that the offeree has not previously served notice of acceptance, the offeror may withdraw the offer or change its terms to be less advantageous to the offeree without the permission of the court.

(7) The offeror does so by serving written notice of the withdrawal or change of terms on the offeree.

Time when a Part 36 offer is made

36.7

(1) A Part 36 offer is made when it is served on the offeree.

(2) A change in the terms of a Part 36 offer will be effective when notice of the change is served on the offeree. G

Acceptance of a Part 36 offer

36.9

(1) A Part 36 offer is accepted by serving written notice of the acceptance on the offeror.

(2) … a Part 36 offer may be accepted at any time (whether or not the offeree has subsequently made a different offer) unless the offeror serves notice of withdrawal on the offeree.

…

Costs consequences following judgment

36.14

(1) This rule applies where upon judgment being entered H

(a) a claimant fails to obtain a judgment more advantageous than a defendant's Part 36 offer; or

(b) judgment against the defendant is at least as advantageous to the claimant as the proposals contained in a claimant's Part 36 offer.

(2) … where rule 36.14(1)(a) applies, the court will, unless it considers it unjust to do so, order that the defendant is entitled to

(a) his costs from the date on which the relevant period expired; and

(b) interest on those costs.

(3) … where rule 36.14(1)(b) applies, the court will, unless it considers it unjust to do so, order that the claimant is entitled to

(a) interest on the whole or part of any sum of money (excluding interest) awarded at a rate not exceeding 10% above base rate for some or all of the period starting with the date on which the relevant period expired; J

(b) his costs on the indemnity basis from the date on which the relevant period expired; and

(c) interest on those costs at a rate not exceeding 10% above base rate.

[4] It can be seen from CPR 36 as a whole, as well as from the extracts cited above, that it contains a carefully structured and highly prescriptive set of rules dealing with formal offers to settle proceedings that have specific consequences in respect of costs in those cases where the offer is not accepted and the offeree fails to do better after a trial. In cases where there has been no CPR 36 offer or a CPR 36 offer has been bettered, the judge has a broad discretion in dealing with costs within the framework provided by CPR 44. CPR 44.3(4) provides that when exercising its discretion as to costs, the court will have regard to the general rule that the unsuccessful party should pay the costs of the successful party, but will also have regard to the conduct of the parties and any payment into court or admissible offer to settle made by one or other party that falls outside the terms of CPR 36. In seeking to settle the proceedings, therefore, parties are not bound to make use of the mechanism provided by CPR 36, but if they want to take advantage of the particular consequences for costs and other matters that flow from making a CPR 36 offer, in respect of which the court's discretion is much more confined, they must follow its requirements. K L

[5] CPR 36 is drafted as a self-contained code. It prescribes in some detail the manner in which an offer may be made and the consequences that flow from accepting or failing to accept it. In some respects, those consequences reflect broadly the approach the court might be expected to take in respect of costs; in others, they do not. For example, CPR 36.14(3) allows the court to award a claimant that has obtained a judgment at least as advantageous as its offer interest on the sum for which it has obtained judgment at an enhanced rate of up to 10% over base rate, costs on the indemnity basis and interest on those costs at an enhanced rate as well.

[6] Basic concepts of offer and acceptance clearly underpin CPR 36, but that is inevitable given that it contains a voluntary procedure under which either party may take the initiative to bring about a consensual resolution of the dispute. Such concepts are part of the landscape in which everyone conducts their daily life. It does not follow, however, that CPR 36 should be understood as incorporating all the rules of law governing the formation of contracts, some of which are quite technical in nature. Indeed, it is not desirable that it should do so. Certainty is as much to be commended in procedural as in substantive law, especially perhaps in a procedural code that must be understood and followed by ordinary citizens who wish to conduct their own litigation. In my view, CPR 36 was drafted with these considerations in mind and is to be read M

and understood according to its terms without importing other rules derived from the general law, save where that was clearly intended.

[7] With these general observations in mind, I turn to consider the two appeals before us.

Gibbon v *Manchester City Council*

[8] In December 2007, Mrs Susan Gibbon was injured when she tripped and fell in a playground for which Manchester City Council were responsible. She brought proceedings against the council, which admitted liability.

[9] On 10 November 2008, the council made a CPR 36 offer to Mrs Gibbon of £1,150 in settlement of her claim. She rejected that offer and, on 18 November, made a CPR 36 offer in response, offering to accept £2,500 plus the repayment of any sums recoverable by the compensation recovery unit and costs.

[10] On 24 November, the council increased their offer to £1,500. The solicitor acting for Mrs Gibbon rejected that offer, indicating that it expected something more generous. However, it did not withdraw the CPR 36 offer made on her behalf on 18 November.

[11] On 7 January 2009, the council increased their offer to £2,500 but, on 18 February, Mrs Gibbon's solicitor again rejected it. Once again, however, it took no steps to withdraw the offer made on 18 November.

[12] On 26 February, the council, mindful of the fact that the offer made on 18 November had not been withdrawn, wrote to Mrs Gibbon's solicitor formally accepting that offer. On receipt of that letter, the solicitor promptly purported to withdraw the offer and again rejected the council's offer of £2,500, but the council were determined to stick to their position. They made an application for a declaration that they were entitled to accept the offer made on 18 November and an order for judgment to be entered to reflect the fact that they had done so.

[13] The application was heard by District Judge Richmond. He held that, since the offer made on 18 November had not been formally withdrawn, the council were entitled to accept it and gave judgment for Mrs Gibbon in the sum of £2,500. He ordered the council to pay her costs up to the date of the offer and Mrs Gibbon to pay the council's costs thereafter.

[14] Mrs Gibbon's appeal was heard by HH Judge Holman. He held that CPR 36 is quite explicit and puts the onus on the offeror to take positive steps to withdraw any existing offer if it does not want it to be available for acceptance. He therefore dismissed the appeal.

[15] Dr Mark Friston, who appeared on behalf of Mrs Gibbon, submitted that although no formal steps were taken by her solicitor to withdraw the CPR 36 offer made on 18 November, its rejection by the council on 24 November 2008 rendered it incapable of acceptance thereafter in accordance with general principles of law. Alternatively, he submitted that the solicitor's letter of 18 February rejecting the council's offer of £2,500 made it quite clear that she was unwilling to accept that amount in settlement of her claim and thus amounted to an implied withdrawal of her CPR 36 offer sufficient to satisfy the requirements of CPR 36.3(7). The offer was therefore no longer open for acceptance.

[16] In my view, attractive though these arguments are, they cannot be reconciled with the clear language of CPR 36 or indeed with the scheme that it embodies. CPR 36.9(2) is quite clear: a CPR 36 offer may be accepted *at any time* unless the offeror has withdrawn the offer by serving notice of withdrawal on the offeree. Moreover, it may be accepted whether or not the offeree has subsequently made a different offer, a provision that is contrary to the general position at common law. The rules state clearly how a CPR 36 offer may be made, how it may be varied and how it may be withdrawn. They do not provide for it to lapse or become incapable of acceptance on being rejected by the offeree. That would be the case at common law, but it is inconsistent with the concepts underlying CPR 36, which proceeds on the footing that the offer is on the table and available for acceptance until the offeror chooses to withdraw it. There are good reasons for that. An offer that appears unattractive when made, and is therefore rejected, may become more attractive as the proceedings progress and the parties reassess the strength of their respective cases. A defendant that chooses to leave its offer on the table may tempt the claimant into accepting it, with the benefit to itself of the consequences for costs of an offer made at an early stage. CPR 36 allows a defendant (or for that matter a claimant) to decide whether to leave its offer open for acceptance or to withdraw it and make another offer later. To import into CPR 36 the common law rule that an offer lapses on rejection by the offeree would undermine this important element of the scheme. It could give rise to disputes concerning whether the offer had been rejected in any given case so as to render it incapable of acceptance. In *Sampla* v *Rushmoor Borough Council* [2008] EWHC 2616 (TCC), Coulson J held, largely for these reasons, that the rejection of a CPR 36 offer does not render it incapable of later acceptance. In my view, he was right to do so.

[17] Nor do I think that the letter of 18 February 2009 can be read as containing an implied withdrawal of the CPR 36 offer. CPR 36.3(7) provides that an offer is withdrawn by serving written notice on the offeree. In my view, that leaves no room for the concept of implied withdrawal; it requires express notice in writing in terms that bring home to the offeree that the offer has been withdrawn. If justification for that requirement is sought, it can be found once again in the need for clarity and certainty in the operation of the CPR 36 procedure. Although the rule does not prescribe any particular form of notice, in order to avoid uncertainty it should include an express reference to the date of the offer and its terms, together with some words making it clear that it is withdrawn. There may, of course, be cases in which the terms of the notice are less clear than might be wished so that there is room for argument concerning whether the notice fulfils the requirements of the rule. However, that is a different question. The letter of 18 February 2009 did not refer to the CPR 36 offer and did not contain any language that might be construed as a notice of withdrawal. In my view, it was quite incapable of constituting a notice of withdrawal of the kind required by CPR 36.3(7).

[18] In some cases, the demands of clarity and certainty in the operation of CPR 36 may appear to produce injustice and I see the force of Dr Friston's submission that Mrs Gibbon's solicitor had made it clear that she was not willing to accept £2,500 to settle her claim. Her difficulty, however, is that a CPR 36 offer had been made on her behalf and had not been withdrawn. It remained on the table and was therefore open for acceptance. We are not concerned with whether she has any ground of complaint against the council for accepting her offer despite the fact that she had made her intentions clear, since no suggestion has ever been made that she has a legitimate complaint about that. In those circumstances, I am of the view that the judge below was right and that this appeal must be dismissed.

LG Blower Ltd v *Reeves*

[19] In January 2006, the appellants, Mr John and Mrs Anne Reeves, entered into a contract with a building contractor, the respondent, LG Blower Specialist Bricklayer, to carry out improvements to their home, including the construction of a two-storey extension, rewiring and the supply and fitting of replacement windows. During the summer of 2006, disputes arose because Mr and Mrs Reeves were dissatisfied with the quality of some of the respondent's work. On 20 July 2006, the respondent sent an invoice for the outstanding amount it claimed to be due and, in mid-August 2006, it stopped work altogether, leaving Mr and Mrs Reeves to complete the job using other contractors.

[20] Mr and Mrs Reeves sought to set off against the respondent's final invoice various amounts that they said represented the loss they had suffered as a result of the respondent's poor workmanship and failure to complete the work. They admitted that part of the money claimed was due and offered to pay the balance. However, the respondent was unwilling to accept that the appellants had any genuine grounds for complaint and was unwilling to accept less than the full amount covered by its invoice. In April 2007, it brought proceedings against Mr and Mrs Reeves in Shrewsbury County Court claiming £15,793.06. They made a counter-claim in the sum of £9,160.60 together with an unquantified claim for damages for breach of contract.

[21] Mr and Mrs Reeves were reasonable people and had tried to reach a settlement even before the matter got to court. In response to the letter before action dated 2 February 2007 claiming (at that time) £15,003.41, they made an offer on 9 February to pay £8,023.14 and tendered a cheque in that amount. However, that was not acceptable to the respondent. The claim form was issued on 30 March and served on 6 April 2007. On 15 May, Mr and Mrs Reeves made a CPR 36 offer of £8,023.14 inclusive of interest in full and final settlement of the claim. I shall refer to this as "the May offer".

[22] Part of the respondent's bill related to plumbing work for which it sought payment of £649.36. On 6 August 2007, Mr and Mrs Reeves paid that item in full, thereby reducing the claim against them.

[23] On 28 August, Mr and Mrs Reeves made another CPR 36 offer, this time of £8,188.38. I shall refer to it as "the August offer". The respondent rejected it.

[24] On 9 November 2007, Mr and Mrs Reeves made another CPR 36 offer, this time of £9,000, but, on 9 January 2008, they withdrew all their offers apart from the May offer.

[25] Further offers, however, followed. On 28 February 2008, Mr and Mrs Reeves "repeated" their offer of £8,188.38, this time inclusive of both interest and costs. There is a dispute concerning whether that was a CPR 36 offer and to what exactly it was intended to refer.

[26] Initially, the respondent had adopted a rather intransigent attitude, but, in the latter part of 2007, it made offers of settlement and thereafter various offers were made by both the respondent and by Mr and Mrs Reeves. Regrettably, however, the parties were unable to reach agreement and the matter went to trial.

[27] On 9 June 2009, District Judge Brown gave judgment for the respondent in the sum of £8,375.94, together with interest and costs. He ordered Mr and Mrs Reeves to pay half of the respondent's costs from 8 January 2008.

[28] Mr and Mrs Reeves considered that the district judge's order was far too favourable to the respondent. In their view, he should have awarded them their costs, or at least made no order as to costs. Their appeal against his order was heard by HH Judge Rubery. They argued that the judgment obtained by the respondent was not materially more advantageous to it than the May offer, especially after taking into account the payment that they had made in respect of the plumbing, and that they should therefore have the benefit of CPR 36. However, the judge did not find it necessary to consider that point. He held that both the May offer and the August offer had been superseded by the offer made on 28 February 2008 and that, since the respondent had emerged as the winner, the district judge had not gone seriously wrong in making his order for costs.

[29] The first question for consideration is whether the judge was right to hold that Mr and Mrs Reeves' earlier CPR 36 offers had been superseded by the offer made on 28 February, since, if he was right about that, none of those earlier offers falls to be considered for the purposes of CPR 36. For the reasons given earlier, I think Mr Thomas Plewman, for the appellant, was right in submitting that CPR 36 is a self-contained code that provides expressly for the manner in which offers may be made, modified and withdrawn and that as such it displaces the ordinary rules of common law. Once made, a CPR 36 offer remains open for acceptance without the need for the court's permission until the start of the trial or its withdrawal in the manner set out in CPR 36.3(7). It follows that I think the judge was wrong to hold that the May and August offers were superseded by the offer made in February 2008. By that letter, Mr and Mrs Reeves repeated their offer of £8,188.38, but they did not expressly withdraw any earlier CPR 36 offer that remained open for acceptance. Accordingly, the May offer was unaffected by it.

[30] The position at the end of the trial, therefore, can be summarised as follows. On 9 January 2008, Mr and Mrs Reeves had revoked all existing CPR 36 offers other than the May offer, which accordingly remained for consideration. On 28 February, they had repeated their offer of £8,188.38, inclusive of interest and costs. Mrs Frances Pigott, for the respondents, submitted that that should be understood to be a reference to the May offer, but reading the letter as a whole I do not think that can be right. The reference in an earlier paragraph to an offer of £8,188.38 suggests that the intention was to repeat or reinstate the August offer, but in slightly less advantageous terms, since the sum offered was to include costs. I agree with Mr Plewman, however, that whatever its intention it did not constitute a CPR 36 offer because it did not comply with the requirements of CPR 36.2(2)(b) or (c). The only CPR 36 offer that the district judge had to consider, therefore, was the May offer.

[31] That makes it unnecessary to decide whether, under CPR 36, a later offer in different terms automatically varies a previous offer or whether a party can make a number of different offers, all of which are concurrently open for acceptance. However, the point is one of some interest and importance and I shall therefore express my views on it.

[32] Although, at first sight, it may seem anomalous that a party should be able to make several offers in different terms, all of which may at any one time be capable of acceptance, that does, in my view, reflect both the language and the purpose of CPR 36. As to the language, CPR 36 is quite clear as to the manner in which offers may be made, varied and withdrawn. It does not provide that only one offer may be available for acceptance at any one time, nor does it provide that a later offer is to be treated as varying or revoking a previous offer and it would be inconsistent with the recognition of CPR 36 as a self-contained code to read provisions of that kind into it. The purpose of CPR 36 is to promote settlement by encouraging sensible offers. As Mr Plewman pointed out, every CPR 36 offer carries with it certain consequences for costs linked to the date on which it is made and the financial implications vary as the proceedings progress. For example, an offer of £10,000 made at an early stage may be more valuable than an offer of £12,000 made at a later stage, depending on the amount of costs that have been incurred in the meantime and prevailing rates of interest. There is no reason why a party should not make more than one offer and leave it to the other to decide which, if any, to accept. Or, if it wishes, it may change the terms of the original offer, which then continues to stand in its varied form as from the date it was originally made. I accept that, in some cases, there could be argument concerning whether a later offer was intended to vary an earlier offer or to stand alongside it. The solution, however, is for parties and their legal advisers to follow the requirements of the rules carefully and to make their intentions clear. If they do so, problems of that kind should not arise.

[33] Mr Plewman submitted that the district judge went wrong in this case because he failed to recognise that the effect of Mr and Mrs Reeves paying the plumbing bill was to increase the value of the May offer by £649.36. Once that is recognised, it could not be said that the judgment was more advantageous to the respondent than the May offer. Mrs Pigott, on the other hand, submitted that however one looked at it the respondent had clearly been successful and had obtained a judgment that was worth significantly more than the May offer. Accordingly, the district judge was right to hold that costs were in his discretion and, applying the principles set out in CPR 44, he was entitled to make an order giving the respondent at least part of its costs.

[34] In any consideration of the effect on costs of an offer to settle it is necessary to compare the amount of the offer with the amount for which judgment has ultimately been obtained. Offers under CPR 36 are treated as inclusive of interest: see CPR 36.3(3). However, since interest is normally awarded in respect of most or all of the period for which the successful claimant has been kept out of its money, it is usually simpler to concentrate on the principal amount in dispute when deciding whether the claimant has recovered more than was available to it under the offer. Clearly, the longer proceedings go on, the more the successful claimant can expect to recover by way of interest, but the additional amount reflects nothing more than the additional time spent waiting to obtain judgment and any change in interest rates in the meantime. Accordingly, in cases where a sum is offered under CPR 36 or an offer is expressed to be inclusive of interest, in order to evaluate the offer it is necessary to take into account how long the principal sum has been outstanding and the rates of interest likely to be awarded by the court.

[35] In the course of his judgment, the district judge commented on the offer made by Mr and Mrs Reeves on 9 February 2007 to pay £8,023.14 and observed that it was not quite as close to the eventual judgment as might appear because part of it related to the plumbing bill that they later accepted to be due. He considered the true amount of the offer to be £7,373.78, and it seems clear that he viewed the May offer (which was in the same amount) in the same way.

[36] The respondent eventually obtained judgment in the principal sum of £8,375.94 after deducting the various items of the counterclaim, including the sum of £1,000 awarded to Mr and Mrs Reeves by way of damages for the interference with the use and enjoyment of their home. (The latter claim was added by amendment shortly before the trial began.) At the time it was made, the May offer (which included interest) represented a principal sum of £7,789.46, allowing for interest at 4% for nine months. (Although the district judge awarded interest at a rate of 8%, that was, in my view, considerably higher than could have been expected. However, if one were to deduct interest at that rate, the principal sum would be reduced even further.) Viewed in that light, at the time it was made the May offer fell £1,586.48 short of the amount that the respondent was entitled to recover on its final account and £586.48 short of the amount for which it eventually obtained judgment, even after deducting the sum awarded to Mr and Mrs Reeves for interference with the use and enjoyment of their home.

[37] Since the plumbing bill was still outstanding when the May offer was made, the district judge was wrong to deduct it when considering the value of that offer. The effect of paying the plumbing bill was to reduce the value of the claim against Mr and Mrs Reeves by £649.36, and to that extent it made the May offer, which they left open for acceptance, more attractive. None the less, the fact remains that, at the end of the day, the respondent recovered £661.38 more than the amount of the enhanced May offer, which, allowing for one year's interest at 4% (the plumbing bill having been paid in August 2007), represented a principal sum of only £7,714.56.

[38] That raises the question of whether the respondent failed to obtain a judgment more advantageous than the enhanced May offer within the meaning of CPR 36.14(1)(a). Mr Plewman submitted that, if one applies the principles set out in *Carver* v *BAA plc* [2008] EWCA Civ 412; [2009] 1 WLR 113, it did.

[39] In *Carver*, this court considered the terms of CPR 36.14(1) and concluded that when asking itself whether the judgment was more advantageous to the claimant than the CPR 36 offer the court should take into account all aspects of the case, including emotional stress and financial factors, such as the incurring of unrecoverable costs. It therefore held that the judge below had been right in that case to look at the matter broadly and to take into account that an additional £51 obtained after trial was more than offset by the irrecoverable cost incurred by the claimant in continuing to contest the case for as long as she had. He was also entitled to take into account the added stress to her as she waited for the trial and the stress of the trial process itself.

[40] The decision in *Carver* has been criticised by many commentators, and most recently by Jackson LJ in his *Review of Civil Litigation Costs: Final Report*, on the ground that it introduces an unwelcome degree of uncertainty into the operation of CPR 36. In my view, there is much force in that criticism. Moreover, I do not think that the decision can be confined to cases in which one party has made a CPR 36 offer that is nearly, but not quite, sufficient and the other has rejected it outright, as Jackson J (as he then was) held in *Multiplex Constructions (UK) Ltd* v *Cleveland Bridge UK Ltd* [2008] EWHC 2280 (TCC): see [71]. The decision in *Carver* is binding on us, but it should be recognised that what may be more important than the factors to be taken into account is the weight that is to be attached to them, and that remains a matter for the judge in each case. Moreover, when deciding how much weight to attach to any particular factor, I think it important to see things from the litigant's perspective rather than to be too ready to impose the court's own view of what is and is not to its advantage. That is particularly important when dealing with money claims, both because to recover judgment for more than what was offered is legitimately regarded as success and because a party faced with a CPR 36 offer ought to be entitled to evaluate it by reference to a rational assessment of its own case (including the risk of incurring unrecoverable costs if it presses on). It should not have to make a significant allowance for the court's view of factors that are inherently difficult to value, such as the amount of unrecoverable costs and (even more so) the stress likely to be generated by pursuing the case to judgment. In a case where the offer has been beaten by a very small amount and there is clear evidence that the successful party has suffered serious adverse consequences as a result of pursuing the case to judgment, those factors may be sufficient to outweigh success in pure financial terms, but, in my view, such cases are likely to be rare. In most cases, obtaining judgment for an amount greater than the offer is likely to outweigh all other factors.

[41] In the present case, we do not know on what basis the respondent has funded the litigation, and even if it is to be assumed that it will have incurred some unrecoverable costs, we have no evidence of their amount. Nor is there evidence of any other factors that might be set against the value of success in pure financial terms. In any event, having regard to the size of the claim and counter-claim, the extent to which the judgment exceeded even the enhanced May offer would, in my view, outweigh all but the most powerful factors that might be relied on by way of disadvantage.

[42] The district judge proceeded on the basis that the judgment was more advantageous to the respondent than the CPR 36 offer and, for the reasons that I have already given (which differ to some extent from his), I think that he was right to do so. He was therefore faced with the task of exercising his discretion, taking into account the general rules on costs contained in CPR 44. The district judge was clearly exercised by the fact that for part of the period prior to the trial Mr and Mrs Reeves had acted sensibly and responsibly in seeking to settle the claim by making various offers and by the fact that the respondent had failed to respond in kind. However, a turning point came in January 2008, when Mr and Mrs Reeves withdrew all their existing CPR 36 offers other than the May offer. The district judge regarded that as a serious mistake and inferred that, at that point, they had decided that their set-off should at least equal the claim, as indeed was their case at trial. Several offers were made on both sides thereafter on different terms, but none at a level that made the failure to accept it culpable.

[43] Quite correctly, the district judge took as his starting point the fact that the respondent had been successful, having recovered a substantial proportion of its claim in the face of a defence that, at the start of the trial, denied that anything was due. However, he considered that the respondent's conduct during the early stages of the proceedings, in particular its dismissive response to sensible offers of settlement, called both for an expression of the court's disapproval and for a reward for Mr and Mrs Reeves. None the less, he criticised them for changing their position in January 2008 and deciding to contest the claim in its entirety, which he considered had increased the costs of both parties without any real prospect of success. His solution was to deprive the respondent of its costs for the period up to 8 January 2008 and to order Mr and Mrs Reeves to pay only half of its costs thereafter.

[44] Mr Plewman submitted that the judge's decision fell outside the generous ambit of his discretion on costs. He submitted that he had failed to take into account the fact that the May offer had remained in place after 8 January 2008 and the likelihood that the bulk of the costs had been incurred in the period immediately prior to the trial and at the trial itself.

[45] In my view, the judge's order was well within the ambit of his discretion. For the reasons that I have given, I think that he was right to assess costs in accordance with the principles set out in CPR 44. Once that is accepted, it was for him to assess the conduct of the parties and how to reflect it in his award of costs. His own experience would have told him that the bulk of the costs were likely to have been incurred in the period immediately leading up to and at the trial and it was those costs that, in his view, the claimant had been obliged to incur as a result of the change in attitude on the part of Mr and Mrs Reeves. It is certainly unfortunate that, once again, litigation in respect of a relatively modest claim has led to the parties' incurring substantial costs, but the

opportunity to avoid it was there on many occasions if only they had both been willing to grasp it.

[46] In my view, there are no grounds for criticising the district judge's decision and I would therefore dismiss this appeal.

Giving the second judgment, **Carnwath LJ** said:

[47] I agree. I add a brief comment on the judgment of this court in *Carver* v *BAA plc* [2008] EWCA Civ 412; [2009] 1 WLR 113, having regard to the criticisms to which Moore-Bick LJ has referred.

[48] In *Multiplex Constructions (UK) Ltd* v *Cleveland Bridge UK Ltd*, Jackson J, in [71], interpreted [28] to [32] of that judgment as setting out:

> how the court ought to approach the matter in circumstances where: (a) one party has made an offer which was nearly but not quite sufficient; and (b) the other party has rejected that offer outright without any attempt to negotiate.

[49] I agree with Moore-Bick LJ that it is not strictly possible to interpret the judgment so narrowly. In giving the leading judgment (with which the other members of the court agreed), Ward LJ treated the matter as one of construction of the new CPR 36. He saw that as intended to put money and non-money claims on the same footing. In the expression "more advantageous" (applied to both categories), the draftsman had used an "open-textured" phrase, which permitted, in para 30:

> a more wide-ranging review of all the facts and circumstances of the case in deciding whether the judgment, which is the fruit of the litigation, was worth the fight.

The rationale lay in the modern approach exemplified in the CPR, under which compromise was encouraged in the interests both of the parties and of the administration of justice, in para 31:

> Litigation is time-consuming and it comes at a cost, emotional as well as financial. Those are, therefore, appropriate factors to take into account in deciding whether the battle was worth it. Money is not the sole governing criterion.

These comments were expressed as an indication of the correct approach in law, which was then applied to the facts of the particular case in the following paragraph:

> 32. It follows that Judge Knight was correct in looking at the case broadly. He was entitled to take into account that the extra £51 gained was more than offset by the irrecoverable costs incurred by the claimant in continuing to contest the case for as long as she did. He was entitled to take into account the added stress to her as she waited for the trial and the stress of the trial process itself. No reasonable litigant would have embarked upon this campaign for a gain of £51.

[50] For my part, with respect, I share the difficulty of some commentators in seeing how the court can sensibly make a judgment of such factors as pre-trial stress or other personal factors, as part of an objective assessment of overall advantage. However, although we are bound to hold that such "emotional" factors *may* be relevant, the practical application of that approach must be considered in context. The sole issue for the court in *Carver* was whether the judge had erred in law in taking such factors into account on the particular facts of that case. As I read the decision, it was because the potential gain was so small in that case, that the court felt able to uphold the judge's approach.

[51] Accordingly, the judgment in *Carver* should not be interpreted as opening the way to a wide ranging investigation of emotional and other factors in every case, even where the financial advantage is significant. I agree with Moore Bick LJ that in most cases success in financial terms will be the governing consideration. Thus, although Jackson J's interpretation of *Carver* may not be supportable in strict legal terms, the practical effect of the proper approach to the exercise of the court's discretion is likely to achieve the same result in most cases.

Sir Anthony May P said:

[52] I agree that each of these appeals should be dismissed for the reasons given by Moore-Bick LJ.

Appeals dismissed.

REAL PROPERTY AND CONVEYANCING

Weir *v* Area Estates Ltd

A

COURT OF APPEAL

[2010] EWCA Civ 801

20 July 2010

CARNWATH, MOORE-BICK AND WILSON LJJ

Estates Gazette 23 October 2010

[2010] 42 EG 106

B

Sale of land — Contract — Rescission Land Registration Act 2002 — Sale of registered freehold land on full title basis under general condition 4.3(a) — Special condition requiring purchaser to accept leasehold interest shown in schedule to freehold title determined by surrender — Surrender ineffective owing to pending bankruptcy petition against lessee at relevant time — Whether special condition precluding reliance on continuance of lease as ground for purchaser to rescind contract — Whether lease a technical conveyancing defect insufficient to justify rescission

In January 2008, the respondent contracted to purchase a registered freehold property from the appellant for £400,000; a deposit of £40,000
C
was paid. The contract included the full title guarantee contained in condition 4.3(a) of the general conditions of sale and provided for vacant possession to be given on completion. The schedule to the freehold title showed the existence of a lease, granted in 2004 for a term of nine years. However, extra special condition 7 (ESC 7) of the contract stated that the lease had determined by operation of law in August 2006 and required the purchaser to accept that position without requiring further proof or raising any objection or requisition in that regard.

The lessee had purportedly surrendered the lease in August 2006. However, the surrender was ineffective, by virtue of section 284 of the
D
Insolvency Act 1986, because of a pending bankruptcy petition against the lessee that had been registered against the leasehold title. The appellant served notice to complete on the respondent in March 2008. The respondent purported to rescind the contract on the ground of the continued existence of the lease. In May 2008, the appellant secured a disclaimer of the lease from the lessee's trustee in bankruptcy.

Summary judgment was given for the respondent on a claim for the return of his deposit. The judge held that: (i) the appellant had no defence based on section 284(4)(*a*) of the 1986 Act; (ii) ESC 7 was not an exception to the full title guarantee in general condition 4.3(a) but affirmed it in the particular respect of the lease; and (iii) the continuance
E
of the lease was not a purely technical conveyancing defect: see [2010] 1 EGLR 77; [2010] 10 EG 121. The appellant appealed on the second and third points. The respondent contended that a vendor could not rely on a condition such as ESC 7 if it was misleading or there had been less than full disclosure. He also submitted that ESC 7 was void, under section 42(3) of the Law of Property Act 1925, as a stipulation in a contract for the sale of land to the effect that no objection should be taken on account of an outstanding legal estate.

***Held*: The appeal was dismissed.**

(1) Rather than simply affirming the "full title guarantee" basis of the sale, ESC 7 seemed to be designed to transfer risk to the purchaser in respect of the lease by drawing specific attention to it, expressing the vendor's position in respect of it and requiring F
the purchaser to accept that position without objection. Since, by general condition 4.3(a), the purchaser was to be taken to have actual knowledge of inclusions on the register, he could not claim to be unaware of the potential problem created by the bankruptcy petition. The condition requiring vacant possession on sale was general in nature and did not necessarily override the specific provision under ESC 7 in respect of the lease.

(2) A condition that purports to exclude the vendor's ordinary obligation to give good title will not always be effective to cover the vendor in respect of a defect in title of which it is aware. The knowledge that the vendor's solicitor should have had, and which G
the vendor would have had if properly advised, may be treated as equivalent to actual knowledge in that regard. Although ESC 7 drew the respondent's attention to the potential issue of the lease, it also contained an unqualified assertion that the problem no longer existed. A properly advised vendor would or should have known that this was untrue. It did not assist the appellant that the respondent might also have had the means of knowledge of the bankruptcy petition through his assumed notice of the contents of the register, since knowledge of the petition alone was inconclusive and did not rule out a valid surrender with the court's consent. By the unqualified terms of the representation, the appellant had H
impliedly represented that it had taken reasonable steps to confirm its accuracy. (3) The existence of the lease could not be treated as a technical conveyancing defect insufficient to justify rescission. The appellant could not show that, as at the date for completion, there had been no risk of the trustee in bankruptcy successfully asserting that the lease subsisted. To terminate the lease would have required a successful application to the court seeking ratification of the surrender or a successful request to the trustee in bankruptcy to disclaim. The outcome of neither had been a foregone conclusion at the relevant time because neither was in the appellant's sole control. Thus, the facts and circumstances of the case did not J
compel the court to conclude beyond reasonable doubt that, at the relevant date, the purchaser had not been at risk of a successful assertion of an incumbrance against him.

***Per curiam*: It was likely that the reference to "outstanding legal estate" in section 42(3) of the 1925 Act was intended to refer to the quality of the interest to be transferred, rather than to any potential incumbrance. There was no obvious reason, as a matter of legislative policy, for introducing a strict anti-avoidance provision relating only to those incumbrances that were legal estates under the scheme of the 1925 Act.**

The following cases are referred to in this report. K

Banister, Re; *sub nom Broad* v *Munton* (1879) LR 12 Ch D 131, CA
Becker v *Partridge* [1966] 2 QB 155; [1966] 2 WLR 803; [1966] 2 All ER 266, CA
Cumberland Consolidated Holdings Ltd v *Ireland* [1946] KB 264; [1946] 1 All ER 284, CA
MEPC Ltd v *Christian-Edwards* [1981] AC 205; [1979] 3 WLR 713; [1979] 3 All ER 752, HL
Weir v *Area Estates Ltd* [2010] 1 EGLR 77; [2010] 10 EG 121, Ch

This was an appeal by the appellant, Area Estates Ltd, from a decision of Mr Robin Knowles CBE QC, sitting in the Chancery Division of the

A High Court, giving summary to the respondent, Graham Weir, on a claim for the return of a deposit paid under a contract for the sale and purchase of land.

Mark Loveday (instructed by Romain Coleman) appeared for the appellant; Jane Evans-Gordon (instructed by Tarbox Robinson & Partners, of Northwood) represented the respondent.

Giving judgment, **Carnwath LJ** said:

[1] Area Estates (the vendor) appeals against the order of Mr Robin Knowles QC, sitting as a deputy High Court judge, dated 18 December 2009, granting Mr Graham Weir (the purchaser) summary
B judgment on his claim to rescind a contract for the sale of a property at 22-24 St Andrew Street, Hertford (the property).

Factual background

[2] On 29 January 2008, the purchaser contracted with the vendor to purchase the registered freehold property at a price of £400,000 subject to a £40,000 deposit. The property had been subject to a lease in favour of a Mr Airey, granted on 28 April 2004 for a term of nine years. The existence of the lease was apparent from the schedule to the freehold title. However, extra special condition 7 (ESC 7) of the contract of sale provided:

C The Lease dated 28 April 2004 and referred to on Title Number HD431124 determined by operation of law on 31 August 2006. The Buyer shall accept the position and shall not be entitled to require any further proof of the determination or raise any objection or requisition with regard thereto.

[3] Unfortunately, the statement that the lease had "determined by operation of law on 31 August 2006" was not correct. What had happened was that the lessee, Mr Airey, had moved out of the property purporting to surrender the lease. However, that purported surrender was ineffective in law because he was at the time subject to a bankruptcy petition, presented on 8 August 2006. The consequence was that any disposition of property was void unless made with the consent of the
D court: see section 284 of the Insolvency Act 1986 (the 1986 Act). The petition was duly registered against the leasehold title, under section 86 of the Land Registration Act 2002, on 11 August 2006. A bankruptcy order was made on 11 October 2006, and was registered against the leasehold title on 2 November 2006.

[4] Notice to complete was served on 7 March 2008. In response, the purchaser sought to rescind the contract because of the encumbrance in the form of the leasehold interest. On 16 May 2008, the vendor secured a disclaimer of the lease from Mr Airey's trustee in bankruptcy. By a claim issued on 4 June 2008, the purchaser sought the return of the £40,000 deposit plus interest and by application dated 5 March 2009
E applied for summary judgment. The application came before the deputy judge on 22 October 2009.

[5] The vendor advanced in substance three defences:

(i) section 284(4)(*a*) of the 1986 Act, which provides an exception for property "received before the commencement of the bankruptcy in good faith, for value and without notice that the petition had been presented";

(ii) condition ESC 7;

(iii) that the defect was a "technical conveyancing defect".

[6] The judge rejected all three defences and, by an order dated 18 December 2009, ordered the repayment of the deposit with interest
F and costs. Permission to appeal was granted by Lloyd LJ.

Issues on the appeal

[7] At the start of the hearing, Mr Mark Loveday, for the vendor, indicated that he was no longer pursuing the first issue. This leaves two issues:

(i) whether condition ESC 7 prevents the purchaser from relying on the continued existence of the lease as a ground for rescission; and (ii) if not, whether the existence of the lease should be treated as a "technical conveyancing defect" insufficient to justify rescission.

[8] I remind myself that this is an application for summary judgment and that we must take the facts at their most favourable from the vendor's point of view. Thus, we must assume that, at the time of the contract, the vendor had acted in good faith and in the belief that the G
lease had come to an end by operation of law.

Condition ESC 7

Judge

[9] The judge looked at ESC 7 in the context of the other provisions of the contract. He referred in particular to general condition 4.3:

Unless otherwise stated in the special conditions the seller sells with full title guarantee except that:

(a) all matters recorded in registers open to public inspection are to be treated as within the actual knowledge of the buyer...

and ESC 1: H

Vacant possession will be given on completion...

[10] He held, in [22]:

(1) On its true construction Special Condition 7 required Mr Weir to accept as factually accurate the proposition that the Lease had determined by operation of law on 31 August 2006.

(2) This requirement was not an exception to the "full title guarantee" basis of the sale, under General Condition 4.3(a). Rather, it was an affirmation of it in this particular.

(3) Consistently with the requirement, Special Condition 7 ruled out Mr Weir's being entitled to require any further proof of the determination, or to J
raise any objection or requisition with regards to it.

(4) Although the Land Register and the Land Charges Register (both registers under General Condition 4.3(a)) revealed the Lease and the bankruptcy petition, and did not mention any determination of the Lease, the position remained that Mr Weir was contractually required to accept that there had been a determination. It transpired that there had not.

[11] He rejected an argument that the purchaser (by clause 4.3(a)) was fixed with notice of the bankruptcy petition, and therefore able to work out the true position. He did not see how that could help the vendor on the question of whether it was able "to convey at completion the unencumbered title they had contracted to convey": see [24]. K

Submissions

[12] Before us, Mr Loveday argued that the judge's interpretation of ESC 7 was too narrow. The vendor is permitted to protect itself against the possible uncertainty of the title, particularly with regard to the inevitable uncertainty surrounding the concept of surrender by operation of law. The plain meaning of the condition was that the risk of such uncertainty was transferred to the purchaser, who could not complain about an incumbrance on the title even though the surrender was found to be void.

[13] For the purchaser, Ms Jane Evans-Gordon submitted that the condition did not prevent the purchaser from objecting if, as it emerged, L
the lease had not in fact been surrendered. Its effect was only that he was not entitled to require further proof from the vendor or to take objection with regard to the lack of such proof. This was confirmed by ESC 1, which required "vacant possession" on completion and was inconsistent with the existence of someone with a better title to immediate occupation.

[14] Further, she submitted, even if the condition would otherwise be construed as the vendor proposes, that construction cannot overcome two obstacles, one derived from statute and one from case law:

(i) section 42(3) of the Law of Property Act 1925 (the 1925 Act),
which would make a condition as so construed void; and M

(ii) the principle (established by cases such as *Re Banister* (1879) LR 12 Ch D 131) that such a condition cannot be relied on if it is misleading or there has been less than full disclosure: see *Emmet and Farrand on Title* (2010), in para 5.008.

Discussion

[15] I have not found this an easy question. I do not share the judge's confidence that the condition is to be read as simply affirming the "full title guarantee" basis of the sale. I see some force in Mr Loveday's contention that it is designed to transfer risk to the purchaser: to draw specific attention to the issue of the lease, express the vendor's position in respect of it and require the purchaser to accept that position without

A objection. Furthermore, since the purchaser was taken to have actual knowledge of anything in the register (condition 4.3(a)), he could not claim to be unaware of the potential problem created by the bankruptcy petition. Although I agree with Ms Evans-Gordon that, taken on its own, the condition requiring "vacant possession" would be inconsistent with the existence of an occupational lease, even if the premises were in fact vacant at the time of sale (see, for example, *Cumberland Consolidated Holdings Ltd* v *Ireland* [1946] KB 264, at pp270-271), that general provision does not necessarily override the specific provision made by ESC 7 in respect of the lease.

[16] It is therefore necessary to consider the two points presented by
B Ms Evans-Gordon as obstacles to the effectiveness of ESC 7. Neither is entirely straightforward. I will deal with the case law principle first, since the ground is much better trodden.

[17] It is well established by authorities going back over a century and a half that a condition that purports to exclude the vendor's ordinary obligation to give good title may not be always be effective. However, the cases do not speak with one voice as to the test to be applied. The important issue, for present purposes, is the necessary state of mind of the vendor. Does the principle apply only when he had actual knowledge of the defect or is it enough that he had the means of knowledge? *Banister*, as I read it, was a case in which the representation
C in the conditions was contradicted by facts of which the vendor had actual knowledge: see at p142, *per* Sir George Jessel MR:

> conditions which… contained representations, either expressed or implied, which are untrue, and were untrue to the knowledge of the vendor at the time those conditions were framed.

[18] The passage in *Emmet and Farrand* to which Ms Evans-Gordon referred, has some useful illustrations from the authorities. I note the following, commenting on the effect of a condition that the property is sold "subject to any general incumbrances or defects in title which may exist": see para 5.008:

D

> Further, such a condition will apparently not be sufficient where there are incumbrances or defects of which the vendor ought reasonably to have known (*Heywood* v *Mallalieu* (1883) 25 Ch D 357, where the vendor's solicitor would have discovered the existence of an easement if he had made fuller enquiries into certain claims).

[19] Reference is made in the same passage to *Becker* v *Partridge* [1966] 2 QB 155, in which the contract for sale of an underlease provided that the vendor's title "has been accepted… and the purchaser shall raise no requisition or objection thereon". This was held to be ineffective when it emerged that there had been breaches of covenant
E in the superior lease giving grounds for forfeiture. Although the vendor did not have actual knowledge, he was held to have had "constructive notice" because his solicitor had neglected to inspect the superior lease as he would have been entitled to do when taking the underlease.

[20] There is a clear statement of principle to similar effect in para 15-072 of *Megarry and Wade: The Law of Real Property* (7th ed), commenting on the effect of a general exclusion such as "subject to any existing rights and easements of whatever nature":

> However, it is a fundamental rule of equity that the vendor cannot rely on such a condition to cover a latent defect in title of which he knew or ought to
> F have known.

A number of cases are referred to in the footnote, including *Becker*. (Reference is also made to an interesting article by Mr Charles Harpum, editor of the current *Megarry and Wade*, in [1992] CLJ: *Exclusion Clauses and Contracts for the Sale of Land.*

[21] For present purposes, the proposition as stated by *Megarry and Wade* seems to me to be sufficiently supported by the decision in *Becker*, which is binding on us. In that case, as I have said, the exclusion was held not to protect the vendor against defects that his solicitor could and should reasonably have discovered. (I note that the facts and, in particular, the basis of the finding of "constructive notice" are more fully stated in the All England report: [1966] 2 All ER 266, at p269.) However, in holding that the exclusion was ineffective, the judgment G was clear: see at pp171G-172A:

> There is no doubt that by a clearly drawn condition put in the contract by a vendor who acts in good faith, and disclosing a possible defect in the title, the purchaser may be compelled to accept the title offered by the vendor. But the vendor must have disclosed the defects of which he knew. In this case the vendor did not know of the breaches which would give rise to forfeiture. But he ought to have known that such breaches might exist.

Thus, the knowledge that the vendor's solicitor should have had, and which the vendor would have had if properly advised, was treated as being equivalent to actual knowledge. The present case seems to me H indistinguishable. The vendor, properly advised, could and should have known that the purported surrender, unsupported by any order of court, was ineffective.

[22] I accept that general statements, such as that in *Megarry and Wade*, need to be read in the context of the particular clause in question. In this case, the exclusion is more specific than those considered in most of the cases, in that it is directed at the particular issue of the lease. However, this works both ways. On the one hand, it may be said to have drawn the purchaser's attention to the potential issue of the lease. On the other hand, it did not simply alert him to a problem; it contained an unqualified assertion that the problem had ceased to exist. A properly J advised vendor would or should have known that the statement was untrue. Nor does it assist the vendor, in my view, that the purchaser may also have had the means of knowledge through his assumed notice of the contents of the register. Mere knowledge of the bankruptcy petition was not conclusive since it did not rule out a valid surrender with consent of the court. In any event, by the unqualified terms of the representation, in my view the vendor impliedly represented that he had taken reasonable steps to confirm its accuracy.

[23] This conclusion makes it unnecessary to reach a firm view of the effect of section 42(3). However, I will comment briefly on what appears to be a surprisingly novel proposition (suggested, it seems, by an enterprising pupil in Ms Evans-Gordon's chambers). The subsection K reads as follows:

> A stipulation contained in any contract for the sale or exchange of land made after the commencement of this Act, to the effect that an outstanding legal estate is to be traced or got in by or at the expense of a purchaser *or that no objection is to be taken on account of an outstanding legal estate*, shall be void.

(Emphasis added.)
Ms Evans-Gordon submitted that, under the second part of the subsection, ESC 7 is ineffective, in so far as it purports to rule out any objection on account of the existence of the lease (being a legal estate). L

[24] The subsection has attracted very little attention remarkably little if it has the effect for which she argued. It does not appear to have been cited in any reported cases. It attracts little comment in any of the modern textbooks; it is not even mentioned in the current edition of *Megarry and Wade*. According to a contemporary textbook (*Dart's Treatise on the Law and Practice relating to Vendors and Purchasers of Real Estate* (8th ed) 1929, vol 1, at pp150-151), section 42(3) marked a shift from previous practice:

> A condition binding a purchaser, if he considered the legal estate outstanding, to be at the expense of getting it in, was, formerly, not uncommon, and was held to throw on him the risk of making out in whom the legal estate was vested; but by [s 42(3)] such a condition is nugatory. M

In *Wolstenholme & Cherry* (12th ed) 1932) (which is often a useful guide to the thinking of the original draftsmen), there is the following note:

> a condition that a legal estate should be left outstanding would have meant that the vendor had a bad title. It does not mean that the vendor cannot sell subject to a legal term, a legal mortgage, a legal rentcharge or a legal easement…

[25] Mr Loveday suggested that the most likely explanation for the provision is that it was related to the simplification of legal and equitable estates that was a central feature of the 1925 Act reforms, but of little practical concern today. He referred by analogy to section 39(2),

A which as part of the elaborate transitional provisions in the first Schedule, provided for the discharging or getting in of "outstanding legal estates".

[26] Whether or not it is so limited, it seems to me more likely that the reference to an "outstanding legal estate" was intended to refer to the quality of the interest to be transferred, rather than of any potential incumbrance. I find it hard to see any reason, as a matter of legislative policy, for the introduction of a strict avoidance provision relating only to those incumbrances that happen to be "legal estates" under the 1925 Act scheme. That seems to me consistent with the note in *Wolstenholme & Cherry*. Fortunately, I need not explore this topic further for the B purposes of this case.

Technical defect

[27] I can deal with this point shortly, because I agree entirely with the judge. He said:

> 26. In my judgment this submission is unsustainable. Either a successful application to the Court seeking ratification of the surrender would be required, or a successful request to Mr Airey's trustee in bankruptcy to disclaim. The outcome of neither can have been a foregone conclusion at the time. Neither was in Area Estates' sole power, control or grant. Area Estates had no right that C would enable it to insist on ratification or disclaimer.
>
> 27. Indeed it is hard to see why the Court would have ratified a surrender by Mr Airey when a bankruptcy petition was pending. As for disclaimer, the outcome of the request would depend on the trustee's assessment at the time whether, in one form or another, there was value to be had for the bankrupt estate from the Lease.
>
> 28. Of course a trustee's assessment will vary from case to case. Even if it is impossible that an assessment by the trustee in the present case would have concluded that there was value, the Property was still subject to a lease and Area Estates had no right to remove it. I cannot treat the presence of a subsisting lease encumbering freehold property sold with vacant possession as a "technical conveyancing defect".

[28] Mr Loveday sought comfort in the judge's acknowledgement D of the "powerful and… commercially persuasive" picture arising from the facts of the case, in which the premises had stood empty for more than a year with no rent being paid and the trustee would be only too pleased to rid himself of the liability. He also referred to the various cases mentioned in *Emmet and Farrand*, in para 5.005, which, it is said, have left the test as one of "uncertain degrees".

[29] However, like the judge, I think that the best guide is to be found in the leading case *MEPC Ltd* v *Christian-Edwards* [1981] AC 205, in which Lord Russell of Killowen, at p220C-D said:

> [I]f the facts and circumstances of a case are so compelling to the mind of the E court that the court concludes beyond reasonable doubt that the purchaser will not be at risk of a successful assertion against him of the incumbrance, the court should declare in favour of a good title shown.

[30] The judge commented:

> 34. There is no prospect in the present case of such a conclusion being appropriate. For all the argument about degree of risk, on which views will differ, Area Estates simply cannot show that as at the date for completion there was no risk of the trustee in bankruptcy asserting successfully that the Lease subsisted. The present case is a far cry from that before the House of Lords, where it was concluded that there was no risk of a claim for specific performance being upheld on the basis of a contract some 70 years old which F had been abandoned some 50 years before.

I agree.

[31] For these reasons, I would dismiss the appeal.

Moore-Bick LJ said:

[32] I agree.

Wilson LJ said:

[33] I also agree.

Appeal dismissed.

Pittack *v* Naviede

CHANCERY DIVISION

[2010] EWHC 1509 (Ch)

24 June 2010

MR MARK HERBERT QC, SITTING AS A DEPUTY HIGH COURT JUDGE

Estates Gazette 30 October 2010

[2010] 43 EG 100

Sale of land — Subsales — Standard conditions of sale — Contract for sale of residential property — Purchaser seeking to effect subsale by direct transfer from vendor to subpurchaser — Vendor refusing to transfer to party other than purchaser — Vendor forfeiting purchaser's deposit for non-completion — Whether subsale contrary to clause 1.5 of standard conditions (4th ed) prohibiting transfer of benefit of contract — Whether purchaser entitled to return of deposit

In July 2007, contracts were exchanged on a sale by the defendant to the claimant of a leasehold residential property for £2.7m. The contract incorporated the standard conditions of sale (4th ed), clause 1.5 of which prohibited the purchaser from transferring "the benefit of the contract". Prior to completion, the claimant had found subpurchasers to whom he proposed to subsell the property for the same price that he had agreed to pay, with the two contracts to be completed by a single transfer from the defendant to the subpurchasers. However, he did not inform the defendant of this at the time. The defendant became aware of the subsale arrangements early in June 2008, shortly before the date for completion of the sale to the claimant. He insisted that he would transfer the property only to the claimant. Thereafter, the subpurchasers withdrew from the subsale, on which contracts had not been exchanged. The claimant therefore lacked the necessary funds with which to complete the purchase from the defendant.

The defendant served a notice to complete, with which the claimant did not comply. The claimant instead purported to rescind the contract, claiming that the defendant was in breach of contract for improperly refusing to execute a subsale transfer. Thereafter, the defendant also purported to rescind, in reliance on the claimant's failure to complete, and forfeited the claimant's deposit.

The claimant brought proceedings to recover the deposit, contending that his failure to complete had been caused by the defendant's wrongful refusal to transfer to the subpurchasers. He also alleged a failure on the part of the defendant to provide the necessary licence to assign from the landlord by the relevant date. The defendant argued that he had been entitled to refuse a transfer to a subpurchaser because a subsale breached clause 1.5 of the standard conditions.

***Held*: The claim was allowed. (1) An assignment of the benefit of a contract is not the same as a requirement to effect the conveyance in favour of a third party. In the absence of a contractual provision to the contrary, be it express or implied, a purchaser is entitled to require the vendor to transfer the property on completion to a third party of its own choosing. It can therefore effect a subsale and require the vendor to execute a conveyance or transfer directly in favour of the subpurchaser. The standard conditions of sale (4th ed) do not alter that position under the general law. Unlike the standard commercial property conditions of sale (2nd ed) that apply to commercial property sales, the standard conditions for residential conveyancing do not expressly prohibit subsales; nor are there any grounds for implying such a prohibition. A vendor, when adopting the non-commercial standard terms, may include an explicit exclusion of subsales to the same effect as that in the commercial standard terms. If it does not do so, however, subsales will not be precluded under clause 1.5 of the standard conditions. Accordingly, the claimant had been entitled to rescind the contract of sale following the defendant's refusal to transfer to a subpurchaser. (2) On a sale of a leasehold interest, clause 8.3 of the standard conditions obliges the vendor to obtain a licence to assign only to the original purchaser. It is the responsibility of that purchaser to obtain consent for any assignment to a subpurchaser. That position is not inconsistent with the purchaser being entitled to nominate its subpurchaser as transferee, provided that it obtains the necessary additional consent. The defendant had failed to procure a written licence to assign to the claimant three working days before the contractual completion date, such that the claimant had been entitled to rescind on that ground also, pursuant to clause 8.3.3 of the standard conditions. The claimant was entitled to the return of his deposit.**

The following cases are referred to in this report.

Aubergine Enterprises Ltd v *Lakewood International Ltd* [2002] EWCA Civ 177; [2002] 1 WLR 2149

Curtis Moffat Ltd v *Wheeler* [1929] 2 Ch 224

Earl of Egmont v *Smith*; *Smith* v *Earl of Egmont* (1877) LR 6 Ch D 469, Ch

This was the hearing of a claim by the claimant, Daniel Pittack, against the defendant, Muhammad Naviede, for the return of a deposit paid under a contract for the sale of land.

Timothy Dutton (instructed by Lucas McMullan Jacobs) appeared for the claimant; the defendant appeared in person.

Giving judgment, **MR MARK HERBERT QC** said:

[1] This is an action concerning property conveyancing. It raises a question of construction on the standard conditions of sale (4th ed), namely whether a clause excluding the transfer of the benefit of the contract operates to exclude subsales. There is no doubt that the parties entered into a contract, and no doubt that the contract has been rescinded. The only question is whether it was the purchaser or the vendor who was entitled to rescind, and the answer to that question determines which of them is entitled to the deposit. Neither side claims damages.

[2] Both parties are individuals, and the property in question is a townhouse in Regents Park held on a long lease from the Crown Estate Commissioners. The claimant is the purchaser, Mr Daniel Andrew Pittack, and he appeared before me by Mr Timothy Dutton of counsel. The defendant is Mr Muhammad Maximilian Naviede, and he appeared before me in person.

[3] On the purchaser's side, I heard evidence from Mr Pittack himself, from his personal assistant Ms Sarabjeet Howard-Brown (also known as Ms Sally Howard) and from his solicitor Mr Garry Stephen Lucas, of Lucas McMullan Jacobs. I also read a short witness statement from Mrs Asha Narang, who was at one time intending to be a subpurchaser of the property. On the vendor's side, I heard evidence from Mr Naviede

himself and also from Mr Emanuel Langley, a business associate of Mrs Narang's son. Mrs Narang is in India, and her statement was admitted under the Civil Evidence Act 1968. All the other witnesses were cross-examined. I may say that although oral evidence took up more than five hours of the court's time, it contained little of value that was not clear from the documents.

Facts

[4] The basic facts are that, in July 2007, Mr Pittack and Mr Naviede exchanged contracts to buy and sell a leasehold residential property and the contractual completion date was originally in July 2008 but, in the event, was brought forward to 20 June 2008. During 2008, Mr Pittack formed the intention of subselling the property, but no contracts for the subsale were ever exchanged and, on 16 June, the subpurchasers withdrew. Mr Pittack had no means immediately available to complete the contract himself, and he did not do so. He claims, however, that Mr Naviede was already in breach of contract in two ways, namely that: (i) Mr Naviede had wrongly refused to transfer the property to any subpurchasers; and (ii) he had not received the landlord's licence to assign the lease to Mr Pittack in due time. He claims that each of these alleged breaches entitled him to rescind and recover the deposit. If Mr Pittack is wrong on both those points, it is obvious that he himself was in breach and Mr Naviede can keep the deposit.

[5] In more detail, the facts are these. A lease of the property was granted by the Crown Estate Commissioners to Mr Naviede in 2002 for a term of more than 105 years, and the lease is therefore a "new tenancy" for the purposes of the Landlord and Tenant (Covenants) Act 1995 (the 1995 Act). This means, among other consequences, that, under section 5 of the Act, an assignment of the lease by Mr Naviede would release him from liability under the tenant's covenants in the lease, even though he is the original lessee.

[6] On 25 July 2007, after several months of precontract negotiations, Mr Pittack and Mr Naviede exchanged contracts for the sale of the property to Mr Pittack for £2.7m. The contract incorporated the standard conditions of sale (4th ed). Clause 1.5 provides:

> 1.5 Assignment
> The buyer is not entitled to transfer the benefit of the contract.

Clause 8.3 dealt with obtaining consent where the contract was for the assignment of a lease, as was the case. For this purpose, "consent" means consent in the form that satisfies the requirement to obtain it, and clause 3.29.3(c) of the lease, read with clause 1.16, provides for written consent, signed by or on behalf of the landlord, such consent not to be unreasonably withheld. Clause 8.3.2 requires: (a) the seller to apply for the consent at his own expense and to use all reasonable efforts to obtain it; and (b) the buyer to provide all information and references reasonably required. Clause 8.3.3 provides:

> 8.3.3 Unless he is in breach of his obligation under condition 8.3.2, either party may rescind the contract by notice to the other party if three working days before completion date (or before a later date on which the parties have agreed to complete the contract):
> (a) the consent has not been given, or
> (b) the consent has been given subject to a condition to which a party reasonably objects.
> In that case, neither party is to be treated as in breach of contract and condition 7.2 applies.

Clause 7.2 provides:

> 7.2 Rescission
> If either party rescinds the contract:
> (a) unless the rescission is a result of the buyer's breach of contract the deposit is to be repaid to the buyer with accrued interest
> (b) the buyer is to return any documents he received from the seller and is to cancel any registration of the contract.

[7] There were also special conditions, including the following:

(a) The contractual date for completion was fixed at 25 July 2008, a whole year after the date of the contract, but the vendor was given the right to bring the completion date forward by serving three months' notice.

(b) The deposit was to be held by the solicitor as the vendor's agent, so that the vendor had the use of it straightaway.

(c) Against those conditions, which were of benefit to the vendor, he agreed not to increase the borrowing secured on the property without the buyer's consent.

(d) Special condition 9 provides:

> The Buyer will join in any Licence to Assign issued by the Landlord and will enter into any covenants and adhere to any conditions reasonably required by the Landlord. The Buyer will also produce to the Landlord such references as the Landlord shall require.

[8] On 31 July 2007, soon after exchange, Lucas McMullan Jacobs asked Mr Naviede's solicitor Zatman & Co whether it was contacting the landlords' agent to find out their requirements in respect of the licence to assign. Zatman & Co responded rather slowly, perhaps understandably so in view of the long completion date. On 21 September 2007, it did indeed write to the Crown Estate and, on 28 September 2007, Cluttons responded on its behalf enclosing a printed form of application incorporating a request for the information required from an assignor. Some 12 weeks later, on 17 December 2007, Zatman & Co forwarded that to Lucas McMullan Jacobs and, on 2 January 2008, the latter replied that it was obtaining the references and information required and that it would forward these when they were to hand. In the end, that did not literally happen, as will appear.

[9] Some weeks after exchange of contracts, Mr Pittack considered that he might want to borrow money to effect the purchase and through his personal assistant, Ms Howard, asked Mr Naviede for a surveyor to have access to the property for the purpose of an inspection. This was refused, despite Ms Howard speaking on several occasions to Mr Naviede himself and also to his associate, Mr Steve Lavin. Mr Pittack then began considering selling on the property to one or more subpurchasers.

[10] One potential subpurchaser, although not the first, was Mrs Narang. By the beginning of April 2008, Lucas McMullan Jacobs was in correspondence with Goldkorn Matthias Gentle, which wrote that it had been instructed by Mrs Narang and Mr Langley as co-purchasers. The proposal was that the subpurchasers would buy the property for the same price that Mr Pittack had agreed to pay, namely £2.7m, and Lucas McMullan Jacobs intended that the two contracts would be completed by a single transfer from Mr Naviede to the subpurchasers. It did not, however, inform Mr Naviede or his solicitor of that intention at that time.

[11] On 20 March, Mr Naviede's solicitor duly gave notice bringing the contractual date for completion forward to 20 June 2008.

[12] From that point, matters moved somewhat more quickly. During May, there was correspondence between Goldkorn Matthias Gentle and Lucas McMullan Jacobs with regard to the proposed subpurchase, and references were obtained for Mr Langley, as was a bank reference for Mrs Narang. On 6 June, Lucas McMullan Jacobs sent a draft transfer to Zatman & Co for approval, but this was for a transfer to Mr Pittack alone. However, Lucas McMullan Jacobs had also been communicating with Cluttons, as agent for the Crown Estate and, by 10 June, a draft licence to assign to Mrs Narang and Mr Langley was in its final form, and Pemberton Greenish, acting for the Crown Estate, had written that it did not expect the Crown to want Mr Pittack to be a party to the licence.

[13] On 11 and 12 June 2008, there was a comparative flood of correspondence that is not easy to analyse because it does not all appear in strict chronological order in the trial papers. First, Cluttons must have given Zatman & Co the names of the intended subpurchasers, and Mr Naviede's view on potential subsales was revealed to those acting for the Crown Estate. On 11 June, Zatman & Co wrote to Cluttons and Pemberton Greenish to say that it had come to its attention that Lucas McMullan Jacobs had been corresponding directly with it in connection with a licence to assign to Mr Langley and Mrs Narang, and that Mr Naviede did not know Mr Langley and Mrs Narang because

A he had contracted to sell the property to Mr Pittack. To Pemberton Greenish, it added that Mr Naviede had notified it that he was not prepared to be a party to a licence to anyone other than Mr Pittack. On 12 June, Cluttons replied to Zatman & Co confirming that Cluttons and Pemberton Greenish had been contacted by Lucas McMullan Jacobs in connection with the assignment and that it had received satisfactory references for Mr Langley and Mrs Narang. It also wrote that it had advised Pemberton Greenish not to action the assignment any further until the matter was clarified. Pemberton Greenish wrote that it had been instructed to take no further action on the matter.

[14] Also on 12 June, Lucas McMullan Jacobs wrote to
B Zatman & Co that Mr Pittack may exchange contracts to sell the property prior to completion, and enclosed a second draft transfer, this being for a transfer direct from Mr Naviede to the subpurchasers.

[15] On the same day, Zatman & Co first replied by fax to Lucas McMullan Jacobs' letter of 6 June. It approved the form of the draft transfer (which was expressed as a transfer to Mr Pittack). As to the licence to assign, it wrote that it did not understand why Lucas McMullan Jacobs had written to Cluttons and Pemberton Greenish, pointing out that the contract provides for this to be done by the vendor. The fax also mentioned that Mr Pittack had not provided the references requested on 17 December 2007. Still, on 12 June, Lucas McMullan
C Jacobs replied to the fax of that day, writing that it did not understand Zatman & Co's objection to its approach to the Crown's agent directly, and that this was now academic because the Crown was prepared to grant the licence. At around the same time, Zatman & Co faxed a short second letter stating that the contract was not assignable and that Mr Naviede was not prepared to transfer the property to anyone other than Mr Pittack. Lucas McMullan Jacobs therefore added a postscript to its reply, to the effect that Mr Pittack was not assigning the contract, and that he was entitled to direct Mr Naviede to transfer the property to a third party because the contract did not contain a prohibition on subsales.

D [16] On 13 June, Zatman & Co repeated Mr Naviede's insistence that he would transfer only to Mr Pittack and that Mr Pittack was obliged to provide references for himself. At the same time, it spoke to Cluttons by telephone, which told it that, since Mr Pittack was a current lessee of the Crown Estate, references for him may not be required. Zatman & Co faxed Cluttons to ask it to confirm this in writing.

[17] The next working day was Monday 16 June. Zatman & Co telephoned Cluttons again and, after receiving confirmation that the Crown Estate would not require references for Mr Pittack, asked Cluttons to arrange for a licence to assign to be prepared authorising a transfer to Mr Pittack. Cluttons duly replied, stating that Pemberton
E Greenish had been instructed to continue with the granting of the licence in favour of Mr Pittack. That letter was sent by post, and there is no evidence that it was either delivered by hand (to Zatman & Co in Manchester) or transmitted to it or anyone else by fax or by e-mail.

[18] Also on 16 June, Goldkorn Matthias Gentle e-mailed Lucas McMullan Jacobs to say that its clients were not proceeding with the purchase, and it confirmed this by letter the following day. Mrs Narang and Mr Langley gave differing reasons for the withdrawal: she that Mr Naviede was not co-operating with the landlords, he that Mr Naviede had not allowed access to the property (and it is the case that Mrs Narang and Mr Langley had only ever seen the outside).
F However, the reasons do not matter. The fact is that no contracts for any subsale were exchanged. I should record that Mr Langley gave evidence before me that he never intended to be a co-purchaser himself, despite having apparently authorised Goldkorn Matthias Gentle to represent to all relevant parties that he was. This casual attitude to accuracy raised doubts concerning his evidence generally, but in the end nothing turned on it.

[19] Mr Lucas was away from the office for the entire week beginning 16 June, and an assistant solicitor, Mr Mughal, was holding the fort at Lucas McMullan Jacobs. On 19 June, Mr Mughal e-mailed Zatman & Co that Mr Pittack would not be completing the purchase because Mr Naviede had improperly refused to execute a subsale transfer and had not provided a licence to assign. Zatman & Co replied by fax, repeating that Mr Naviede was obliged to transfer the property only to G
Mr Pittack (referring to para 14.3.1 of the Law Society's *Conveyancing Handbook*), and stating that the reason why the licence to assign had not been provided was that Mr Pittack had failed to provide references. Even so, it had found that the Crown Estate would grant the licence, and the letter enclosed a draft licence that needed only the addition of Mr Pittack's full address.

[20] During 19 June, there was some attempt to negotiate terms for a supplemental agreement, but this did not result in a concluded agreement. Instead, on 20 June, Zatman & Co served a standard-form notice to complete, giving 10 working days to complete by 4 July and making time of the essence. H

[21] After that date, there was no attempt on the part of Mr Pittack to complete the purchase. Instead, on 1 July, Lucas McMullan Jacobs wrote a formal letter that reasserted the argument that clause 1.5 does not exclude direct-transfer subsales, so that Mr Naviede's refusal to transfer on those terms was improper, and gave notice rescinding the contract on the basis of Mr Naviede's repudiatory breach and, alternatively, because the licence to assign was not available by the required date. On 7 July, Zatman & Co replied, relying on Mr Pittack's failure to complete on 4 July in accordance with the notice to complete, and rescinded the contract.

Clause 1.5 of the standard conditions (4th ed) J

[22] At the root of this case is the true construction of clause 1.5 of the standard conditions, which provides that a purchaser is not entitled to transfer the benefit of the contract to a third party.

[23] On behalf of the purchaser Mr Pittack, Mr Dutton said that it means what it says but no more, and that it does not prevent subsales. He also pointed to special condition 9, which would be irrelevant with regard to a new lease to which the 1995 Act applies except in the case of a subsale completed by a direct transfer to the subpurchaser. This implies, he said, that the possibility of a direct transfer to a subpurchaser was impliedly acknowledged in the terms of the contract.

[24] According to this argument, what clause 1.5 prevents is K
the actual transfer to a third party of the benefit of the contract, by which, after due notice to the vendor, the assignee takes the place of the original purchaser and can maintain a claim against the vendor itself. The general law provides that the purchaser is generally entitled to obtain a transfer to itself or as it directs, and clause 1.5 does not remove that right. If this is right, then when, on 12 and 13 June 2008, Mr Naviede refused outright to transfer the property to anyone other than Mr Pittack, that amounted to an improper repudiation of the contract and Mr Pittack was entitled to rescind on the strength of that repudiation, as he did on 1 July, and will be able to recover the deposit. For this purpose, said Mr Dutton, it is immaterial that Mr Langley and L
Mrs Narang withdrew from the purchase on 16 June.

[25] Mr Dutton showed me the corresponding clause in the standard commercial property conditions of sale (2nd ed). I shall call these the "commercial" standard terms to distinguish them from the "non-commercial" standard terms that were incorporated into the Pittack-Naviede contract. Clause 1.5 of the commercial standard terms has a different heading and an extra subclause:

> 1.5 Assignments and sub-sales
>
> 1.5.1 The buyer is not entitled to transfer the benefit of the contract.
>
> 1.5.2 The seller may not be required to transfer the property in parts or to M
> any person other than the buyer.

That clause makes it clear that subsales are not authorised, so that the purchaser is not entitled to nominate a third party to take a direct transfer from the vendor. Mr Dutton argued that, in the absence of clause 1.5.2 or its equivalent from the non-commercial standard terms, there is nothing in those terms explicitly to exclude subsales, and there is no reason to imply such an exclusion; special condition 9 even points the other way.

Subsales in general

[26] The first step is to establish what the position would be in the absence of clause 1.5. Some of the modern textbooks on land law and

A conveyancing do not deal with subsales at all. Nor has Mr Dutton's research discovered recent judicial authority on the matter. What he does point to is a *dictum* of Sir George Jessel MR in *Earl of Egmont* v *Smith* (1877) LR 6 Ch D 469. That case primarily concerned the purchaser's ability to insist on the property being conveyed to him in separate lots, but the relevant clause also mentioned subsales. Clause 6 of the contract provided:

> The purchaser shall be at liberty to require the vendors to convey the property to him, or as he shall direct, in such portions and by such separate conveyances as he shall think fit; provided that the purchaser shall bear and pay all additional expense whatever incurred or sustained by the vendors in consequence thereof;
> B and if the vendors or their solicitor shall enter, at the request or on behalf of the purchaser, into any negotiation or communication with, or furnish any abstracts, evidence, or information to any sub-purchaser or sub-purchasers from him, the same shall be without prejudice in any way to the vendor's rights and the purchaser's liabilities under this agreement, and shall not be deemed to be a recognition of such sub-purchaser or sub-purchasers, or a waiver or abandonment of any of the vendor's rights under this agreement; and the vendors shall not be required to execute any conveyance of any part of the premises except on payment of the entirety of the purchase-money and all interest, if any, due thereon, and all costs and expenses payable by the purchaser under this agreement or otherwise.

C [27] Sir George Jessel MR said this about that clause, at p474:

> As I understand that condition, it is no more than the law would require. I take it that in no case can a vendor object to convey the sold property in parcels on receiving the whole purchase-money. Whether he could or could not object to convey it at various times is another matter, but that he could not object to convey on his being paid the additional expense occasioned by his joining in several conveyances instead of one, I am quite clear.
>
> An ordinary contract of sale is not only to convey to the purchaser, but to convey as the purchaser shall direct. Whether or not he could be asked to keep his legal estate for a long period, and convey it in portions at various times is a question perhaps deserving of more consideration. But this condition does not exceed the ordinary law; it gives the vendors the additional expense; it
> D gives them the whole purchase-money, and is not to operate as a waiver or abandonment of any of their rights.

[28] Mr Naviede submits that this should be read in a narrow sense, limited by the context, namely that a purchaser can direct the vendor to convey the land, to the purchaser itself, in whatever lots it requires. However, Mr Dutton said that it should be taken as meaning also that a purchaser is entitled to direct the vendor to transfer the property to a third party, such as a subpurchaser, on completion. He pointed out that the wider reading was adopted by no less an authority than *Williams on Vendor and Purchaser* (4th ed, 1936), at pp641-642:

> E On the side of the grantee or grantees under the conveyance, the purchaser himself is in general the only necessary party. But… the vendor's obligation is to execute a conveyance of the land sold to the purchaser, or as he shall direct. The purchaser is therefore entitled to require the conveyance to be made to some other person or persons than himself, or to himself and others, and the vendor is bound to assure the lands sold accordingly. It appears that in such case the vendor may in general demand that the purchaser, with whom alone he has contracted, shall be made a party to the conveyance in order to testify that he has directed the conveyance to be made to a stranger to the contract and that the vendor has duly performed his part of the contract by complying with this direction. But if the purchaser should have made an absolute assignment of all his interest in the contract, and the assignee has given notice of the assignment
> F and is willing to take upon himself the whole burden of the original contract and prove his title by assignment from the purchaser, then it seems that the vendor ought to complete the contract with the assignee alone, without requiring any further concurrence of the purchaser.

[29] That passage does not use the word subsale or subpurchaser, but to my mind the meaning is clear enough. In the absence of a provision to the contrary, express or implied, a purchaser may effect a subsale and require the purchaser to execute a conveyance or transfer in favour of the subpurchaser direct; although the original purchaser may be required to participate for the reason stated. There is also a clear distinction between an assignment of the benefit of the contract and a requirement to effect the conveyance in favour of a third party.

[30] A similar question arose in *Curtis Moffat Ltd* v *Wheeler* [1929] G
2 Ch 224, which also concerned a contract for the sale of leasehold property having a prohibition on assignment without the landlord's consent. The landlord was not prepared to grant a licence to assign to the plaintiff, being a company, but it was willing to grant a licence in favour of an individual as nominee for the company. The plaintiff requested an assignment in favour of a nominee, and it was held that the vendor was not entitled to refuse. After referring to authorities in which a lessor that has entered into a contract to *grant* a lease to a particular person can be compelled to grant the lease to a solvent nominee, Maugham J applied the same principle to the case of an *assignment* of a lease, at p237: H

> In my opinion, treating the matter as one of substance, the principle of these cases should be followed in the case of a sale of leaseholds where, as here, the contract was not entered into upon considerations personal to the purchaser. The restriction on assignment was so qualified that the defendant was always able to compel the [landlord] to assent to an assignment to any proper person, and no difficulty has in fact arisen on this head.

The reference to the restriction on assignment being qualified was a reference to a proviso that consent should not be withheld to an assignment or underletting of the premises to a respectable and responsible person. J

[31] The conclusion that I draw from these passages is that, in the absence of provision to the contrary, a buyer is entitled to require the vendor to transfer the property on completion to a third party of its choosing. In my view, this applies similarly to the sale of leasehold property, including cases where the prohibition on assignment without consent (if any) cannot be unreasonably withheld, although in this last case the fact may be that the landlord withholds its consent reasonably, in which case other considerations arise.

Effect of clause 1.5

[32] It remains to decide what effect clause 1.5 of this particular contract has on that general statement of law. K

[33] Mr Dutton was prepared to acknowledge a potential difficulty with his construction in the context of a lease where the vendor has covenanted not to assign the lease without the consent of the landlord. That was the case here. The relevant clause is clause 8.3, the important parts of which I have quoted in [6] above. The essence is that the purchaser is required to provide references and other information as required by the landlord, and the vendor is then required to obtain the licence at his own expense. If there is a subsale, in whose favour should the licence to assign be obtained? Moreover, if the purchaser initially intends to take the assignment itself and the vendor obtains a licence to assign to it and then the purchaser contracts to subsell and requires the L
vendor to complete by a direct transfer to the subpurchaser, what is the vendor obliged to do? Does it have to obtain a further licence in favour of the subpurchaser, again at its own expense?

[34] In such a case, it seems to me that unless and until the purchaser informs the vendor of the identity of the subpurchaser, the only obligation on the vendor is to obtain a licence to the original purchaser, and a prerequisite to that is for the purchaser to provide whatever references and other information are required by the landlord. In addition, it is no doubt required under its own contract to obtain the licence to assign to its subpurchaser, subject to the subpurchaser providing its references and other information. Under the main M
contract, it remains the duty of the vendor only to obtain consent for the assignment to the purchaser. The purchaser will have to obtain consent for the assignment to the subpurchaser.

[35] I therefore find that the difficulties under clause 8.3 do not truly help me to construe clause 1.5. Clause 8.3 creates an obligation only to obtain consent for the assignment to the purchaser, but that does not seem to me inconsistent with the purchaser being entitled to nominate its subpurchaser as transferee, provided that the purchaser obtains the necessary additional consent.

[36] Mr Naviede resists all this analysis of clause 1.5. He acknowledges, doubtless in deference to the passage from *Earl of Egmont* on which Mr Dutton relied, and also *Curtis Moffat*, that a

purchaser may require the vendor to transfer the property direct to a nominee of the purchaser, but that this applies only to a true nominee. He acknowledges, too, that there is nothing to stop a purchaser effecting a subsale and arranging for a separate transfer from the purchaser to the subpurchaser, after (perhaps immediately after) the vendor has effected a transfer in favour of the original purchaser. In short, he says, clause 1.5 does not prevent a transfer to a nominee or back-to-back transfers. However, he says that a direct transfer from the vendor to a subpurchaser involves the transfer of the benefit of the contract to the subpurchaser and is therefore prohibited by clause 1.5. On the other hand, he accepts that if he is wrong about clause 1.5 and it does not exclude subsales by direct transfer to the subpurchaser, Mr Pittack was entitled to rescind when he, Mr Naviede, refused to complete in favour of anyone other than Mr Pittack. He made that concession first in the witness box and repeated it in his closing submissions.

[37] The most strongly expressed ground for his hostility to a substituted transferee was that a direct transfer would, or might, avoid stamp duty land tax (SDLT) and that the vendor should not be compelled to be party to a document effecting any such transaction. He also pointed out that the property is a listed building and that he was apprehensive (if I may paraphrase his concerns) about being made personally liable for breaches of the regulations committed in future by successors in title.

[38] I am unimpressed by those reasons. As to SDLT, either a direct subsale does avoid the imposition of two charges to tax or it does not, and there are provisions in the Finance Act 2003 (as amended) determining that question. To my mind, there is therefore nothing improper in a purchaser and subpurchaser seeking to effect the subsale by a direct transfer and no impropriety in the vendor acceding to such a request, even if it is not required to do so. As to the rules relating to listed buildings, a vendor either is or is not exposed to possible fraudulent claims by successors in title alleging that defaults of their own were effected by it, but that does not seem to me to be affected one way or the other by the subsale being effected by a direct transfer as opposed to back-to-back transfers.

[39] Mr Naviede has also seized on the qualification expressed by Maugham J in the passage cited from *Curtis Moffat*, in [30] above, that the freedom to substitute a subpurchaser of a lease applies only if "the contract was not entered into upon considerations personal to the purchaser". He points to the prohibition in special condition 9 against him borrowing more than a stated amount. However, that is not, in my view, a consideration that is personal to the purchaser. If anything, it is a consideration that is personal to the vendor, but in reality it is neither. It is a prohibition inserted by agreement, in recognition of the long completion date and the deposit being available to the vendor. The term was designed to ensure, as Mr Pittack stated in his evidence, that the property would not become overburdened with debts secured on it, a circumstance that might have hindered the vendor in satisfying its obligation to discharge incumbrances on completion. It is a term of benefit to any transferee. I do not regard it as a consideration that is personal to the individual purchaser. In any case, what Maugham J referred to was a *contract* (not individual terms in a normal contract) made for personal considerations.

[40] Even so, Mr Naviede has respectable support for his argument, namely the *Law Society's Conveyancing Handbook* (16th ed). Part B.14 of this handbook is headed "Sub-sales", and para 14.3 states (in part) as follows:

> 14.3 Contractual restriction on sub-sales
>
> 14.3.1 Where a buyer intends to sell the property to a third party by way of sub-sale he must ensure that his existing contract to purchase the land contains no restriction which would prevent the sub-sale. Standard Condition 1.5 and Standard Commercial Property Condition 1.5 both state that the buyer is not entitled to transfer the benefit of the contract. As a result the seller cannot be forced to transfer the title to anyone other than the buyer named in the contract. This provision will not prevent sub-sales of freeholds nor sub-assignments of leases. However, the buyer will have to complete the second transaction separately and subsequent to his own purchase. The provision does however provide protection on the grant of a lease as the seller (landlord) cannot be forced to grant the lease to anyone other than the buyer (tenant) in the contract. A seller should resist a buyer's request to remove Condition 1.5 from the contract, because its presence ensures that the seller will take the benefit of the covenants given by the first tenant on grant of a lease.
>
> 14.3.2 If a restriction against sub-sales is included in the buyer's contract, he will be able to contract to resell the land to a third party, but will have to complete this second transaction separately and subsequently to his own purchase.
>
> ...
>
> 14.3.4 Where there is a restriction on sub-sales in the contract it may be possible for the buyer to assign the benefit of his contract with the seller to the third party and then to complete the purchase as nominee for the third party.

[41] To the same effect is para 16.9 of Frances Silverman's *The Standard Conditions of Sale: A Conveyancer's Guide* (7th ed); unsurprisingly so, given that Mrs Silverman is, I am told, also the author of the Law Society's *Conveyancing Handbook*.

[42] If it is correct that clause 1.5 of the non-commercial standard terms excludes subsales, then I agree with the passage just cited. However, I regret that I find Mrs Silverman's commentary unsatisfactory because it does not mention, let alone emphasise or explain, the difference between the texts of the commercial and non-commercial standard terms. It asserts, but does not explain, that the short-form clause 1.5 of the non-commercial standard terms does prohibit subsales, and at the same time it points out that subsales completed by back-to-back transfers cannot be prevented. I certainly accept that subsales completed by back-to-back transfers cannot in practice be prevented, however clause 1.5 is construed and whichever version of the standard terms is in issue. The vendor in such a case is not being asked to do anything but execute a transfer to the contractual purchaser subject to receiving the balance of the sale price. However, I find difficulty with two remaining questions:

(1) If a direct transfer to a nominee is permitted in the sense that the purchaser can insist on it, as Mr Naviede accepts, how is it that a direct transfer to a subpurchaser is not also permitted?

(2) If a subsale completed by back-to-back transfers is permitted, so that *ex hypothesi* it does not amount to a transfer of the benefit of the contract, exactly how and when does a subsale completed by direct transfer become a transfer of the benefit of the contract so as to be excluded by clause 1.5?

[43] I do not find satisfactory answers to those questions. The most promising answer at first sight, preserving some consistency, is that on reflection a vendor cannot be required to effect a direct transfer in favour of a nominee, and that every subsale does, on analysis, amount to the transfer of the benefit of the contract. After all, clause 1.5 is not couched in technical terms, and it might therefore be right to read it in a non-technical way that does equate the substitution of a subpurchaser with a transfer of the benefit of the contract. If correct, that would allow a vendor to refuse any demand by the purchaser to execute a transfer in favour of a third party. It would nevertheless have to recognise that the vendor cannot in practice prevent back-to-back transfers.

[44] However, in the end, I am persuaded that such a construction of clause 1.5 is not justified. The *Egmontdictum* and particularly the passage from *Williams on Vendor and Purchaser* that I have cited in [28] above, show that there are two potential difficulties for a vendor and that they are different although perhaps often connected. One difficulty is the purchaser assigning the benefit of the contract to a third party, thus exposing the vendor to unwanted contractual obligations owed to the assignee. The draftsmen of the standard terms (in both versions) evidently regard it as commonplace for a vendor to want to exclude that difficulty. The other difficulty is a subsale not including a transfer of the benefit of the contract, and the draftsmen of the commercial standard terms evidently regard it as commonplace for a vendor of commercial property to want to exclude that possibility too. However, the draftsmen of the non-commercial standard terms have not visibly reached the same conclusion with regard to residential conveyancing.

[45] In my judgment, Mr Dutton's approach to clause 1.5 is correct. It is possible to include an exclusion of subsales, as clause 1.5.2 of the commercial standard terms does, and it is open to any vendor adopting the non-commercial standard terms to include an explicit exclusion

to the same effect. However, if that explicit exclusion is not present, clause 1.5 of the non-commercial standard terms should, in my view, be read as not excluding subsales. It is significant that the commercial standard terms are different, and the Law Society's *Handbook* and the *Conveyancer's Guide* mentioned above are unfortunate or even misleading in failing to draw a distinction between the two.

Licence to assign

[46] Mr Dutton has also relied on a second argument, added by a relatively late amendment to his pleading, based on the licence to assign. Under clause 8.3.2, the buyer is required to provide all information and references reasonably required and, under clause 8.3.3, it is the obligation of the vendor to procure the licence. If the licence has not been received three working days before the contractual date for completion, the purchaser is entitled to rescind and recover the deposit, unless it is itself in breach of its obligations under clause 8.3.2. The contractual completion date was 20 June 2008, so the requirement was to "receive" the licence by the close of business, or possibly before midnight on 16 June: see *Aubergine Enterprises Ltd* v *Lakewood International Ltd* [2002] EWCA Civ 177; [2002] 1 WLR 2149, in [46]. Accepting that no particular form may be needed, it nevertheless does need to be in writing, and it needs to be received. Mr Naviede relies on the letter from Cluttons dated 16 June, mentioned in [17] above. That letter might indeed have been adequate confirmation that the licence would formally be given, but there is no evidence that it was received until 17 June at the earliest. In particular, there is no evidence that it was received by fax or e-mail on 16 June. There was also an earlier draft of a licence to be issued in favour of Mr Pittack, but that was clearly no more than a draft.

[47] I have therefore [had] to conclude that Mr Naviede did not receive a written licence to assign three working days before 20 June. The factual question therefore is whether Mr Pittack had provided "all information and references reasonably required". On that question, the Crown Estate did not require references from Mr Pittack, and there is no question of it requiring any other information not already provided, except perhaps for Mr Pittack's address (which I regard as trivial). Mr Pittack was therefore not in breach, and he was entitled to rescind.

Conclusions

[48] For both those reasons, I have come to the conclusion that Mr Pittack was entitled to rescind and, not being himself in breach of the contract, is entitled to the return of the deposit with interest.

[49] I reach that conclusion without a great deal of satisfaction. Mr Pittack strikes me as fortunate in the extreme, having failed to prepare for the possibility of his intended subpurchasers withdrawing, that Mr Naviede reacted so uncompromisingly to the suggestion of a subsale. Mr Dutton described Mr Naviede as "too clever by half" in his attempt to box Mr Pittack into a corner. That phrase has a pejorative connotation that I do not endorse, even though by refusing access to the property Mr Naviede had acted consciously in making things difficult for Mr Pittack. The fact is that a substantial degree of mistrust had arisen on both sides. I would characterise Mr Naviede merely as unfortunate that he objected to the subsale before the subpurchasers withdrew. As for the licence to assign, the fact is that Mr Naviede never made a formal application for the licence, relying on Mr Pittack's failure to provide him with references. He was again unfortunate that Mr Pittack himself had provided what Cluttons needed. Even so, by the time he discovered the position on 16 June, he could have obtained written confirmation from Cluttons by fax or e-mail that day, and I can only suppose that the urgency of the matter was not then appreciated.

[50] Mr Pittack's pleadings include a claim for a lien over the property to enforce the return of the deposit. However, this was not pressed at the hearing, and indeed Mr Dutton's only submission (if that is the word) on this point was that he handed up copies of two pages from *Megarry & Wade: The Law of Real Property* (7th ed), containing para 15056, a paragraph with a plethora of authorities, none of which I was invited to read, and none of which Mr Naviede had any proper opportunity to answer. I make no order on this part of the claim.

Claim allowed.

SUPPLEMENT
CASES NOT REPORTED IN “ESTATES GAZETTE”

CONSTRUCTION

Anglian Water Services Ltd *v* Laing O'Rourke Utilities Ltd

TECHNOLOGY AND CONSTRUCTION COURT

[2010] EWHC 1529 (TCC)

25 June 2010

EDWARDS-STUART J

Construction — NEC Engineering and Construction Contract — Arbitration — Notice — Validity of service — Housing Grants, Construction and Regeneration Act 1996 (HGCRA) — Parties entering into contract in terms of NEC Engineering and Construction Contract (2nd ed) 1995 — Contract specifying address for service of notice to refer adjudicator's decision to arbitration — Employer faxing notice to contractor's solicitor — Solicitor confirming receipt of notice and forwarding it to relevant individuals at employer company — Whether arbitration provisions in clause 93.1 contrary to Part II of HGCRA — Whether notice an effective communication under clause 13 of the contract — Whether appropriate to extend time for service

The claimant engaged the defendant, under a contract incorporating the terms of the NEC Engineering and Construction Contract (2nd ed) 1995, to design and construct several tanks at a sewage works. The contractual adjudication provisions did not comply with Part II of the Housing Grants, Construction and Regeneration Act 1996 (the HGCRA) because they did not provide for a party to refer a dispute to adjudication "at any time", as required by section 108, and they were accordingly replaced by the adjudication provisions in the Scheme for Construction Contracts. Clause 93.1 of the contract further provided that a party that was dissatisfied with an adjudicator's decision could refer the dispute to arbitration by serving notice on the other party within four weeks of notification of the decision, which notice would, by clause 13.2, take effect when received at the last address notified by the recipient for receiving communications or, if none was notified, at the recipient's address as stated in the contract data.

In March 2010, the claimant faxed such a notice to the defendant's solicitor. The solicitor replied by e-mail on the same day confirming receipt of the fax and then forwarded the notice to the defendant's relevant personnel. The four-week time limit for serving the notice expired on the following day. The defendant subsequently argued that the claimant had failed to serve a valid notice within the time limit because the notice had been sent to the wrong address.

The claimant applied to the court for: (i) a declaration that it had validly, and in time, notified its intention to refer a dispute to arbitration; alternatively (ii) an extension of time for taking that step pursuant to section 12 of the Arbitration Act 1996. An issue arose as to whether the arbitration clause in the contract was ineffective for non-compliance with Part II of the HGCRA; the claimant argued that it was, since the right to refer a dispute to adjudication under section 108 was intended to be permissive, leaving open the possibility of some other dispute resolution, and was not complied with if the contractual terms made an adjudication mandatory prior to the final determination of any dispute, as was the case where it was a precondition of a referral to arbitration.

Held: **The claim was allowed. (1) A contract that obliges a party to refer a dispute to adjudication before it can pursue that dispute either by litigation or arbitration does not fetter the right to refer a dispute to adjudication at any time. The fact that a party cannot begin proceedings by way of arbitration in respect of any particular dispute until it has obtained a decision from an adjudicator on that dispute has no bearing on its ability to choose when to refer a dispute to adjudication. Although a contractual provision to that effect will fetter a party's ability to start proceedings in the court or by way of arbitration because it cannot do so without first having referred the dispute to adjudication, the HGCRA does not entitle parties to start court or arbitration proceedings at any time and such a requirement cannot be read into its provisions. Furthermore, clause 93.1 does not form part of the adjudication provisions of the contract and is not one of the provisions that falls to be deleted as a result of the incorporation of the adjudication provisions of the scheme into the contract. Clause 93.1 and the Scheme for Construction Contracts are not incompatible or mutually exclusive and can operate together. Accordingly, clause 93.1 was a valid and surviving provision. (2) The commercial purpose of clause 13.2 of the contract is to enable each party to work on the basis that all communications relating to the contract will be channelled through one particular office, so enabling every incoming document to be properly filed and its arrival properly recorded. The clause is intended to fix the moment in time at which a communication takes effect. It would be unsatisfactory if in any case involving a dispute over this, the parties had to investigate the circumstances in which the communication was made and received in order to determine whether the mode of delivery was as good as, or better than, that prescribed by clause 13.2. Accordingly, compliance with the mode of delivery specified in clause 13.2 is the only means of achieving effective delivery of a communication under the contract because the communication takes effect only when a copy is received at the prescribed address. The fact that a notification is received in time by the relevant personnel cannot trump a failure to give notice in accordance with the contract. (3) However, on the evidence, the claimant had sent its notice to the correct address. The defendant's solicitor had agreed in correspondence that it would, on behalf of the defendant, accept service of any documentation relevant to the adjudication. Accordingly, the address of the defendant's solicitor had been the "last address notified" for the receipt of such communications. There is no reason why an address cannot be notified under clause 13.2 in respect of a particular type of communication but not any other. The claimant's notice of its intention to refer to arbitration a dispute decided by the adjudicator was a dispute relevant to the adjudication because, in the absence of such a notice served within the four-week time limit, the adjudicator's decision would become finally binding on the parties. A valid notice of intention to refer served in time was therefore relevant to the adjudication because it prevented the adjudicator's decision from becoming final. The claimant was therefore entitled to the declaration sought. (4) Had the claimant's notice not been validly served, it would have been appropriate to extend time for service under section 12 of the 1996 Act. It was relevant in that connection that the notice had been received in time by the defendant's relevant personnel and that the conduct of the**

those acting for the defendant, in giving unqualified confirmation of safe receipt, was a material contributing cause of the claimant's failure also to serve the documents at the defendant's office.

The following cases are referred to in this report.

Harbour & General Works Ltd v *Environment Agency* [2000] 1 WLR 950; [2000] 1 All ER 50; [2000] 1 Lloyd's Rep 65

Lantic Sugar Ltd v *Baffin Investments Ltd* [2009] EWHC 3325 (Comm); [2010] 2 Lloyd's Rep 141

Manchester Diocesan Council of Education v *Commercial & General Investments Ltd* [1970] 1 WLR 241; [1969] 3 All ER 1593; (1970) 21 P&CR 38, Ch

Spectra Pty Ltd v *Pindari Pty Ltd* [1974] 2 NSWLR 617

Yates Building Co Ltd v *RJ Pulleyn & Sons (York) Ltd* [1976] 1 EGLR 157; (1975) 237 EG 183, CA

Yuanda (UK) Co Ltd v *WW Gear Construction Ltd* [2010] EWHC 720 (TCC); [2010] 1 CLC 491

This was the hearing of a claim by the claimant, Anglian Water Services Ltd, against the defendant, Laing O'Rourke Utilities Ltd, for declaratory relief as to the valid and timely service of a notice to refer a dispute to arbitration under a construction contract or, alternatively, for an extension of time for such service.

Andrew Wales and Sarah Martin (instructed by CMS Cameron McKenna LLP) appeared for the claimant; David Streatfeild-James QC (instructed by Pinsent Masons LLP, of Bristol) represented the defendant.

Giving judgment, **Edwards-Stuart J** said:

Introduction

[1] This is the hearing of an arbitration claim by which the claimant, Anglian Water Services Ltd (AWS), applies for:

[1.1] a declaration that it validly notified its intention to refer a dispute to arbitration by 23 March 2010 and therefore in time; alternatively

[1.2] an extension of time for taking that step until 30 March 2010 or such other date as the court may permit, pursuant to section 12 of the Arbitration Act 1996 (the 1996 Act).

[2] By a contract that incorporated the terms of the second edition, 1995, of the NEC Engineering and Construction Contract (ECC), AWS engaged the respondent, Laing O'Rourke Utilities Ltd (LOR), to design and construct a number of tanks at AWS's Saltfleet sewage treatment works (SSTW).

[3] It was common ground that the adjudication provisions in the contract did not comply with Part II of the Housing Grants, Construction and Regeneration Act 1996 (the HGCRA) with the result that the adjudication provisions in the Scheme for Construction Contracts (the scheme) replaced the adjudication provisions in the contract. This was for the simple reason that the provisions in the contract did not permit a party to the contract to refer a dispute to adjudication "at any time", as section 108 of the HGCRA requires.

[4] Clause 93.1 of the core clauses provided for arbitration in the following terms (the reference to *tribunal* in the clause means arbitration and words in italics are defined in terms in the contract):

> If after the *Adjudicator*
> • notifies his decision or
> • fails to do so
>
> within the time provided by this contract a Party is dissatisfied, that Party notifies the other Party of his intention to refer the matter which he disputes to the *tribunal*. It is not referable to the *tribunal* unless the dissatisfied Party notifies his intention within four weeks of
> • notification of the *Adjudicator's* decision or
> • the time provided by this contract for this notification if the *Adjudicator* fails to notify his decision within that time
>
> whichever is the earlier. The *tribunal* proceedings are not started before Completion of the whole of the *works* or earlier termination.

[5] It is alleged by LOR that AWS failed to serve a valid notice of dissatisfaction under clause 93.1 within four weeks of a decision of the adjudicator, Dr Robert Gaitskell QC. The failure did not consist of a failure to send the document to LOR within the prescribed four weeks, but a failure to send it to what LOR contends to be the correct address for service.

[6] LOR relies on clause 13 of the core clauses, the relevant parts of which provide as follows:

> Communications 13
>
> 13.1 Each instruction, certificate, submission, proposal, record, acceptance, notification and reply which this contract requires is communicated in a form which can be read, copied and recorded. Writing is in the *language of this contract.*
>
> 13.2 A communication has effect when it is received at the last address notified by the recipient for receiving communications or, if none is notified, at the address of the recipient stated in the Contract Data.
>
> 13.3 If this contract requires the *Project Manager*, the *Supervisor* or the *Contractor* to reply to a communication, unless otherwise stated in this contract, he replies within the *period for reply.*
>
> …
>
> 13.7 A notification which this contract requires is communicated separately from other communications.

[7] I will have to go into the facts in some detail later in this judgment, but what happened in outline was that at 16.19 on 22 March 2010, AWS's solicitor, CMS Cameron McKenna (CMS), sent a letter by fax to LOR's solicitor, Pinsent Masons (PM), that was a notice of dissatisfaction with the decision of Dr Gaitskell issued on 24 February 2010, together with a separate notice to refer a dispute to arbitration. This was acknowledged by an e-mail at 17.03 the same day from Mr Andrew Shelling, of PM, saying: "I confirm safe receipt of your fax sent earlier today."

[8] PM then forwarded the documents to its clients, and it is not in dispute that they went straight into the hands of the two individuals at LOR who were responsible for handling the dispute with AWS. The four-week time limit for serving the notice of dissatisfaction expired at the end of the following day (or, possibly, the day after).

[9] Having received the acknowledgement from PM, the partner and assistant at CMS — thinking that all was well — took no further steps in respect of further service of the documents. However, one week later, on 29 March 2010, CMS received a letter from PM saying that it would take instructions from its clients as to whether it was instructed to accept service and asking CMS to confirm that the relevant notice under clause 93.1 of the contract had been served on its clients in accordance with the contract.

Issues

[10] These can be summarised as follows:

(1) is clause 93.1 incompatible with Part II of the HGCRA and of no effect? If not,

(2) was delivery of the notice of dissatisfaction to PM within the four-week period an effective communication of that notification under the contract?; and if not

(3) since the notification of dissatisfaction was received by the relevant individuals at LOR within the four-week period, did that of itself make it an effective communication under clause 13?

(4) if each of the above issues is answered in favour of LOR, should AWS be given an extension of time for service of the notice of dissatisfaction under section 12 of the 1996 Act?

[11] Since the notice of dissatisfaction was subsequently served on LOR at the address given in the contract on 30 March 2010, that is the date to which AWS seeks an extension of time (if it needs it).

[12] I will deal with the issues in the order in which I have set them out.

Validity of clause 93.1

[13] The principal point made by Mr Andrew Wales, who appeared for AWS together with Ms Sarah Martin, was that since section 108(1) of the HGCRA confers on a party to a construction contract the right to refer a dispute arising under the contract for adjudication under a procedure complying with the section, that must carry with it a right not to refer a dispute for adjudication but to refer it to some other method of dispute resolution.

[14] In para 62.3 of his skeleton argument, Mr Wales put the point this way:

> A contract which requires a mandatory adjudication prior to the chosen means of final determination is non-compliant with the HGCRA. The adjudication for which the Scheme provides will not (as required) be permissive in nature if contract terms remain effective to make an adjudication mandatory prior to the final determination of any dispute.

[15] This is a short submission, but unfortunately I cannot give it such a short answer. The HGCRA provides that a contract must entitle each party to a construction contract to refer a dispute to adjudication at any time and it provides that the decision of the adjudicator will be binding until the dispute is finally determined by legal proceedings, or arbitration if that is what the contract provides, although the parties may agree to accept the decision of the adjudicator as finally determining the dispute.

[16] The adjudication provisions in the scheme, which as I have already said have been incorporated into this contract, entitle any party to the contract to give written notice of its intention to refer any dispute arising under the contract to adjudication. However, as I have already noted, it is provided by section 108(2) of the HGCRA that the contract must enable a party to give notice *at any time* of its intention to refer a dispute to adjudication. However, the HGCRA is silent on the question of whether a referral to adjudication should be a precondition of starting proceedings in the courts or by way of arbitration, as appropriate. A contract that obliges a party to refer a dispute to adjudication before it can pursue it by either litigation or arbitration does not, in my view, impose any fetter on the right to refer a dispute to adjudication at any time. However, it does prevent a party from starting proceedings in the courts or by way of arbitration at any time because it cannot do so without having first referred the dispute to adjudication.

[17] The HGCRA does not provide that a party to a contract has a right to start proceedings, either by way of litigation or arbitration, at any time, and I cannot see any room for reading such a requirement into its provisions as a matter of construction. However, that is effectively what Mr Wales is seeking to do: he just puts the point in a much more subtle way. His argument, when looked at more broadly, amounted to this: that Part II of the HGCRA entitles any party to the contract to give notice of its intention to refer any dispute arising under the contract to *arbitration at any time*. This it clearly does not do. Accordingly, I consider that there is nothing in the point that clause 93.1 cannot survive the provisions of Part II of the HGCRA.

[18] However, Mr Wales had a further way of putting the point, albeit in a slightly different way. What he said is that "the requirement for a mandatory adjudication prior to arbitration deprives the parties of the right under section 108(2)(a) to be able to give notification at any time": see para 62.4 of the skeleton. The fallacy in this argument is that it confuses relative time with absolute time. Under the HGCRA and the scheme, a party is entitled to refer a dispute to adjudication at any time from day one of the contract. Provided that there is a legitimate dispute, that right subsists (at least, in England and Wales, in practice if not in theory) up to the time its cause of action becomes statute-barred. The fact that a party cannot start proceedings by way of arbitration in respect of any particular dispute until it has obtained a decision from an adjudicator on that dispute has no bearing whatever on its ability to choose when to refer a dispute to adjudication.

[19] It is only if one asks the question "can a dispute be referred to adjudication before or after it has been referred to arbitration?" that there will be a restricted answer. However, one obvious reason for the restricted answer is that once the dispute has been decided by arbitration there can be no dispute to refer to adjudication. The dispute will have been finally determined by the arbitrator's award. If there is an adjudication of a dispute, that adjudication must, by definition, always precede the determination of the same dispute by arbitration.

[20] So this just brings me back to the submission that the real effect of clause 93.1 is that it prevents either party to the contract from referring a dispute to arbitration at any time. However, for the reasons that I have already given, unless this is contrary to the HGCRA, the point goes nowhere.

[21] However, Mr Wales had another string to his bow. He submitted that the importation of the scheme into the contract means that all the adjudication provisions of the scheme apply in place of the contractual provisions covering the same ground in the contract: see *Yuanda (UK) Co Ltd* v *WW Gear Construction Ltd* [2010] EWHC 720 (TCC)*, a decision of mine (which I understand may be going to the Court of Appeal, but not on this point). Although it is not difficult to identify the adjudication provisions of the scheme that are to be imported into the contract, because they are contained in Part I of the scheme under the heading "Adjudication", it may not be so easy to identify the provisions in the contract that they are to replace.

[22] Section 9 of ECC option C is headed "Disputes and termination". Clause 90 is headed "Settlement of disputes"; clause 91 is headed "The adjudication"; clause 92 is headed "The *Adjudicator*"; and clause 93 is headed "Review by the *tribunal*". On the face of it, therefore, these clauses, as they stand, suggest that clauses 90-92 inclusive are concerned with adjudication and clause 93 is concerned with arbitration.

[23] However, some of these clauses were amended by the incorporation of addendum Y(UK)2 to the second edition of the NEC form of contract. Clause 90 was deleted in its entirety and replaced by a new clause 90, with the heading "Avoidance and settlement of disputes"; clause 91 was amended by replacing the side heading "The adjudication" with the heading "Combining procedures", replacing clause 91.1 with a new clause and including certain minor amendments to the text of clause 91.2. In addition, there were also some minor amendments to the text of clause 92, but the heading of that clause was not changed. There were no amendments to clause 93.

[24] Unfortunately, this particular paper trail does not stop there because there was then a further amendment to clause 90 by option Z. By this amendment, clause 90 was deleted in its entirety and replaced by new provisions in four subclauses, 90.1 to 90.4. Effectively, these provided that, in the first instance, any dissatisfaction or dispute under the contract was to be submitted to and settled by the project team, and failing resolution there, to a body called the partnership board. In the event that the partnership board was unable to settle the dissatisfaction or dispute, the matter was to be submitted to and settled by the adjudicator. There were various consequential provisions in addition.

[25] However, in the light of these amendments, the long and the short of all this is that clause 90 sets out initial steps of a dispute resolution process that had to be followed before a dispute or dissatisfaction could be referred to adjudication; clause 91 deals with the situation where a matter in dispute involved a subcontractor; and clause 92 deals with the adjudication process itself.

[26] By contrast, clause 93.1 deals with the situation where the adjudicator either made his decision within the time required by the contract or failed to do so within that time. In either event, a dissatisfied party was then entitled to refer the matter in dispute to arbitration provided that it notified its intention to do so within four weeks of the notification of the adjudicator's decision or, if he failed to give a decision, within four weeks of the date by which he should have given it. In addition, the clause provides that the arbitration proceedings themselves were not to be started before completion of the works. Clause 93.2 deals with the powers and jurisdiction of the arbitrator.

[27] Paragraph 19(1) of the scheme sets out the time limits within which an adjudicator must reach his decision. Paragraph 19(2) provides that if the adjudicator fails, for any reason, to reach his decision in accordance with those time limits, any of the parties to the dispute may serve a fresh notice of adjudication and continue with a new adjudicator.

[28] In my view, para 19 of the scheme and clause 93.1 of the contract can be operated together and are not mutually exclusive. If they are to be read together (which Mr Wales disputed), they produce a situation

---

* Editor's note: Reported at [2010] 1 CLC 491

A in which, if the adjudicator fails to give a decision within the required time, a party can refer the dispute to arbitration under clause 93.1 provided that it notifies its intention to do so within four weeks of the date by which the adjudicator should have notified his decision. If it fails or chooses not to do this, it can start a fresh adjudication with a new adjudicator under para 19 of the scheme. I accept that although the wording of clause 93.1 appears to be mandatory in its terms (in that it appears to be saying that if a party is dissatisfied with the failure to make a decision it *must* notify the other party of its intention to refer the dispute to arbitration), it provides no remedy at all if the dissatisfied party does not act within the four-week time limit. It does not give
B that party the residual right to refer the dispute to a fresh adjudicator. I have to confess that the task of construing the provisions in this form of contract is not made any easier by the widespread use of the present tense in its operative provisions. No doubt this approach to drafting has its adherents within the industry, but, speaking for myself and from the point of view of a lawyer, it seems to me to represent a triumph of form over substance. Paragraph 19(2) of the scheme, by contrast, is clearly drafted and provides no time limit within which the referral to a new adjudicator must take place.

[29] The real difficulty, in my judgment, with the submission that clause 93.1 must be included within the provisions of the contract
C that are replaced by the scheme is that it removes the operative part of the arbitration clause. Although the arbitration clause is the entirety of clause 93, not just clause 93.1, the operative part of the clause is contained only in clause 93.1. If clause 93.1 is removed without an equivalent replacement (and there is no such replacement in the scheme), one is left with a clause that defines the role of the arbitrator but provides no entitlement to refer a dispute to arbitration or any machinery by which to do so. Even if clause 93.2 were to be read as conferring a right to refer a dispute to arbitration, it would seem to be a right to refer the dispute to arbitration at any time because the clause is silent on any restriction as to when a dispute can be referred to arbitration. This would not sit easily with the adjudication provisions
D in the scheme.

[30] Although I accept that the coexistence of clause 93.1 and the scheme is not an entirely happy one, it is workable for the reasons that I have given. If, on the other hand, clause 93.1 is to be deleted, it would be open to question whether there was a surviving right to refer a dispute to arbitration at all.

[31] Standing back and looking at the matter more broadly, it seems to me that clause 93 does not form part of the adjudication provisions of the contract. Although it is relevant to the adjudication provisions, because non-compliance with its requirements would leave
E the adjudicator's decision as finally binding, it is both in form and substance concerned with the provision of a right to refer a dispute to arbitration.

[32] For these reasons, therefore, I conclude that clause 93.1 is not incompatible with the scheme and is not one of the adjudication provisions of the contract that falls to be deleted as a result of the incorporation of the adjudication provisions of the scheme into the contract. The result is that it is a valid and surviving provision.

Was delivery of the notice of dissatisfaction to PM on 22 March 2010 effective?

F [33] This issue raises, I think, three further sub-issues:

(1) Is compliance with clause 13.2 of the contract the only means of rendering effective a communication required by the contract?

(2) As at 22 March 2010, what was the "last address notified" by LOR for receiving communications required by the contract, or at least communications relevant to the adjudication?

(3) If the answer to issue (2) is PM's office in Bristol for the purpose of communications relevant to the adjudication, was the notice of intention to refer the dispute to arbitration a communication relevant to the adjudication?

Is compliance with clause 13.2 the only means of achieving effective communication?

[34] The argument, attractively put by Mr Wales, is that LOR is G
reading clause 13.2 as though it said that a communication has effect *only* when received at the last address notified. That, he submitted, is a restrictive reading that is not justified as a matter of construction, and he cited various authorities in support of this proposition. Further, he submitted that such a construction would be contrary to the commercial purpose of the clause, since it would preclude the parties from communicating in ways that must have been contemplated at the time the contract was made. For example, communications that are physically handed over at a meeting on site (the site not being the last address notified) and recorded in the minutes. In short, he submitted, clause 13.2 stipulates the effectiveness of a communication H
received in the stated way, but it does not preclude the effectiveness of communications made in other ways.

[35] Mr David Streatfeild-James QC, who appeared for LOR, submitted that clause 13.2 must be complied with in order to make a communication effective. He submitted that there is a very good reason for this. He reminded me of numerous provisions in the contract that require notification of a particular thing or state of affairs that then has certain contractual repercussions. For example, within clause 13 there is clause 13.3, which requires the project manager and others to reply to a communication within the *period for reply* (which is 14 days). A failure to do so is a compensation event under clause 60.1. In order for the J
project manager to be able to do this, submitted Mr Streatfeild-James, there must be certainty as to when the relevant communication was received.

[36] Although he accepted that it may be cumbersome, and in some ways unsatisfactory, Mr Streatfeild-James submitted that the contract requires every communication to be served at the notified address in order to be effective. It is dangerous, he submitted, to construe clause 13.2 solely in the light of the events that have given rise to the present application: clause 13.2 applies to all communications that may take place throughout the contract, and they take a wide variety of forms. K

[37] I now turn to the authorities relied on by Mr Wales. The first was *Manchester Diocesan Council of Education* v *Commercial & General Investments Ltd* [1970] 1 WLR 241. The council wanted to sell some school buildings and conditions of sale, incorporating a form of tender, were issued on their behalf by their surveyor. Condition 4 provided that the party whose offer was accepted should be notified by letter sent to it by post at the address given in its tender and that every letter should be "deemed to have been received in due course of post". The council wanted to accept the defendant's offer and their surveyor wrote to the defendant's surveyor — not the defendant — stating that the offer had been accepted. A dispute arose as to whether there was a binding L
contract. It was held that where an offeror prescribes a particular mode of acceptance, but does not stipulate that acceptance in that mode only will be binding, acceptance may be communicated in any other mode not less advantageous to the offeror. Since condition 4 was included in the offer at the instance of the council, strict compliance with it could be waived by the council, provided that the defendant was not thereby adversely affected. Accordingly, it was held that there was a binding contract. At p246C, Buckley J said:

> Where, however, the offeror has prescribed a particular method of acceptance, but not in terms insisting that only acceptance in that mode shall be binding,
> I am of opinion that acceptance communicated to the offeror by any other mode M
> which is no less advantageous to him will conclude the contract.

[38] The next case was *Spectra Pty Ltd* v *Pindari Pty Ltd* [1974] 2 NSWLR 617. A clause containing an option to renew a lease concluded as follows: "Should the Lessee decide to exercise this option it shall give to the Lessor notice in writing to be sent by prepaid registered mail to the Lessor's last known place of residence of its desire so to do not later than (blank) calendar months prior to the expiration of the term hereby granted."

[39] The option was exercised by ordinary post instead of registered post, and it was contended that this was a fatal defect. The judge, Wootten J, said that it appeared that the notice of exercise of the option

A was in fact sent by the plaintiff to the defendant prior to the expiry of the term and was received by the defendant in the normal course of post. He said that the question in the case was whether the clause should be construed as a stipulation that prepaid registered mail is to be not merely a sufficient method of giving notice but the sole and essential method of doing so. His conclusion, at p623A, was:

> in the absence of a very clear indication of a contrary intention, it would not be reasonable to construe a provision for service by registered mail as excluding the giving of notice by other equally expeditious means which do in fact result in the actual receipt of the notice by the offeror, eg personal delivery or unregistered mail, although of course in the latter event the offeree will run the
> B risk of non-delivery.

[40] The third case relied on by Mr Wales was *Yates Building Co Ltd* v *RJ Pulleyn & Sons (York) Ltd* [1976] 1 EGLR 157*. That was another case of an option to purchase land, and the relevant clause said that the option could be exercised by notice in writing "to be sent by registered or recorded delivery post to the registered office of Pulleyns or the offices of their said solicitors".

[41] At pp157J-158A, Lord Denning MR said:

> It seems to me that this depends on the construction of the option clause. The option is an offer: an irrevocable offer. When a person makes an offer,
> C he does sometimes prescribe the method by which it is to be accepted. If he prescribes it in terms which are mandatory or obligatory, the acceptance is only good if it complies with the stated requirements. Thus in the present case the notice of acceptance *must* be in writing, and *must* be given to Pulleyns or to Pulleyns' solicitors, and *must* be given between April 6 1973 and May 6 1973. But the question is whether the words "such notice to be sent by registered or recorded delivery post" are mandatory or directory. That test is used by lawyers in the construction of statutory instruments, but it can also be used in the construction of other documents. The distinction is this: a mandatory provision must be fulfilled exactly according to the letter, whereas a directory provision is satisfied if it is in substance according to the general intent… In applying this rule of construction, you must look to the subject-matter, consider the object to be fulfilled, and then see whether the provision must be fulfilled strictly to the
> D letter or whether the substance of it is enough.

[42] The acceptance was not sent by registered or recorded delivery post, but it arrived in time. Unsurprisingly, perhaps, Lord Denning held, as a matter of interpretation, that if the letter did reach the sellers in time, it was a valid exercise of the option. Orr and Scarman LJJ agreed, although Scarman LJ said that he was not convinced that the term "directory" has any application to the field of contract. He considered that contractual provisions were either obligatory or permissive.

[43] Mr Streatfeild-James did not seek to challenge the principle established by these cases; however, he submitted that it was necessary
E to look at this particular clause in the context of this particular widely used standard form of contract.

[44] An obvious difference between this clause and the clauses in the three cases to which I have referred is that this clause is intended to be of general application to a large number of different types of communications. In each of the cases cited, it was clear that the method of service adopted caused no detriment to the other party, and indeed may even have been advantageous. The problem here is that there is no way of knowing, as a matter of general application, whether service on a particular individual responsible for the matter concerned would in fact be more advantageous to the relevant party than service in the
F prescribed manner. One can readily envisage situations where that may not be the case.

[45] It seems to me that the probable commercial purpose of the clause is to enable each party to the contract to work on the basis that all communications in respect of the contract will be channelled through one particular office, with the obvious advantage of enabling every incoming document to be properly filed and its arrival properly recorded. I think that, as a specialist judge in this field, I can probably take judicial notice of the fact that for the best part of a decade or so electronic document management systems for providing a database of all communications generated in the course of the contract have G
been employed in many substantial civil engineering and construction projects. Service through one designated office may be more cumbersome, but it does enable proper records to be kept. It is then for the project manager or co-ordinator, or a particular member or members of its staff, to ensure that incoming documents are then copied to all those individuals with an interest in seeing them.

[46] Turning to the wording of the clause, it seems to me that clause 13.2 is there to fix the moment in time when a communication takes effect. For the reasons given by Mr Streatfeild-James, in the case of certain types of communication the date of receipt will trigger the start of the period in which a response or action is required. The answer H
to the rhetorical question asked by Mr Wales — what is wrong with a document being handed over in a meeting? — is nothing, but I consider that the contract requires that a copy of the document should be sent also to that party's prescribed address.

[47] It would be unsatisfactory, in my view, if in any case in which there was a dispute about the time at which a communication took effect, the parties had to investigate the circumstances in which the communication was made and received in order to determine whether the mode of delivery was actually as good as, or better than, the mode of delivery prescribed by clause 13.2. Apart from anything else, there might well be legitimate room for disagreement as to whether the J
actual mode of service was an improvement on the prescribed mode of service.

[48] I accept the point forcefully made by Mr Wales that, in the present case, the notice of dissatisfaction found its way to the relevant individuals in LOR within the prescribed time and almost certainly sooner than if it had been sent to what LOR contends was the last notified address, namely the office in St Neots. However, that can provide no basis for reaching any general assumption in the context of a project such as this about whether delivery to the individuals concerned with any particular communication is in general going to be more advantageous to the receiving party than delivery to the K
prescribed address.

[49] For these reasons, I conclude that compliance with the mode of delivery specified in clause 13.2 is the only means of achieving or securing effective delivery of a communication under the contract because the communication takes effect only when it is received at the prescribed address.

[50] The consequence of this conclusion is that I reject the submission that the fact that notification was received within time by the relevant personnel at LOR effectively trumps the failure to give notice in accordance with the contract. On this aspect, I would echo the words of Gross J, albeit made in a different context and to a different L
end, in *Lantic Sugar Ltd* v *Baffin Investments Ltd* [2009] EWHC 3325 (Comm)*, in [40], namely:

> If a claimant is required to serve X and, mistakenly purports to serve Y, the mere fact that Y informs X of the purported service so that X knows of it, cannot convert Y's receipt of the documents into good service upon X.

[51] However, the fact that notice of AWS's intention to refer the dispute to arbitration was brought to the attention of the relevant officers or directors of LOR within the four-week time limit is clearly a relevant factor if it becomes necessary to consider the application under M
section 12 of the 1996 Act to extend time.

As at 22 March 2010, what was the "last address notified" by LOR?

[52] Mr Streatfeild-James submitted that the notification of a change of address under clause 13.2 is itself a communication that has to be sent to "the last address notified" and does not take effect unless this happens. Mr Wales's response to this was that the communication of a change of address was not a communication that "the contract requires" within the meaning of clause 13.1.

* Editor's note: Also reported at (1975) 237 EG 183

* Editor's note: Reported at [2010] 2 Lloyd's Rep 141

[53] In support of this submission, Mr Wales argued that a party did not have to change its address or give an address that was different from that stated in the contract data, and so the notification of a different address was not a communication that the contract required. It seems to me that precisely the same argument could be raised, for example, in respect of an instruction to search for defects under clause 42. The supervisor does not have to give such an instruction — indeed, it would be possible for the works to be carried out and completed without the need for such an instruction ever arising. However, it is plain, in my view, that if such an instruction is given, it has to be communicated in the manner prescribed by clause 13.2. Accordingly, on this point, I accept the submissions of Mr Streatfeild-James.

[54] The next point is to consider which, if any, was the "last address notified" by LOR to AWS for the receipt of relevant communications as at 22 March 2010. I say "relevant communications" because I see no reason why an address should not be notified under clause 13.2 in respect of a particular type of communication but not others. For example, I do not see anything in the wording of clause 13.2 to prevent LOR from notifying an address solely for the purpose of, say, receiving certificates issued under the contract.

[55] Whether this conclusion is correct may not matter because I was told that the contract works had been completed and that the only activity that was continuing under the terms of the contract was the dispute resolution process that has given rise to this application. As Mr Wales put it in his skeleton argument, "the dispute was all that remained to the life of the contract". I did not understand this to be disputed.

[56] By a letter dated 29 February 2008, addressed to "Laing O'Rourke" (that is, not specifically to LOR, although there could have been no doubt that LOR was the identity of the intended recipient) at the St Neots office, CMS notified LOR that it had been instructed by AWS in respect of the recovery of the losses caused by the collapse of reactor 2 at SSTW. After setting out the details of the claim and making some proposals for the way forward, the letter concluded: "We look forward to hearing from you as to your availability for a without prejudice meeting in the week commencing 7 April."

[57] Mr Wales submitted that if it was necessary for AWS to give notice of an address for receiving communications in respect of this dispute, the letter of 29 February 2008 did this. I accept this submission. Clause 13.2 does not require a notification to be in any particular form, the only requirement is that for it to be effective it must be sent to the last address notified by the recipient for receiving communications. It appears to be common ground that the St Neots office was the "last address notified" on behalf of LOR at the time this letter was sent. Although CMS's letter does not say in terms that communications relating to the dispute are to be sent to its office, that, in my view, is implicit given the terms of the letter as a whole and the concluding sentence. That sentence, in particular, makes it clear that the requested response should go to CMS.

[58] On 11 April 2008, PM wrote to CMS saying that it was instructed by "Laing O'Rourke" in connection with CMS's letter of 29 February 2008. The letter did not address any of the substantive issues in the dispute but merely the proposals put forward by CMS as to the way forward. The letter concluded by saying that PM would be grateful if CMS "would note our interest in this matter" and asking CMS to confirm that its client would be prepared to defer the issue of any proceedings until it had had sight of LOR's proposals as to a timetable for without prejudice discussions. I do not consider that the request to note PM's interest amounted to a notification of an address for the delivery of communications under clause 13.2. I did not understand Mr Wales to make any serious suggestion to the contrary.

[59] On 14 November 2008, PM wrote to CMS enclosing a report prepared by Arup and setting out a summary of LOR's position. The letter concluded by saying that LOR denied that it had any liability to AWS, as alleged in the letter of 29 February 2008. After some further correspondence between the respective solicitors, on 11 December 2009 CMS wrote to PM enclosing "by way of service" a notice of adjudication in respect of the dispute. The letter stated that a copy of the notice of adjudication had also been served on PM's clients, and it was in fact copied (as it stated at its foot) to Mr Paul Grammer, who was the European commercial construction director of Laing O'Rourke and was the person who had primary conduct of the dispute on behalf of LOR. The same day, PM replied by an e-mail from Mr Adam Harris, the partner at PM who had the conduct of the case, stating "Safe receipt of the Notice and correspondence is acknowledged".

[60] PM replied to this letter on 15 December 2009 and made various proposals for what it described as a "sensible adjudication timetable". The letter then addressed what were described as some matters of housekeeping, and it is perhaps worth quoting the last three paragraphs of the letter in full:

> The first matter is the identity of the parties. The contract for the Saltfleet facility was signed by Laing O'Rourke Utilities Limited. There is no company called Laing O'Rourke Limited. Consequently, as matters currently stand we do not believe that the notice settled by leading and junior counsel is, as served, fit for its purpose.
>
> The second matter may be of limited importance if a timetable and procedure can be agreed. It is common ground that the adjudication provisions of the contract fall away, therefore any adjudication will be subject to the Scheme and not to the ICE adjudication procedure as your covering letter suggests.
>
> We would hope that we can take a practical approach to these matters to enable an adjudication on the issues to proceed against the suggested timetable.

[61] Since there was no direct indication in this letter that PM was instructed to accept service of documents in respect of the adjudication, CMS replied on the following day, 16 December 2009, by e-mail in the following terms:

> We enclose a copy of our clients revised adjudication notice (together with covering letter), service copies of which will be served directly on your clients *as required by the contract.*
>
> We would be grateful if you that (sic), in future, service can be effected directly on your offices. In the absence of such agreement our client will serve the referral notice and supporting documents on your clients who will, in turn, have to arrange for these documents to be forwarded to your firm.

(Emphasis added.)

[62] In spite of the missing words, the request contained in the second paragraph of this e-mail was perfectly clear. There was no immediate response. On 17 December 2009, CMS sent a further e-mail to PM setting out the agreed timetable for the adjudication and referring to the fact that agreement had been reached on a standstill for purposes of limitation. The e-mail continued:

> In the meantime... we should be grateful if you would confirm by return (a)... and (b) that your firm will accept service of the referral notice and any other documentation relevant to the adjudication.

[63] Around 20 minutes later, Mr Harris replied by e-mail with the single word "Confirmed". In other words, PM had agreed that it would accept service of any documentation relevant to the adjudication on behalf of LOR. I am quite satisfied that this was a sufficient notification under clause 13.2 that PM's Bristol office would from then on be the address for service of communications relevant to the adjudication. Accordingly, I decide this issue in favour of AWS.

Was the notice of intention to refer the dispute to arbitration a communication relevant to the adjudication?

[64] On 18 December 2009, Dr Gaitskell wrote to the parties noting the agreed directions and suggesting a date for a hearing if one were required. The letter directed (or asked) the parties to "use email for speed, but with each recipient immediately confirming receipt by emailing back". On 24 February 2010, Dr Gaitskell sent his decision to the parties. Accordingly, the latest date for either party to serve a notice on the other of its intention to refer the matter to arbitration was 23 March 2010 or, possibly, 24 March 2010 (it probably does not matter which).

[65] On 22 March 2010, CMS sent a fax to PM at 16.19, the cover sheet of which stated "Notice to Refer a Dispute to Arbitration". The letter that followed immediately was in the following terms:

> We refer to the Decision of Mr Robert Gaitskell QC in the adjudication of the above matter, dated 24 February 2010. Needless to say our clients are dissatisfied with the Decision and we hereby put you on notice that they wish to refer the matter to arbitration in accordance with the ICE Arbitration Procedure (1997). We duly enclose by way of service a Notice to Refer.

The notice to refer followed in the remaining pages.

[66] At 17.03 on 22 March 2010, Mr Andrew Shelling, of PM, replied by e-mail, copied to Mr Adam Harris, the partner responsible at PM, saying: "I confirm safe receipt of your fax sent earlier today." It is relevant to note that, by this e-mail, Mr Shelling was confirming receipt of both the notice of intention to refer a dispute to arbitration, which was required by clause 93.1, and the notice to refer. The latter was not required by clause 93.1, but there was no reason why it should not have been sent at the same time since the works had been completed. The letter from CMS did not say that it enclosed the notice of intention to refer a dispute to arbitration by way of service; that was said only with specific reference to the notice to refer.

[67] The first issue raised by this exchange is whether the letter giving notice of AWS's intention to refer the dispute to arbitration was a communication "relevant to the adjudication" and was therefore one that PM had confirmed could be served on PM instead of LOR.

[68] In my view, this is essentially a matter of impression. Although I have already concluded that clause 93 did not form part of the adjudication provisions of the contract, it does not follow that a notice of intention to refer to arbitration a dispute decided by the adjudicator is not a communication relevant to the adjudication. In my view, it is. In the absence of a notice of intention to refer served within the four-week period, the adjudicator's decision becomes finally binding on the parties. A valid notice of intention to refer served in time is therefore relevant to the adjudication because it prevents the adjudicator's decision from being final. Although it is not part of the adjudication process itself, in my judgment it is relevant to the adjudication for this reason. I therefore reject the submission of Mr Streatfeild-James to the contrary.

[69] Accordingly, I decide this issue also in favour of AWS, with the result that AWS is, in my judgment, entitled to the declaration that it seeks. However, in case I am wrong in any of my conclusions so far, I will go on to consider the remaining issues.

Was the communication effective because it was received by the relevant individuals?

[70] A few minutes after Mr Shelling sent his e-mail to CMS on 22 March 2010, Mr Harris sent an e-mail to his clients, Mr Grammer and Mr Clive McKenzie. The latter was Laing O'Rourke's general counsel and the company secretary of LOR, and he was based at Laing O'Rourke's Dartford office. I should set out the text of this e-mail in full:

> I attach a notice of arbitration that has been served by Anglian. There is a 4 week limit under the contract for the service of this notice to preserve the right to arbitrate. The Adjudicator's decision was dated 24 February. We calculate that effective notice has to be received by close of business 25 March at the latest. Fax is an acceptable method of communication. We have simply acknowledged receipt (which they will know from the fax confirmation sheet anyway).
>
> …
>
> They may well have sent a notice direct to you. If they have not and do not do so within the contractual period, issues will arise as to service given the requirements of the contract. We have not indicated that we have instructions to accept service of notices under the contract and so service on us is unlikely to be effective service. The provisions of the contract require it to be served at the place notified or if no notice has been given at the St Neots address given in the contract. It would be sensible for people to be watching out for anything coming in over the next few days.

[71] Apart from appreciating the brevity and clarity of the advice given, several points can be made concerning this e-mail, which was obviously privileged (but PM has been given limited authority to waive privilege in respect of this document). First, it is clear that PM accepted that AWS still had plenty of time in which to serve the notice of intention to refer on the St Neots office if it saw fit to do so. Second, for the reasons that I have already given, I consider that Mr Harris was not correct to say that PM had not indicated that it had instructions to accept service of notices under the contract. It had, in my judgment, indicated that it had had instructions to accept service of documents relevant to the adjudication. Third, PM had been careful not to say anything more to CMS beyond acknowledging safe receipt of the notice of intention to refer and the notice to refer.

[72] I should say at once that I do not consider that Mr Harris can be criticised for taking the line that he did. His primary duty was to his client and it might have felt legitimately aggrieved if he had sold the pass on a time-bar issue. He had to tread the delicate line between not consciously misleading his opponents while at the same time preserving his own client's rights. Mr Harris had to keep on one side of a fine line: I find that he just kept on the right side of it. This does not mean that CMS may not have been induced to take no further action in respect of the notices as a result of Mr Shelling's e-mail, but simply that Mr Harris and Mr Shelling did not act with the intention of deliberately misleading CMS.

[73] Mr Harris has prepared and served a witness statement in response to this application. In that witness statement (in para 33), he notes, correctly, that as at 17 December 2009 CMS was aware that PM did not have instructions to accept service and that it was aware of the contractual requirement for service at LOR's last notified address (being the St Neots office).

[74] Mr Harris accepts, of course, that he confirmed that PM would accept service of documentation relevant to the adjudication, but he maintains that none of that documentation was required by the contract to be the subject of a notice within the meaning of clause 13 of the contract. He says that his instructions from LOR did not extend to notices required under the contract or to documents that were not relevant to the adjudication. I accept his evidence concerning the extent of his instructions, although whether any documentation relevant to the adjudication should have been the subject of a notification that complied with clause 13 of the contract is a matter for the court. It will be apparent from what I have already said that I do not agree with Mr Harris in so far as the notice of intention to refer is concerned.

[75] However, for present purposes, that does not matter. I find that Mr Harris did not have authority, actual or ostensible, to accept service of the notice to refer and CMS had no grounds for assuming before 22 March 2010 that he might have such authority. This is because the notice to refer was not a document relevant to the adjudication. However, nothing turns on that because service of the notice to refer itself was not subject to the four-week time limit in clause 93.1.

[76] Since I accept that the only authority in respect of service given to Mr Harris was to accept service of documents relevant to the adjudication if, contrary to my earlier finding, the notice of intention to refer was not a document relevant to the adjudication, Mr Harris did not have authority to accept service of it and did not in fact do so. As I have already noted, the letter of 22 March 2010 that was sent by fax did not say that the notice of intention to refer was being sent by way of service: it just said that CMS was putting PM on notice that AWS wanted to refer the matter to arbitration. In spite of the persuasive submission of Mr Wales, I do not accept that the consequence of not following the contractually prescribed method of service can be overcome by the fact that the documents were in fact passed on to the people who ultimately needed to receive them. As I have already held, the contract states that service takes effect when received at LOR's last notified address (or the address in the contract data, if not changed). However, for the reasons that I have already given, the letter of 22 March 2010 was validly sent to PM because it was a document relevant to the adjudication.

Application under section 12 of the 1996 Act

[77] Both parties have very properly reminded me of the fact that section 12 of the 1996 Act was intended to introduce a higher threshold for an applicant seeking an extension of time for starting arbitral proceedings than the test that existed under the Arbitration Act 1950. This reflects the advice of the departmental advisory committee that, in principle, party autonomy should prevail and the jurisdiction of the court to interfere in the bargain made by the parties should be limited.

[78] The application is made under section 12 of the Act, which provides that the court may extend the time for starting arbitral proceedings where an arbitration agreement provides that a claim shall be barred unless the claimant takes, within a time fixed by the agreement, a step to begin proceedings. Section 12(3)(b), which is relied on here, provides:

> (3) The court shall make an order only if satisfied —
> (a)…
> (b) that the conduct of one party makes it unjust to hold the other party to the strict terms of the provision in question.

[79] Mr Wales submitted that, in this case, the conduct of PM, acting on behalf of LOR, in acknowledging receipt of the letter giving notice of AWS's intention to refer the dispute to arbitration and the notice to refer, was a material or effective cause of CMS's failure to take any further steps, which could easily have been taken, to serve the relevant documents on LOR's office in St Neots during the following day, 23 March 2010.

[80] It is common ground that the test to be applied is that identified by Colman J in *Harbour & General Works Ltd* v *Environment Agency* [2000] 1 Lloyd's Rep 65, at p73, where he said:

> For it to be held that the conduct of one party makes it unjust to hold the other party to the strict terms of the time bar, there must, in my judgment, at the very least be conduct which is proved somehow to have led the claimant to omit to give notice in time.

[81] I do not propose to go into the facts of that case because, in my judgment, they provide very little assistance on the question that is before me.

[82] Mr Wales relied strongly on the decision of Gross J in *Lantic Sugar*, a case to which I have already referred briefly. For present purposes, I can simplify the facts of that case as follows. A dispute arose between the consignees of a cargo of raw cane sugar and the vessel's owners. Any claims under the bill of lading were subject to a 12-month time bar that was ultimately extended up to and including 12 March 2009. On around 11 March 2009, there was a telephone conversation between a Mr Behr, representing the consignees, and a Mr Morland, who was acting on behalf of the owners, P&I Club. A few days earlier, on 6 March 2009, Mr Behr had sent a notice for arbitration to the club, for the attention of Mr Morland. The case for the owners was that during the telephone conversation on 11 March, Mr Morland had informed Mr Behr that he was taking instructions on the fax (which had sent the notice for arbitration) but that, pending express instructions, the club did not have authority to accept service on behalf of the owners.

[83] The evidence of Mr Behr was that he did not remember Mr Morland telling him that the club was not authorised to accept service of the notice of arbitration on behalf of the owners, but equally he did not remember being told expressly that the club did have such authority. When, the following day, Mr Morland sent an e-mail to Mr Behr saying that he had previously told him by telephone that the club had received the notice of arbitration but that the club was not able to accept/acknowledge service, Mr Behr's reaction was that that was "absolutely untrue". The judge accepted Mr Behr's account. Thereafter, having received the e-mail from Mr Morland on 12 March 2009, Mr Behr did everything he could to serve the notice for arbitration, but the registered office of the owners' agent appeared to be in the Marshall Islands and it was simply not possible to effect service before the deadline because of the differences in the time zones.

[84] In respect of the law, Gross J, in [46], described the relevant test under section 12(3)(b) in the following terms:

> Here, therefore, the court can *only* extend time if the conduct of [the owners] or the Club (acting on [the owners'] behalf) makes it unjust to hold the Claimants to the time bar. Some conduct must be shown that is causative of the failure to comply with the time bar or related to the injustice which would arise if relief is not granted: *Thyssen* v *Calypso* [2000] 2 Lloyd's Rep 243, at [24 to 25]; *Harris, Planterose & Tecks, The Arbitration Act 1996, A Commentary* (4th ed), at p81; that conduct, however, need not amount to an estoppel or something akin to it (*Harris et al, ibid*). Mere silence on the part of the party seeking to rely on the time bar, or failure to alert the party seeking the extension to the need to comply with the time bar, cannot make the barring of the claim unjust: *Harbour & General* v *Environment Agency* [2000] 1 Lloyd's Rep 65, at pp73 (Colman J) and 82 (CA).

(Gross J's emphasis.)

[85] Applying this test, the conclusion of Gross J was in the following terms:

> [52] Pulling the threads together, the course taken by the 11th March conversation meant that if a point was to be taken on the Club lacking authority to accept service of the 6th March Notice, it was only fair that something should then have been said. However inadvertent, the failure to do so, against a background of the Club taking instructions on the Notice, was misleading — and none the less so because its effect was to reinforce Mr Behr's own error as to the Club's authority to accept service. It is plain from Mr Behr's actions that the moment he did realise that there was a difficulty with serving the Club, he sought with determination and speed to serve [the owners] directly or through other agents (as in the Marshall Islands) with authority to accept service. Had Mr Morland said on the 11th March that which he later claimed to have said (and which he did say on the 12th March, upon receipt of the 12th March Notice), it is at least probable that the matter was still curable and that proper service would have been achieved in time… That conduct on the part of the Club contributed to, even if it was not the sole cause of, the Claimants' failure to comply with the time limit in question. In my view, on the particular facts of this case, s12(3)(b) of the Act is satisfied. That being so, in all the circumstances (as already set out) and in the exercise of my discretion under s12(1) of the Act, I am amply persuaded that the just course is to extend time.

[86] I must now turn back to the facts. On behalf of AWS, witness statements were served by Ms Belinda Schofield, the partner with conduct of the case, and Mr Neil Winterbourne, her assistant. It was Mr Winterbourne who sent the fax to PM dated 22 March 2010 enclosing the notice of intention to refer. He also was the person to whom the e-mail in response from Mr Shelling was sent. In his witness statement, he said:

> I was conscious at the time that this was a confirmation of receipt by someone whom I knew was involved in the matter and in terms which did not give any cause for concern. When Belinda Schofield came into the office on the morning of 23 March 2010, she asked whether I had heard anything in response to the Notice sent the previous day. I confirmed that we had and, in fact, read out to her the e-mail which Mr Shelling had sent the previous evening. Belinda Schofield commented on the phraseology used; along the lines that it was the same as that required by the adjudicator. If Mr Shelling had given any indication or hint that a point might be taken on service of either the Notice to Refer or the notice of intention, or even if Mr Shelling had simply enquired whether the notice of intention had been sent directly to LOR, I would have noted that and reacted to it immediately.
>
> Tuesday 23 March 2010 was the last day of the four week period after notification to the parties of Dr Gaitskell QC's decision. However, I took no further steps on that day because, having received Pinsent Masons' e-mail on 22 March in the terms in which it was received, I believed I had satisfied the requirements of the contract by the fax and letter I had sent to Pinsent Masons the day before.

[87] In her witness statement, Ms Schofield said the following:

> With reference to our e-mails of 16 December 2009 and 17 December 2009 regarding service of documents in the future on Pinsent Masons (NAW 2/p27 and p51), our intention was clear: we wanted to obtain comfort from Pinsent Masons that they had authority to accept service of all documents and pleadings relating to the adjudication. They so confirmed and it put this issue in my mind beyond doubt.
>
> …
>
> Following discussion with Leading and Junior Counsel, it was felt that the contract was not clear as to whether a notice of AWS's intention to refer to arbitration was separate and distinct from a Notice to Refer to Arbitration. We therefore decided that when informing LOR of AWS's intention to refer to arbitration we would also serve a Notice to Refer to Arbitration. Counsel drafted such a Notice to Refer on 22 March 2010.
>
> I was mindful that insofar as clause 93 of the contract applied, AWS had until 23 March 2010 to serve notice of its dissatisfaction with the adjudication award and its intention to refer to arbitration. I was therefore concerned to ensure that we should get the notice delivered to Pinsent Masons before that date, as the address to which I understood the notification should properly be given.

> On the morning of 22 March, having checked the draft Notice to Refer and the draft letter to go to Pinsent Masons, I asked Neil Winterbourne to let me know when these had gone to Pinsent Masons. As this was the penultimate day for service within the time period, I wanted to know that the letter and enclosure was delivered to Pinsent Masons. I left the office shortly after midday for a meeting in our Bristol office. Neil confirmed to me shortly after 5.00pm that the Notices had been sent.
>
> The next morning I enquired of Neil whether he had received any response from Pinsent Masons acknowledging receipt. I remained anxious that we had effected delivery so close to the deadline. I did not focus at the time on the fact that because we had sent the notice by fax, a transmission report would have been generated. Even if I had focused on that fact, I believe I would have wanted comfort that it had made its way from the electronic fax system to those at Pinsent Masons with conduct of the matter. Neil confirmed that he had received a response, and read the e-mail from Mr Shelling of 22 March 2010 to me from his screen. I was assured by Pinsent Masons' response; I noted that the acknowledgement used the same phrase to acknowledge receipt and service of all the documents and pleadings in the adjudication…
>
> Further, I also noted that there was no reference to Pinsent Masons taking instructions. That phrase would have been a surprise to me because the notification we had sent was not something that called for any action or response on the part of LOR. It would have prompted me to check the position regarding service of the notice and if no immediate comfort could have been obtained that Pinsent Masons were not taking a point on service, given the deadline, I would have insisted that the notification was sent directly to LOR.
>
> Although I believed prior to the fax of 22 March that the offices of Pinsent Masons were the correct address for service, I am clear in my own mind that Mr Shelling's e-mail which Neil Winterbourne read out to me, influenced me in dispelling my lingering concern to know that the notification had been properly given and confirmed that this view was the right view — as a result of which no further steps were taken. The lingering concern I had was not a specific doubt — I was not taking a gamble on getting service right. It was a concern just in case any point we had not considered was raised and particularly so given that we were up against the time limit.

[88] When considering the position of Ms Schofield and Mr Winterbourne, I consider that it is most important to remember two things. First, they were well aware of the contractual requirement for service of documents at the contract address, or the last notified address, because that is what they had done in December 2009. The whole point of the requests to PM to confirm that it had instructions to accept service of documents relating to the adjudication was to avoid having to send them to St Neots (only for them to be forwarded to PM and Laing O'Rourke's Dartford office). Second, there was no cost or difficulty involved in serving the relevant documents on LOR at its St Neots office: it would have taken a secretary a matter of a few minutes. No one in Ms Schofield's position would have failed to take such a simple precaution unless she was confident that it was not necessary. I find that that was Ms Schofield's state of mind.

[89] I consider that it is abundantly clear that CMS did not serve the documents on LOR's St Neots office because it thought that it had achieved effective service when it sent the documents to PM and received unqualified confirmation of safe receipt. The service and acknowledgement of these documents followed exactly the same pattern as had taken place on several previous occasions when no question of the validity of the service had arisen.

[90] It seems to me that it could be forgiven for assuming that if any point was going to be taken concerning service, particularly when the notice to refer had been sent expressly "by way of service", PM would have qualified its acknowledgement by saying something about taking instructions from its clients. I find that but for the terms in which PM acknowledged service of those documents, CMS would have sent them the following day to LOR's St Neots office. When it had doubts on previous occasions, that is what it did, and there is no other rational explanation for the fact that it did not do so on this occasion. However, I repeat my earlier observation that I do not consider that Mr Harris, at PM, acted with any intention to mislead CMS. He felt that his duty to his clients was to say as little as possible and he knew that he was dealing with sophisticated solicitors who could reasonably be expected to know what they were doing.

For these reasons, I find that the conduct of those acting for LOR was a material contributing cause to the fact that the notice of intention to refer the dispute to arbitration was not sent to St Neots on 23 March 2010 before the deadline expired. Given that the notice was in fact received in time by Mr Grammer and Mr McKenzie, who were the two people at LOR who needed to see them, I would regard it as unjust to hold AWS to the strict provisions in respect of service if it was in fact in breach of those provisions (contrary to my earlier finding). This seems to me to be a paradigm case for an extension of time under section 12 of the 1996 Act and, if it is required, I would have no hesitation in exercising my discretion in favour of granting the necessary extension of time.

Claim allowed.

LANDLORD AND TENANT

HOUSING ACTS

Manchester City Council *v* Pinnock

SUPREME COURT

[2010] UKSC 45

3 November 2010

LORD PHILLIPS, PRESIDENT, LORD HOPE, DEPUTY PRESIDENT, LORD RODGER, LORD WALKER, LADY HALE, LORD BROWN, LORD MANCE, LORD NEUBERGER AND LORD COLLINS

Landlord and tenant — Housing Act 1996 — Demoted tenancy — Possession order — Secure tenancy demoted on grounds of anti-social behaviour by tenant's family members — Possession notice served by local authority landlords — Review panel upholding notice — Section 143D of 1996 Act — Application to court for possession order — County court conducting judicial review of panel's decision and granting possession order — Extent of county court's review jurisdiction — Article 8 of European Convention on Human Rights — Whether section 143D(2) compatible with Article 8

Since 1978, the appellant had lived with his partner, with whom he had five children, in a property leased from the respondent local authority. His secure tenancy was demoted in 2005 on an application by the respondents, under section 82A of the Housing Act 1985, as amended by the Anti-Social Behaviour Act 2003, on the ground of anti-social behaviour by his partner and children. In 2008, shortly before the demotion order was due to lapse, the respondents served a notice, under section 143E of the Housing Act 1996, of their intention to seek possession on the ground of further alleged anti-social behaviour by two of the appellant's sons. The appellant exercised his right to a review but the review panel upheld the notice.

The respondents applied to the court for a possession order under section 143D of the 1996 Act. In those proceedings, the appellant sought to challenge the factual basis on which the respondents had sought possession and on which the panel had upheld that decision. He further contended that a possession order would breach his rights under Article 8 of the European Convention on Human Rights. The county court judge considered that his role was limited to conducting an ordinary judicial review of the respondents' decision and did not extend to resolving factual disputes or arguments based on Article 8. He found that the respondents' decision was rational and made an order for possession.

The appellant's appeal against that decision was dismissed by the Court of Appeal, which took an even more limited view of the county court's role. It held that, on the express wording of section 143D(2), the county court could not review the substance or rationality of the decision to seek possession but was obliged, on the express wording of section 143D(2) of the 1996 Act, to grant the possession order sought so long as the correct procedure under sections 143E and 143F had been followed: see [2009] EWCA Civ 852; [2010] 1 WLR 713. The appellant appealed. The issues on the appeal related to the compatibility of the demoted tenancy regime in the 1985, 1996 and 2003 Acts with Article 8 of the Convention.

***Held*: The appeal was dismissed.**

(1) The jurisprudence of the European Court of Human Rights (ECtHR) establishes that any person at risk of being dispossessed of his home at the suit of a local authority should, in principle, have the right to raise the proportionality of the measure and have that matter determined by an independent tribunal in the light of Article 8, even if his right of occupation under domestic law has come to an end, and that a judicial procedure that is limited to addressing the proportionality of the measure through the medium of traditional judicial review will be inadequate because it is not appropriate for resolving sensitive factual issues. Although the Supreme Court is not obliged to follow every decision of the ECtHR, it would be wrong not to follow a consistent line of decisions that is not inconsistent with a substantive or procedural aspect of domestic law and the reasoning in which does not overlook or misunderstand some argument or point of principle. Accordingly, where a court is asked to make an order for possession of a person's home at the suit of a local authority, the court must have the power to assess the proportionality of making the order and, in making that assessment, to resolve any relevant dispute of fact. Although section 143D(2) of the 1996 Act would exclude such a power if it were construed in the traditional way, it is possible, under section 3(1) of the Human Rights Act 1998, to construe it in a Convention-compliant manner. The requirement to follow the procedure under sections 143E and 143F includes an inherent requirement to act lawfully in following that procedure, including a duty on the local authority in domestic law to act rationally and to investigate the facts fairly when deciding whether to bring possession proceedings against a demoted tenant, as well as a duty under Article 8 to consider proportionality. Section 143D(2) should therefore be read as allowing a demoted tenant to argue in its defence that the procedure has not been lawfully complied with, either because its express requirements have not been observed or the rules of natural justice have been infringed, and as enabling the court to consider that matter having regard to the occupier's Article 8 rights. Since ECtHR jurisprudence requires the court considering such a challenge to have the power to make its own assessment of any relevant disputed facts, the traditional review powers of the court should be expanded so as to permit it to carry out that exercise. Section 143D(2) should be read as allowing the court to exercise the powers that are necessary to give effect to any Article 8 defence that the tenant raises in the possession proceedings, including the power to reconsider the facts by hearing evidence and forming its own view. A tenant can therefore raise proportionality issues in the possession proceedings even if they are in the county court, and the county court can exercise the powers necessary for that purpose. So construed, the demoted tenancy regime in the 1996 Act is compatible with Article 8. (2) Where possession is sought, the question is always whether eviction is a proportionate means of achieving a legitimate aim. If an Article 8 point is raised by a residential occupier in proceedings, that point should first be considered summarily and if, as will often be the case, the court is satisfied that the point cannot succeed, even if the facts relied on are made out, it should be dismissed. In most cases in which a residential occupier has no contractual or statutory protection and the local authority are entitled to possession under domestic law, there will be a strong case for the proportionality of a possession order. In the instant case, it had been proportionate to make an order for possession such that it was unnecessary to remit the case for further determination.

The following cases are referred to in this report.

Belchikova v *Russia* 2408/06 unreported 25 March 2010
Blečić v *Croatia* 59532/00 [2006] ECHR 207; (2005) 41 EHRR 13, ECtHR
Connors v *United Kingdom* 66746/01 [2004] 4 PLR 16; [2004] HLR 52, (2005) 40 EHRR 9, ECtHR
Ćosić v *Croatia* unreported 15 January 2009
Cumming v *Danson* [1942] 2 All ER 653
Di Palma v *United Kingdom* 11949/86 (1986) 10 EHRR 149
Doherty v *Birmingham City Council*; *sub nom Birmingham City Council* v *Doherty* [2008] UKHL 57; [2009] 1 AC 367; [2008] 3 WLR 636; [2009] 1 All ER 653; [2009] 1 P&CR 11; [2008] HLR 45
Harrow London Borough Council v *Qazi*; *sub nom Qazi* v *Harrow London Borough Council* v *Quazi* [2003] UKHL 43; [2004] 1 AC 983; [2003] 3 WLR 792; [2003] 4 All ER 461; [2004] 1 P&CR 19; [2003] 3 EGLR 109; [2003] HLR 75
Kay v *Lambeth London Borough Council*; *Gorman* v *Lambeth London Borough Council*; *Constantine* v *Lambeth London Borough Council*; *Barnett* v *Lambeth London Borough Council*; *Cole* v *Lambeth London Borough Council*; *Dymny* v *Lambeth London Borough Council*; *Price* v *Leeds County Council* [2006] UKHL 10; [2006] 2 AC 465; [2006] 2 WLR 570; [2006] 4 All ER 128; [2006] 2 P&CR 25; [2006] HLR 22
Kay v *United Kingdom* 37341/06 unreported 21 September 2010
Manchester City Council v *Cochrane* [1999] 1 WLR 809; (1999) 31 HLR 810, CA
Manchester City Council v *Pinnock* [2009] EWCA Civ 852; [2010] 1 WLR 713; [2010] 3 All ER 201; [2010] HLR 7
McCann v *United Kingdom* 19009/04 [2008] 2 EGLR 45; [2008] 28 EG 114; [2008] HLR 40; (2008) 47 EHRR 40; [2008] LGR 474, ECtHR
Mullen v *Salford City Council*; *Powell* v *Hounslow London Borough Council*; *Hall* v *Leeds City Council*; *Frisby* v *Birmingham City Council*; *Mushin* v *Manchester City Council* [2010] EWCA Civ 336; [2010] HLR 35
Paulić v *Croatia* 3572/06 unreported 22 October 2009
Pyx Granite Co Ltd v *Ministry of Housing and Local Government* [1960] AC 260; [1959] 3 WLR 346; [1959] 3 All ER 1; (1959) 58 LGR 1; 10 P&CR 319, HL
R v *Horncastle*; *R* v *Blackmore*; *R* v *Carter*; *R* v *Maquis*; *R* v *Graham* [2009] 2 AC 373; [2010] 2 WLR 47; [2010] 2 All ER 359
R (on the application of Ullah) v *Special Adjudicator*; *Do* v *Immigration Appeal Tribunal* [2004] UKHL 26; [2004] 2 AC 323; [2004] 3 WLR 23; [2004] 3 All ER 785
Wandsworth London Borough Council v *Winder (No 1)* [1985] AC 461; [1984] 3 WLR 1254; [1984] 3 All ER 976; 83 LGR 143, HL
Zehentner v *Austria* 20082/02 unreported 16 July 2009

This was an appeal by the appellant, Cleveland Pinnock, from a decision of the Court of Appeal dismissing his appeal against a decision of HH Judge Holman, sitting in Manchester County Court, granting a possession order to the respondents, Manchester City Council, in respect of residential property under section 143D of the Housing Act 1996.

Richard Drabble QC and James Stark (instructed by Platt Halpern, of Manchester) appeared for the appellant; Andrew Arden QC and Jonathan Manning (instructed by the legal department of Manchester City Council) appeared for the respondents; Daniel Stilitz QC and Ben Hooper (instructed by the Treasury Solicitor) appeared for the Secretary of State for Communities and Local Government, as intervener; Jan Luba QC (instructed by the Equality and Human Rights Commission) represented the Equality and Human Rights Commission, as intervener.

Giving the first opinion, **Lord Neuberger** said:

[1] This is the judgment of the court, to which all members have contributed.

[2] The principal questions in this appeal are: (a) whether Article 8 of the European Convention on Human Rights (the Convention) requires a court, which is being asked to make an order for possession under section 143D(2) of the Housing Act 1996 (the 1996 Act) against a person occupying premises under a demoted tenancy, to have the power to consider whether the order would be "necessary in a democratic society"; and (b) if so, whether section 143D(2) is compatible with Article 8 of the Convention. In the result, the court answers both questions in the affirmative: the first in [22] to [54]; and the second in [65] to [107] below.

[3] The appeal concerns a tenancy granted by a local authority, but observations relating to local authority landlords in this judgment apply equally to other social landlords to the extent that they are public authorities under the Human Rights Act 1998 (the HRA).

[4] On the other hand, we should emphasise at the outset that nothing in this judgment is intended to bear on cases where the person seeking the order for possession is a private landowner. We briefly explain why in [50] below.

Background to the appeal: Secure and demoted tenancies

[5] Most residential occupiers of houses and flats owned by local authorities are "secure tenants" under Part IV of the Housing Act 1985 (the 1985 Act). By virtue of section 84 of the 1985 Act, a secure tenant cannot be evicted unless the landlord establishes to the satisfaction of the court: (a) that one of the grounds specified in Schedule 2 to the 1985 Act (for example, non-payment of rent or nuisance to neighbours) exists; and (b) except in some specified categories of case where suitable alternative accommodation is available, that it is reasonable to make an order for possession against the tenant.

[6] Even where the landlord establishes that these two requirements are satisfied, the court has a wide discretion, under section 85 of the 1985 Act, as to what order to make. It may refuse to make any order, it may adjourn the proceedings, it may make an outright possession order that takes effect on a specific day or it may make a suspended possession order that will not take effect so long as, for instance, the tenant pays the rent or creates no nuisance.

[7] The secure tenancy regime was originally introduced by the Housing Act 1980 (the 1980 Act), but its provisions were consolidated by Part IV of the 1985 Act. Certain types of tenancy are excluded from this regime, and they are set out in Schedule 1 to the 1985 Act. Subsequently, amendments were made to the regime, most relevantly for present purposes by the 1996 Act and the Anti-social Behaviour Act 2003 (the 2003 Act).

[8] The 2003 Act inserted a new section 82A into the 1985 Act. This section gives the court the power to make a "demotion order" in respect of a secure tenancy. A demotion order results in a tenancy ceasing to be a secure tenancy and becoming instead a "demoted tenancy". Section 82A(4) states that such an order may be made only if: (a) the tenant (or someone living with him) has engaged, or has threatened to engage, in (i) "housing-related anti-social conduct" (as defined in section 153A of the 1996 Act) or (ii) conduct that consists of or involves using the "premises for unlawful purposes" (as explained in section 153B of the 1996 Act); and (b) "it is reasonable to make the order". Section 82A makes it clear that the demoted tenancy is a new tenancy. The terms of the previous tenancy as to rent are carried across into the new tenancy but the demotion results in much-reduced rights of security of tenure for the tenant.

[9] Somewhat confusingly, the provisions dealing with the operation of the demoted tenancy regime were inserted as Chapter 1A of Part V of the 1996 Act. Subsection (1) of section 143B of that Act explains that if a tenancy is demoted, the demotion will last for a year, unless the landlord brings possession proceedings within that year. If such proceedings are brought within the year and are not determined before the year's end, the demoted tenancy continues until the proceedings are determined. If such proceedings are brought within the year and an order for possession is made, the tenancy ends. If no such proceedings are brought, or they are brought and they fail, then, at the end of the year, the demoted tenancy will become a secure tenancy.

[10] Subsection (1) of section 143D of the 1996 Act states that a landlord can bring a demoted tenancy to an end only by obtaining an order for possession from the court. Since it is central to the present appeal, section 143D(2) must be quoted in full:

> The court must make an order for possession unless it thinks that the procedure under sections 143E and 143F has not been followed.

[11] The effect of section 143E of the 1996 Act is that before issuing possession proceedings against a demoted tenant, a local authority landlord must serve a notice (a "notice") informing him of: (a) the fact

that they have decided to seek possession; (b) the reasons why; (c) the date after which the proceedings will be issued; (d) the tenant's right to request a review of the landlord's decision (a "review"); and (e) where to get legal advice. Section 143F of the 1996 Act entitles the tenant, within 14 days of the notice, to request a review, in which case the local authority landlords are obliged to carry out a review that complies with regulations made by the Secretary of State under section 143F(3) and (4) and then to inform the tenant of the outcome. Such regulations have been made in the Demoted Tenancies (Review of Decisions)(England) Regulations 2004 (SI 2004/1679).

[12] Section 143N of the 1996 Act states that the county court has jurisdiction "to determine questions arising", and "to entertain proceedings brought", under, *inter alia*, sections 143B-143F, even if the only relief sought is a declaration.

[13] The procedures of the demoted tenancy regime are closely based on those of a regime first introduced by Chapter 1 of Part V of the 1996 Act. It enabled a local authority to grant a tenancy under which a new tenant had a one-year probationary period before becoming a secure tenant. During that first year, the tenancy is an "introductory tenancy". The procedure governing the landlords' right to claim possession during that probationary period is contained in sections 127, 128, and 129 of the 1996 Act, whose provisions are, *mutatis mutandis*, virtually identical to sections 143D, 143E and 143F respectively.

Procedural background to the appeal

[14] In November 1978, Manchester City Council (the council) granted Mr Cleveland Pinnock a tenancy of a house at 65 Meldon Road, Longsight (the property), where he has lived ever since with his partner, Ms Christine Walker, and, from time to time, with all or some of their five children. In March 2005, the council applied to Manchester County Court for an order for possession of the property or, in the alternative, a demotion order in respect of Mr Pinnock's secure tenancy. Each of these claims was based on the contention that all of Mr Pinnock's children and Ms Walker (but not Mr Pinnock) had been guilty of serious anti-social behaviour in breach of the covenants in Mr Pinnock's tenancy. The proceedings came before Mr Recorder Scott Donovan, who heard considerable evidence and argument over a total of six days. In part, the length of the proceedings was due to the council amending their case in respect of the relief that they were seeking.

[15] The recorder gave a full judgment on 8 June 2007. He concluded that a large number of serious allegations against Ms Walker and Mr Pinnock's children were well founded. He nevertheless decided that it would be "truly draconian" to make an order for possession "bearing in mind the length of the tenancy and Mr Pinnock's blameless life looked at from his own lack of direct involvement in criminal activity". However, he went on to say that "[a]pplying the criteria of 'reasonableness', I am satisfied that a demotion of tenancy order is the most appropriate order and that compliance with the order is entirely within Mr Pinnock's and Christine Walker's own hands".

[16] The demotion order therefore took effect from 8 June 2007. On 6 June 2008, the day before the order would effectively have lapsed, the council served a notice under section 143E, which indicated that possession would be sought. The notice had the effect, pursuant to section 143B, of prolonging the demoted tenancy and of initiating the procedure envisaged in sections 143D, 143E and 143F. The notice sought to justify the projected possession proceedings on the ground of further alleged incidents of anti-social behaviour in the vicinity of the property involving two of Mr Pinnock's sons. Mr Pinnock exercised his right to seek a review, which duly took place before a panel appointed by the council (the panel). In its decision of 3 July 2008, the panel effectively upheld the notice.

[17] The council then issued a claim for possession that came before HH Judge Holman in Manchester County Court. After a two-day hearing, the judge gave a full judgment on 22 December 2008. The upshot of his decision was that he made an outright order for possession of the property. Mr Pinnock appealed to the Court of Appeal, who dismissed his appeal: see [2009] EWCA Civ 852*. Mr Pinnock now appeals to this court.

Issues that arise on this appeal

[18] That simple description of the present proceedings rather masks the important and difficult issues to which they give rise. Those issues are apparent from the clear and careful judgments of Judge Holman, in the county court, and of Stanley Burnton LJ (with whom Mummery and Lloyd LJJ agreed) in the Court of Appeal. Mr Pinnock wanted to challenge the factual basis on which the council had decided to seek possession and the panel had decided to uphold the decision. He also contended that the making of an order for possession would violate his Article 8 Convention rights.

[19] Judge Holman concluded that his role in this case was, as he put it in [60], "limited to conducting a conventional judicial review" of the council's decision to bring the possession proceedings, and that his remit did not extend to "resolv[ing] factual disputes". In particular, he could not entertain any argument based on Article 8. Having accepted that he could review the council's decision to bring and maintain the possession claim on normal judicial review principles, the judge concluded that the council's decision to prosecute the claim was rational. He accordingly made an outright order for possession.

[20] Stanley Burnton LJ agreed in the result, but although largely agreeing with Judge Holman's analysis, he thought that the county court's role was even more limited. He said, in [50]:

> Section 143D of the 1996 Act restricts the county court to considering whether the procedure under sections 143E and 143F has been followed. If the court concludes the procedure has not been followed, it will not make an order for possession. If it has been followed, it must make the order. I emphasise the word procedure. The court's review is limited to matters of procedure, and the county court cannot review the substance or rationality of the landlord's decision, or whether or not it is consistent with the tenant's or other occupiers' Convention rights.

Stanley Burnton LJ none the less went on, helpfully, to consider whether he would have agreed with the judge's conclusion that the council's decision to maintain a claim for possession was rational. He concluded that it was; indeed, he thought that the judge had taken rather a restrictive view of the relevant evidence that the council could have taken into account.

[21] This appeal gives rise to four main issues, of increasing specificity. The first is whether the jurisprudence of the European Court of Human Rights (the ECtHR) requires that before making an order for possession of property that consists of a person's home pursuant to a claim made by a local authority (or other public authority), a domestic court should be able to consider the proportionality of evicting that person from his home under Article 8, and, in the process of doing so, to resolve any relevant factual disputes between the parties. We deal with that question in [22] to [54] below and answer it in the affirmative. The second issue ([55] to [64] below) is what this conclusion means in practice in respect of claims for possession, and related claims, in respect of residential property. The third issue ([65] to [107] below) is whether the demoted tenancy regime in the 1985, 1996 and 2003 Acts can properly be interpreted so as to comply with the requirements of Article 8 or whether at least some aspects of that regime are incompatible with the occupier's Article 8 Convention rights. The fourth issue ([108] to [132] below), which requires a fuller consideration of the facts of this case, is how the appeal should be disposed of in the light of the answers on the first three issues.

First issue: What does the Convention require of the courts?

Nature of the issue

[22] So far as relevant, Article 8 of the Convention provides:

> 1. Everyone has the right to respect for… his home….
>
> 2. There shall be no interference by a public authority with the exercise of this right except such as is in accordance with the law and is necessary in a democratic society in the interests of… the economic well-being of the country,

---

* Editor's note: Reported at [2010] 1 WLR 713

for the prevention of disorder or crime… or for the protection of the rights… of others.

It is also appropriate to refer to Article 6, which, so far as relevant, provides:

1. In the determination of his civil rights and obligations… everyone is entitled to a fair… hearing… by an independent and impartial tribunal established by law.

[23] The argument on behalf of Mr Pinnock is as follows.

(a) Where the person seeking possession is a "public authority", a court invited to make an order for possession of a person's home must be satisfied that Article 8 is complied with.

(b) Accordingly, in the present proceedings, Judge Holman had to satisfy himself that the order for possession that he was being invited to make complied with Article 8.

(c) Article 8, when read together with Article 6, required the judge, as the relevant independent tribunal, to be satisfied that the order for possession would: (i) be "in accordance with the law"; and (ii) be "necessary in a democratic society", that is, that it would be proportionate.

(d) The order for possession was "in accordance with the law" since it was made pursuant to the provisions relating to demoted tenancies in the 1985 and 1996 Acts, which are in principle unobjectionable under Article 8.

(e) However, Mr Pinnock was not given the opportunity to raise with the court the question of whether the order for possession was, in all the circumstances of this case, proportionate. Therefore, Article 8 was violated.

(f) Further, in order to determine proportionality, the court should have had power to resolve for itself any issues of fact between the council and Mr Pinnock and to form its own view of proportionality, rather than adopting the traditional judicial review approach taken by the judge.

(g) Either the legislation should be interpreted to have the effect contended for in points (e) and (f) or this court should make a declaration of incompatibility.

[24] The issues identified in the argument for Mr Pinnock are by no means novel. It is therefore necessary for the court to look briefly at the decisions of the House of Lords that deal with them and then, in a little more detail, at the relevant decisions of the ECtHR.

House of Lords cases

[25] In three relatively recent cases, the House of Lords held that it was not open to a residential occupier, against whom possession was being sought by a local authority, to raise a proportionality argument under Article 8. In other words, the House rejected points (e) and (f) in the outline of the argument for Mr Pinnock in [23] above. Point (g) therefore did not arise. For this reason, the Court of Appeal and Judge Holman were bound to come to the conclusions that we have summarised in [19] and [20] above.

[26] The three decisions of the House of Lords are *Harrow London Borough Council* v *Qazi* [2003] UKHL 43; [2004] 1 AC 983*, *Kay* v *Lambeth London Borough Council* [2006] UKHL 10; [2006] 2 AC 465, and *Doherty* v *Birmingham City Council* [2008] UKHL 57; [2009] 1 AC 367. In each of them, the defendants were residential occupiers of properties owned by a local authority, but, for one reason or another, they were not secure tenants and, having had any right to continue to occupy the respective properties brought to an end in accordance with domestic law, they were trespassers. So, in accordance with domestic law, the defendants could raise no defence to the local authority's claim for possession. In each case, however, the defendants contended that they should be able to rely on the argument that, even though they were trespassers with no defence to a claim for possession under domestic law, they had the right to have the proportionality of the loss of their home taken into account by virtue of their Article 8 Convention rights. No disrespect is intended to the impressive and careful reasoning in those three decisions when we say that, for present purposes, it is unnecessary to consider them in any detail.

[27] In *Harrow* and *Kay*, albeit in each case by a bare majority, the House decided that "a defence which does not challenge the law under which the possession order is sought as being incompatible with Article 8 but is based only on the occupier's personal circumstances should be struck out": *Kay*, at pp516-517, in [110], *per* Lord Hope of Craighead, with whom Lord Scott of Foscote, Baroness Hale of Richmond and Lord Brown of Eaton-under-Heywood agreed. This observation applied to claims against trespassers, just as much as to claims against current or former tenants or licensees. At the end of the same paragraph, Lord Hope explained that, following *Wandsworth London Borough Council* v *Winder (No 1)* [1985] AC 461, in principle it would be open to a defendant "to challenge the decision of a local authority to recover possession as an improper exercise of its powers at common law" on the traditional judicial review ground "that it was a decision that no reasonable person would consider justifiable".

[28] In *Doherty*, the law as stated in [110] of *Kay* v *Lambeth* was substantially reaffirmed. On the Article 8 point, Lord Mance, however, dissented, in [132], and Lord Walker of Gestingthorpe displayed less than whole-hearted enthusiasm in [107] to [108]. The law on the judicial review point was affirmed by Lord Hope, Lord Walker and Lord Mance in [56], [123] and [157] respectively. Nevertheless, in the light of the developments in the Strasbourg jurisprudence that we describe below, the House developed the law by acknowledging that the traditional approach to judicial review would have to be expanded, particularly to permit the court to make its own assessment of the relevant facts: see especially at p416, in [68], *per* Lord Scott, and at p443, in [138], *per* Lord Mance.

[29] In both *Harrow* and *Kay*, Lord Bingham of Cornhill (dissenting along with Lord Steyn in the former case and with Lord Nicholls of Birkenhead and Lord Walker in the latter) accepted that it should be open, as a matter of principle, to a residential occupier, against whom a local authority are seeking possession, to raise an Article 8 proportionality argument based on the facts of the particular case. However, in *Harrow*, in [25], Lord Bingham said that if this was right "the occasions on which a court would be justified in declining to make a possession order would be very highly exceptional". He effectively repeated this view in *Kay*, at pp491-492, in [29], where he suggested that only in "rare and exceptional cases" would an Article 8 proportionality challenge "not be futile".

Strasbourg jurisprudence

[30] Mr Pinnock contends that, exceptionally, it is appropriate for this nine-judge court to depart from the majority view in these cases because there is now a consistent series of decisions of the ECtHR that unambiguously supports the minority view in the earlier House of Lords decisions, and there is no good reason not to follow that series of decisions. We must therefore examine them.

[31] In *Connors* v *United Kingdom* 66746/01 (2005) 40 EHRR 9, gypsies had initially been permitted to locate their caravan on a piece of land owned by a local authority, but their right of occupation was brought to an end because the local authority considered that they were committing a nuisance. The local authority then successfully brought summary proceedings for possession on the ground that they were trespassers and had no right to remain in occupation of the land. Before the First Section of the ECtHR, the gypsies successfully contended that the proceedings violated their rights under Article 8.

[32] Although the local authority's decision to evict the gypsies was susceptible to judicial review, the ECtHR considered, in [92], that this procedure was insufficient to satisfy the requirements of Article 8 because "the local authority was not required to establish any substantive justification for evicting [the gypsies] and on this point judicial review could not provide any opportunity for an examination of the facts in dispute between the parties".

[33] In a passage, in [81] to [83], that has often been quoted verbatim in subsequent decisions, the ECtHR said:

* Editor's note: Also reported at [2003] 3 EGLR 109

A [81] An interference will be considered "necessary in a democratic society" for a legitimate aim if it answers a "pressing social need" and, in particular, if it is proportionate to the legitimate aim pursued. While it is for the national authorities to make the initial assessment of necessity, the final evaluation as to whether the reasons cited for the interference are relevant and sufficient remains subject to review by the Court for conformity with the requirements of the Convention.

[82] In this regard, a margin of appreciation must, inevitably, be left to the national authorities… This margin will vary according to the nature of the Convention right in issue, its importance for the individual and the nature of the activities restricted, as well as the nature of the aim pursued by the restrictions. … Where general social and economic policy considerations have arisen in the
B context of Article 8 itself, the scope of the margin of appreciation depends on the context of the case, with particular significance attaching to the extent of the intrusion into the personal sphere of the applicant.

[83] The procedural safeguards available to the individual will be especially material in determining whether the respondent state has, when fixing the regulatory framework, remained within its margin of appreciation. In particular, the Court must examine whether the decision-making process leading to measures of interference was fair and such as to afford due respect to the interests safeguarded to the individual by Article 8.

[34] In *Blečić* v *Croatia* 59532/00 (2005) 41 EHRR 13, the First Section of the ECtHR held that there had been no violation of the
C applicant's Article 8 rights in circumstances where her protected tenancy of her home had been terminated by the Croatian court on the ground that she had ceased to occupy it for 10 months during 1991-92. Her case was that it had been her home since 1953, and that her absence had been attributable to armed conflict in Dalmatia, but it was held that it had been her "personal decision… to leave". The ECtHR said, in [65]:

> State intervention in socio-economic matters such as housing is often necessary in securing social justice and public benefit. In this area, the margin of appreciation available to the State in implementing social and economic policies is necessarily a wide one. The domestic authorities' judgment as to
> D what is necessary to achieve the objectives of those policies should be respected unless that judgment is manifestly without reasonable foundation. Although this principle was originally set forth in the context of complaints under Article 1 of Protocol No 1… the State enjoys an equally wide margin of appreciation as regards respect for the home in circumstances such as those prevailing in the present case, in the context of Article 8. Thus, the Court will accept the judgment of the domestic authorities as to what is necessary in a democratic society unless that judgment is manifestly without reasonable foundation, that is, unless the measure employed is manifestly disproportionate to the legitimate aim pursued.

The case then went to the Grand Chamber, which held that, *ratione*
E *temporis*, the court had had no jurisdiction to hear it. The Grand Chamber said nothing, however, to cast doubt on what the First Section had said in the passage that we have quoted: see [2006] ECHR 207.

[35] In *McCann* v *United Kingdom* 19009/04 (2008) 47 EHRR 40*, the county court made an order for possession against a man who occupied his home as a joint tenant with his estranged wife on the ground that the tenancy had been determined by a notice to quit that she had served at the request of the local authority landlord and without reference to her husband. The ECtHR (Fourth Section) rejected the contention that the reasoning in *Connors*, in [81] to [83], was "confined only to cases involving the eviction of gypsies or cases where the applicant sought to challenge the law itself rather than its application in
F his particular case": see [50]. The court continued:

> The loss of one's home is the most extreme form of interference with the right for respect for the home. Any person at risk of an interference of this magnitude should in principle be able to have the proportionality of the measure determined by an independent tribunal in the light of the relevant principles under [Article 8], notwithstanding that, under domestic law, his right of occupation has come to an end.

[36] In [54], the ECtHR considered and rejected the contention that "the grant of the right to the occupier to raise an issue under Article 8 would have serious consequences for the functioning of the system or G for the domestic law of landlord and tenant", citing and confirming the view of Lord Bingham in *Kay*, at pp491-492, in [29], to the effect that "only in very exceptional cases" could "an applicant… succeed in raising an arguable case which would require a court to examine the issue". The court also said that "in the great majority of cases, an order for possession could continue to be made in summary proceedings".

[37] In *Ćosić* v *Croatia* 28261/06 unreported 15 January 2009, the Croatian state had obtained an order evicting the applicant from her home, which she had ceased to have any right to occupy as a matter of domestic law. After quoting and considering, in [81] to [83], the ECtHR (First Section) pointed out, in [21], that the national courts had H based their decision "exclusively on the [domestic] applicable laws" and had "thus confined themselves to finding that occupation by the appellant was without legal basis [and] made no further analysis as to the proportionality of the measure to be applied against the applicant". The court immediately went on to say that the Convention required that the eviction order was "proportionate… to the legitimate aim pursued" and that "no legal provision of domestic law should be interpreted and applied in a manner incompatible with Croatia's obligations under the Convention".

[38] In [22] and [23], the ECtHR concluded that Article 8 had been violated because "the applicant [had not been] afforded [the] J possibility" of having "the proportionality and reasonableness of the measure [viz, an order of court evicting her from her home] determined by an independent tribunal in the light of the relevant principles under Article 8".

[39] In *Zehentner* v *Austria* 20082/02 unreported 16 July 2009, the ECtHR (First Section) had to consider the effect of Article 8 in the context of an order evicting the applicant from her home following a "judicial sale", after the making of the Austrian equivalent of a charging order. The procedural circumstances were rather unusual, but the court held, in [54], that "the judicial sale and the applicant's eviction are to be seen as a whole". Importantly, for present purposes, in [52] to [59], the K court reaffirmed the approach in *Connors* and *McCann*. In particular, the court also stated, in [59], that a person at risk of eviction from his home should "be able to have the proportionality of the measure determined by an independent tribunal in the light of the relevant principles under Article 8".

[40] In *Paulić* v *Croatia* 3572/06 unreported 22 October 2009, the ECtHR (First Section) cited *McCann* and reiterated, in [43], "that any person at risk of an interference with his right to home should in principle be able to have the proportionality and reasonableness of the measure determined by an independent tribunal… notwithstanding that, under domestic law, he or she has no right to occupy a flat". The court L went on to explain that this right "does not arise automatically", but only if the issue is raised with the court by the person concerned.

[41] Finally, there is *Kay* v *United Kingdom* 37341/06 unreported 21 September 2010, in which the ECtHR (Fourth Section) gave its judgment after the conclusion of the oral argument in this case. We then received written submissions on the decision from the parties. In that case, the application was made to the Strasbourg court by the unsuccessful appellants in *Kay* v *Lambeth*. They had no security of tenure in their homes and their defences to claims for possession brought by the local authority based on the contention that it was disproportionate to deprive them of their homes in the light of Article 8 had been struck out. After M carefully considering the various views expressed in the House of Lords in *Kay* v *Lambeth* and *Doherty*, and the relevant decisions of the Court of Appeal, the ECtHR stated, in [65] to [68], that the principles laid down in *Connors* and *McCann* applied.

[42] The ECtHR then stated, in [73]:

> The Court welcomes the increasing tendency of the domestic courts to develop and expand conventional judicial review grounds in the light of Article 8. A number of their Lordships in *Doherty* alluded to the possibility for challenges on conventional judicial review grounds in cases such as the applicants' to encompass more than just traditional *Wednesbury* grounds (see Lord Hope at paragraph 55; Lord Scott at paragraphs 70 and 84 to 85; and Lord Mance at paragraphs 133 to 135 of the House of Lords judgment).

---

* Editor's note: Also reported at [2008] 2 EGLR 45; [2008] 28 EG 114

However, notwithstanding these developments, the Court considers that at the time that the applicants' cases were considered by the domestic courts, there was an important distinction between the majority and minority approaches in the House of Lords, as demonstrated by the opinions in *Kay* itself. In *McCann*, the Court agreed with the minority approach [in *Kay* v *Lambeth*] although it noted that, in the great majority of cases, an order for possession could continue to be made in summary proceedings and that it would be only in very exceptional cases that an applicant would succeed in raising an arguable case which would require a court to examine the issue.

[43] Accordingly, in the next paragraph of its judgment, the ECtHR concluded:

> 74. In conclusion, the *Kay* applicants' challenge to the decision to strike out their Article 8 defences failed because it was not possible at that time to challenge the decision of a local authority to seek a possession order on the basis of the alleged disproportionality of that decision in light of personal circumstances. Accordingly, for the reasons given in *McCann*, the Court concludes that the decision by the county court to strike out the applicant's Article 8 defences meant that the procedural safeguards required by Article 8 for the assessment of the proportionality of the interference were not observed. As a result, the applicants were dispossessed of their homes without any possibility to have the proportionality of the measure determined by an independent tribunal. It follows that there has been a violation of Article 8 of the Convention in the instant case.

[44] The ECtHR was therefore saying that, in so far as the law had subsequently been developed in *Doherty*, this development could not be relied on by the UK in *Kay* v *United Kingdom.*

Conclusion on the first issue

[45] From these cases, it is clear that the following propositions are now well established in the jurisprudence of the ECtHR:

(a) Any person at risk of being dispossessed of his home at the suit of a local authority should in principle have the right to raise the question of the proportionality of the measure and to have it determined by an independent tribunal in the light of Article 8, even if his right of occupation under domestic law has come to an end: see *McCann*, in [50]; *Ćosić*, in [22]; *Zehentner*, in [59]; *Paulić*, in [43], and *Kay* v *United Kingdom*, in [73] to [74].

(b) A judicial procedure that is limited to addressing the proportionality of the measure through the medium of traditional judicial review (that is, one that does not permit the court to make its own assessment of the facts in an appropriate case) is inadequate because it is not appropriate for resolving sensitive factual issues: see *Connors*, in [92]; *McCann*, in [53]; *Kay* v *United Kingdom*, in [72] to [73].

(c) Where the measure includes proceedings involving more than one stage, it is the proceedings as a whole that must be considered in order to see if Article 8 has been complied with: see *Zehentner*, in [54].

(d) If the court concludes that it would be disproportionate to evict a person from his home notwithstanding the fact that he has no domestic right to remain there, it would be unlawful to evict him so long as the conclusion obtains, for example, for a specified period or until a specified event occurs or a particular condition is satisfied.

Although it cannot be described as a point of principle, it seems that the ECtHR has also franked the view that it will be only in exceptional cases that Article 8 proportionality would even arguably give a right to continued possession where the applicant has no right under domestic law to remain: see *McCann*, in [54]; *Kay* v *United Kingdom*, in [73].

[46] We have referred in a little detail to the ECtHR jurisprudence. This is because it is important for the court to emphasise what is now the unambiguous and consistent approach of the ECtHR, when we have to consider whether it is appropriate for this court to depart from the three decisions of the House of Lords.

[47] As we have already explained, the House of Lords decisions have to be seen against the backdrop of the evolving Strasbourg jurisprudence. So, for instance, the first of the House of Lords decisions, *Harrow*, came before any of the ECtHR judgments. *Kay* v *Lambeth* was decided after *Connors*. However, viewed without the benefit of subsequent ECtHR jurisprudence, the reasoning in *Connors* could have been interpreted as applying only to gypsies. Indeed, one point made on the applicant's behalf was that gypsies occupying sites owned by local authorities were not given any rights of security of tenure, unlike occupiers of flats or houses owned by local authorities, who were secure tenants. Although *McCann* had been decided by the time of *Doherty*, it would have been inappropriate for a five-judge court, at least in the particular circumstances, to depart substantially from the decision of the seven-judge court in *Kay*. Importantly, the judgments in *Ćosić*, *Zehentner*, *Paulić* and *Kay* v *United Kingdom* were all given after the last of the three House of Lords decisions.

[48] This court is not bound to follow every decision of the ECtHR. Not only would it be impractical to do so, it would sometimes be inappropriate, since it would destroy the ability of the court to engage in the constructive dialogue with the ECtHR that is of value to the development of Convention law: see, for example, *R* v *Horncastle* [2009] UKSC 14; [2010] 2 WLR 47. Of course, we should usually follow a clear and constant line of decisions by the ECtHR: see *R (on the application of Ullah)* v *Special Adjudicator* [2004] UKHL 26; [2004] 2 AC 323. However, we are not actually bound to do so or (in theory at least) to follow a decision of the Grand Chamber. As Lord Mance pointed out in *Doherty*, in [126], section 2 of the HRA requires our courts to "take into account" ECtHR decisions, not necessarily to follow them. Where, however, there is a clear and constant line of decisions whose effect is not inconsistent with some fundamental substantive or procedural aspect of our law, and whose reasoning does not appear to overlook or misunderstand some argument or point of principle, we consider that it would be wrong for this court not to follow that line.

[49] In the present case, there is no question of the jurisprudence of the ECtHR failing to take into account some principle or cutting across our domestic substantive or procedural law in some fundamental way. That is clear from the minority opinions in *Harrow* and *Kay* v *Lambeth*, and also from the fact that our domestic law was already moving in the direction of the European jurisprudence in *Doherty*. Even before the decision in *Kay* v *United Kingdom*, we would in any event have been of the opinion that this court should now accept and apply the minority view of the House of Lords in those cases. In the light of *Kay* v *United Kingdom*, that is clearly the right conclusion. Therefore, if our law is to be compatible with Article 8, where a court is asked to make an order for possession of a person's home at the suit of a local authority, the court must have the power to assess the proportionality of making the order and, in making that assessment, to resolve any relevant dispute of fact.

[50] We emphasise that this conclusion relates to possession proceedings brought by local authorities. As we pointed out in [4] above, nothing that we say is intended to bear on cases where the person seeking the order for possession is a private landowner. Conflicting views have been expressed both domestically and in Strasbourg on that situation. In *Harrow*, the views of Lord Bingham and Lord Steyn, in [23] and [26], can be contrasted with the view of Lord Hope, in [52]. In *Belchikova* v *Russia* 2408/06 unreported 25 March 2010, the application was held to be inadmissible, but the ECtHR (First Section) seems to have considered that Article 8 was relevant, even when the person seeking possession was a private sector landowner. Presumably, this was on the basis that the court making the order was itself a public authority. However, it is not clear whether the point was in contention. In the rather older admissibility decision of *Di Palma* v *United Kingdom* 11949/86 (1986) 10 EHRR 149, in paras 155 to 156, the Commission seems to have taken a different view, but the point was only very briefly discussed. No doubt, in such cases Article 1 of the First Protocol to the Convention will have a part to play, but it is preferable for this court to express no view on the issue until it arises and has to be determined.

Exceptionality

[51] It is necessary to address the proposition that only in "very highly exceptional cases" will it be appropriate for the court to consider a proportionality argument. Such a proposition undoubtedly derives support from the views expressed by Lord Bingham, and has been

referred to with apparent approval by the ECtHR in more than one case. Nevertheless, it seems to us to be both unsafe and unhelpful to invoke exceptionality as a guide. It is unhelpful because, as Lady Hale pointed out in argument, exceptionality is an outcome and not a guide. It is unsafe because, as Lord Walker observed in *Doherty*, in [122], there may be more cases than the ECtHR or Lord Bingham supposed where Article 8 could reasonably be invoked by a residential tenant.

[52] We would prefer to express the position slightly differently. The question is always whether the eviction is a proportionate means of achieving a legitimate aim. Where a person has no right in domestic law to remain in occupation of his home, the proportionality of making an order for possession at the suit of the local authority will be supported not merely by the fact that it would serve to vindicate the authority's ownership rights. It will also, at least normally, be supported by the fact that it would enable the authority to comply with their duties in respect of the distribution and management of their housing stock, including, for example, the fair allocation of their housing, the redevelopment of the site, the refurbishing of substandard accommodation, the need to move people who are in accommodation that now exceeds their needs and the need to move vulnerable people into sheltered or warden-assisted housing. Furthermore, in many cases (such as this appeal) other cogent reasons, such as the need to remove a source of nuisance to neighbours, may support the proportionality of dispossessing the occupiers.

[53] In this connection, it is right to refer to a point raised by the Secretary of State. He submitted that a local authority's aim in wanting possession should be a "given", which does not have to be explained or justified in court, so that the court will be concerned only with the occupiers' personal circumstances. In our view, there is indeed force in the point, which finds support in Lord Bingham's comment in *Kay* v *Lambeth*, at p491, in [29], that to require the local authority routinely, from the outset, to plead and prove that the possession order sought is justified would, in the overwhelming majority of cases, be burdensome and futile. In other words, the fact that the authority are entitled to possession and should, in the absence of cogent evidence to the contrary, be assumed to be acting in accordance with their duties, will be a strong factor in support of the proportionality of making an order for possession. However, in a particular case, the authority may have what they believe to be particularly strong or unusual reasons for wanting possession, for example that the property is the only occupied part of a site intended for immediate development for community housing. The authority could rely on that factor, but would have to plead it and adduce evidence to support it.

[54] Unencumbered property rights, even where they are enjoyed by a public body such as a local authority, are of real weight when it comes to proportionality. So, too, is the right, indeed the obligation, of a local authority to decide who should occupy their residential property. As Lord Bingham said in *Harrow*, at p997, in [25]:

> [T]he administration of public housing under various statutory schemes is entrusted to local housing authorities. It is not for the court to second-guess allocation decisions. The Strasbourg authorities have adopted a very pragmatic and realistic approach to the issue of justification.

Therefore, in virtually every case in which a residential occupier has no contractual or statutory protection and the local authority are entitled to possession as a matter of domestic law, there will be a very strong case for saying that making an order for possession would be proportionate. However, in some cases, there may be factors that would tell the other way.

Second issue: Application of this conclusion in general

[55] The conclusion that, before making an order for possession, the court must be able to decide not only that the order would be justified under domestic law but also that it would be proportionate under Article 8(2) to make the order presents no difficulties of principle or practice in respect of secure tenancies. As explained above, no order for possession can be made against a secure tenant unless, *inter alia*, it is reasonable to make the order. Any factor that has to be taken into account, or any dispute of fact that has to be resolved for the purpose of assessing proportionality under Article 8(2), would have to be taken into account or resolved for the purpose of assessing reasonableness under section 84 of the 1985 Act. Reasonableness under that section, as with proportionality under Article 8(2), requires the court to consider whether to order possession at all, and, if so, whether to make an outright order rather than a suspended order, and, if so, whether to direct that the outright order should not take effect for a significant time.

[56] Moreover, reasonableness involves the trial judge "tak[ing] into account all relevant circumstances... in... a broad, common-sense way": *Cumming* v *Danson* [1942] 2 All ER 653, at p655, *per* Lord Greene MR. It therefore seems highly unlikely, as a practical matter, that it could be reasonable for a court to make an order for possession in circumstances in which it would be disproportionate to do so under Article 8.

[57] The implications of Article 8 being potentially in play are much more significant where a local authority are seeking possession of a person's home in circumstances in which domestic law imposes no requirement of reasonableness and gives an unqualified right to an order for possession. In such a case, the court's obligation, under Article 8(2), to consider the proportionality of making the order sought does represent a potential new obstacle to the making of an order for possession. The wide implications of this obligation will have to be worked out. As in many situations, that is best left to the good sense and experience of judges sitting in the county court.

[58] The present appeal involves a type of case that arises relatively rarely, namely a claim for possession against a demoted tenant, and we heard relatively little in the submissions about other types of case. When it comes to possession proceedings, a demoted tenant is unusual in two respects: (a) he has already been the subject of proceedings that have resulted in the loss of statutory protection; and (b) he will have been given notice of the grounds on which possession is being sought, and an opportunity to challenge those grounds.

[59] The conjoined appeals in *Mullen* v *Salford City Council* [2010] EWCA Civ 336*, which are due to be heard by this court later this month, involve possession orders made in different and more common circumstances, namely the introductory tenancy regime (under Chapter 1 of Part V of the 1996 Act) and the homelessness regime (under Part VII of the 1996 Act). Those appeals may therefore provide a more appropriate vehicle for the giving of general guidance. Moreover, in respect of the homelessness regime, this court will be able to consider whether any guidance can usefully be given to local authorities as to what course to take before seeking possession in cases where there is no provision for the kind of procedure envisaged in sections 143E and 143F of the 1996 Act. In the light of our decision in the present appeal, the lawyers preparing for those appeals will have the opportunity to give particular attention to these aspects of the matter.

[60] Nevertheless, certain general points can be made, even at this stage.

[61] First, it is only where a person's "home" is under threat that Article 8 comes into play, and there may be cases where it is open to argument whether the premises involved are the defendant's home (for example, where very short-term accommodation has been provided). Second, as a general rule, Article 8 needs to be considered by the court only if it is raised in the proceedings by or on behalf of the residential occupier. Third, if an Article 8 point is raised, the court should initially consider it summarily, and if, as will no doubt often be the case, the court is satisfied that even if the facts relied on are made out the point would not succeed, it should be dismissed. Only if the court is satisfied that it could affect the order that the court might make should the point be further entertained.

[62] Fourth, if domestic law justifies an outright order for possession, the effect of Article 8 may, albeit in exceptional cases, justify (in ascending order of effect) granting an extended period for possession, suspending the order for possession on the happening of an event or even refusing an order altogether.

---

* Editor's note: Reported at [2010] HLR 35

[63] Fifth, the conclusion that the court must have the ability to assess the Article 8 proportionality of making a possession order in respect of a person's home may require certain statutory and procedural provisions to be revisited. For example, section 89 of the 1980 Act limits the period for which a possession order can be postponed to 14 days or, in cases of "exceptional hardship", 42 days. Moreover, some of the provisions of CPR 55, which appear to mandate a summary procedure in some types of possession claim, may present difficulties in respect of cases in which Article 8 claims are raised. Again, we say no more on the point, since these aspects were not canvassed on the present appeal to any significant extent, save in respect of the legislation on demoted tenancies that we are about to discuss under the third issue.

[64] Sixth, the suggestions put forward on behalf of the Equality and Human Rights Commission, that proportionality is more likely to be a relevant issue "in respect of occupants who are vulnerable as a result of mental illness, physical or learning disability, poor health or frailty", and that "the issue may also require the local authority to explain why they are not securing alternative accommodation in such cases" seem to us well made.

Third issue: Application of this conclusion to demoted tenancies
Introductory

[65] As explained above, where an order for possession is made against a demoted tenant such as Mr Pinnock, the court is involved at two different stages. The first stage, which arises if the landlord decides to apply for a demotion order, requires the court to decide whether to make such an order. The second stage, which arises if the landlord decides to make an application for an order for possession while the demotion order applies, requires the court to decide whether to make an order for possession. Each stage involves a significant and direct assault on the tenant's right to occupy his home, and therefore engages Article 8.

[66] So far as the first stage is concerned, before making a demotion order, the court must consider for itself the factual basis for making such an order. Moreover, the court can make such an order only once it is satisfied that: (a) the facts that it investigates and determines justify the order under section 82A(4)(a); and (b) it is reasonable to make the order under section 82A(4)(b). I therefore find it impossible to conceive of circumstances in which the requirements of Article 8 would not be satisfied by the plain words of the relevant statutory provisions.

[67] Greater problems arise, however, when one turns to the second stage, where, as in this case, the court is asked to make an order for possession against a demoted tenant.

Proper interpretation of section 143D(2) of the 1996 Act

[68] The first argument raised against the conclusion that the county court judge who is asked to make an order for possession under section 143D(2) can carry out his own Article 8 assessment of the proportionality of making such an order arises from the wording of the subsection. We have quoted it in [10] above. The provision requires the court to make an order for possession "unless it thinks that the procedure under sections 143E and 143F has not been followed". If one construes that section in accordance with the traditional approach to interpretation, it is hard to see how the court could have the power either to investigate for itself the facts relied on to justify the decision to seek possession or to refuse to make an order for possession if it considered that it would be disproportionate to do so.

[69] Therefore, there is obvious force in the point that, in the absence of any Article 8 Convention right, section 143D(2) would limit the court to satisfying itself that the procedural requirements of sections 143E and 143F had been complied with. Stanley Burnton LJ took that view in the Court of Appeal. At any rate, absent the HRA, the purpose of section 143D appears to be to deprive the courts of almost any ability to stand in the way of a landlord that had decided to seek possession against a demoted tenant.

[70] However, since the Convention requires the court to have the power to consider the proportionality under Article 8 of making a possession order at the instance of a local authority in respect of a person's home, the effect of section 3(1) of the HRA is that section 143D(2) should be read as not excluding that power, if at all possible. Accordingly, it is necessary to examine the issue rather more critically.

[71] Clearly, the local authority, when deciding to bring possession proceedings against a demoted tenant under section 143E, and any panel reviewing that decision under section 143F, have a duty in domestic law to act rationally and to investigate the relevant facts fairly, as well as a duty under Article 8 to consider proportionality, which includes investigating the relevant facts.

[72] Rightly, in our view, it is common ground that a court has jurisdiction, under normal judicial review principles, to satisfy itself that the local authority and panel have indeed acted reasonably and have investigated the relevant facts fairly when deciding to bring possession proceedings. From this, it must follow that any decision by the local authority to continue possession proceedings is similarly susceptible to judicial review. At the same time, it is right to emphasise that it would almost always require a marked change of circumstances following a panel's decision to approve the proceedings before an attempt could properly be made judicially to review the continuance of proceedings that were initially justified.

[73] In our judgment, once it is accepted that it is open to a demoted tenant to seek judicial review of a landlord's decision to bring and continue possession proceedings, it inevitably follows that, as a generality, it is open to a tenant to challenge that decision on the ground that it would be disproportionate and therefore contrary to Article 8. Further, as we saw in [31] to [43] above, the ECtHR jurisprudence requires the court considering such a challenge to have the power to make its own assessment of any relevant facts that are in dispute. We have already pointed out, in [28] above, that Lord Scott and Lord Mance in particular reached this conclusion in *Doherty*, in [68] and [138]. The ECtHR acknowledged this development in *Kay* v *United Kingdom*, in [73]. In these circumstances, we are satisfied that, wherever possible, the traditional review powers of the court should be expanded so as to permit it to carry out that exercise.

[74] In summary, where it is required in order to give effect to an occupier's Article 8 Convention rights, the court's powers of review can, in an appropriate case, extend to reconsidering for itself the facts found by a local authority, or indeed to considering facts that have arisen since the issue of proceedings, by hearing evidence and forming its own view.

[75] Much the more difficult question, however, is whether it is possible to read and give effect to section 143D(2) in a way that would permit the county court judge to carry out this exercise. As we have pointed out in [69] above, the purpose of the subsection appears to be to ensure that the court makes an order for possession in all cases except where it thinks that the procedure under sections 143E and 143F has not been followed. In other words, the purpose is to ensure that the court does nothing more than check whether the procedure has been followed. It could therefore be argued that holding that the court could assess the proportionality of the local authority's decision to bring and to continue the possession proceedings would go against the whole import of the section and would amount to amending rather than interpreting it.

[76] We have come to the conclusion that we should reject that argument.

[77] In our view, if the procedure laid down in sections 143E or 143F has not been lawfully complied with, either because the express requirements of that section have not been observed or because the rules of natural justice have been infringed, the tenant should be able to raise that as a defence to a possession claim under section 143D(2). After all, the tenant's argument in such circumstances would be within the scope of the ambit of section 143D(2), namely that "the procedure under sections 143E and 143F has not been [lawfully] followed", since lawfulness must be an inherent requirement of the procedure. It must equally be open to the court to consider whether the procedure has been lawfully followed, having regard to the defendant's Article 8 Convention rights and section 6 of the HRA.

[78] This approach is borne out by section 7(1) of the HRA, which, so far as relevant, provides:

> A person who claims that a public authority has acted (or proposes to act) in a way which is made unlawful by section 6(1) may —
>
> …
>
> (b) rely on the Convention right or rights concerned in any legal proceedings…

By virtue of this provision, an occupier who is the defendant in possession proceedings in the county court and who claims that it would be incompatible with his Article 8 Convention rights for him to be put out of his home must be able to rely on those rights in defending those proceedings. This approach fits with the observation of the ECtHR in *Paulić*, in [43], that the court need consider proportionality only if it is raised by the person whose Article 8 rights are said to be infringed.

[79] We therefore consider that section 143D(2) should be read as allowing the court to exercise the powers that are necessary to consider and, where appropriate, to give effect to, any Article 8 defence that the defendant raises in the possession proceedings.

[80] This approach to the interpretation of section 143D(2) also goes a long way towards disposing of Mr Arden's argument that even if Article 8 required this kind of review, the county court does not have jurisdiction to carry it out. So, he suggested, the issue would have to be referred to the High Court, where it would presumably be assigned to the Administrative Court. In effect, section 7(1)(b) confers the necessary jurisdiction on county court judges when it is necessary for them to deal with a defence that relies on an alleged breach of the defendant's Article 8 Convention rights.

[81] The same conclusion can be justified on the rather wider basis that where a tenant contends that the decision of a local authority landlord to issue or indeed to continue possession proceedings can in some way be impugned, the tenant should be entitled to raise that contention in the possession proceedings themselves, even if they are in the county court. This seems to us to follow from the decision of the House of Lords in *Wandsworth*, as cited and approved in the present context in *Kay* v *Lambeth*, in [110], and again in *Doherty*, in [56], [123] and [157]: see [28] above. This approach also derives strong support from the observations of Lord Bingham in *Kay* v *Lambeth*, in [30].

[82] This second reason involves disapproving part of the reasoning of the Court of Appeal in *Manchester City Council* v *Cochrane* [1999] 1 WLR 809, by which, understandably, the Court of Appeal in this case appears to have regarded itself as bound. In *Manchester City Council*, the Court of Appeal held that an introductory tenant could not raise a defence to a claim for possession when that defence was based on the contentions that: (a) there had been no breaches of the tenancy agreement (the substantive ground relied on by the council for bringing the instant proceedings); (b) the relevant regulations had not been complied with; and (c) there had been a failure to comply with the rules of natural justice in the conduct of the review by the panel.

[83] As a result of our conclusion on the first issue on this appeal, Article 8 would require the court to be able to consider the facts, as well as proportionality, for itself. However, even in the absence of Article 8, a court would have had power to consider whether a reasonable local authority and panel could have reached the conclusion that such breaches existed. Similarly, a court would have had power to consider whether the relevant regulations had been followed, and whether the rules of natural justice had been followed. The question is whether that court could be the court hearing the possession claim, given that it is (virtually always) the county court.

[84] In *Manchester City Council*, three reasons were given for concluding that the defences sought to be raised could not be pursued in the county court. The first was that section 127(2) of the 1996 Act, which is in similar terms to section 143D(2), required the court to make an order for possession: see pp818G-820B. That is, in substance, the view that we have rejected in [76] to [79] above.

[85] The second reason for the Court of Appeal's conclusion in *Manchester City Council* was based on the contrast between section 127(2) and section 204 of the 1996 Act, in Part VII of the 1996 Act that is concerned with homelessness: see pp820B-C and 821H-822A.

[86] However, like Lord Fraser of Tullybelton in *Wandsworth*, at p510A-B, we would adopt the principle stated by Viscount Simonds in *Pyx Granite Co Ltd* v *Ministry of Housing and Local Government* [1960] AC 260, at p286, that a citizen's "recourse to Her Majesty's courts for the determination of his rights is not to be excluded except by clear words".

[87] The third reason for the Court of Appeal's decision in *Manchester City Council* is discussed at pp820C-821E, and relies on the presumption that possession claims against demoted tenants could be procedurally derailed if tenants could raise public law points in the course of the possession proceedings. We do not consider that this presumption is correct. Indeed, the ability of a tenant to delay possession proceedings by raising a public law point would be greater if such points had to be taken in separate proceedings in the High Court.

[88] For these reasons, we are satisfied that we should apply the approach of the House of Lords in *Wandsworth*. This permits us to confirm our earlier conclusion that section 143D(2) should be read as allowing the county court to exercise the powers that are necessary to consider and, where appropriate, to give effect to any Article 8 defence that the defendant raises in possession proceedings brought in that court.

Section 17(1)(a) of the Crime and Disorder Act 1998

[89] A further difficulty that is said to stand in the way of the conclusion that the county court judge can carry out a proportionality exercise is based on section 17(1) of the Crime and Disorder Act 1998, which provides: "Without prejudice to any other obligation imposed on it, it shall be the duty of each authority to which this section applies to exercise its various functions with due regard to the likely effect of the exercise of those functions on, and the need to do all that it reasonably can to prevent, (a) crime and disorder in its area…"

[90] This section, Mr Arden rightly submitted on behalf of the council, applied when, *inter alia*, a local authority were exercising their function as the landlords of their housing stock. He then went on to submit that the council's duty under section 17 could conflict with their duty, by virtue of the occupier's Article 8 Convention rights, to consider whether it would be proportionate to bring or continue possession proceedings against him. For instance, bringing such proceedings might be a reasonable way of preventing crime and disorder in the authority's area, even though bringing those proceedings would be disproportionate when viewed in the particular context of the individual concerned.

[91] In our view, this argument is devoid of substance. In the first place, section 17 begins with the qualifying words: "Without prejudice to any other obligation imposed on it…". Therefore, if the effect of the HRA is to impose an obligation on a local authority landlord to consider proportionality under Article 8 before embarking on possession proceedings against a demoted tenant, section 17 is not inconsistent with and does not undermine that obligation. As far as the county court is concerned, in so far as it is to be treated as reviewing the local authority's decision to bring proceedings, the same point applies, and, in so far as it is to be treated as carrying out its own assessment, nothing in section 17 impinges on it.

[92] Second, section 17 requires a local authority to exercise their functions, paying "due regard" to the need to prevent crime and disorder. The section imposes no absolute obligation on an authority to do everything to reduce crime and disorder, irrespective of other persons' rights or of their own other duties, and it would be very surprising if it did. Accordingly, the furthest this point goes is to suggest that a local authority, when deciding to bring possession proceedings against a demoted tenant, should take into account their duty under section 17, as well as the Article 8 Convention rights of the tenant and any other Convention rights that may be in play. That would also be a factor to be taken into account by the panel when reviewing the local authority's decision. Similarly, it would be a factor for the county court judge to take into account when considering whether the local authority had acted proportionately.

A *Section 6(2) of the HRA*

[93] We have concluded that section 143D(2) of the 1996 Act can be read and given effect compatibly with the Article 8 Convention rights of any occupiers of local authority housing. So, no question of the application of section 6(2) of the HRA arises in that respect.

[94] Nor, indeed, did Mr Arden argue that section 6(2) would make it lawful for the local authority to disregard the occupier's Article 8 Convention rights when deciding whether to bring possession proceedings against him. On the contrary, it was rightly common ground that a local authority must take into account a demoted tenant's Article 8 rights when taking possession proceedings under the 1996 Act.
B The same applied to a panel reviewing that decision.

[95] However, as is plain from the speeches of Lord Walker and Lord Mance in *Doherty*, in [110] to [113] and [141] to [153], respectively, two passages in [86] and [114] in the speech of Lord Hope in *Kay* v *Lambeth*, could be interpreted as indicating that section 6(2) did apply to the local authority's decision as to whether to bring possession proceedings in the circumstances of those cases so as to make it lawful for the authority to disregard the occupier's Article 8 Convention rights.

[96] The absence of any real debate on the point makes the present case an unsuitable vehicle for any wide-ranging discussion of
C section 6(2). Nevertheless, we think it right to confirm that, in our view, the subsection has no application to the decision of a local authority as to whether to bring or continue possession proceedings against demoted tenants.

[97] Section 6 of the HRA provides:

> (1) It is unlawful for a public authority to act in a way which is incompatible with a Convention right.
> (2) Subsection (1) does not apply to an act if —
> (a) as the result of one or more provisions of primary legislation, the authority could not have acted differently; or
> (b) in the case of one or more provisions of, or made under, primary
> D legislation which cannot be read or given effect in a way which is compatible with the Convention rights, the authority was acting so as to give effect to or enforce those provisions.

[98] Plainly, subsection (2)(a) applies only where legislation imposes a duty to act. That provision is not relevant for present purposes since the local authority are under no statutory duty that would compel them to take possession proceedings against a demoted tenant.

[99] The statutory provision that relates to the bringing of possession proceedings against demoted tenants is section 143D(1) of the 1996 Act. It provides:

> E (1) The landlord may only bring a demoted tenancy to an end by obtaining —
> (a) an order of the court for the possession of the dwelling-house, and
> (b) the execution of the order.

In addition, as already explained, section 143E provides for the local authority to give notice of their decision to apply for an order for possession of the tenant's house. Section 143F provides for the panel to review that decision at the request of the tenant.

[100] It does not particularly matter on which of these provisions we choose to concentrate. However, it can be assumed, for the purposes of the argument, that when a local authority landlord decide to bring
F possession proceedings against a demoted tenant, the authority are "acting so as to give effect to" section 143D(1) within the meaning of section 6(2) of the HRA.

[101] Then, the only question is whether section 143D(1) can be given effect in a way that is compatible with the demoted tenant's Article 8 Convention rights. If so, section 6(2) does not apply. For the reasons that we have already set out in detail, the answer is that section 143D(1) can unquestionably be given effect in a way that is compatible with the demoted tenant's Article 8 Convention rights. Most obviously, the local authority will give effect to section 143D(1) in a way that is compatible with those rights when they bring proceedings that are proportionate because the demoted tenant has, for instance, continued to act in a manner that causes a nuisance to his neighbours. That being so, section 6(2) of the HRA has no application to the G
decision of a local authority to bring or continue possession proceedings against a demoted tenant.

[102] In these circumstances, *mutatis mutandis*, the conclusion of Lord Mance in *Doherty*, at p447, in [153], applies in the present situation:

> Accordingly, a local authority which fails to take into account Convention values when deciding whether or not to give any and if so what length of notice to quit cannot, in my opinion, be said to be "acting so as to give effect to or enforce" statutory provisions which are incompatible with the Convention rights.

[103] As the council accept, the result therefore is that section 6(2) H
does not make it lawful for a local authority to fail to consider whether it would be proportionate to bring or continue such proceedings. The same must apply to a local authority's decision to take possession proceedings against other occupiers who are not secure tenants.

Conclusion on the third issue

[104] We are accordingly of the view that a county court judge who is invited to make an order for possession against a demoted tenant pursuant to section 143D(2) can consider whether it is proportionate to make the order sought, and can investigate and determine any issues of fact relevant for the purpose of that exercise. It follows that the demoted J
tenancy regime in the 1996 Act is compatible with Article 8.

[105] Having said that, there are two further points that we should make.

[106] First, as already observed, there appears to be no express fetter on the nature of the grounds that a local authority can invoke for seeking possession against a demoted tenant. It would seem that, as in this case, local authorities seeking possession against demoted tenants in practice normally rely on repetitions of the type of incidents that gave rise to the demotion order. It may well be that the nature of the grounds on which possession can be sought against demoted tenants is limited in that way, as a matter of law. However, that would involve implying some sort of K
limitation into the statute, since there is no express provision that would prevent a local authority relying on, say, the fact that they have a more deserving potential occupier of the premises in question. We say no more on the matter since the point does not arise in this appeal, and it was not the subject of any argument.

[107] Second, we have expressed reservations concerning the view that, in respect of possession claims generally, Article 8 will assist an occupier only in "highly exceptional" circumstances. However, there are two features of possession claims under section 143D that enable us to express such a view in respect of these claims. First, the court will already have decided that it was reasonable (and therefore L
proportionate under Article 8) to have made the demotion order, largely removing the tenant's security of tenure. The court will have done this less than two years (and, no doubt, in some cases less than one year) before it is being asked to make an order for possession. The two sets of proceedings must be viewed as a whole for the purposes of Article 8: see *Zehentner*, in [54] (quoted in [39] above). This highlights the fact that although Article 8 is still engaged at the second, possession order stage, it would be difficult for the tenant successfully to invoke it, given that its requirements had been satisfied at the first, demotion order stage. Second, as with introductory tenancies, the tenant will have been given the local authority's reasons for deciding to seek possession. So M
he will have had the opportunity to challenge the decision and to have that challenge considered by the panel.

Fourth issue: Application of these conclusions to the facts of this case

[108] For the reasons already explained, neither Judge Holman nor the Court of Appeal thought that they had jurisdiction either to consider whether the making of an order for possession in this case was "necessary in a democratic society" under Article 8 (that is, whether it was proportionate to evict Mr Pinnock and Ms Walker from the property) or to resolve any disputes of fact between the parties in respect of that issue. Since we have concluded that the judge had jurisdiction to deal with both matters, there are two alternative courses that we could

now take: we could address the proportionality issue ourselves or we could remit the issue to Manchester County Court.

[109] If we can take the former course, we should do so: it is more than three years since the demotion order was made in respect of Mr Pinnock's tenancy and more than two years since these possession proceedings were started against him. Before they reached this court, they had already taken up three days of court time (plus the six hearing days before the recorder who made the demotion order). However, we can determine the issue of Article 8 proportionality only if we can do so without needing to hear further evidence.

[110] In order to consider which course to take, we must set out the relevant circumstances in a little more detail.

[111] After stating that, as the tenant, he was "responsible for the behaviour of every person (including children) living in or visiting [his] home", Mr Pinnock's tenancy agreement contained covenants to the effect that neither he nor anyone residing with him would "cause a nuisance, annoyance or disturbance to any other person" or would "harass any other person". Examples were given in the agreement of possible breaches of these covenants. They included "offensive drunkenness" and "doing anything that interferes with the peace, comfort or convenience of others".

[112] The events that led the recorder to conclude that a demotion order was justified were many and serious. In very summary terms, anti-social behaviour injunctions (under section 152 of the 1996 Act) had been granted against Ms Walker and one of Mr Pinnock's sons, Clive, in 2003. Ms Walker had gone on to breach the injunction. Anti-social behaviour orders had been granted against another son, Devon, in 2002, and against his twin sons, Orreon and Orraine, in 2004. Each of them had been breached. Further, each of the five children had appeared before the criminal courts, where they had been convicted of a variety of offences, including a racial Public Order Act offence, driving while disqualified and blackmail. The last of these involved the obtaining of some £1,000 by repeated, almost daily, threats of violence against a 16-year-old youth. In a schedule to his judgment, the recorder listed no fewer than 32 crimes or serious nuisances that were committed by Ms Walker and Mr Pinnock's five children between 1992 and 2006.

[113] There are some differences among the allegations relied on by the council in their notice served on 6 June 2008 under section 143E, the allegations relied on by the panel that carried out the review pursuant to section 143F, the allegations relied on by Judge Holman and the allegations relied on by the Court of Appeal. In our view, however, the Court of Appeal's analysis of the relevant allegations was clearly correct.

[114] We would make three comments in respect of that analysis.

[115] First, there is nothing in the statutory provisions relating to the demoted tenancy regime that limits the particular grounds on which a local authority can rely when deciding to issue possession proceedings against a demoted tenant. Subject to the possible type of limitation discussed in [106] above, we see no reason to restrict those grounds in a particular case, save by reference to rationality in domestic law and proportionality in the light of the Convention. This view is not based only on a reluctance to imply words or conditions into statutory provisions. It is also based on the point that, by demoting a tenancy, a court has decided that the tenant has forfeited any statutory protection for at least a year, and it seems wrong to imply a degree of protection back into the statute unless it is necessary to do so, for example because the Convention requires it. We are therefore satisfied that a local authority are not limited to relying on matters that amount to breaches of the tenancy in question in order to justify a decision to issue and continue a claim for possession against a demoted tenant.

[116] Second, the panel should be able to take into account all the available information when it assesses the justification for, and proportionality of, the local authority issuing a claim for possession against a demoted tenant. It seems obvious that before the panel the tenant could raise events that happened after the notice, and it is hard to see why the same should not apply to the landlords. In any event, if the tenant raises his Article 8 Convention rights as a defence to possession proceedings, the court must consider all relevant issues. These must include a matter that arose after the date of the notice. We are therefore satisfied that it is open to the panel and to the court hearing the possession claim to take into account grounds that are not contained in the notice.

[117] Third, we can see no reason why the fact that a notice contains a bad reason should destroy the landlords' right to seek possession unless, for instance, the bad reason somehow infects the good faith of the landlords.

[118] On that basis, the following matters are relied on as supporting the council's decision to bring and maintain the possession proceedings against Mr Pinnock.

[119] First, on 22 September 2007, Mr Clive Pinnock resisted arrest at the property and ran off. In due course, he was convicted of resisting or obstructing a constable in the execution of his duty. Although this conduct was obviously an annoyance for the police officers involved, there was no evidence that it caused nuisance or annoyance to neighbours. Therefore, it may not have constituted a breach of the tenancy agreement. Nevertheless, it was plainly relevant to the housing management functions of the council. Further, as Stanley Burnton LJ pointed out, this behaviour was *capable* of causing nuisance or annoyance to any person.

[120] Second, Mr Devon Pinnock pleaded guilty to causing death by dangerous driving and driving a vehicle while disqualified and uninsured on 18 January 2008. A young woman died and two others were seriously injured in the incident that occurred 1.55 miles from the property.

[121] It is relevant to mention that Ms Walker blamed the police for the incident and did not accept that Devon was in any way responsible. As Stanley Burnton LJ said, this "bore on whether she and Mr Pinnock were able and willing to exercise parental control over their children who lived at, or visited, the property so as to bring their anti-social behaviour to an end".

[122] Third, in February 2008, Mr Orreon Pinnock committed a burglary of premises a few minutes' walk from the property, an offence that also involved an assault on a woman. He was convicted of this offence after the service of the notice, but the judge and the Court of Appeal rightly held that this was a relevant factor when deciding whether to issue and prosecute possession proceedings against Mr Pinnock.

[123] Judge Holman thought that the second incident (but only in so far as it involved Ms Walker blaming the police and excusing Devon) and the third incident could be relied on by the council as a ground for justifying their claim for possession against Mr Pinnock. He therefore concluded, in [70], that "there was material before the review panel… entitling it to uphold the decision to terminate". To much the same effect, Stanley Burnton LJ said that he could "not see any basis for a finding that the review panel's decision was one that no reasonable person could consider appropriate, and if the judge had had jurisdiction to review that decision I would have upheld his decision to uphold it": see [67].

[124] Mr Pinnock's case is that it would be disproportionate to evict him (now a pensioner) and Ms Walker (still in employment) from their home of more than 30 years, given that none of their five children lives with them and that there have been no further incidents since February 2008. In this connection, Mr Richard Drabble QC made a number of points that he said the panel, Judge Holman and the Court of Appeal had failed to take into account. None of the matters relied on, he said, constituted a breach of the tenancy agreement. Ms Walker had committed no nuisance, offence or harassment since 2003 and there was no suggestion that she or Mr Pinnock is likely to commit any nuisance or crime in the future. As for the children save for Orreon, they did not reside in the property at the date of the offences relied on. Further, any crime or nuisance that they might commit in the area in the future could not be attributable to the tenancy continuing because they do not live in the property. For the same reason, Mr Pinnock could not be treated as responsible for their behaviour. Moreover, there was no evidence that the children were or would be particularly drawn to the area by their parents living at the property. In any event, other remedies such as

anti-social behaviour orders and injunctions and orders excluding the children from the area under section 153C of the 1996 Act, would be more effective deterrents.

[125] We see the force of these points. However, unless there is some dispute of fact that needs to be resolved, we are not persuaded that this is a case in which the occupiers of the property have any real prospect of successfully relying on Article 8 proportionality or, indeed, on the contention that the decision of the council to issue and maintain possession proceedings against them was unreasonable.

[126] The history of crime, nuisance and harassment on the part of those living at the property in the period leading up to the demotion order made in June 2007 was extraordinary in its extent and persistence. Were it not for Mr Pinnock being innocent of any such conduct on his own account, we doubt whether the recorder would have thought it right to refuse the council's original claim for possession. As it was, he made it clear that the demotion order represented what was very much a last chance for Mr Pinnock (and for Ms Walker).

[127] Despite this being their last chance, as we have explained, there were incidents at or near the property. Clive resisted arrest at the property and ran away from the police (of which he was convicted). Devon caused death by dangerous driving in the vicinity of the property and then ran away (of which he was convicted). Ms Walker refused to accept that Devon was in any way responsible for this and instead blamed the police. Orreon committed a burglary near the property (and was later imprisoned). In short, there were three serious incidents in a year: one in the property, two in its immediate vicinity. Mr Pinnock's children were responsible for all of them. Moreover, there is every sign that Ms Walker, at least, has learnt nothing. All this happened under the shadow of a demotion order.

[128] The argument that none of the children lives in the property any longer is of scant assistance to Mr Pinnock since his case is that none of them has lived there since the demotion order was made. Even if that is true, it is clear that the children visit the property and, unfortunately, when they do they appear to commit crimes and make a nuisance of themselves in the vicinity. Furthermore, there is no guarantee that at least some of the children will not stay at the property on a temporary, intermittent or permanent basis. For the council to evict Mr Pinnock on such grounds may well seem to him harsh. However, in the light of the history, the demotion order, the interests of their neighbours and the council's right and duty to manage and allocate their housing stock, the decision cannot be characterised as unreasonable or disproportionate.

[129] If some of the children did in fact live in the property, Mr Pinnock has been dishonest, and the council's case is even stronger. In this connection, it should be said that there is good reason to think that Devon did live in the property. Mr Pinnock's evidence to the panel was that Devon had moved out around five years earlier, but that evidence had been given to and rejected by the recorder, not least because Devon had given the property as his address to the criminal courts in July 2005, June 2006, and March 2007. When he appeared in court on the charge of causing death by dangerous driving in January 2008, he again gave his address as the property. The only new evidence before the panel disputing his residence at the property was a statement by the mother of his girlfriend. However, she too said that he had ceased to live in the property four or five years previously. It is thus hard to see how any tribunal could conclude that he did not reside there, but, as the judge said, the panel ducked the issue of residence in their written decision.

[130] The fact that some (or even all) of the grounds justifying the rationality and proportionality of the council's decision to seek possession may not have involved any breach of the tenancy agreement does not give rise to a problem. There is no requirement in the 1996 Act that they should, and, as already mentioned, there is no warrant for implying any such requirement into the statute. The fact that Mr Pinnock may not be responsible for the incidents is not of great significance: the order for possession was not sought or made to punish him. The fact that there may be other remedies to deal with the children is also of little force: rather than seeking Asbos or Asbis to keep them out of the vicinity, it is scarcely irrational or disproportionate to decide to remove their parents, whom they undoubtedly visit, even if (which is an unresolved issue) they do not live with them.

Conclusion

[131] In these circumstances, it is unnecessary to remit this case for the question of proportionality to be determined. The only issues of fact that are in dispute are whether Devon lived in the property at the time he caused death by dangerous driving and whether Clive's resisting his arrest actually caused any nuisance locally. For the reasons given in [128] above, it is unnecessary to decide whether Devon was living in the property at the relevant time. Equally, it is unnecessary to establish whether Clive's action actually resulted in a nuisance: as Stanley Burnton LJ said, it is sufficient that he resisted arrest and that this could have caused a nuisance.

[132] We shall accordingly dismiss the appeal and uphold the order for possession made against Mr Pinnock, albeit for reasons that are rather different from those of Judge Holman and the Court of Appeal. Mr Pinnock is and was entitled to an opportunity of having the proportionality of the measure determined by a court and, if necessary for that purpose, of having any relevant issue of fact resolved. That right was not acknowledged by the courts below (for wholly understandable reasons). We have, however, afforded him the opportunity to have the proportionality of the possession order considered. Having considered the issue, we are satisfied that it was proportionate to make the order, irrespective of the truth relating to the two possible issues of fact between Mr Pinnock and the council.

Appeal dismissed.

LEGITIMATE EXPECTATION

R (on the application of Grimsby Institute of Further and Higher Education) *v* Chief Executive of Skills Funding (formerly the Learning and Skills Council)

ADMINISTRATIVE COURT

[2010] EWHC 2134 (Admin)

12 August 2010

HH JUDGE PETER LANGAN QC, sitting as a deputy High Court judge

Legitimate expectation — Statutory body — Building grant — Two-stage process for obtaining grants involving approval in principle followed by application for approval in detail — Further education institute applying for grant towards construction of new building — Funding body giving approval in principle — Approval in detail later refused owing to lack of funds — Whether legitimate expectation that approval in detail would be granted or that certain procedures would be followed

The defendant was the statutory successor to the Learning and Skills Council (LSC), a body that was responsible for securing higher education provision in England. Part of LSC's functions included the making of grants to colleges in order to fund capital expenditure. The system for dealing with grant applications involved two stages: an approval in principle (AiP) followed by the submission of a further application for approval in detail (AiD). The claimant further education institute had applied to the LSC for a grant in connection with the construction of a new building and had received AiP in 2008. It subsequently became apparent that the LSC had sufficient funds to make a grant to only a small number of the schemes for which it had granted AiP. The claimant's scheme was not among that number and AiD was refused in 2009.

The claimant brought a claim for judicial review against the defendant, contending that it was in breach of a twofold legitimate expectation that: (i) once AiP had been granted, the application would be dealt with in accordance with the LSC's usual and known procedures; and (ii) that body would be funded and organised in a manner that enabled it to meet commitments given at the AiP stage. The claimant argued that the expectation in question had arisen from implied representations made by the LSC in its *Capital Handbook*, which set out the procedure for applying for grants and the criteria to be met, and in meetings and correspondence between representatives of the claimant and the LSC. The claimant sought to recover £3,728,452 that it had spent on working up its application to the point at which it applied for AiD.

***Held*: The application was dismissed. (1) Save in an exceptional case, where a legitimate expectation is founded on a representation, that representation must be unambiguous. If a past practice is relied on, there must have been a specific undertaking to an individual group whereby the continuance of that practice would be assured. Cases of legitimate expectation not founded on a representation will be highly exceptional and must satisfy the test of conspicuous unfairness. Even cases based on an abuse of power are properly characterised as being founded on a general representation that a public authority will not act so unfairly that its conduct amounts to an abuse of power. The requirements for legitimate expectation in public law are separate from private law doctrines such as estoppel, and detrimental reliance, although relevant, is not essential; however, the court's approach to the interpretation of statements will be the same. Maladministration alone is not a ground for judicial review; the question will always be whether the public body has acted unlawfully. In deciding what if any relief should be granted, the court should take into account: (i) whether the decision being challenged is in the macro-political field; (ii) whether it involves social or political value judgments as to priority of expenditure; and (iii) the nature and clarity of the promise or prior practice in question. Where the legitimate expectation is not merely procedural in nature but is an expectation that a claimant will receive a substantive benefit, the determination of the appropriate remedy should not concentrate on the understandable disappointment of the claimant to the exclusion of wider considerations. (2) The things said and done on behalf of the LSC had contained no implicit representation that the claimant's application for AiD would be dealt with on its merits without regard to the financial state of the LSC. There had been no unambiguous statement, either in the *Capital Handbook* or in meetings and correspondence, whether individually or taken together, that were AiP to be granted, the consideration of a detailed application for AiD would be confined to the inherent merits of the application and the availability of funds would thereafter be irrelevant. Moreover, throughout the process of applying for AiD, the claimant's officers had not relied on any such assumption but had appreciated that they were proceeding at the risk of failure. (3) The claimant could not establish any practice-based legitimate expectation. There had been no specific undertaking to colleges that any practice of confining the consideration of an application for AiD to the merits of the application would continue. (4) Further, although the outcome of the AiD application was a considerable disappointment for the claimant and left it with substantial unreimbursed expenditure, the facts fell far short of what was required to make out a rare case of conspicuous unfairness on the part of a public authority. That result, although unfortunate, could not be characterised as the product of an abuse of power by the LSC. (5) Although the LSC's decision had not left the claimant without any public law rights, and it was entitled to due consideration of its request for reimbursement, along with such requests made by other colleges, that right had been satisfied in all the circumstances.**

The following cases are referred to in this report.

Associated Provincial Picture Houses Ltd v *Wednesbury Corporation* [1948] 1 KB 223; [1947] 2 All ER 680; (1947) 45 LGR 635, CA

R (on the application of Association of British Civilian Internees (Far East Region)) v *Secretary of State for Defence* [2003] EWCA Civ 473; [2003] QB 1397; [2003] 3 WLR 80

R (on the application of Bancoult) v *Secretary of State for Foreign and Commonwealth Affairs (No 2)* [2008] UKHL 61; [2009] 1 AC 453; [2008] 3 WLR 955; [2008] 4 All ER 1055

R (on the application of Bibi) v *Newham London Borough Council*; R (on the application of Al-Nashed) v *Newham London Borough Council* [2001] EWCA Civ 607; [2002] 1 WLR 237; (2001) 33 HLR 84, CA

R (on the application of Niazi) v *Secretary of State for the Home Department*; *R (on the application of Bhatt Murphy (a firm))* v *Independent Assessor* [2008] EWCA Civ 755

R (on the application of Reprotech (Pebsham) Ltd) v *East Sussex County Council* [2002] UKHL 8; [2003] 1 WLR 348; [2002] 4 All ER 58; [2002] 2 PLR 60; [2003] 1 P&CR 5; [2002] JPL 82

R v *Inland Revenue Commissioners, ex parte Unilever plc*, *R* v *Inland Revenue Commissioners, ex parte Matteson's Walls Ltd* [1996] STC 681

R v *Inland Revenue Commissioners, ex parte MFK Underwriting Agents Ltd* [1990] 1 WLR 1545; [1990] 1 All ER 91, QBD

R v *North and East Devon Health Authority, ex parte Coughlan* [2001] QB 213; [2000] 2 WLR 622; [2000] 3 All ER 850, CA

R v *Secretary of State for Education and Employment, ex parte Begbie* [2000] 1 WLR 1115, CA

R v *Secretary of State for the Home Department, ex parte Brind* [1991] 1 AC 696; [1991] 2 WLR 588; [1991] 1 All ER 720, HL

Rowland v *Environment Agency* [2003] EWCA Civ 1885; [2005] Ch 1; [2004] 3 WLR 249; [2004] 2 Lloyd's Rep 55

This was the hearing a claim by the claimant, Grimsby Institute of Further and Higher Education, against the defendant, the Chief Executive of Skills Funding, formerly the Learning and Skills Council, for judicial review, on the ground of breach of legitimate expectation, of a decision to withhold final approval for a capital grant.

Peter Knox QC and Ximena Montes-Manzano (instructed by the legal department of Grimsby Institute of Further and Higher Education) appeared for the claimant; John Randall QC and David Warner (instructed by the legal department of the Young People's Learning Agency) represented the defendant.

Giving judgment, **HH Judge Peter Langan QC** said:

Introduction

[1] This case has to do with the concept of legitimate expectation in public law.

[2] At the material time, the Learning and Skills Council (the LSC) was responsible for securing the provision of further education (other than higher education) in England. The defendant is the statutory successor to the LSC, but it will be convenient to use the acronym LSC to refer indifferently to the defendant and the LSC itself. As part of the carrying out of its functions, the LSC made grants to colleges in order to fund capital expenditure.

[3] Grimsby Institute of Further and Higher Education (the institute) applied to the LSC for a grant towards the erection of a new building. The system operated by the LSC for dealing with applications in respect of schemes that had, as the institute's project had, a capital value of more than £10m involved two distinct stages. These were known as approval in principle (AiP) and approval in detail (AiD). The institute obtained AiP for its application. It failed thereafter to obtain AiD, not because of any lack of merit in the scheme presented, but because the LSC had run out of funds. It transpired that the LSC had given AiP to so many applications, without regard to the fact that it had insufficient money to fund them all, that only a small number of the schemes could be given AiD. The institute was not one of that number.

[4] The institute had spent a substantial amount of money in working up its application to the point at which it applied for AiD. Some of this expenditure was reimbursed by the LSC: the greater part, a sum put by the institute at £3,728,452, was not.

[5] It is not part of the institute's case that it should have been granted AiD for its scheme. The case advanced is that the institute had a twofold legitimate expectation. First, that once AiP had been granted, the application would be dealt with in accordance with the usual and known procedures of the LSC. Second, that the LSC would be funded and organised in a manner that enabled it to meet commitments given at the AiP stage. It is now clear that AiD will never be given to the institute's scheme. The institute seeks compensation for the non-fulfilment of the expectation in the form of reimbursement of the expenditure that it had incurred.

[6] The answer of the LSC to this claim is, put very shortly, that the institute falls a long way short of what is required to make good a public law case founded on legitimate expectation.

[7] This has been, as judicial review claims go, a heavy case. There is a massive amount of documentary material before the court. Witnesses were cross-examined on their statements. The submissions of counsel, both written and oral, were lengthy. The hearing lasted some six days. It should be possible in this judgment, particularly in the light of the way in which the case was presented on behalf of the institute in closing submissions, to deal more shortly with the facts than at one time I had anticipated. This will, however, be possible only if I begin by rehearsing at some length, first, the statutory provisions relating to the LSC and its functions and, second, the structure of the LSC and the procedures that it followed. The facts will then appear in what will, I hope, be a comprehensible statutory and administrative context.

[8] I am grateful to all the counsel engaged in the case for their considerable assistance.

LSC: Statutory provisions

[9] The LSC was established by section 1 of the Learning and Skills Act 2000 (the Act) as a body of between 12 and 16 members, with functions expressly limited to England (a separate council for Wales was established by section 30 of the Act).

[10] The main duties of the LSC were set out in section 2(1) of the Act. The LSC:

> must secure the provision of proper facilities for —
>
> (a) education (other than higher education) suitable to the requirements of persons who are above compulsory school age but have not attained the age of 19,
>
> (b) training suitable to the requirements of such persons,
>
> (c) organised leisure-time occupation connected with such education, and
>
> (d) organised leisure-time occupation connected with such training.

[11] Section 3(1) of the Act provided that the LSC must secure the provision of reasonable facilities (which is a concept that differs from that of proper facilities (see sections 2(2) and 3(2)) for the same descriptions of education, training and organised leisure-time for persons who have attained the age of 19.

[12] By section 5(1)(a) of the Act, the LSC was empowered to secure the provision of financial resources to "persons providing or proposing to provide post-16 education or training". Section 6 of the Act empowered the LSC to impose conditions on the provision of financial resources.

[13] Section 19(1) of the Act required the LSC to "establish a committee (to be called a local learning and skills council) for each area of England specified by the Secretary of State". Each local council were to perform in respect of their area such of the duties of the LSC as the LSC specified, and might exercise in relation to their area such of the powers of the LSC as the LSC specified: see section 20(1).

[14] Finally, section 27 dealt with grants to the LSC, and, so far as is material to this case, made these provisions:

> (1) The Secretary of State may make grants to the Council of such amounts and subject to such conditions as he thinks fit…
>
> (2) The conditions may set the Council's budget for any of its financial years.

LSC: Structure and procedures

Structure of the LSC

[15] At national level, the LSC was headed by the council itself. Among the committees of the LSC was a national capital committee, the function of which was to consider large-scale capital projects and make recommendations on them to the council.

[16] Prior to 2008, the LSC had nine regional boards and 47 local councils to which regional and local functions were delegated. The local councils were dissolved on 21 February 2008. The regional boards were replaced by regional councils on 1 September 2008. It was the role of a regional council, assisted by a regional capital advisory group, to identify, develop and recommend major capital projects. The institute lay within the LSC's Yorkshire and Humber region, which was itself divided into four areas. The institute was within the Humber area.

A *Financing of the LSC*

[17] At the period with which I am concerned, government expenditure was managed by way of triennial comprehensive spending reviews. Each year in or around the month of December, the LSC received a grant letter from the Secretary of State. This informed the LSC of the amount that it would receive from government for the year starting on the following 6 April. At the start of a comprehensive spending review period, the letter would provide a firm figure for the grant for the first year of the triennium and indicative figures for the second and third years. One of the LSC's witnesses, Mr Philip Head, said that, in accordance with this practice, in December 2004 the LSC
B had been given a firm figure for 2005-06 and indicative figures for 2006-07 and 2007-08.

Grant applications: Capital Handbook

[18] The procedure for applying for grants was set out in the LSC's *Capital Handbook*. I have the second edition, which was published in November 2006. I do not have the edition that was extant at the beginning of the story with which I am concerned. That does not matter. It is common ground that the institute was aware, from October 2006, of the imminent publication of the second edition, and also that the basic division of the application process into having to obtain AiP before
C seeking AiD was carried over from the first edition.

[19] As will appear later in this judgment, the contents of the *Capital Handbook* form an important plank in the institute's case and it is accordingly necessary to look at three parts of the *Handbook* in some detail. First, there is some general, introductory material, through which I was taken when the case was opened. Second, and relied on in the closing submissions for the institute, there are passages in the body of the *Handbook* that deal with applications for AiP and AiD and with project fee support. Third, and similarly relied on in closing submissions, there is Annex A to the *Handbook*, which is entitled "Capital Projects Criteria".

D *Handbook: Introductory material*

[20] The *Handbook* begins, in para 1.1, by stating its purpose, which is to update:

> the arrangements for the administration, assessment and determination of applications to the Learning and Skills Council (LSC) for consent and capital support. It consolidates all general capital guidance provided by the LSC for further education (FE) sector providers.

[21] The *Handbook* then proceeds, in a "Capital overview", to refer to the FE White Paper *Raising Skills, Improving Life Chances,* which "sets out the Government's expectation that the LSC's capital resources will be directed to implement the FE component of the 14-19 vision in
E each locality".

> 2.1... In this context, the LSC is seeking to both challenge and support the FE sector to develop world-class buildings for world-class teaching and learning…
>
> …
>
> 2.3 Based on the additional capital funding for 2008-10 announced in the 2005 budget, if this level of funding continues and most of it continues to be directed to the FE sector, the LSC estimates that the renewal and modernisation of the FE estate could be substantially completed by 2013-2014.

[22] There are then sections dealing with national and regional
F capital strategies (para 3) and with provider property strategies (para 4). As regards the latter:

> 4.1 The LSC expects all FE colleges to produce a college property strategy on a three-year basis. In particular, all colleges submitting a capital project application will need to have an agreed property strategy in place in order for the application to be approved…

Handbook: AiP, AiD and fee support

[23] Applications for AiP and AiD are dealt with in para 7 of the *Handbook*.

> 7.1 The LSC will continue to operate two levels of capital project approvals for FE and 16-19 capital funding: approval in principle (AiP) and detailed approval. Prior to preparing and submitting an application, all providers should consult with the LSC local partnership team, LSC regional provider financial G
> management team and LSC RPA [regional property adviser] to discuss the key education, finance and property aspects of the project.
>
> 7.2 All colleges and other qualifying providers with proposals that have a total capital cost of more than £10 million will normally be required to make an AiP application. Once the AiP application has been determined by the LSC, a detailed application will be worked up and submitted. All proposals with estimated costs of more than £10 million will normally be considered by the LSC at these two stages. All applications that have a total capital cost of less than £10 million will normally be considered once by the LSC as a detailed application.
>
> 7.3 The LSC will only accept applications when the college or other
> qualifying provider has reached a certain stage in the project design and H
> preparation process. In the case of AiP applications, the proposal must be developed approximately to the Royal Institute of British Architects (RIBA) design stage C. In summary, the proposal should include the following:
>
> • A project brief
>
> • Floor plans and proposed elevations
>
> • An estimate of construction and other relevant costs
>
> • A clear indication of town planning issues including the identification of any onerous Section 106 agreements.
>
> 7.4 In the case of detailed applications, the proposal must be developed in the design stage to beyond RIBA stage D. This means that the proposal should include the following:
>
> • A fully completed project brief J
>
> • A detailed proposal showing spatial arrangements, material and appearance, as well as a cost estimate
>
> • Evidence of full planning permission
>
> • A breakdown of tendered and other relevant costs.
>
> 7.5 All capital applications should use the appropriate forms, contained within Annex E…

[24] Paragraph 8 of the *Handbook* deals with feasibility and project fee support. Feasibility fee support is available towards the costs of preparing feasibility studies for projects with estimated capital costs of more than £5m. Costs are supported to a maximum of £200,000 on a
matched funding basis: this means that the LSC will pay up to £100,000 K
so long as a similar sum is paid by the applicant. The paragraph goes on to deal with project fee support.

> **AiP-project fee support**
>
> 8.3 In addition to the feasibility fee support outlined above, the LSC has now introduced project fee support arrangements where colleges can apply to the LSC for grant support towards the fee costs associated with developing and preparing capital project applications, both to the AiP stage and then to the detailed stage, and on a basis suitable for LSC determination. The purpose of these arrangements is to encourage colleges to bring forward more accurate and better founded applications by accessing grant funding for capital projects from
> the LSC at an earlier stage in the capital process, and to help reduce the financial L
> burden on colleges of preparing capital projects.
>
> 8.4 In order to develop a project proposal to an AiP stage, colleges are likely to expend circa 15 per cent of the consultants' total projected fees (usually up to 10 per cent of total project costs). Colleges can apply for project fee support on a matched funding basis towards maximum costs of £500,000 to take an application up to the AiP stage. This is not additional funding as such but is really an advance of support that would have been payable later with the aim of helping to ease cash flow and incentivising better applications. …
>
> **Detailed-project fee support**
>
> 8.5 In order to develop a detailed project application, colleges are likely to have expended circa 45 per cent of the consultants' total projected fees.
> Colleges, that have approved AiP applications, can apply for project fee support M
> to enable them to develop the project to a stage where it can be considered as a detailed application. The amount of project fee support is determined by the rate of the indicative level of grant support the LSC will have agreed for the AiP application. The maximum allowable fee cost is up to £1 million. For example, this means that if a college spends £600,000 on fees between the AiP and detailed stage and the AiP was agreed at an indicative rate of 40 per cent, the college would be entitled to claim a grant of up to £240,000. As with the AiP project fee support, this is not additional funding but an advance of support to ease college cash flow…

Handbook: Capital projects criteria

[25] Annex A to the *Handbook* is divided into four sections: Stage 1: Validation Criteria; Stage 2: Application to the Learning and Skills

Council; Stage 3: Selection of Proposals by the Learning and Skills Council; and Stage 4: Moderation Arrangements.

[26] Stage 1: Validation Criteria: Before applying to the LSC for support, a provider is expected to apply specified validation criteria to its own proposals (paras 1 to 3). It is also expected to be in a position to provide some of the funding of its project:

> 4 Given the scarcity of capital funds, a provider will be expected to contribute as much as it reasonably can to the costs of its projects.
>
> 5 The LSC's maximum contribution towards the costs of a project will usually be no more than the sum the LSC considers necessary for the provider to be able to undertake the project. A provider will need to demonstrate the need for a particular level of grant support in confirming the affordability of the project.
>
> 6 An appropriate figure should be provisionally confirmed with the LSC's regional PFM [provider finance manager] director before proposals are finalised. Under the stage 3 criteria (see below), preference may be given to providers applying for less than the maximum available LSC support should there be competition for funds. …

Providers are "required to demonstrate that they have diligently examined whether better value for money might be obtained by private sector investment" or through a public/private partnership: see para 9.

[27] Stage 2: Application to the LSC: This section of Annex A lists the factors that applicants are advised to consider in developing a capital project. They are: the delivery plan; quality issues; property indicators; the economic appraisal; borrowing criteria; financial health criteria; and race relations issues.

[28] Stage 3: Selection of Proposals by the LSC: This section begins with para 23:

> The LSC will consider all proposals that pass the first two stages of the process against the seven selection criteria described at paragraphs 24-31.

The selection criteria are: the contributions that the project would make to the provision of proper and reasonable facilities for further education (FE) and work-based training in the area; quality assessments of the existing provision; economy (on a cost-per-square-metre comparison with similar constructions in the same area); value for money; efficient space utilisation; projects required to meet legal requirements, such as health and safety legislation (which will be given preference); and financing (preference is to be given to public-private financed projects and loan-financed projects, in that order).

[29] Stage 4: Moderation Arrangements: This section starts with para 32:

> Local LSCs will determine the priority of competing capital applications according to the extent that they will contribute to meeting the LSC's local and national targets and objectives. In the event of applications for capital grant exceeding the funds available, the LSC will put in place national moderation arrangements to determine the relative priority to be given to competing applications.

Narrative

[30] I now turn to the history of this case.

Background

[31] The institute was founded in 1944. Like other colleges in the FE sector, it was from its inception part of the educational provision made by the relevant local authority. In April 1993, pursuant to the Higher and Further Education Act 1992, such colleges became independent corporate bodies, operating on a not-for-profit basis. Council land and buildings occupied by the colleges were taken over by the new corporations from the local authorities. In 2007-08, there were 366 FE colleges in England: there are somewhat fewer now because of mergers.

[32] In 1993, much of the infrastructure of the colleges was in poor condition. In the 2005 Budget, the government announced "a step change in capital investment in the FE sector". Existing plans would enable government support for investment in the FE estate to rise to £250m in 2007-08, while "Budget 2005 announces that an additional £350m of capital investment will be made over 2008-09 to 2009-10 to support the longer-term transformation of the further education sector".

[33] By 2005, the institute had around 20,000 students, the majority of whom were studying on part-time courses. The main campus was at Nuns Corner, in Grimsby. As was the case with so many FE colleges, the institute's buildings left a good deal to be desired, both in their general condition and as regards what is considered appropriate in a modern learning environment. In discussions with local officers of the LSC, it was acknowledged by the LSC that the institute had not over the years done particularly well in the allocation of capital grants from the LSC. At the same time, it was made clear to the institute that, in order to obtain large-scale funding, a college must come up with a detailed masterplan, that is, an estates strategy for a period of five years or so, which would demonstrate that any physical development would cohere with the college's educational plans and vision.

From masterplan to AiP

[34] On 6 July 2005, the institute applied to the LSC for feasibility fee support "to employ consultants to undertake a 10-year masterplan/vision for estates rationalisation". Fee costs of the consultants were to come to £47,829 and support was requested at 50% of this figure. On 28 October 2005, the LSC approved the application. The consultant engaged by the institute for the purpose of preparing the masterplan was Dyers.

[35] By May 2006, North East Lincolnshire Council and Franklin Sixth Form College were on the scene, the former as the education authority that were responsible for Hereford School in Grimsby. The idea that was being pursued was the comprehensive redevelopment of all three establishments on a single site so as to form a "learning village" or "learning corridor". There were various meetings at which representatives of the institute, the local authority, the Sixth Form College and the LSC were present and exchanged ideas about the scheme.

[36] For the most part, it is not necessary for me to examine these meetings. There were, however, three meetings in 2006 on which particular reliance was placed in submissions made on behalf of the institute. These took place on 18 August, 15 September and 13 October 2006. Since I intend to deal only with uncontroversial matters in this part of the judgment, I shall at this stage do no more than refer to the place of each meeting and the persons present.

[37] The meeting of 18 August 2006 took place at the institute. There were 14 persons present. Among them were: Professor Daniel Khan, who was then the principal of the institute; Mr Adrian Clarke, the associate principal (strategic services and resourcing); Mr Michael Green FRICS, the regional property adviser for the Yorkshire and Humber region of the LSC; and representatives of Franklin College, the local authority and the institute's professional advisers. Someone who was not at this meeting was Mr Ray Ellis, the deputy principal, who had the lead role in the project team at the institute.

[38] On 22 August 2006, the institute placed an advertisement in the *Official Journal of the European Union* regarding the obtaining of professional services for the project.

[39] The next important meeting took place on 15 September 2006 at the institute. There were 12 people (together with a 13th person recorded as being "in attendance") at the meeting. They included, from the institute, Professor Khan and Mr Ellis; from the LSC, Mr Bob Flockton, who held the post of partnership director; representatives of Franklin College and the local authority and professional advisers.

[40] Around this time, Dyers produced its "Property Strategy and Ten Year Masterplan". As might be anticipated, this is a lengthy document. It presented the institute with four options, with option 4 being recommended as the preferred option. Option 1 was "Do nothing"; option 2 was "Do minimum", that is, simply the outstanding planned maintenance. Option 3, which was called "Combined new build HE and FE facility", in fact comprised four sub-options, 3A, 3B and so on. What these had in common was that each included some elements of refurbishment and some elements of new building on the Nuns Corner site. Option 4 was "Complete new build with contribution to HE and Learning Corridor objectives" and was just that: 46,680m^2 of

A purpose-built accommodation, including the provision of a nursery and a linked university centre facility.

[41] By this stage, North East Lincolnshire Council and Hereford School appear to have left the stage, and they do not appear again in the story.

[42] Following the production of the masterplan, there were further meetings involving representatives of the institute and Franklin College and of the LSC. Ms Jane Lyon, the area director for the Humber area of the LSC, wrote to Mr Philip Jenkinson, the chair of the governors of the institute. The letter is undated, but it must have been written after the masterplan had been seen by Ms Lyon and prior to the meeting of
B 26 September 2006, to which I will come shortly, because the letter was referred to at that meeting.

[43] Ms Lyon set out the main elements of Dyers' option 4, noting that it embodied "a significant change to the original proposals but one which still supports the concept of a learning corridor". She continued:

> After a further meeting with Peter [Newcome, the principal of Franklin College] to share his views, the LSC felt it was important at this early stage to inform you of our support, in principle, and its further exploration.
>
> As with any project of this magnitude, there will be many aspects to resolve. However, the potential benefits for learners in North East Lincolnshire are seen as considerable.
>
> C From the LSC funding perspective, there is an opportunity for FE colleges to obtain substantial support that is unlikely to be available again for some considerable time. Inevitably, both colleges will need to consider the financial implications but the LSC are hopeful that the Governing Bodies will recognise the benefits.
>
> We do not want this to be a distraction from moving the overall scheme forward but we felt we need to determine, at an early stage, whether this radical plan is worthy of further investigation. If it is, then we need to move speedily into the next stage and prepare a feasibility study as soon as possible.
>
> We appreciate this will require a degree of trust and mutual understanding requiring the best endeavours of both Corporations to make it happen. If you need any further information from the LSC then do let us know, we
> D are more than happy to give any presentations to the governing body if that is appropriate.

[44] On 26 September 2006, there was a meeting of the corporation, or governing body, of the institute. Mr Ellis is recorded in the draft minutes (I have not seen an approved version) as saying that "the LSC have embedded themselves in the process". The minutes say that "[t]here was general support for the preferred option in terms of a joint campus with Franklin College". Then:

> The Corporation **APPROVED** the 10 year estates master plan and the submission of the detailed business case for the preferred option.
>
> E Lawrence Hart [of AA Projects, one of the professional advisers] advised that the next stage was to have discussions with the national LSC, so that the Institute could understand from an early state whether they supported the numbers. He advised that the meeting should take place within 4-6 weeks.

[45] The third of the meetings as to which there is controversy was of a very different kind from the other two. It took place on 13 October 2006 at the Cedar Court Hotel, in Bradford, and was organised by the Yorkshire and Humber region of the LSC for an invited audience of college principals and chairs of governors or their representatives. The agenda was headed "Building for the Future: Yorkshire and the Humber Regional Capital Strategy Event". The speakers included
F Mr Alastair Grindlay, the LSC's capital programme manager. Mr Green was among the many LSC officers who were present. The institute was represented by Mr Ellis and Mr Clarke.

[46] It is now possible to speed up the narrative somewhat.

[47] After the meeting of 13 October 2006, the institute started the process of selecting architects, main contractors and the various professionals (quantity surveyors and environmental consultants) who would be required for the early stages of the project.

[48] On 3 November 2006, the institute lodged with the LSC an application for fee support up to AiP stage. On the application form, the estimated project cost was shown as £113,134,340, the total fee cost up to AiP as £2m and the LSC contribution requested as £500,000.

[49] Around the end of 2006 and the beginning of 2007, the institute G
appointed Tribal Property Ltd as project manager for the development.

[50] At some stage, Franklin College decided to pursue its own application separately from that of the institute.

[51] On 9 February 2007, the LSC informed the institute that fee support of £250,000 had been approved.

[52] Formal appointments of professionals were then made.

[53] On 17 July 2007, the institute submitted its application for AiP. The original application was revised in October 2007 (the precise date is uncertain), and in the revised form the project cost was shown as £118.699m before inflation and £133,656,525 after inflation. The contribution requested from the LSC was £96,003,221 or 81% of the H
pre-inflation cost. This percentage is referred to as the "intervention rate". The balance was to be provided by a contribution from the institute (£16.666m) and other public sector grants (£6m).

[54] There was then a process of discussion between the LSC's regional finance officers and representatives of the institute, and the application wound its way through the various stages required in order to obtain AiP. The application was considered by the LSC's Humber capital committee on 12 November 2007, by the regional capital committee on 27 November 2007 and by the national capital committee on 13 December 2007. At each stage, the project was recommended for approval to the next tier in the hierarchy, but with an intervention rate J
of 80%.

[55] The national council of the LSC considered the application at a meeting of 23 January 2008. There were nine proposals in all before the meeting, but because the papers for the meeting had not been circulated in sufficient time for consideration by the members, no firm decisions were reached. Instead, the chair and chief executive were given delegated authority to determine the applications, subject to comments that they might receive from members.

[56] Thereafter, the decision made was to grant AiP to the institute's application, but at a slightly lower intervention rate of 79%. The news was communicated in a letter of 31 January 2008 from Mr Philip Head, K
director of infrastructure and property services, to Professor Khan. I should set out two paragraphs in full:

> 4 I confirm that in order to assist in marketing to, and negotiating with potential developers and partners, without prejudice to the determination of a detailed application in due course, that the Council has agreed that the project proposal would meet the Council's projects criteria. The Council has also agreed that budgetary provision equivalent to 79% of the estimated total project cost (£93,749,000) should also be made for the project proposal for a period of up to twelve months pending receipt of a detailed application.
>
> …
>
> 5 I must inform you, however, that the Council's in principle support L
> for these project proposals is subject to the condition that the project should be implemented in phases, broadly equal in terms in floor space and costs for which individual detailed applications should be made. To clarify this point, the assumption is that the college should have successfully completed phase 1 before implementing phase 2 and any further phases. The grant support intervention rate will be calculated for each phase of the proposal as and when they come forward to the Council for determination.

From AiP to detailed application

[57] Most of the year 2008 was taken up with moving the application from AiP to a state in which it could be submitted for AiD. By May 2008, Davis Langdon LLP had replaced Tribal Property as project manager M
(although the individual responsible, Mr Robert Hardy, moved from Tribal Property to Davis Langdon, thus providing continuity); BAM Construction Ltd was appointed as main contractor and work by the institute and its advisers towards preparation of the detailed application was well under way.

[58] In June 2008, a serious problem raised its head. It will be recalled that AiP had been given on the footing that the project would proceed in two distinct and roughly equal phases. It became apparent to the institute and those advising it that this two-phase implementation would have serious disadvantages. The principal disadvantages were: (i) an escalation in cost to a figure well in excess of that for which AiP had been given (the main contractor put the cost of a two-phase

construction at £160m); (ii) disruption during the second phase (because the first phase would in effect amount to the construction of one-half of a single building); (iii) the risk that AiD might not be obtained for the second phase; and (iv) the risk that contractors would have the institute "over a barrel" on pricing for the second phase because of the need for continuity.

[59] This matter was considered at a meeting at the LSC's offices in Coventry on 29 July 2008. The meeting was, contrary to the case originally advanced by the institute, convened at the request of the institute, not that of the LSC. The persons present included Mr Ellis and Mr Clarke from the institute, and Mr Green and Mr Grindlay of the LSC. All of these apart from Mr Grindlay, who is seriously ill, gave oral evidence. The account of the meeting that I find to be reliable is that which was given by Mr Green.

[60] The upshot of the meeting was that the institute was to revise its plans so as to submit a single-phase project for AiD. Mr Green said, during cross-examination:

> They had a good case, but they would have to come up with figures. We would support the proposal for a single AiD, if there was a real prospect of getting AiD. Mr Grindlay said that, if a sufficient case was put up, he would support it. He did say that we, as officers, would have a difficult time at the National Capital Committee. I told Mr Hardy [this must have been after the meeting, as Mr Hardy was not there] that the departure from AiP was a concern.

[61] It is, I think, a fair reading of the evidence as a whole that proceeding with a single-phase application for AiD was something that had the support, but the cautious and qualified (particularly as to cost) support, of the responsible officers at the LSC. On the institute's side, there was a recognition that AiD could not be taken for granted and that the situation after the change to a single-phase bid was less certain than it had been previously. Thus, in the risk register that was compiled on a monthly basis by Davis Langdon, one finds, as at 30 June 2008, that the risk that the LSC does not continue with a positive approach to the project is put at 50%, but by 30 October 2008, that risk factor has increased to 0.8 (the equivalent of 80%).

[62] In the meantime, quite significant enabling works were being undertaken on-site. These included the conversion of the institute's business solutions centre to workshops, the construction of a new brick workshop and the provision of a temporary car park. The approval of these works, together with the incurring of professional fees, was said by Mr Ellis, in an e-mail of 10 September 2008, to "have tested the risk appetite of our Governing Body".

[63] The detailed application on the single-phase basis was submitted to the LSC on 27 October 2008. The project cost was £154.068m before inflation and £161,772,260 after inflation. The contribution sought from the LSC was £135.080m, which represented an intervention rate of 85%.

[64] From the point of view of the LSC, the figures in this application were, as Mr Green said in evidence, "seriously too high". They were high in a double sense: they exceeded both the level of grant that was anticipated by the LSC and they were much greater than the amounts on the basis of which the AiP had been given. This view was made clear to the institute, which set about revising the detailed application, which was then resubmitted on 27 or 28 November 2008. The figures had come down: project cost, £140,028,661 before inflation and £146,754,863 after inflation; contribution sought, £115.936m; intervention rate, 79%.

[65] The institute had been working towards having the detailed application considered at a meeting of the national council of the LSC on 17 December 2008. The application would first have to pass through a meeting of the regional capital committee earlier in the month. As was made clear by Mr Green in an e-mail of 3 December 2008 to Professor Khan, the late submission of the revised application had made this timescale one that could not be met. It was then hoped to have the papers ready for the meeting of the regional capital committee fixed for January 2009.

[66] The papers were not, I think, ready in time: but it would have made no difference to the outcome if they had been. The balloon was about to go up.

Collapse of funding

[67] The national council of the LSC met on 17 December 2008 as planned. The papers prepared for the meeting and the minutes are documents of considerable length. Put very shortly, the LSC found that it was running out of money. It did not have sufficient funds to grant AiP or AiD to applications that, on their merits, might have been successful. There had, as matters were later described in a draft report of the committee on public accounts, been "a catastrophic mismanagement of the LSC capital budget during 2008 and neglect of oversight by those in senior positions in the LSC".

[68] The following decisions were taken at the meeting of 17 December 2008. Decisions on all applications listed for AiP or AiD at the meeting of 17 December were deferred to the meeting of 4 March 2009. There would be a general deferral of the determination of all capital applications pending a review of the capital programme. The outcome of this review would be considered at the meeting of 4 March 2009. The capital committee meetings for January and February 2009 were cancelled, and there was to be no consideration of projects by regional capital committees. At the meeting of 4 March 2009, the council would confirm the timescales and the terms on which pending applications might be considered. Communications about all this "needed to be urgently and sensitively handled".

[69] On 9 January 2009, there was a meeting in Bradford between representatives of the LSC and of the institute. It is sufficient to say that the funding difficulties of the LSC and the concerns of the institute over any delay in the funding of the project were the main subjects for discussion.

[70] Work continued, probably in hope rather than in expectation, on a further revision to the detailed application with a view to reducing costs. The application in its final form was submitted on around 25 February 2009. The cost of the project before inflation had come down to £138,948,385; after inflation, the cost was £139.950m; the contribution sought from the LSC was £106m; the intervention rate was in the region of 76%.

[71] At the meeting of the national council on 4 March 2009, it became apparent that the situation faced by the LSC was even more serious than had been anticipated. Seventy-nine projects had received AiP, with £2.7bn required from the LSC: the extent by which the grants of AiP exceeded funds that could be made available at the AiD stage is apparent from the fact that, as will be seen, a few months later, AiD could be given to only 13 of these schemes. The meeting was told that Sir Andrew Foster had been asked by the Secretary of State to carry out a review, and in the meantime the LSC was working on a consultation exercise to identify a strategy for prioritising all future projects.

[72] On 16 March 2009, Ms Lyon wrote to Professor Khan to inform him of the current situation, and told him that, pending the completion of the consultation exercise, "as previously advised, your college should continue to avoid incurring any further costs in developing its project".

[73] On 1 April 2009, Sir Andrew Foster's *A Review of the Capital Programme in Further Education* was presented to the House of Commons by the Secretary of State. This contained a damning indictment of financial management at the LSC. The implementation of the laudable policy to transform the FE estate "did not include a robust financial strategy or a regional or national approach to prioritisation". Overcommitment to projects should have been detected sooner and, on the basis of "plenty of experience in the public sector", might have been anticipated so that "the surge [in take-up] could have been mitigated and managed". There was a "champagne moment culture", in which AiP was celebrated "as the point of no return". It seemed to be normal practice for projects not to be rejected at the detailed stage, provided that costs stayed in line with those that had received AiP.

[74] On 21 April 2009, Professor Khan wrote to the chief executive of the LSC. He set out the "severe consequences" for the project of the

A "current suspension of capital projects and the lack of clear timescales going forward". He said that the institute "need[ed] to recoup from the LSC the development costs of the project". He enclosed a breakdown, which showed costs of £4,768,452, incurred under eight headings: architects; project managers; quantity surveyors; building contractors; mechanical and electrical consultants; structural and civil consultants; enabling works; and "other (including planning)". From the total, Professor Khan deducted grants of £790,000 and £250,000 received from the LSC, leaving a balance (for which he enclosed an invoice) of £3,728,452. I should say that I have been unable to trace documents relating to the larger of the two grants, but the receipt of that sum by the B institute is not a matter of controversy.

[75] On 24 April 2009, the chief executive wrote to Professor Khan. The letter does not appear to be a direct reply to Professor Khan's letter of 21 April, but rather a standard letter designed to be sent to all FE college heads. As regards projects awaiting approval, PricewaterhouseCoopers had been appointed to assist in the preparation of new criteria for prioritisation. As regards costs, once the LSC had "a full analysis of the costs and their nature" it would "be able to identify those cases where a contribution to such costs from the LSC may be justified".

[76] On 27 May 2009, the institute sent the pre-action protocol letter to the LSC.

C [77] On 19 June 2009, the LSC replied. There are three points of particular significance in the letter. First, the institute's application for AiD would be considered in accordance with the LSC's new prioritisation process. Second, the institute had been responsible for delay in submitting the detailed application, by reason of the change in the scheme and the costs being in excess of those for which AiP had been given. It was "not until February 2009 that the cost of the scheme had reached a reasonable level". Third, the failure to reimburse costs over and above the payments already made was not unreasonable.

[78] The LSC, with the assistance of PricewaterhouseCoopers, then proceeded to the assessment of applications so as to select those D that should be supported with additional capital funding that had been provided by HM Treasury. The details of the assessment exercise have not been examined in the course of the hearing. The competitors were those applications that were "shovel-ready", that is to say, where work could begin within three months. They were then awarded points under various criteria, and the 13 colleges with the highest overall scores were allocated funds. The institute missed, but only by a narrow margin, coming into the top 13. The institute was informed of the fact that it had been unsuccessful by a letter of 26 June 2009.

[79] As a result, the project for the redevelopment of the Nuns Corner site is, for all practical purposes, dead.

E [80] There remained the question of the institute's wasted expenditure. The LSC sent out a questionnaire on costs incurred by applicants for grants whose cases would not be going any further. Two hundred-and-nine colleges returned the questionnaire duly completed. In round figures, the total amount spent by colleges on costs relating to capital projects that did not obtain AiP or AiD was £340m; £94m was due to the colleges under the established arrangements for fee support; the gap between the two figures was £246m. The latter sum was simply something that the LSC could not afford.

[81] The decision eventually taken by the LSC was to reimburse those colleges that would face possible insolvency if they were not F reimbursed. There were 41 such colleges, and between them these received payments came to £44m.

[82] I was told that the institute is the only college that has challenged the decision not to reimburse it in full.

Law

Preliminary

[83] Mr Randall put before the court nine propositions of law, which provided him with headings for what I found an impressive and comprehensive review of the subject of legitimate expectation in public law. In this section of the judgment, it will be convenient to set out those propositions, together with Mr Knox's response and, where this seems to be required, some observations of my own.

First proposition G

[84] "Save in an exceptional case, a legitimate expectation founded on a representation requires that representation to be clear and unambiguous." The authorities relied on are *R* v *Inland Revenue Commissioners, ex parte MFK Underwriting Agents Ltd* [1990] 1 WLR 1545, at p1569G-H and *R (on the application of Association of British Civilian Internees (Far East Region))* v *Secretary of State for Defence* [2003] EWCA Civ 473; [2003] QB 1397, in [72] and [73] (hereafter *ABCIFER*)

[85] Mr Knox accepted this proposition, subject to the qualification that a statement must always be read in context. One has to consider, as was said in the judgment of the Court of Appeal in *ABCIFER*, how "on H a fair reading" a statement "would reasonably have been understood by those to whom it was directed": see [56].

[86] There is no real difference between counsel on this point.

Second proposition

[87] "A legitimate expectation founded on a past practice requires there to have been a specific undertaking to an individual or group whereby its continuance is assured." The authority relied on is *R (on the application of Bhatt Murphy (a firm))* v *Independent Assessor* [2008] EWCA Civ 755: see, in particular, [40] to [46].

[88] Mr Knox did not question the proposition but, rather, turned J it to the advantage of the institute by relying on the following passage from the judgment of Laws LJ in *Bhatt Murphy*.

> 43. Authority shows that where a substantive expectation is to run the promise or practice which is its genesis is not merely a reflection of the ordinary fact (as I have put it) that a policy with no terminal date or terminating event will continue in effect until rational grounds for its cessation arise. Rather, it must constitute a specific undertaking, directed at a particular individual or group, by which the relevant policy's continuance is assured…
>
> 44. I will give two concrete examples from the cases. In *Ex p Khan* [1985] 1 All ER 40 the Home Office promulgated specific criteria for the admission of children into this country for the purposes of adoption here. The appellant K sought entry for his prospective adoptive child. He relied in terms on the published criteria which he fulfilled. But he found his application blocked by a further, unannounced criterion which he did not satisfy. This court allowed his appeal.

"This," Mr Knox said, "is my case."

Third proposition

[89] "Although an abuse of power through acting with conspicuous unfairness can arise without any legitimate expectation being infringed, such cases are highly exceptional." Although this proposition, which was accompanied by a lengthy review of the authorities (*R* v *Inland Revenue* L *Commissioners, ex parte Unilever plc* [1996] STC 681, *ABCIFER* and *Rowland* v *Environment Agency* [2003] EWCA Civ 1885; [2005] Ch 1) was accepted by Mr Knox as being "what the authorities say", I would myself put it slightly differently. It seems to me that abuse of power cases are properly characterised as legitimate expectation cases. The expectation is, however, not based on a specific representation but is of a more general kind, namely that a public authority will not act so unfairly that its conduct amounts to an abuse of power: see *ABCIFER,* in [72], and *Rowland,* in [68]. However, perhaps more importantly, it is common ground that cases of legitimate expectation not founded on a representation are highly exceptional. M

Fourth proposition

[90] "The requirements for legitimate expectation in public law are now sufficiently developed to stand separately from private law doctrines such as estoppel, and should do so, being more sensitive, and tailored to, the particular context of public law." The authorities relied on are *R (on the application of Reprotech (Pebsham) Ltd)* v *East Sussex County Council* [2002] UKHL 8; [2003] 1 WLR 348, in [6] and [33] to [35], and *Bhatt Murphy*, which was previously cited, in [40] and [43].

[91] Mr Knox accepts this proposition, but reminds the court that the moral values of estoppel have been absorbed into public law. Further, although the remedies in private and public law cases are quite

different, the approach of the court to the interpretation of statements is the same.

Fifth proposition

[92] "Legitimate expectation not being the same as estoppel, detrimental reliance is not essential to making it out, though it remains highly relevant." There was extensive citation of authority in support of this proposition (*R* v *Secretary of State for Education and Employment, ex parte Begbie* [2000] 1 WLR 1115, at p1131G, *R (on the application of Bibi)* v *Newham London Borough Council* [2001] EWCA Civ 607; [2002] 1 WLR 237, in [29] to [31], and *R (on the application of Bancoult)* v *Secretary of State for Foreign and Commonwealth Affairs (No 2)* [2008] UKHL 61; [2009] 1 AC 453 in [60] and [135]), which was accepted as correct by Mr Knox.

Sixth proposition

[93] "When allotting an alleged legitimate expectation as between the three *Coughlan* categories, the correct approach is that of the Court of Appeal in *Bibi*."

[94] This proposition requires some exegesis from me. In *R* v *North and East Devon Health Authority, ex parte Coughlan* [2001] QB 213, in [55] to [60], there is in the judgment of the Court of Appeal a somewhat involved discussion of the role of the court in legitimate expectation cases. The starting point is to determine what was the legitimate expectation of a member of the public as to how he would be treated by a public authority. There are at least three possible outcomes of the enquiry. (1) The court might decide that the authority was required to bear in mind only its previous policy or representations giving it the weight it thought right, but no more, before deciding whether to change course. (2) The court might decide that the promise or practice induced a legitimate expectation of being consulted before a particular decision is taken. (3) The court might decide that a lawful promise or practice has induced a legitimate expectation of a benefit that is substantive.

[95] Once it has been decided into which category a legitimate expectation falls, the role of the court differs according to the category. In category (1), the court is limited to reviewing the decision on conventional *Wednesbury* grounds: *Associated Provincial Picture Houses Ltd* v *Wednesbury Corporation* [1948] 1 KB 223. In category (2), the court has to determine whether the decision was procedurally fair, and will require the opportunity for consultation to be given unless there is an overriding reason to depart from it. In category (3), the court will have to decide whether to frustrate the expectation is so unfair as to amount to an abuse of power and, when necessary, to decide whether there is a sufficient overriding interest to depart from what has been promised.

[96] *Coughlan* was decided in 1999. *Bibi* came before the Court of Appeal two years later. As I read the latter decision, it is dealing not so much (as is suggested by the sixth proposition) with the allocation of cases between the *Coughlan* categories but, rather, with the resolution of cases that fall within category (3), where the relevant legitimate expectation is that of a substantive, and not merely a procedural, benefit. *Bibi* qualifies *Coughlan* in two respects. First, doubt is expressly cast on the approach that suggests that the question for the court in a category (3) case is whether the authority, by reneging on its promise, was acting so unfairly as to be an abuse of power. This question provides an uncertain guide because a major part of the problem in legitimate expectation cases "is that it is often not adequate to look at the situation of the disappointed promisee" apart from the situations of the promisor and (sometimes) of many other persons to whom promises have been made: see [33] to [37]. Second, doubt is cast by implication on what was said in *Coughlan* concerning the remedy in category (3) cases. The judgment of the court, which was handed down by Schiemann LJ, contains the following passages:

> 40 The court has two functions assessing the legality of actions by administrators and if it finds unlawfulness on the administrators' part, deciding what relief it should give. It is in our judgment a mistake to isolate from the rest of administrative law cases those which turn on representations made by authorities. The same constitutional principles apply to the exercise by the court of each of these two functions.
>
> 41 The court, even where it finds that the applicant has a legitimate expectation of some benefit, will not order the authority to honour its promise where to do so would be to assume the powers of the executive. Once the court has established such an abuse it may ask the decision taker to take the legitimate expectation properly into account in the decision making process.
>
> ...
>
> 43 While in some cases there can only be one lawful ultimate answer to the question whether the authority should honour its promise, at any rate in cases involving a legitimate expectation of a substantive benefit, this will not invariably be the case.

[97] Mr Knox was prepared to accept in general terms that the judgment in *Bibi* provided the road along which the court now has to go: but he emphasised that *Bibi* should not be considered in isolation from other cases on legitimate expectation, and he drew my attention to two other passages in the judgment in *Bibi*:

> 39 But, on any view, if an authority, without even considering the fact that it is in breach of a promise which has given rise to a legitimate expectation that it will be honoured, makes a decision to adopt a course of action at variance with that promise then the authority is abusing its powers.
>
> ...
>
> 59 But when the authority looks at the matter again it must take into account the legitimate expectations. Unless there are reasons for not giving effect to those legitimate expectations then effect should be given to them.

Seventh proposition

[98] "(1) It is clear that in non-ECHR/EU cases, the test of proportionality has not been substituted for the *Wednesbury* principle. (2) That being so, many of the comments about balancing the public and private interests in *Coughlan* have to be read with caution." I have divided this proposition into two parts in order to take account of Mr Knox's response.

[99] Mr Knox accepted subproposition (1) as correct. Several authorities were cited by Mr Randall: the point was perhaps most clearly articulated in 1991 in the decision of the House of Lords in *R* v *Secretary of State for the Home Department, ex parte Brind* [1991] 1 AC 696, particularly in [762] and [763]. Mr Knox, however, rejected sub-proposition (2), on the basis that *Coughlan* had been approved in and survived *Bibi*. I have already referred to the *Coughlan* categories. The controversy here is focused on category (3), where a promise resulting in a substantive legitimate expectation has been broken. It was common ground in *Coughlan* that the public authority could break its promise (to permit the claimant to remain in a particular care home for the rest of her life) if, and only if, an overriding public interest required this. The court said at [52]:

> Both [counsel] adopted the position that, while the initial judgment on this question has to be made by the health authority, it can be impugned if improperly reached. We consider that it is for the court to decide in an arguable case whether such a judgment, albeit properly arrived at, strikes a proper balance between the public and the private interest.

[100] I think that so far as the application of the law to the facts of the present case is concerned, this debate may well be an academic one. I will therefore deal with it briefly. I have already commented on the relationship between the decisions in *Coughlan* and *Bibi*. I agree with Mr Randall, but only to this limited extent: although a striking of the balance between the public and the private interest may be required in some cases, *Bibi* provides a reminder that the court should not concentrate on the understandable disappointment of the claimant to the exclusion of wider considerations.

Eighth proposition

[101] "The fact that maladministration has occurred is not a ground for judicial review. The question is only has the public body acted unlawfully?"

[102] Mr Knox agreed with this proposition, subject to the qualification that although maladministration is not a ground for judicial review, "it is still relevant in deciding whether it is fair to allow a public body to change its practices". I have to say that I do not really understand the qualification. It seems to me, with respect, to come close to contradicting the main proposition, at any rate in legitimate

A expectation cases. The question must always be whether the act of the public body is unlawful.

Ninth proposition

[103] "In deciding what, if any, relief should be granted, the court will take into account (a) whether the decision challenged is in the macro-political field, and/or (b) involves social or political value judgments as to priority of expenditure, and/or (c) the nature and clarity of the promise or prior practice in question." There was, once again, extensive citation of authority: see (a) *Begbie*, at p1131C-D, and *Bibi*, in [23]; for (b) *Bibi*, in [64]; for (c) *Bancoult*, in [182].

B [104] This ninth proposition is, I think, incontrovertible. Mr Knox limited his response to saying (and I agree) that it is simply not realistic to characterise the decisions in this case as having been made in the macro-political field.

Discussion

Claim as pleaded

[105] Formally, the challenge made in these proceedings is to the decision, which was conveyed to the institute by the letter of 26 June 2009, that its application for funding was not going to proceed to AiD. As I have already stated, the claim has effectively turned into one for wasted expenditure, so that the true challenge is to the decision not to C reimburse the institute in full. The heart of the institute's case is to be found in para 28 of the amended grounds of claim:

> By so encouraging the Institute, the LSC implicitly represented to it that if and when it obtained AiP, (a) its application for AiD would be dealt with and approved in the usual way (subject to satisfying the LSC of relevant criteria) and (b) the LSC was and would continue to be properly financed and organised, so that it could deal with (and approve) such application in the usual way. Accordingly, the Institute had a legitimate expectation to this effect. Had it not been for this encouragement, the Institute would not have undertaken a project of this magnitude.

D [106] This formulation was criticised by Mr Randall on the basis that one could not derive from it any clear meaning, phrases such as "the usual way", "properly financed" and "properly organised" being shot through with ambiguity. In my judgment, the criticism is not a sound one. The meat of the alleged representation, for it seems to me that one is looking at a single representation expressed in alternative ways, is that the institute's application for AiD would be dealt with on its merits and without regard, as Mr Knox put the matter early in his closing submissions, to the financial state of the LSC.

[107] Whether any such representation was implicit in things said or done on behalf of the LSC is of course quite a different matter.

E *Alleged representations*

[108] In his closing submissions, Mr Knox set out the representations on which the institute was relying. They fell into two main groups, of which the second was further subdivided. (1) There were representations made in the institute's *Capital Handbook*. Even if these stood alone, the institute would, on Mr Knox's submissions, be entitled to succeed. (2) There were representations made otherwise than in the *Handbook*. These were: (i) representations made prior to 13 October 2006, namely (a) on 18 August 2006, (b) on 15 September 2006 and (c) in the undated letter from Ms Lyon that was produced at the meeting of 26 September 2006; (ii) representations made on 13 October 2006; F (iii) representations that were (a) implicit in the dealings of the LSC thereafter and (b) particularly in the AiP letter of 31 January 2008.

[109] I will consider these groups of alleged representations in turn.

Group 1 representations: Capital Handbook

[110] The parts of the *Handbook* that Mr Knox treated as central to his submission are paras 7.3 (the stage to which a project must be developed before it can be considered for AiP), 8.5 (fee support leading to the application for AiD) and, more particularly, Annex A. There is within Annex A, at para 6, the reference to competition for funds and, at paras 23-31, the criteria for the selection of projects. Finally, there is para 32 of Annex A, the reference to moderation arrangements to determine the relative priority to be given to competing applications G where insufficient funds are available to meet all applications.

[111] The thrust of Mr Knox's submissions was that Annex A was limited to the stage at which AiP was being sought. A finding to this effect is vital to the case resting on Group 1 representations because, if Annex A extends to the AiD stage, para 32 opens the door to the refusal of an application on the footing that the LSC has insufficient funds to meet all applications that have received AiP. There was in Mr Knox's submission no suggestion in the *Handbook* that once AiP had been obtained there would be any competition for grants. Two of the selection criteria, those relating to preference being given to projects that are put forward in order to meet legal requirements and that are to H be financed on a public-private basis or on loan, supported the view that the annex applied only up to AiP.

[112] Mr Randall disagreed. He pointed out that guidance note A, which is part of the *Handbook*, required the same application form to be used for applications for both AiP and AiD and that almost all the requirements set out in the guidance note applied indifferently to both types of application. As regards criteria according to which applications would be selected, the only published criteria were those set out in Annex A. If they applied only up to AiP, there would be no criteria applicable at the AiD stage, and that was something that must be highly improbable. Even if one did not have the express provision as J to moderation arrangements, it would be unthinkable that a public body with manifold calls on its funds was somehow debarred from exercising a choice when the funds available turned out to be insufficient to meet all those calls.

[113] In my judgment, the submissions made by Mr Randall are to be preferred. For the reasons advanced by him, I do not accept that Annex A is to be restricted in the manner suggested by Mr Knox: and as soon as one treats the moderation arrangements as applicable at AiD stage, the implied representation on which reliance is placed melts away. Representations, even implied representations, have to be clear and unequivocal. Nothing in the *Handbook* seems to me to K come anywhere close to implying that, once AiP has been obtained, the availability of funds to the LSC becomes irrelevant.

Group 2(i) representations: Representations prior to 13 October 2006

Representation 2(i)(a): Meeting on 18 August 2006

[114] The relevant pleading is in para 5B(1) of the amended grounds of claim.

> At a meeting on 18 August 2006, Mr Green of the LSC informed the representatives from the Institute (Professor Khan, Mr Adrian Clarke and others), from Hereford School and Franklin Sixth Form College: (a) that the L LSC fully supported the masterplan; (b) that a "total new rebuild should be considered as an option for all parties"; (c) that the LSC had an additional £350 million over the next three financial years. (The LSC relies on the Institute's own minute of this meeting).

[115] The three points recorded in para 5B(1) as having been made by Mr Green come verbatim from the minute referred to. Mr Green in evidence accepted the accuracy of the minute.

[116] I have real difficulty in grappling with this part of the institute's case: (a) is a representation of the LSC's attitude or collective state of mind and is not said to be inaccurate, (b) is a suggestion, or recommendation, as to how matters might or should proceed, and (c) is M a representation of fact and is not said to be inaccurate. I am simply unable to derive from these statements, individually or taken together, an implied representation to the effect that if an application for a capital grant were to be made by the institute, and if (at some undefined time in the future) AiP were to be obtained, the availability of funds to the LSC would thereafter be irrelevant in determining an application for AiD. The institute has, as I see matters, got to get as far as that to establish a case on implied representation and it gets nowhere near doing so.

Representation 2(i)(b): Meeting on 15 September 2006

[117] The pleading is in para 5B(3) of the amended grounds of claim:

At a meeting on 15 September 2006 to discuss option 4, attended by Professor Khan, Mr Ellis, and Mr Clarke for the Institute, by Mr Bob Flockton for the LSC, and representatives from the North East Lincolnshire Council and Dyers (who had drawn up the masterplan):

(a) There was discussion about the financial backing for the project and the requirement for it to be considered within "two year window of opportunity with the LSC";

(b) Mr Bob Flockton confirmed that the LSC had agreed the project in principle and to explore what level of funds would be available to support the development costs.

[118] The statements are recorded in a minute taken for the institute. The witnesses from the institute confirmed the accuracy of the minute. No one who attended the meeting on behalf of the LSC gave evidence. Mr Green made the observation that he would have thought that the phrase "two year window of opportunity" would be more likely to have been used by a representative of the institute rather than by an officer of the LSC.

[119] My difficulty with this part of the case is the same as that which I had with the previous representation. This meeting took place at an early stage. Discussions were still exploratory in nature. The LSC was, as the witnesses from the institute stated (and as I accept), adopting a positive and encouraging attitude to the project. That, however, is the beginning and end of the matter. It is, in my judgment, once again simply impossible to translate these statements into the kind of implied representation that is embodied in para 28 of the amended grounds of claim.

Representation 2(i)(c): Ms Lyon's undated letter

[120] I have precisely the same problem as with representations 2(1)(a) and 2(1)(b). The letter is expressly supportive of the institute's plans, albeit cautiously so ("we need to determine… whether this radical plan is worthy of further investigation"). However, the letter, read as a whole, is tentative in nature and, once again, falls far short of being able to bear the weight that Mr Knox seeks to place on it.

Group 2(ii) representations: Meeting on 13 October 2006

[121] This was the large gathering at Bradford attended by representatives of many FE colleges and of the LSC. It is said in para 6 of the amended grounds of claim that "[t]he attendees were briefed on the process for applications and then the LSC staff positively encouraged the colleges to put in binds for large capital developments as the LSC was said to have *significant under spend*" (italics in original).

[122] There has been controversy over what was said at the meeting, but, on reviewing my notes of the evidence, some of the points of difference between the two sides seem to have disappeared. Thus, Mr Ellis, for the institute, was unable to say firmly that phrases such as "once in a lifetime opportunity" and "significant under spend" were used at this (rather than at some other) meeting. On the other hand, Mr Green, for the LSC, accepted that colleges were invited to come back to the LSC with proposals, and he also agreed that the LSC was in effect informing colleges that they had a once-in-lifetime opportunity to obtain funding.

[123] Mr Green was an impressive witness, who gave his evidence in a measured and transparently honest manner. Mr Ellis was argumentative and, on one other significant part of the history, has been shown up as unreliable. Mr Clarke's recollection was exposed in cross-examination as being patchy and uncertain. The only safe course (which I adopt without hesitation) is therefore to accept what Mr Green said. I will set out the bulk of my notes of the relevant part of his cross-examination:

> I can't recall "world class buildings" being mentioned at the meeting. We did not encourage [colleges] to come back with mega-schemes.
>
> Colleges were not asked for big buildings. They were asked to go away, look at their estates and come back with proposals.
>
> I can't recall the LSC using the phrase "once in a lifetime." I think "support" rather than "encourage" was the word used. It was a positive message.
>
> The extra £350m over 3 years was not mentioned, but there was a slide showing the money available.
>
> I don't think [LSC officers] said there was an under spend. They did say that there were insufficient projects coming forward so that Yorkshire and Humberside were not doing well compared to other regions. Yorkshire and Humberside were missing out on the opportunity to get [funds] from the national budget.
>
> I don't recall it being said that there would be competition for AiD among projects which have AiP.

[124] It is common ground that the information imparted by the LSC at the meeting included an explanation of the procedure for obtaining capital grants, with express mention of the AiP and AiD stages. On the basis of Mr Green's evidence, the college representatives were told that funds were available and that they should go away and think about applying for grants for capital projects. However, I find it impossible to derive from anything that was said, or from the positive tone of the meeting, a representation of the kind relied on. It must not be forgotten that the representation pleaded in para 28 of the amended grounds of claim comes to this: that, in the event of AiP being granted, there would be excluded from further consideration by the LSC both the totality of funds available to the LSC and other calls on those funds. Once again, I find that there are light years between matters established as having been said or done, on the one hand, and what the institute seeks to infer from those matters, on the other hand.

Group 3 representations: Representations after 13 October 2006
Representation 3(i): "Implicit in all other dealings"

[125] This is far too vague and unparticularised. It is true that the officers of the LSC adopted a positive attitude towards the application for AiD. It is clear from the evidence that they regarded it as part of their job to assist an applicant (if necessary, by constructive criticism) to work up a project to a state in which it was suitable for submission to the relevant decision-making bodies at the LSC. Indeed, the officers appear to have been unwilling to bring a project to committee unless it was such as they felt able to recommend for approval. However, none of this seems to me to be apt to convey to the mind of the applicant and, in particular, the institute, the message that consideration of a detailed application would be confined to the inherent merits of the application.

Representation 3(ii): Letter of 31 January 2008

[126] This letter was put forward by Mr Knox as coming close to an express representation. He attached particular importance to the passage in which Professor Khan was informed that budgetary provision for 79% of the cost of the project was being made for a period of up to 12 months pending receipt of the detailed application. This was, Mr Knox submitted, tantamount to saying that "the money will be there if you satisfy the criteria".

[127] This letter had not been pleaded as a representation in the amended grounds of claim and was not dealt with, or complained about, by any of the witnesses for the institute. It raised its head only in Mr Knox's cross-examination of one of the witnesses for the LSC towards the end of the case. I appreciate the force of Mr Randall's objection that "it is grossly unfair to bounce the defendant into a legitimate expectation argument based on this letter".

[128] I do not think that Mr Randall needed to be too concerned. If the information regarding budgetary provision stood alone, it would be arguable that it could be read as conveying to the reader the message that funds had been ring-fenced and that there was something close to a guarantee of availability. However, the passage has to be read in context. It follows immediately on a sentence that informs the institute that AiP has been granted "without prejudice to the determination of a detailed application in due course". There is, once again, a total absence of words from which one could fairly infer the exclusion from consideration by the LSC of financial matters external to the application.

[129] Mr Randall put his point succinctly:

> The sentence does not support the paragraph 28 representation. It is a late thought to salvage a fatally flawed case.

I agree.

Conclusion on representations

[130] The case for the institute was put primarily on the basis of legitimate expectation based on implied representations. For the reasons set out in the foregoing paragraphs, I do not think that the para 28 representations can, on any objective basis, be regarded as implicit in the words and actions relied on by the institute. There was, in my judgment, nothing in the nature of a clear and unambiguous statement of the kind required to found a legitimate expectation (the first proposition of law discussed above).

[131] Mr Randall made several points that demonstrated to my mind that the evidence given on behalf of the institute was inconsistent with the case presented on its behalf.

[132] First, none of the witnesses for the institute said that he understood the para 28 representation to have been implicit in the words relied on as constituting that representation. The statements now relied on never seem to have registered contemporaneously with the institute's own officers.

[133] Second, although detrimental reliance is not essential to making out a case of legitimate expectation, it is relevant (the fifth proposition of law), yet reliance is not actually claimed by any of the witnesses for the institute.

[134] Third, there was throughout the process of applying for AiD an appreciation on the part of the officers of the institute that they were proceeding at risk of failure. It would be remarkable if, after the risk came to fruition, a claim based on legitimate expectation of success were to be established.

Claim based on practice

[135] There was not, in my judgment, anything in the nature of a specific undertaking to colleges that a practice (if it existed) of confining consideration of an application for AiD to the merits of that application would continue (see the second proposition). The real weakness of the institute's case here lies in the impossibility of finding in the *Capital Handbook* or elsewhere any assurance of the kind that would be required to found a practice-based legitimate expectation.

Claim based on conspicuous unfairness

[136] This is an alternative way of putting the case for the institute. I dealt with the relevant legal principles earlier (the third proposition). There can be no doubt as to three matters: a case of legitimate expectation not founded on a representation must satisfy the test of conspicuous unfairness; such cases are exceptional; and whether a case falls within this exceptional category is entirely dependent on the facts.

[137] In his closing submissions, Mr Knox put the conspicuous unfairness claim in this way:

> My clients were encouraged to apply for a very substantial grant in circumstances in which it was known that they would incur very substantial fees. The fees were bound to be far greater than the subsidised amount [reference was made to the *Capital Handbook*]. They have not been accused of extravagance. It is unfair to turn them down at the last moment on the "no money" basis… The fault lay with the LSC, not with us.

[138] It is of course understandable that the outcome of the application for AiD was a considerable disappointment to the institute. Further, the institute has been left with substantial unreimbursed expenditure. What happened cannot be dismissed as "just one of those things". However, in my judgment, the facts of the case fall a long way short of what is required to make out, as can be done only in a rare case, conspicuous unfairness on the part of a public authority. There was no departure by the LSC from any published practice or concession. The institute acted throughout with what Mr Randall aptly called "an active appreciation of risk". The institute was not left entirely out of pocket, but received the maximum contribution to costs available under the guidelines in the *Capital Handbook*. The institute's request for the additional amount now claimed was not dismissed out of hand, but was considered together with similar requests from many other colleges. The result for the institute is most unfortunate, but it cannot be characterised as the product of an abuse of power by the LSC.

True analysis

[139] The debacle that gave rise to this litigation did not, in my judgment, leave the institute without any rights in public law. It must, I think, have had a right to due consideration of its request for reimbursement along with the requests made by other colleges that returned the questionnaire regarding costs to the LSC. However, that right has been satisfied: the request was considered and the decision to apply such funds as were available to meeting the expenses incurred by colleges that were facing insolvency cannot be stigmatised as irrational.

Disposal

[140] It follows from what I have said that the claim for judicial review must be dismissed.

Quantum

[141] In case the litigation goes to appeal and my decision is held to be wrong, I should say a little about the relief that I would have granted if I had held the case of the institute to be sound in principle.

[142] The very short statement of expenditure put in evidence by the institute would not be an adequate foundation for an order for payment of the sum alleged to be due. There are three matters that require attention. First, the claim is unvouched. Second, it seems to me that expenditure incurred prior to AiP should be quantified and excluded. Third, it may be that the institute has derived value from some of the enabling works carried out after AiP and, if so, that there should be some deduction in respect of betterment. I would therefore have directed, in the absence of agreement between the parties, a short hearing at which to determine any outstanding issues on quantum.

Envoi

[143] I do not pretend to have dealt with every point that was canvassed either in evidence or in submissions. That is, in part, the result of a determination on my part to avoid missing the wood from the trees. It is all too easy to fall into that trap in a case such as this, which did not merely suffer from documentary overload but was not simplified (I put the matter gently) from the making of quite substantial additions to the court papers as the hearing progressed. I have tried, and I hope succeeded, not to omit consideration of anything of real importance.

Application dismissed.

PROCEDURE

R (on the application of Garner) *v* Elmbridge Borough Council

COURT OF APPEAL

[2010] EWCA Civ 1006

29 July 2010

LLOYD, RICHARDS AND SULLIVAN LJJ

Civil practice and procedure — Protective costs order (PCO) — Judicial review — Planning permission — Directive 83/337/EEC — Application for PCO in connection with judicial review of planning decision — Whether claimant required to show that issues raised of general public importance and public interest requiring their resolution in accordance with principles in* R (on the application of Corner House Research) *v* Secretary of State for Trade and Industry *— Whether PCO necessary to comply with requirement in article 10a of directive that procedure to challenge legality of decision to which directive applying not to be prohibitively expensive — Whether issue of prohibitive expense depending on financial means of individual claimant

In June 2009, the respondent council granted planning permission for a comprehensive redevelopment of Hampton Court station and adjoining land. The redevelopment site was close to Hampton Court Palace, a scheduled ancient monument and a Grade I listed building. The appellant, who had previously worked for Historic Royal Palaces as an architect specialising in the conservation of historic buildings, applied for permission for a judicial review of that decision and for a protective costs order (PCO) in respect of such proceedings.

The application was initially refused, but on a renewed application the parties agreed to a "rolled up" hearing dealing with both permission to apply for judicial review and the substantive hearing. However, the judge refused to make a PCO on the ground that the issues raised were not of general public importance and the public interest did not require their resolution, pursuant to the principles in *R (on the application of Corner House Research)* v *Secretary of State for Trade and Industry* [2005] EWCA Civ 192; [2005] 1 WLR 2600. He further rejected the appellant's contention that a PCO should be granted pursuant to the requirement in article 10a of Council Directive 83/337/EEC (the EIA Directive) that members of the public concerned with a sufficient interest should have access to a review procedure that was not prohibitively expensive. He held that there was before him insufficient evidence of the appellant's financial resources on which to determine whether the proceedings would be prohibitively expensive for him.

The appellant appealed. He contended that: (i) article 10 of the EIA Directive, which incorporated the principles set out in the Aarhus Convention and had direct effect in domestic law, did not require that the issues raised in a challenge to an environmental decision should be of general public importance or that the public interest should require their resolution; and (ii) the issue of prohibitive expense should be determined objectively, by reference to whether the potential costs would be prohibitively expensive to an ordinary member of the public without requiring public disclosure of an individual claimant's financial affairs.

Held: The appeal was allowed. (1) The _Corner House_ principles relating to PCOs, although settled, must be applied flexibly. Furthermore, the EIA Directive has direct effect in domestic law, and in cases to which it applies the _Corner House_ principles must be interpreted and applied, and if necessary modified, in such a way as to secure compliance with the directive. In an article 10a case, there is no justification for applying the _Corner House_ requirements that the issues should be of general public importance and that the public interest should require their resolution. Both the Aarhus Convention and the EIA Directive are based on the premise that it is in the public interest for there to be effective public participation in the decision-making process in significant environmental cases. An important component of that public participation is that the public should be able to ensure, through an effective review procedure that is not prohibitively expensive, that such environmental decisions are taken lawfully. Accordingly, under community law it is a matter of general public importance that those environmental decisions to which the directive applies are taken in a lawful manner. If there is an issue with that, the general public interest requires that it should be resolved in an effective review process. The respondents' decision to grant planning permission for the redevelopment was a decision to which the directive applied. The appellant was a member of the public concerned, in that he had an interest in the decision-making process in respect of applications for planning permission that affected the setting of Hampton Court. Such a finding was consistent with the underlying objective of article 10a of giving the public wide access to justice. Although the question of whether the appellant had a "sufficient interest" to bring the proceedings for the purposes of section 32(1) of the Senior Courts Act 1981 remained to be considered at the rolled-up hearing, it was at least arguable that he did. Accordingly, the judge had not been entitled to reject the appellant's application for a PCO on the ground that the issues raised were not issues of public importance that the public interest required to be resolved. (2) The judge's approach to the issue of prohibitive costs, although consistent with _Corner House_ principles, was not consistent with article 10a. A purely subjective approach to the issue of prohibitive expense is not consistent with the objectives underlying the directive. Even if it is permissible or necessary to have regard to an individual claimant's financial circumstances, the underlying purpose of the directive would be frustrated if the court were entitled to decide the matter solely by reference to the claimant's means, without considering whether the potential costs would be prohibitive for an ordinary member of the public. The more intrusive the investigation into the means of those who seek PCOs and the more detail that is required of them, the more likely it becomes that there will be a chilling effect on the willingness of ordinary members of the public to challenge the lawfulness of environmental decisions. There was evidence that a PCO was necessary if the instant proceedings were not to be prohibitively expensive. (3) Where a PCO is granted in favour of a claimant, the imposition of some reciprocal limit on a respondent's liability for costs is not necessarily inconsistent with article 10a. The requirement that the review procedure should not be prohibitively expensive applies to both parties engaging in it. The limit on the liability of each party does not necessarily have to be the same, but it should properly reflect the disparity of resources. The imposition of a reciprocal cap was fair and reasonable in all the circumstances

A **of the instant case. A PCO of £5,000, with a reciprocal limit of £35,000 on the respondents' liability, was reasonable.**

The following cases are referred to in this report.

Commission of the European Communities v *Ireland* C-427/07 [2010] Env LR 8

Edwards v *Environment Agency; sub nom R (on the application of Edwards)* v Environment Agency (No 2) [2006] EWCA Civ 877; [2007] JPL 82

Marleasing SA v *La Comercial Internacional de Alimentación SA* C-106/89 [1990] ECR I-4135, ECJ

Morgan v *Hinton Organics (Wessex) Ltd* [2009] EWCA Civ 107; [2009] 2 P&CR 4; [2009] JPL 1335

B *R (on the application of Buglife: The Invertebrate Conservation Trust)* v *Thurrock Thames Gateway Development Corporation* [2008] EWCA Civ 1209

R (on the application of Corner House Research) v *Secretary of State for Trade and Industry* [2005] EWCA Civ 192; [2005] 1 WLR 2600; [2005] 4 All ER 1

R (on the application of Garner) v *Elmbridge Borough Council* [2010] EWHC 567 (Admin)

This was an appeal by the appellant, Keith Garner, from a decision of Nicol J, sitting in the Administrative Court, refusing an application for a protective costs order in respect of an application for judicial review of a decision of the respondents, Elmbridge Borough Council,
C to grant planning permission for the redevelopment of land.

Richard Drabble QC, David Smith and Dr David Wolfe (instructed by Richard Buxton Environmental & Public Law, of Cambridge) appeared for the appellant; James Findlay QC and Robert Williams (instructed by the legal department of Elmbridge Borough Council) appeared for the respondents; Jeremy Hyam (instructed by Leigh Day & Co) represented WWF (UK) and Friends of the Earth, as interveners.

Giving judgment, **Sullivan LJ** said:

Introduction

[1] This is an appeal against the order made on 3 March 2010 by
D Nicol J, in so far as that order dismissed the appellant's application for a protective costs order (PCO) and ordered that he should pay the respondents' costs of the PCO application, summarily assessed in the sum of £3,000.

[2] There is also an application for permission to appeal against a further order made by Nicol J on 27 April 2010 on the papers, refusing permission to add Keith Garner Ltd as an additional claimant and refusing an application for a PCO in respect of both that company and a Mr Gerald Macaulay, who was granted permission to be added as a second claimant. He has so far declined to take up that permission because he was not granted a PCO.

E

Background

[3] On 16 June 2009, the respondent granted planning permission to the interested parties for a comprehensive redevelopment of Hampton Court station and adjoining land at Hampton Court Way, East Molesey (the site). The site is on the opposite bank of the River Thames from Hampton Court Palace. The palace is a scheduled ancient monument and a Grade I listed building. The appellant lives in London SW11. He is neither a local resident nor is he a local elector. He does, however, have a long-standing interest in Hampton Court Palace; he is an architect and his practice specialises in the conservation of historic buildings. He
F worked for Historic Royal Palaces (HRP) from 1994 to 2004. During that time, one of his duties was to advise the director of Hampton Court Palace on planning applications affecting its setting. He advised that objection should be made to a number of developments over the years that would have had an adverse effect on the setting of the palace. The applications for those developments were refused.

[4] After leaving HRP, the appellant maintained his interest in the palace. In 2007, he objected to an earlier scheme for the redevelopment of the site. This proposal became known as "the boathouse scheme" because of the architectural treatment of the hotel that was proposed within the scheme. The appellant did not object to the architectural treatment but to the scale of the proposed development and the effect of a redevelopment on the site of that scale on the setting of the palace. In the present scheme, for which planning permission was granted G
on 16 June 2009, the architectural treatment has been changed. It is known as the "classical scheme". The appellant says in his second witness statement that he was not sent a copy of the notice that the respondents sent to objectors to the boathouse scheme informing them of the new application for the classical scheme, but, in any event, when the appellant learnt of the new proposal he did not think it necessary to resubmit his objection because his objection in 2007 to the scale of the proposed redevelopment was, in his judgment, equally applicable to both the boathouse scheme and the classical scheme, since the scale of the proposed redevelopment on the site remained the same. The only material difference was the architectural treatment, to which he had H
not objected.

[5] The appellant's judicial review claim form was filed on 14 September 2009. The grant of planning permission was challenged on three principal grounds, namely that: (i) the respondents had failed to give a summary of their reasons for granting planning permission; (ii) they had failed to address or to apply the statutory requirement that, when deciding whether planning permission should be granted, special regard should be had to the desirability of preserving the setting of listed buildings; and (iii) they had not applied the sequential test set out in PPS 25 for development proposals in a flood plain. The claim included an application for a PCO. J

[6] In a witness statement in support of the claim, Ms Lisa Foster, a Californian lawyer employed by the appellant's solicitor, explained why the appellant was seeking a PCO. She said that although the appellant's funds were limited, he had agreed to be responsible for lodging the proceedings so that they could be brought within the three-month time limit, but:

> Given his individual limit on funds, he could not and would not afford to take on the risk of fighting a judicial review without costs protection. If the PCO is not granted, I am instructed that he would have no option but to withdraw from the proceedings or find a substitute claimant.

K

[7] She said that the appellant's solicitor's firm, which, I would add, has extensive experience in environmental judicial reviews, generally estimates each side's costs in a one-day straightforward judicial review, where permission to apply for judicial review is granted on the papers, at around £15,000 plus VAT. However, if there are complications, for example if permission is refused on the papers and the application has to be renewed at an oral hearing, the costs can easily double; so the appellant, if unsuccessful, would be facing a costs liability of £30,000 if the case was straightforward or £60,000 if it proved to be complicated. The appellant could not afford such sums and, realistically, she said £1,000 to £1,500 could be raised from the local community.

[8] The respondents and the two interested parties filed L
acknowledgements of service. They contended that the grounds of challenge had no merit, but they also contended that the appellant did not have a sufficient interest to bring the proceedings and that the proceedings had not been brought promptly. They also opposed the grant of a PCO.

[9] It is relevant to note that both the respondents and the second interested party claimed the costs of their acknowledgements of service. The amounts claimed were £6,080 and £8,839.60 respectively: a total of £14,918 before the papers had even gone before a judge.

[10] On 9 December 2009, Mr George Bartlett QC, sitting as a
deputy High Court judge, considered the matter on the papers and M
refused permission to apply for judicial review. He said:

> There was a failure to give a summary of the reasons for granting permission in accordance with Article 22(1) of the Town and Country Planning Act (General Permitted Development) Order 1995, as the defendant accepts. Such a failure may well not justify the grant of permission where it is otherwise apparent that the planning authority have taken into account and weighed properly the relevant policies and other material considerations. Here, however, the Council were required to have special regard to the desirability of preserving the setting of Hampton Court Palace, and, in view of the duty, an important issue was whether the river frontage of the site should be kept free of substantial development. There is in any view a clearly arguable case, as evidenced by the officer's report, that, while detailed consideration was given to the design of

A the proposed buildings, the Council failed to apply the statutory requirement, of which there is no mention in the report, and failed to address this important issue, which the report does not discuss, by reference to it.

There is no reasonably arguable case on the application of PPG 25, in my view. This matter, which was not apparently of concern to the claimant at any earlier stage, was dealt with in paragraphs 8.4.1 to 8.4.6 of the officer's report in a way that appears to me to address properly both the sequential test and the criteria in the exception test.

Had the claimant objected to the application for planning permission, he would have had sufficient standing. On what is not simply a local issue but one of national significance because of the importance of Hampton Court Palace, his interest, as a person concerned with the protection of historic buildings, B and the Palace in particular, and as one whose views appear to accord with those of many individuals and groups, including Historic Royal Palaces, would have qualified him to bring proceedings. I would also have considered it an appropriate case for a protective costs order. But he did not object to the application, despite the fact that the Committee resolution in December 2008 to grant permission must have alerted him to the need for the Council, when determining the applications in June 2009, to consider the issue of keeping the river frontage substantially free of development. Not having sought to influence the decision, I do not think that he has sufficient standing.

In addition the application was not made promptly, in that over five weeks elapsed before he sought legal advice and a further five weeks elapsed before the pre action protocol was issued. It appears that delay is financially prejudicial C to the charity the Royal Star and Garter.

Despite the importance of the issue that the claimant seeks to bring before the court I do not think in these circumstances that permission should be granted.

[11] Mr Bartlett did not think it was appropriate to award costs. On 16 December 2009, the appellant gave notice of renewal of his claim for permission to apply for judicial review. After a certain amount of procedural toing and froing, the position that was reached before Nicol J on 2 March 2010 was that the parties were agreed that there should be a "rolled up" hearing of the renewed application of permission to apply for judicial review, to be followed by the substantive hearing if D permission were to be granted, and that this rolled-up hearing should be listed with reasonable expedition. As Nicol J said, that left the question of whether a PCO should be granted.

[12] In his judgment dated 3 March 2010 ([2010] EWHC 567 (Admin)), Nicol J explained, in [1.12]:

The first interested party (Gladedale) indicated that it would not be seeking an order that the claimant pay its costs even if the judicial review application was dismissed. So far as its costs were concerned, the application for a protective costs order was therefore academic. For the defendant and the first interested party (Network Rail Infrastructure, hereafter "Network Rail") the issue was potentially a real one. They opposed the making of a PCO. In the alternative, E they asked that, if one was to be made, an order should also be made limiting their costs liability in the event that the judicial review application should be successful.

[13] In [1.13], Nicol J said:

For the principles to be applied in relation to a PCO, both parties referred me to *R (Corner House Research)* v *the Secretary of State for Trade and Industry* [2005] 1 WLR 2600. At paragraph 74 Lord Phillips MR said this:

"74. We would therefore restate the governing principles in these terms:

1. A protective costs order may be made at any stage of the proceedings, on such conditions as the court thinks fit, provided that the court is satisfied that:

(i) the issues raised are of general public importance;

F (ii) the public interest requires that those issues should be resolved;

(iii) the applicant has no private interest in the outcome of the case;

(iv) having regard to the financial resources of the applicant and the respondent(s) and to the amount of costs that are likely to be involved it is fair and just to make the order;

(v) if the order is not made the applicant will probably discontinue the proceedings and will be acting reasonably in so doing.

2. If those acting for the applicant are doing so pro bono this will be likely to enhance the merits of the application for a PCO.

3. It is for the court, in its discretion, to decide whether it is fair and just to make the order in the light of the considerations set out above.

[14] In [1.15], Nicol J set out the relevant provisions of article 10a of Directive 83/337/EEC:

Member states shall ensure that, in accordance with the relevant national legal system, members of the public concerned: G

(a) having a sufficient interest, or alternatively,

(b) maintaining the impairment of a right, where administrative procedural law of a Member State requires this as a precondition,

have access to a review procedure before a court of law or another independent and impartial body established by law to challenge the substantive or procedural legality of decisions, acts or omissions subject to the public participation provisions of this Directive.

[…]

Any such procedure shall be fair, equitable, timely and not prohibitively expensive.

H

[15] Having noted that the respondents and the second interested party had agreed that the appellant had no private interest in the outcome of the case, the judge said that the other *Corner House** conditions were all in issue. He considered the two questions of whether the issues raised were of general public importance and whether the public interest required them to be resolved together, and concluded in [1.27] of the judgment that:

the claimant's challenge is not one where the issues in the case are of general public importance and which the general public interest requires to be resolved.

J

[16] The judge then turned to the questions:

1.28. Having regard to the financial resources of the applicant and the respondent(s) and to the amount of costs that are likely to be involved, is it fair and just to make the order; and if the order is not made will the applicant probably discontinue the proceedings and be acting reasonably in so doing?

[17] He answered those questions in [1.29] by saying:

1.29. Even if I was wrong about the general public importance of the case, I am quite clear that the claimant has not satisfied me of these next matters.

[18] The judge said that so far as the appellant's resources were concerned, the only evidence that he had was Ms Foster's witness statement. Having referred to the passages that I have mentioned, Nicol J continued in [1.31] to [1.36]: K

[1.31] In his witness statement the claimant said only this:

"During this period from 1 September I then had to consider if my funding would permit me to proceed and whether I could raise a community fund to help depending on Elmbridge's response to the pre action letter. Once we had Elmbridge's response on the 7th, I decided I would act individually as the claimant had set up [a] fund which local residents could contribute to. Because I was concerned about an adverse cost award in the event the claim was unsuccessful, it has been explained to me that I am acting on a matter of great public interest, in a situation where the HRP hasn't acted although the advice of counsel is that the permission is potentially unlawful." L

[1.32] In my judgment, the defendant and Network Rail were entitled to observe that the lack of information about the claimant's resources was striking. Ms Foster's statement is vague in the extreme. It entirely lacks the detail which would allow the defendant and Network Rail to test it, or for the court to determine its accuracy. Even these general comments about the claimant's means are presented in just her witness statement. They are not endorsed, let alone elaborated on, in the claimant's own witness statement beyond the even vaguer statement that he was "concerned about an adverse costs order". The point as to the paucity of evidence on this issue was taken clearly in the acknowledgments of service. Yet there has been no evidence subsequently to fill that gap. The claimant has provided a second witness statement which M speaks to his standing to bring these proceedings, but that, too, is silent on the question of his resources.

[1.33] Mr Barnes, the Director of Conservation and Learning at HRP, has provided a witness statement in which he says that the claimant has his organisation's full moral and some limited financial support. The amount is not specified in the evidence, but the claimant's reply to the acknowledgment of service says that it amounted to £5,000. In the course of submissions, I was told that the claimant had contributed £2,000 to the fund for costs, and other donations had amounted to about £1,630.

---

* Editor's note: *R (on the application of Corner House Research)* v *Secretary of State for Trade and Industry* [2005] EWCA Civ 192; [2005] 1 WLR 2600

[1.34] The claimant submits that Article 10a of the Directive requires that legal proceedings to challenge environmental decisions should not be "prohibitively expensive". I do not think that that takes the matter any further. It is impossible to tell whether the proceedings would be "prohibitively" expensive unless there is information about the resources which the claimant would have available to fund them. That evidence is simply lacking.

[1.35] Because I have insufficient evidence as to the claimant's financial position, it is also not possible for me to determine whether it would be reasonable for him to discontinue the proceedings if a PCO is refused. Of course, in a general sense, legal proceedings can be very expensive. But generalities of that kind are not much use. They have to be related to the financial position of the particular claimant.

[1.36] The insufficiency of evidence as to the claimant's financial resources, in my judgment, is a clear reason why the application for a PCO must fail.

[19] The appellant applied for permission to appeal on 24 March. I refused permission to appeal on consideration of the papers, observing, *inter alia*, that this was a kind of satellite litigation in respect of case-management decisions that the Court of Appeal had deprecated in *R (on the application of Buglife: The Invertebrate Conservation Trust)* v *Thurrock Thames Gateway Development Corporation* [2008] EWCA Civ 1209. The appellant then renewed his application for permission to appeal, and at an oral hearing before Munby LJ on 10 June he granted permission to appeal, saying, *inter alia*, that:

I am left with the uncomfortable feeling that the significance of Article 10a and the centrality of the arguments based on Article 10a has been more prominent in the submissions put before me today by Mr Drabble than had been the case before either Nicol J or Sullivan LJ…

[20] Munby LJ said that:

If this was a case which had to be determined purely according to domestic *Corner House* principles, it seems to me that, for all the reasons given by Sullivan LJ, there is no conceivable basis upon which this court should be invited to intervene. But the fact is, as is common ground before me today, that this is a case in which there is a directly applicable European directive and it was arguable that Nicol J was wrong in his approach in terms of law and principle to Article 10a.

Parties' submissions

[21] On behalf of the appellant, Mr Richard Drabble QC submitted that Directive 2003/35/EC, which amends Council Directive 85/337/EEC (the EIA Directive, hereinafter referred to as "the directive"), and 96/61/EC (the IPPC Directive) by, *inter alia*, the insertion of article 10a, has incorporated the principles in the Aarhus Convention into a directive that has direct effect in domestic law. This places the UK under an obligation to ensure that the procedure for challenging the lawfulness of the planning permission in the present case is not "prohibitively expensive". It is not sufficient to rely on an exercise of judicial discretion after the event: see *Commission of the European Communities* v *Ireland* C-427/07*, in [92] to [94]. A PCO must be made in order to avoid the chilling effect of an open-ended exposure to liability for the respondents' and the two interested parties' costs. He submitted that there is no requirement in article 10a, nor indeed in Aarhus, that the issues raised in challenge to an environmental decision should be of "general public importance" or that the public interest should require that those issues are resolved. In applying Aarhus principles, article 10a is based on the premise that public involvement in decision making in the environmental field is in the public interest, and that it is in the public interest that members of the public concerned, having a sufficient interest, should be able to ensure, through an effective review procedure, that environmental decisions are lawfully taken. Alternatively, he submitted that the judge had failed to apply the public importance and public interest conditions in *Corner House* with sufficient flexibility.

[22] In *Morgan* v *Hinton Organics (Wessex) Ltd* [2009] EWCA Civ 107†, Carnwath LJ, giving the judgment of the court, reviewed the relevant authorities relating to PCOs. He drew the threads together in [47] of the judgment:

[47] It may be helpful at this point to draw together some of the threads of the discussion, without attempting definitive conclusions:

(i) The requirement of the Convention that costs should not be "prohibitively expensive" should be taken as applying to the total potential liability of claimants, including the threat of adverse costs orders.

(ii) Certain EU Directives (not applicable in this case) have incorporated Aarhus principles, and thus given them direct effect in domestic law. In those cases, in the light of the Advocate-General's opinion in the Irish cases, the court's discretion may not be regarded as adequate implementation of the rule against prohibitive costs. Some more specific modification of the rules may need to be considered.

(iii) With that possible exception, the rules of the CPR relating to the award of costs remain effective, including the ordinary "loser pays" rule and the principles governing the court's discretion to depart from it. The principles of the Convention are at most a matter to which the court may have regard in exercising its discretion.

(iv) This court has not encouraged the development of separate principles for "environmental" cases (whether defined by reference to the Convention or otherwise). In particular the principles governing the grant of Protective Costs Orders apply alike to environmental and other public interest cases. The *Corner House* statement of those principles must now be regarded as settled as far as this court is concerned, but to be applied "flexibly". Further development or refinement is a matter for legislation or the Rules Committee.

(v) The Jackson review provides an opportunity for considering the Aarhus principles in the context of the system for costs as a whole. Modifications of the present rules in the light of that report are likely to be matters for Parliament or the Civil Procedure Rules Committee. Even if we were otherwise attracted by Mr Wolfe's invitation (on behalf of CAJE) to provide guidelines on the operation of the Aarhus convention, this would not be the right time to do so.

(vi) Apart from the issues of costs, the Convention requires remedies to be "adequate and effective" and "fair, equitable, timely". The variety and lack of coherence of jurisdictional routes currently available to potential litigants may arguably be seen as additional obstacles in the way of achieving these objectives.

[23] Turning to the second reason why Nicol J refused the application for a PCO, Mr Drabble submitted that an approach that placed the onus on the individual claimant to demonstrate, by the production of detailed evidence as to his means, that the proceedings would be prohibitively expensive for him was not article 10a-compliant. Article 10a required systemic compliance with the requirement that the review procedures provided by the member state should not be prohibitively expensive. The obligation on the member state is to ensure that members of the public concerned with a sufficient interest have access to a procedure that is not prohibitively expensive for them. Thus, the procedure must be one that, viewed objectively, is not so expensive as to deter an ordinary member of the public concerned. The test is not whether the particular member of the public who happens to want to access the review procedure would find it prohibitively expensive to do so.

[24] In any event, he submitted that the prospective costs of what was agreed by 2 March 2010 to be a one-and-a-half day rolled-up hearing would be prohibitively expensive for an "ordinary" member of the public, and there was nothing to contradict the appellant's evidence that he would not be able to afford them. Mr Macaulay had been denied a PCO by Nicol J because he was not prepared to make a public disclosure of his assets. However, on the basis of such disclosure as Mr Macaulay had made, it should have been concluded that although relatively fortunate, he is still a person of relatively modest means. As Mr Macaulay put it in his witness statement:

We live off primarily my company pension although I also have a state pension. I also hold an investment portfolio which reflects the savings of my life's work. I consider myself to be a normal citizen who worked hard for over 40 years and I am now able to enjoy a comfortable retirement with moderate good health. I could not put our home or investments at risk to join this case since I have no significant assets or income other than as stated above.

[25] A little later on in his witness statement he put it perhaps a little more graphically, when he said:

* Editor's note: Reported at [2010] Env LR 8

† Editor's note: Reported at [2009] 2 P&CR 4

… I am not remotely in the league of rumoured wealth referred to at the hearing before Nicol J for example pertaining to the owner of a well known department store or certain footballers.

[26] Mr Drabble submitted that the prospect of a public investigation into one's financial means, with a wholly uncertain outcome, would be bound to have a chilling effect on the willingness of members of the public to challenge environmental decisions.

[27] The respondents and the two interested parties all appeared at the hearing before Nicol J. At the hearing before Munby LJ, the respondents and the second interested party were represented. At the hearing before us only the respondents were represented. On their behalf, Mr James Findlay QC submitted that the judge had been right to refuse to grant a PCO and right to refuse to join the additional claimants. He submitted that in order to take advantage of article 10a, the appellant had to be both a member of the public concerned (which he was not) and have a sufficient interest (which was a matter to be decided at the rolled-up hearing, which, it is hoped, will be heard in October). The appellant's company was in no better position than the appellant, although Mr Findlay had accepted that, as a local resident, Mr Macaulay would have a sufficient interest in the matter. He submitted that the *Corner House* principles remained binding on this court; flexibly applied they were not necessarily inconsistent with article 10a. The public importance and public interest tests were not inconsistent with the broader framework of the directive. As for the requirement that the procedure should not be prohibitively expensive, there was no warrant in the directive for a wholly objective approach, regardless of the means of the individual claimant. The court was entitled in an article 10a case to have regard to an individual claimant's circumstances, and the evidence before Nicol J was not such as to enable him to reach a conclusion that the proceedings would be prohibitively expensive for either the appellant or for Mr Macaulay.

[28] Applying *Corner House* principles — in so far as it was necessary to do so to comply with article 10a — the court still had to consider: (a) should there be a PCO?; and (b) if the answer to that question is yes, in what sum should the PCO be made? The second question, he submitted, could not be answered without knowledge of the individual claimant's means. If we were to reject those submissions and a PCO were to be granted, he submitted that the court should also cap the costs that could be recovered from the respondents should the appellant be successful. In open correspondence, the respondents had made an offer to agree to a PCO on the basis that it would limit their costs claim to £5,500 in exchange for a reciprocal cap on their liability in the sum of £35,000.

[29] This was rejected by those acting on behalf of the appellant. Mr Drabble submitted that in an article 10a case there was no reason to impose additional limitations on the remuneration of a claimant's lawyers that would have the practical effect of either penalising a successful claimant in costs or requiring his legal advisers to subsidise the litigation. The ordinary costs provisions were sufficient to ensure that only reasonable and proportionate costs could be recovered.

[30] *Corner House* was not concerned with environmental litigation of this kind, and the reciprocal costs-capping provisions referred to in that case were dealing with a different position, which is now addressed by CPR 44.

[31] For the record, I should mention that the second interested party made it clear, in a letter dated 11 June, that if a PCO were ordered, it would not claim its costs on the basis that the appellant, if successful, would not seek his costs against the second interested party. I should also mention that we gave WWF (UK) and Friends of the Earth (two NGOs concerned with environmental matters) permission to intervene by way of written submissions. We were assisted by those submissions and by brief oral submissions from Mr Jeremy Hyams on the matters of general interest raised by this appeal.

Discussion and conclusions

[32] It is unnecessary to rehearse the authorities that deal with the application of *Corner House* principles. The threads are drawn together in *Morgan* (see above). Although the principles must be applied flexibly, they are settled so far as this court is concerned. However, this court has not had to consider whether those principles do comply with the requirements of article 10a in a case where the directive applies. It is common ground that the directive has a direct effect in our domestic law. In such a case, the Court of Appeal recognised in *Morgan* (see [47(ii)]) that some more specific modification of our domestic costs rules may be required.

[33] There is no dispute that the decision to grant planning permission in the present case is a decision to which the directive applies. The council required that an environmental impact assessment (EIA) should accompany the planning application. It seems to me, therefore, that we must modify the *Corner House* conditions in so far as it is necessary to secure compliance with the directive, but only in so far as it is necessary to secure such compliance.

[34] Approaching the matter in that way, I do not accept Mr Findlay's submission that the appellant is not "a member of the public concerned" for the purposes of article 10a. Article 3 adds to the definitions in the directive a definition of "the public" and "the public concerned". The definitions are as follows:

4. The "public" means one or more natural legal persons and, in accordance with national legislation or practice, their associations, organisations or groups;

5. "The public concerned" means: the public affected or likely to be affected by, or having an interest in, the environmental decision-making procedures referred to in Article 2(2); for the purposes of this definition, non-governmental organisations promoting environmental protection and meeting any requirements under national law shall be deemed to have an interest.

[35] Even if the appellant is not "affected or likely to be affected by the decision to grant planning permission", he does, in my judgment, have an interest in the decision-making process in respect of applications for planning permission that affect the setting of Hampton Court Palace. In reaching that conclusion, I bear in mind that article 10a leaves member states to decide what constitutes a sufficient interest, provided that they do so "consistently with the objective for giving the public concerned wide access to justice". If the underlying objective is to give the public wide access to justice, it seems to me that the appellant at least does have an interest. I have summarised his involvement over many years with applications that affect the setting of Hampton Court Palace; these include his objection to the boathouse scheme on the ground that a redevelopment of such a scale on the site would adversely affect the setting of the palace. The appellant has explained why, having made that objection, he did not think it was necessary for him to object to the classical scheme because it was, in his view, objectionable in principle on exactly the same grounds. As the appellant said in para 13 of his second witness statement:

Thus, over the course of 13 years, leading up to the council's decision on the [classical scheme] in December 2008, I had detailed meetings and discussions with Elmbridge Borough Council, English Heritage, the Thames Landscape Strategy, Railtrack, local residents and a succession of developers and their professional advisers and have made numerous written submissions both on behalf of the HRP and, more recently, in my own name.

[36] The question of whether the appellant has a "sufficient interest" for the purposes of section 31(3) of the Senior Courts Act 1981 is one of the matters to be considered at the rolled-up hearing. I realise that Mr Bartlett refused permission to apply for judicial review on the basis that "not having sought to influence the decision" the appellant did not have sufficient standing, but it is important to bear in mind that when he did so he did not have the benefit of the fuller explanation of the appellant's involvement in earlier schemes affecting the setting of Hampton Court Palace. That is contained in the appellant's second witness statement.

[37] On the basis of that further information it is, in my judgment, at the very least arguable that the appellant does have a sufficient interest for the purposes of section 31(3). That is the only test that we have in the UK for deciding whether a member of the public should be able to access the review procedure to challenge the lawfulness of environmental decisions.

[38] It is unnecessary to consider whether there is a different test for a sufficiency of interest for the purposes of article 10a and whether such a test, if it were to be more stringent, would offend the principle of equivalence because if the question of whether the member of the public concerned does have a sufficient interest is to be resolved in the review procedure itself, as it is in the present case, the procedure for resolving that issue must be article 10a-compliant and must not be prohibitively expensive. I would add that, to the extent that there is any doubt as to whether the appellant has a sufficient interest, there are powerful reasons for permitting Keith Garner Ltd and Mr Macaulay to be joined as claimants, since the former, we are told, wrote a letter objecting to the application for planning permission that is under challenge in these proceedings and the latter is a local resident.

[39] Turning then to the two grounds on which Nicol J refused a PCO, I accept the appellant's submission that in an article 10a case there is no justification for the application of the issues of "general public importance"/"public interest requiring resolution of those issues" in the *Corner House* conditions. Both Aarhus and the directive are based on the premise that it is in the public interest that there should be effective public participation in the decision-making process in significant environmental cases (those cases that are covered by the EIA and IPPC Directives); and an important component of that public participation is that the public should be able to ensure, through an effective review procedure that is not prohibitively expensive, that such important environmental decisions are lawfully taken. In summary, under community law it is a matter of general public importance that those environmental decisions subject to the directive are taken in a lawful manner, and if there is an issue as to that, the general public interest does require that that issue be resolved in an effective review process. The *Corner House* principles are judge-made law and in accordance with the *Marleasing** principle those judge-made rules for PCOs must be interpreted and applied in such a way as to secure conformity with the directive.

[40] It follows that the judge was not entitled to reject the appellant's application for a PCO on the basis that the issues raised were not of general public importance that the public interest required to be resolved.

[41] I turn to the second reason why the judge refused to grant a PCO: that it was impossible to tell whether the proceedings would be "prohibitively expensive" unless there were detailed information concerning the appellant's resources to fund the proceedings.

[42] This raises an important issue of principle. Should the question of whether the procedure is or is not prohibitively expensive be decided on an "objective" basis, by reference to the ability of an "ordinary" member of the public to meet the potential liability for costs, or should it be decided on a "subjective" basis, by reference to the means of the particular claimant, or on some combination of the two bases?

[43] In an ideal world, I would have preferred to defer taking a decision on such an important issue of principle until after the findings of the Aarhus Convention Compliance Committee as to whether our domestic costs rules are Aarhus-compliant are published (it is anticipated that the committee's draft findings will be published within the next two months or so) and until after we know whether the European Commission will accept or reject the UK's response to the Commission's reasoned opinion, announced in a press release dated 18 March 2010, in which the Commission is contending that the UK is failing to comply with the directive because challenges to the legality of environmental decisions are prohibitively expensive.

[44] I am also mindful of the fact that we were told that an appeal to the Supreme Court is pending against the decision of Mrs Registrar di Mambro and Master O'Hare on preliminary issues that arose in the detailed assessment of the bills of costs lodged by the respondents in *Edwards* v *Environment Agency* [2006] EWCA Civ 877†. They decided to disallow any costs that would be prohibitively expensive and, in deciding whether the costs were prohibitively expensive, they adopted an objective test but said that they should also have regard to other matters, including the parties' financial resources: see [18] to [19] of the decision on the preliminary issues. Thus, it seems likely that before too long this issue of principle will have been considered at the highest level by the Aarhus Compliance Committee, the European Commission and the Supreme Court.

[45] Sadly, it is not an ideal world; it is now more than a year since the planning permission was granted, and we are still no nearer to deciding whether it was lawfully granted. The parties anticipate that the rolled-up hearing will be listed in October. This term ends tomorrow, so we must reach a decision today as to whether the judge was wrong to refuse to grant a PCO.

[46] Whether or not the proper approach to the "not prohibitively expensive requirement under Article 10a" should be a wholly objective one, I am satisfied that a purely *subjective* approach, as was applied by Nicol J, is not consistent with the objectives underlying the directive. Even if it is either permissible or necessary to have some regard to the financial circumstances of the individual claimant, the underlying purpose of the directive to ensure that members of the public concerned having a sufficient interest should have access to a review procedure that is not prohibitively expensive would be frustrated if the court were entitled to consider the matter solely by reference to the means of the claimant who happened to come forward, without having to consider whether the potential costs would be prohibitively expensive for an ordinary member of "the public concerned".

[47] In the present case, there was evidence that without a PCO the liability and costs of an unsuccessful appellant were likely to be prohibitively expensive to anyone of "ordinary" means. Although, in the event, Mr Bartlett did not award costs at the refusal of permission to apply for judicial review stage, the costs claimed by the respondents and the second interested party at that preliminary stage, when the matter was simply being considered on the papers and before there had been any hearing, had reached nearly £15,000.

[48] Following the hearing before Nicol J, the respondents applied for and were awarded £3,000 costs. It must be remembered that that was a hearing simply to determine whether there should be a PCO. At that hearing, the two interested parties did not ask for their costs, but they did not give any undertaking that they would not be claiming their costs at the rolled-up hearing that would be considering the substantive issue over one-and-a-half days.

[49] There was no evidence before the judge to suggest that the appellant's solicitor's estimate of a likely costs liability of £60,000 plus VAT if the case were not straightforward was an overestimate. If anything, the sum that was claimed by the respondents and the second interested party for their acknowledgements of service alone suggested that this figure was an underestimate. This was not a straightforward case, as evidenced by the refusal of permission to apply for judicial review, the participation of two interested parties in addition to the appellant and the respondents, the agreement that there should be a rolled-up hearing and the fact that that hearing was listed not for one day but for one-and-a-half days. Even though an unsuccessful claimant would not as a general rule be ordered to pay two sets of costs, the possibility that the appellant might have to pay all or part of the costs of one or both of the two interested parties, as well as the costs of the respondents, could not be ruled out. Clearly, the second interested party considered that there was a real prospect that it might be awarded its costs. That, after all, was the reason why it objected to the grant of a PCO. The fact that such an award might not be made as an exercise of judicial discretion was insufficiently certain for the purposes of article 10a: see *Commission* v *Ireland*.

[50] Against that background, as a matter of common sense, most "ordinary" members of the public, and very many who are much more fortunately placed, would be deterred from proceeding by a potential costs liability, including VAT, that totalled well over double the gross national average wage for a full-time employee (slightly less than £25,500 pa). There is a further aspect to the purely subjective approach that may well have the effect of deterring members of the public from

* Editor's note: Marleasing SA v La Comercial Internacional de Alimentación SA C-106/89 [1990] ECR I-4135

† Editor's note: Reported at [2007] JPL 82

A challenging the lawfulness of environmental decisions, contrary to the underlying purposes of the directive.

[51] Mr Macaulay said that he was unwilling to undergo a means test in a public forum. Applicants for public funding from the Legal Services Commission have to disclose details of their means to the Commission, but they do so in a private process; they do not have to disclose details of their means and personal affairs, for example who has an interest in the house in which they are living, how much it is worth and so forth, to the opposing parties or to the court, in documents that are publicly available and that will be discussed, unless the judge orders otherwise, in an open forum. The possibility that the judge
B might, as an exercise of judicial discretion, order that the public should be excluded while such details were considered would not provide the requisite degree of assurance that an individual's private financial affairs would not be exposed to public gaze if he dared to challenge an environmental decision.

[52] The more intrusive the investigation into the means of those who seek PCOs and the more detail that is required of them, the more likely it is that there will be a chilling effect on the willingness of ordinary members of the public (who need the protection that a PCO would afford) to challenge the lawfulness of environmental decisions.

C [53] For these reasons, the judge's approach to the prohibitively expensive issue in this case, although it was wholly consistent with *Corner House* principles, was not consistent with article 10a. On the evidence before him, including evidence as to the costs that had already been incurred and the claimant's solicitor's estimates of potential costs, a PCO was necessary if these proceedings were not to be prohibitively expensive. What figure, if any, would be appropriate, and should there be some form of reciprocal limit on the respondent's liability for the appellant's costs? So far as these questions are concerned, it seems to me that the respondents' proposals, made in open correspondence, are eminently reasonable, and I am not persuaded by the appellant's submissions that their liability under the PCO should effectively be put
D at nil and that there should be no limit on the amount that the appellant can recover from the respondents. In my judgment, the appellant cannot have it both ways. If the test is not a wholly subjective one and does incorporate an element of objectivity, the requirement in article 10a is that the review procedure shall not be prohibitively expensive, not that it shall not be prohibitively expensive for only one of the parties engaging in the review process.

[54] Although I would accept Mr Drabble's submission that *Corner House* and the reciprocal cost-cap provisions referred to therein were dealing with a different problem, I would also accept Mr Findlay's submission that the imposition of some form of reciprocal limit on a respondent's liability for costs is not necessarily inconsistent with article 10a. Thus, unlike the general public interest condition, which we E are effectively obliged to disapply, this is a feature of *Corner House* that has to be considered on a case-by-case basis.

[55] I would accept that there may well be long-term problems if the courts were consistently to impose reciprocal caps that are "modest" on successful appellants' ability to recover costs from respondents because such a course might well have the effect either that the claimant's lawyer would subsidise the process or that successful claimants would be exposed to prohibitively expensive costs claims from their own lawyers. It seems to me that these are systemic problems that we simply cannot resolve in this appeal, nor should we seek to do so given that the wider issues concerning compliance with article 10a are to be F considered elsewhere, as I have indicated.

[56] The only question, in my judgment, for this court is whether, on the facts available to us, the imposition of a reciprocal cap in the present case would be inconsistent with article 10a. I would answer that question "no". It then falls for consideration as to whether the imposition of a cap would be fair and reasonable in all the circumstances. I would not for a moment suggest that the limit on the liability of each party must necessarily be the same. The limits should properly reflect the disparity of resources, but it does seem to me that the upper limit of £35,000, suggested for the respondents' liability if the appellant is successful, is fair and proportionate and, rather more importantly, it is consistent with G a review process that must not be prohibitively expensive.

[57] Drawing the threads together, since the respondents are still taking the point against the appellant that he does not have a sufficient interest, I would allow the applications by Keith Garner Ltd and Mr Macaulay to be joined. I should mention that at the commencement of this hearing we granted them permission to appeal against the order that had been made by Nicol J.

[58] I would also grant a PCO for the three appellants. When considering what figure would be appropriate, it seems to me that there really is nothing whatsoever to suggest that a figure for the three appellants of £5,000 would be in any way a deterrent to them H continuing with these proceedings. I will therefore make an order in that sum. I would not increase it by reason of the addition of the two appellants; thus, a PCO overall of £5,000; and I would impose, by way of a reciprocal limitation on the respondents' liability, an order that they should not have to pay more than £35,000.

Lloyd LJ said:

[59] I agree.

Richards LJ said:

[60] I also agree.

Appeal allowed.

TRESPASS

Bocardo SA *v* Star Energy Onshore Ltd and another

SUPREME COURT

[2010] UKSC 35

28 July 2010

LORD HOPE, DEPUTY PRESIDENT, LORD WALKER, LORD BROWN, LORD COLLINS AND LORD CLARKE

Trespass — Damages — Petroleum extraction — Licence to search and bore for and extract petroleum granted under Petroleum (Production) Act 1934 — Mines (Working Facilities and Support) Act 1966 — Holders of petroleum licence running oil pipelines beneath land owned by another party — Landowner's permission for pipelines not obtained — Whether landowner having claim in trespass — Correct approach to assessment of damages — Whether principles relevant to compulsory acquisition applying — Appeal and cross-appeal dismissed

The respondents were the successive holders of a petroleum licence, granted under section 2 of the Petroleum (Production) Act 1934, that authorised them to search and bore for and extract petroleum from a natural reservoir of petroleum and petroleum gas. They operated three "deviated" pipelines, drilled by their predecessor, which ran diagonally down from a well-head located close to an estate owned by the appellant. The pipeline extended beneath the appellant's land at depths that varied from 800ft to 2,900ft. The appellant had not given permission for the pipelines, nor had the respondents obtained any ancillary rights in respect of them, with the attendant payment of compensation under section 8(2) of the Mines (Working Facilities and Support) Act 1966.

In 2006, the appellant, having become aware of the existence of the pipelines beneath its land, brought a claim against the respondents for trespass. It contended that the respondents, although owning the oil, had no rights over its land for the purpose of accessing it. The claim was allowed at first instance and substantial damages were awarded, representing a percentage of the value of the extracted oil: see [2008] EWHC 1756 (Ch); [2009] 1 All ER 517. The Court of Appeal upheld the decision on liability but reduced the damages award to £1,000: see [2009] EWCA Civ 579; [2010] Ch 100.

The appellant appealed on the quantum of damages; the respondents cross-appealed on liability. The issues were: (i) on liability, whether the appellant's title to the land extended down to the underground strata through which the pipelines ran; and (ii) on quantum, whether the principles relevant to compulsory acquisition cases applied to section 8(2) of the 1966 Act, including the rule in *Pointe Gourde Quarrying & Transport Co Ltd* v *Sub-Intendent of Crown Lands* [1947] AC 565 that an award of compensation for compulsory acquisition could not include an increase in the value of land that was entirely attributable to the scheme underlying the acquisition.

***Held*: The appeal and the cross-appeal were dismissed.**

(1) The old maxim that the owner of land prima facie owns everything up to the sky and down to the centre of the earth is not obsolete as a proposition of law. Although it has ceased to apply to the use of airspace above a height that might interfere with the ordinary use of land, different considerations apply to the strata beneath the surface of land. The best view is that the owner of the surface of land owns the strata below it, including any minerals to be found there, unless they have been alienated by a conveyance to another party, either at common law or by statute. Since the Crown, through which the respondents held their licence, asserted ownership only of the petroleum and not of the strata in which it was found, the appellant continued to own the strata beneath its land. If possession or a right to possession was a precondition for bringing a claim in trespass, the appellant had met that condition. In the absence of evidence to the contrary, its ownership of the paper title was sufficient for it to be deemed to be in possession of the land. (2) The respondents had no defence to a claim for trespass, either at common law or under the 1934 Act. The common law principle of non-derogation from grant did not apply because the respondents' licence had been granted by the Crown and the appellant was not a party to that arrangement. Nor did the 1934 Act afford a defence because that Act should not be construed as conferring on any party any right that it would not otherwise enjoy to enter onto or interfere with land: see section 10(3). (3) The correct measure of damages for the respondents' trespass was the sum that would have been assessed, on a referral to the court, as proper compensation to the appellant under section 8(2) of the 1966 Act for the grant of the ancillary rights that the respondents were entitled to acquire under the legislation in order to extract the petroleum. The principles relevant to compulsory acquisition apply to the assessment of compensation under section 8(2) because the grant of ancillary rights to a petroleum licence holder under the relevant legislation involves the compulsory acquisition of rights over land; compensation should therefore be assessed on principles that are similar to those applying under other compulsory purchase legislation. (4) (Lord Hope and Lord Clarke dissenting) On an application of ordinary compulsory purchase principles, the appellant was not entitled to assert and benefit from the "key" or "ransom" value of the ancillary rights without which the respondents could not unlock the value of the oil. The relevant "scheme" was the exploitation of the petroleum licence in the specified area. Any value in the ancillary rights required to facilitate that exploitation existed exclusively because of the scheme and thus fell to be disregarded under the *Pointe Gourde* principle. Any special value that applied only to a particular purchaser that had obtained powers of compulsory purchase could not be taken into consideration in setting the price because to do otherwise would be to allow the scheme's existence to enhance the value of the land to be purchased under it. The 1934 Act extinguished whatever pre-existing key value the appellant's land might be thought to have had in the open market and created a new world, in which only the Crown and its licensees had any interest in accessing the oilfield, and in which they had been empowered to do so compulsorily and thus on terms subject to the *Pointe Gourde* approach to compensation.

The following cases are referred to in this report.

Attorney-General v *Blake* [2001] 1 AC 268; [2000] 3 WLR 625; [2000] 4 All ER 385, HL

Batchelor v *Kent County Council* (1989) 59 P&CR 357; [1990] 1 EGLR 32; [1990] 14 EG 129; RVR 181; [1990] JPL 571, CA

Bernstein v *Skyviews & General Ltd* [1978] QB 479; [1977] 3 WLR 136; [1977] 2 All ER 902; [1977] 1 EGLR 96; (1977) 241 EG 917
Bocardo SA v *Star Energy UK Onshore Ltd* [2009] EWCA Civ 579; [2010] Ch 100; [2009] 3 WLR 1010; [2010] 1 All ER 26; [2009] 2 P&CR 23; [2010] RVR 127, [2008] EWHC 1756 (Ch); [2009] 1 All ER 517; [2008] 2 P&CR 23
Boehringer v *Montalto* 142 Misc 560 (1931)
Bowser v *Maclean* (1860) 2 De GF&J 415
BP Petroleum Developments Ltd v *Ryder* [1987] 2 EGLR 233; [1987] RVR 211
Bracewell v *Appleby* [1975] Ch 408; [1975] 2 WLR 282; [1975] 1 All ER 993; [1976] 1 EGLR 119; (1974) 237 EG 731
Chance v *BP Chemicals Inc* 670 NE2d 985 (Ohio 1996)
Chapman Lowry & Puttick Ltd v *Chichester District Council* (1984) 47 P&CR 674; [1984] 1 EGLR 188; 269 EG 955, LT
Commissioner for Railways v *Valuer General* [1974] AC 328; [1973] 2 WLR 1021; [1973] 3 All ER 268, PC (Aus)
Corbett v *Hill* (1870) LR 9 Eq 671, Ch
Didow v *Alberta Power Ltd* [1988] 5 WWR 606
East End Dwellings Co Ltd v *Finsbury Borough Council* [1952] AC 109; [1951] 2 All ER 587; (1951) 2 P&CR 135; 49 LGR 669, HL
Edwards v *Minister of Transport* [1964] 2 QB 134; [1964] 2 WLR 515; [1964] 1 All ER 483; (1963) 62 LGR 223; 15 P&CR 144; 188 EG 1081, CA
Elwes v *Brigg Gas Co* (1886) LR 33 Ch D 562, Ch
Experience Hendrix LLC v *PPX Enterprises Inc* [2003] EWCA Civ 323; [2003] 1 All ER (Comm) 830; [2003] FSR 46
Field Common Ltd v *Elmbridge Borough Council* [2008] EWHC 2079 (Ch); [2009] 1 P&CR 1
Gajapatiraju v *Revenue Divisional Officer, Vizagapatam; sub nom Gajapatiraju* v *Vizagapatam (Revenue Divisional Officer)* [1939] AC 302; [1939] 2 All ER 317; 55 TLR 563, PC
Horsford v *Bird* [2006] UKPC 3; [2006] 1 EGLR 75; [2006] 15 EG 136
Inland Revenue Commissioners v *Clay*; *Inland Revenue Commissioners* v *Buchanan* [1914] 3 KB 466; (1914) 83 LJKB 1425; 111 LT 484; 30 TLR 573
Inverugie Investments Ltd v *Hackett* [1995] 1 WLR 713; [1995] 3 All ER 841; [1996] 1 EGLR 149; [1996] 19 EG 124, PC
Jaggard v *Sawyer* [1995] 1 WLR 269; [1995] 2 All ER 189; [1995] 1 EGLR 146; [1995] 13 EG 132, CA
Kankakee County Board of Review v *Property Tax Appeal Board* 871 NE2d 38 (Illinois 2007)
Kelsen v *Imperial Tobacco Co (of Great Britain and Ireland) Ltd* [1957] 2 QB 334; [1957] 2 WLR 1007; [1957] 2 All ER 343
Kuwait Airways Corporation v *Iraqui Airways (Nos 4 and 5)* [2002] UKHL 19; [2002] 2 AC 883
Logan v *Scottish Water* [2005] CSIH 73; 2006 SC 178
Lucas & Chesterfield Gas & Water Board, Re [1909] 1 KB 16, CA
Lunn Poly Ltd v *Liverpool & Lancashire Properties Ltd* [2006] EWCA Civ 430; [2006] 2 EGLR 29; [2006] 26 EG 210
Mercury Communications Ltd v *London & India Dock Investments Ltd* (1993) 69 P&CR 135; [1994] 1 EGLR 229
Mitchell v *Mosley* [1914] 1 Ch 438, CA
National Carriers Ltd v *Panalpina (Northern) Ltd* [1981] AC 675; [1981] 2 WLR 45; [1981] 1 All ER 161; (1981) 43 P&CR 72, HL
NV de Bataafsche Petroleum Maatschappij v *War Damage Commission* (1956) 23 ILR 810
Orr v *Mortvedt* 735 NW2d 610 (Iowa 2007)
Pell Frischmann Engineering Ltd v *Bow Valley Iran Ltd* [2009] UKPC 45; [2010] BLR 73
Pepper (Inspector of Taxes) v *Hart* [1993] AC 593; [1992] 3 WLR 1032; [1993] 1 All ER 42, HL
Pointe Gourde Quarrying & Transport Co Ltd v *Sub-Intendent of Crown Lands* [1947] AC 565; 43 TLR 486, PC
Porter v *Secretary of State for Transport* [1996] 3 All ER 693; [1996] 2 EGLR 10; [1996] 35 EG 89; [1996] 1 PLR 111; [1996] RVR 195; [1997] JPL 635, CA
Pountney v *Clayton* (1883) LR 11 QBD 820, CA
Powell v *McFarlane* (1979) 38 P&CR 452, Ch
Pye (JA) (Oxford) Ltd v *Graham* [2002] UKHL 30; [2003] 1 AC 419; [2002] 3 WLR 221; [2002] 3 All ER 865; [2003] 1 P&CR 10
R (on the application of Spath Holme Ltd) v *Secretary of State for the Environment, Transport and the Regions; sub nom R* v *Secretary of State for the Environment, Transport and the Regions, ex parte Spath Holme Ltd* [2001] 2 AC 349; [2001] 2 WLR 15; [2001] 1 All ER 195; [2001] 1 EGLR 129; (2001) 33 HLR 31, HL
Rowbotham v *Wilson* (1860) 8 HL Cas 348, QB
Sempra Metals Ltd (formerly Metallgesellschaft Ltd) v *Commissioners of Inland Revenue* [2007] UKHL 34, [2008] 1 AC 561
Sidney v *North Eastern Railway Co* [1914] 3 KB 629, DC
South Eastern Railway Co v *London County Council* [1915] 2 Ch 252
Spirerose Ltd v *Transport for London; sub nom 64-70 Holywell Lane, London, EC2, Re; Transport for London* v *Spirerose Ltd* [2009] UKHL 44; [2009] 1 WLR 1797; [2009] 4 All ER 810; [2009] RVR 225
Stebbing v *Metropolitan Board of Works* (1870) LR 6 QB 37
Stoke-on-Trent City Council v *W&J Wass Ltd (No 1)* [1988] 1 WLR 1406; [1988] 3 All ER 394; (1988) 87 LGR 129, CA
Swordheath Properties Ltd v *Tabet* [1979] 1 WLR 285; [1979] 1 All ER 240; (1978) 37 P&CR 327; [1979] 1 EGLR 58; 249 EG 439, CA
United States v *Causby* 328 US 256 (1946)
Waters v *Welsh Development Agency; sub nom Melville* v *Welsh Development Agency* [2004] UKHL 19; [2004] 1 WLR 1304; [2004] 2 All ER 915; [2004] 2 P&CR 29; [2004] 2 EGLR 103; [2004] RVR 153
Whitwham v *Westminster Brymbo Coal & Coke Co* [1896] 2 Ch 538
Willoughby Hills v *Corrigan* 278 NE2d 658
Wrotham Park Estate Co Ltd v *Parkside Homes Ltd* [1974] 1 WLR 798; [1974] 2 All ER 321; (1973) 27 P&CR 296; 229 EG 617, Ch
WWF World Wide Fund for Nature (formerly World Wildlife Fund) v *World Wrestling Federation Entertainment Inc* [2007] EWCA Civ 286; [2008] 1 WLR 445; [2008] 1 All ER 749

This was an appeal by the appellant, Bocardo SA, from a decision of the Court of Appeal reducing the quantum of an award of damages made by Peter Smith J, sitting in the Chancery Division of the High Court, on a claim for trespass against the respondents, Star Energy Onshore Ltd and Star Energy Weald Basin Ltd, and a cross-appeal by the respondents on the issue of liability.

Jonathan Gaunt QC, Michael Beloff QC and Edward Peters (instructed by Denton Wilde Sapte LLP) appeared for the appellant; Michael Driscoll QC and Ciaran Keller (instructed by Norton Rose LLP) appeared for the respondents; James Strachan (instructed by the Treasury Solicitor) represented the Secretary of State for Energy and Climate Change, as intervener

Giving the first opinion, **Lord Hope** said:

[1] The Palmers Wood oilfield is a naturally occurring reservoir of petroleum and petroleum gas, the north-eastern part of which extends beneath the Oxted Estate of which the appellant Bocardo is the freehold owner. The rest of the oilfield lies under land in different ownerships. Petroleum cannot be recovered from an underground reservoir without the carrying out of works of some kind below the surface of the land. An oil company such as the first respondent, Star Energy Weald Basin Ltd, which has a licence under section 2 of the Petroleum (Production) Act 1934 (the 1934 Act) — now repealed and replaced by section 3 of the Petroleum Act 1998 (the 1998 Act), of which para 4 of Schedule 3 preserves pre-existing licences — to search, bore for and obtain petroleum will have to sink wells into the substratum by means of drilling in order to recover it. It may have to do this by means of wells that are drilled diagonally rather than vertically from the well-head.

[2] A particular feature of this case is that the apex of the oilfield lies beneath Bocardo's land. The most efficient means of recovering the petroleum is to sink a well as close to the apex as possible. If this is not done, and the well is sunk to a point that is substantially below the apex, much of the oil that could otherwise be recovered will be lost. It was for this reason that the respondents' predecessor sunk three wells from the well-head by what is known as deviated or directional drilling from one of the two drilling sites that were created for the extraction of petroleum from the Palmers Wood oilfield. The wells enter the substrata below the Oxted Estate at depths of around 1,300ft, 800ft and 950ft beneath the surface respectively. Two of them are known as PW5 and PW8. They are used to extract petroleum and petroleum gas from the reservoir beneath the Oxted Estate and terminate at around 2,900ft and 2,800ft below the surface of its land respectively. The third, known as PW9, passes through the substrata beneath the Oxted Estate at a depth of approximately 950ft below the surface and ends beyond its perimeter at a point in the reservoir at around 1,400ft below ground level. It is used for injecting water into the oilfield to maximise and speed recovery.

[3] The respondents' predecessor, Conoco (UK) Ltd, did not seek to negotiate any contractual licence or wayleave from Bocardo to drill the wells, lay the casing and tubing within them and extract the petroleum and petroleum gas by this means from the oilfield. Nor did it apply for any statutory right to do this under the Mines (Working Facilities

A and Support) Act 1966 (the 1966 Act) or the Pipelines Act 1962 (the 1962 Act). The respondents in their turn did not seek to do this when they acquired the petroleum production licence from their predecessor. It appears to have been assumed all along that this was not necessary. The evidence at the trial of the respondents' expert was that although deviated or directional drilling has been common industry practice for some years, he was not aware that any onshore oil company had applied for ancillary rights to permit deviated drilling on UK onshore operations. Bocardo was unaware until July 2006 that petroleum and petroleum gas was being extracted by this means from beneath its land.

[4] The issues that this case raises fall into two parts. First, there is
B the question of whether the drilling of the three wells under Bocardo's land was an actionable trespass. Peter Smith J held that it was: see [2008] EWHC 1756 (Ch); [2009] 1 All ER 517. His decision was affirmed by the Court of Appeal (Jacob, Aikens and Sullivan LJJ): see [2009] EWCA Civ 579; [2009] 3 WLR 1010; [2010] Ch 100. Second, if there was an actionable trespass, there is the question of what is the correct measure of damages. The measure that was adopted by the trial judge was rejected by the Court of Appeal, which made a substantial reduction in the award of damages. Bocardo appeals to this court on the damages issue and the respondents cross-appeal on the issue
C of trespass.

(a) Trespass

[5] On 21 July 2006, Bocardo commenced proceedings against the respondents for trespass. The question that this issue raises is whether an oil company that has been granted a licence to search, bore for and get petroleum in the licensed area that is beneath land belonging to another, and drills wells at depth beneath that land in order to recover petroleum from within the licensed area without obtaining the landholder's agreement or an ancillary right under the 1966 Act to do so, is committing a trespass. The respondents accept that if a trespass was committed by drilling the wells in the first place, it will have continued
D until now. In the Court of Appeal, Aikens LJ said that it was logical to examine the question of whether there was a trespass as at July 2000, when, having taken account of the fact that the limitation period under section 2 of the Limitation Act 1980 for a claim in trespass is six years, the cause of action arose: see [48]. However, I agree with him that nothing turns on the precise date at which the issue is considered.

[6] It is common ground that a trespass occurs when there is an unjustified intrusion by one party onto land that is in the possession of another: William Blackstone's *Commentaries on the Laws of England*, vol 3, at p209; *Clerk & Lindsell on Torts* (19th ed) 2006, in para 19.01. It is common ground, too, that Bocardo did not and does not own any
E of the petroleum in the reservoir that is situated beneath its land. Nor does it possess or have the right to possess any of that petroleum. Those rights belonged to the holder of the licence granted by the Secretary of State under section 2 of the 1934 Act, Conoco (UK) Ltd. They now belong to the respondents (the first respondent, Star Energy Weald Basin Ltd) as the original holder's assignees. By virtue of section 1 of the 1934 Act, which vested the property in petroleum existing in its natural condition in strata in Great Britain in the Crown, at no time did Bocardo have any right to search, bore for or get that petroleum from the reservoir beneath its land. Only the Crown or its licensee had the
F right to do so.

[7] The question of whether the drilling of the three wells under Bocardo's land, and the continued presence of the well-casing and tubing within them, was an actionable trespass raises the following issues as to whether: (i) Bocardo's title to the land extends to the strata below the surface through which the three wells and their casing and tubing pass; (ii) possession or a right to possession is a precondition for bringing a claim for trespass and, if so, Bocardo has or is entitled to possession of the subsurface strata through which these facilities pass; (iii) the respondents have a right under the 1934 Act (and subsequently the 1998 Act) to drill and use the three wells and their casing and tubing to extract petroleum from beneath Bocardo's land that gives them a defence to a claim in trespass.

Ownership: How far below the surface? G

[8] There is, of course, nothing new in one person carrying out works under land whose surface is in the ownership or the possession of another. Operations of that kind have been familiar since at least Roman times. They have ranged from great public works such as catacombs, on the one hand, to modest cellars for the storage of wine or other commodities, on the other. What is new is the depth at which the operations that are said to constitute a trespass in this case have been carried out. The advance of modern technology has led to the discovery of things below the surface, and the desire to obtain access to and remove them, that were unimaginable when the depths to which people could go were limited by what manual labour could achieve. H

[9] Bocardo's case is that it is trite law that a conveyance of land includes the surface and everything below it, unless there have been exceptions from the grant such as commonly occur in the case of minerals. The respondents do not dispute this proposition as a general rule that applies where the rights of the surface owner are interfered with. However, they maintain that it does not extend to the depth at which the operations were and are being carried out in this case. The minimum depth was 800ft, while for the most part the depths were greatly in excess of this. Mr Michael Driscoll QC, for the respondents, said that he accepted that, in law, the surface owner owned the substrata to some depth, but not that far. He submitted that the wells and their J tubes and casing did not interfere with or enter onto "land" in any meaningful way. Moreover, the right to search, bore for and get the petroleum was vested in the Crown. Bocardo did not own and had no right to possess the petroleum.

[10] It has often been said that prima facie the owner of the surface is entitled to the surface itself and everything below it down to the centre of the earth: see, for example, *Rowbotham* v *Wilson* (1860) 8 HL Cas 348, at p360, *per* Lord Wensleydale, *Bowser* v *Maclean* (1860) 2 De GF&J 415, at p419, *per* Lord Campbell LC, *Pountney* v *Clayton* (1883) LR 11 QBD 820, at p838, *per* Bowen LJ, *Elwes* v *Brigg Gas Co* (1886) LR 33 Ch D 562, at p568, *per* Chitty J and *National Carriers Ltd* K v *Panalpina (Northern) Ltd* [1981] AC 675, at p708, *per* Lord Russell of Killowen. The proposition that prima facie everything below the surface belongs to the surface owner is often linked to the proposition that everything above it belongs to him too: "everything up to the sky", as Sir William James VC put it in *Corbett* v *Hill* (1870) LR 9 Eq 671, at p673, or "everything under the sky" in the words of Bowen LJ in *Pountney*, at p838. In *Mitchell* v *Mosley* [1914] 1 Ch 438, at p450, Cozens Hardy MR said that the grant of the land includes the surface and all that is supra — houses, trees and the like — and everything that is infra — mines, earth and clay, etc. Agreeing with him, Swinfen Eady and Phillimore LJJ said that this was a recognised rule of law. Plainly, L the source for these remarks was the well-known Latin brocard *cuius est solum, eius est usque ad coelum et ad inferos.*

[11] The soundness of this brocard as a proposition of law was questioned in *Commissioner for Railways* v *Valuer General* [1974] AC 328. The subject of the appeal was a property in the centre of Sydney beneath which there had been extensive excavations to a depth of 40ft or more. The question was how the property was to be valued for rating purposes. The statute proceeded on the basis that it was a parcel of land that had to be valued. The commissioner said that this meant land defined only by vertical boundaries — land *usque ad coelum et ad inferos*, in other words. The valuer-general said that it was M possible to value as land only that which had a recognisable connection with the surface. Otherwise, it had to be valued as stratum, to which special provisions applied. As Lord Wilberforce explained at p351, the question that the valuer-general's argument gave rise to was whether there was a complete dichotomy between land and strata beneath it and, if so, what that dichotomy was. The statutory definition did not answer the question of how, in the context of the legislation, layers defined by horizontal boundaries were to be treated. "It is in relation to this question," he said, at p351C, "that the Latin tag *usque ad coelum et ad inferos* has been introduced and given a prominent place in argument".

[12] Lord Wilberforce did not think much of the brocard, or tag as he called it. As he explained, at p351D-E:

A It is well known that this brocard cannot be traced in the *Digest* or elsewhere in Roman Law. The first recognised appearance is in the 13th century gloss of the Bolognese Accursius upon *Digest* VIII.2.1. It appears there in the form "*cuius est solum eius esse debet usque ad coelum*" (*cf* in the law of Scotland *Stair's Institutions* II.7.7). In the form of a maxim, it only has authority at common law in so far as it has been adopted by decisions, or equivalent authority. The earliest recognition appears to be recorded in *Bury* v *Pope* [1587] Cro Eliz 118 where reference is made to its use Temp Ed I in the form "*cuius est solum, eius est summitas usque ad coelum*," but the context of this statement in the reign of Edward I has not been identified.

Then, after referring to Coke, Litt 4a, which he said contained an
B uncritical adoption of the maxim, and to Blackstone, *Commentaries* II (21st ed) 1844, c2, p18 who followed Coke, he said, at p351G:

> There are a number of examples of its use in judgments of the 19th century, by which time mineral values had drawn attention to downwards extent as well as, or more than, extent upwards. But its use, whether with reference to mineral rights, or trespass in the air space by projections, animals or wires, is imprecise and it is mainly serviceable as dispensing with analysis: *cf Pickering* v *Rudd* (1815) 4 Camp 219 and *Ellis* v *Loftus Iron Co* (1874) LR 10 CP 10. In none of these cases is there an authoritative pronouncement that "land" means the whole of the space from the centre of the earth to the heavens: so sweeping, unscientific and unpractical a doctrine is unlikely to appeal to the common law
> C mind. At most the maxim is used as a statement, imprecise enough, of the extent of the rights, prima facie, of owners of land: Bowen LJ was concerned with these rights when, in a case dealing with rights of support, he said: "Prima facie the owner of the land has everything under the sky down to the centre of the earth": *Pountney* v *Clayton* (1883) 11 QBD 820, 838.

[13] In the Court of Appeal, Aikens LJ, referring to Lord Wilberforce's remarks in *Commissioner for Railways*, said that he had no doubt that Accursius's maxim, or brocard, was not part of English law: see [59]. Asking himself what the general rule is at common law concerning the ownership of the substrata below the surface of land, he said that he found it in *Mitchell*, but shorn of its references to Accursius's maxim.
D In short, he said, the registered freehold proprietor of the surface will also be the owner of the strata beneath the surface of its land, including the entire minerals, unless there has been some express or implied alienation of the whole or a particular part of the strata to another. In his view, in [60], Bocardo's title certainly extended to the strata (other than the petroleum) to be found at the depth of the wells up to 2,800ft below the surface of the Oxted Estate. Precisely how much further into the earth's crust that ownership might go was a question that he did not need to decide. However, if it carried to the centre of the earth landowners, he said, all have a lot of neighbours.

[14] I think, with respect, that Aikens LJ was perhaps a little too
E hasty in asserting that the brocard is not part of English law. It is true that Lord Wilberforce appears to have had little enthusiasm for it. He regarded it as an excuse for dispensing with analysis. However, those remarks were made in a case in which the question was what was meant by the word "land" in the statute. He seems to have been prepared to accept it as having some relevance as a statement, imprecise though it is, of the rights, prima facie, of owners of land: see his reference to Bowen LJ's observation in *Pountney*, at pp820 and 838. Furthermore, although Aikens LJ adopted what Cozens Hardy MR said in *Mitchell*, at p450, was an accurate statement of the law if shorn of his references to Accursius's maxim, it must be acknowledged that it was by
F reference to that maxim that Cozens Hardy MR said what he did. As Lord Wilberforce pointed out, the maxim has authority at common law only in so far as it has been adopted by decisions or equivalent authority. I am inclined to think that the observations by the Court of Appeal in *Mitchell*, seen against the background of various *dicta* in the 19th century cases, including *Pountney*, measure up to that requirement. In the present context, therefore, I believe that the brocard does have something to offer us.

[15] The particular relevance of the brocard to the dispute in this case is that, taken literally, it answers Mr Driscoll's point that the wells in question were too deep for the landowner's interest in its land to be affected. If the brocard is accepted as a sound guide to what the law is, there is no stopping point. This makes it unnecessary to speculate as to how it can be applied in practice as one gets close to the earth's centre. G
The depths to which the wells in question were drilled in this case do not get anywhere near to approaching the point of absurdity. The fact that there were substances at that depth that can be reached and got by human activity is sufficient to raise the question as to who, if anybody, is the owner of the strata where they are to be found. The Crown has asserted ownership of the petroleum, but it does not assert ownership of the strata that surround it. The only plausible candidate is the registered owner of the land above, which is exactly what the brocard itself indicates. Mr Driscoll was unable to point to any contrary authority.

[16] It is perhaps worth looking more closely at the words used by the glossator. The earliest source that we have for them is the *Glossa* H
Ordinaria, which was compiled by Accursius, a professor at the University of Bologna in the 13th century. He set for himself the task of collecting and arranging a vast number of annotations to the *Digest* that had been made by his predecessors in one great work. He supplemented these with annotations of his own. For the most part at least, the authors of these annotations are not identified. The gloss that led to the brocard with which we are all familiar is not attributed to anybody. We have no means of knowing when it was first written down. Francis Lyall in "The maxim *cuius est solum* in Scots Law" [1978] JR 147, at p148, observed that the history of its development is obscure. It may have been one of Accursius's own annotations, but it seems just as likely that it was much J
older. All we can say with confidence is that it was not part of Roman law, but that it had been recognised by 1250 when the *Glossa Ordinaria* was completed.

[17] The wording of the gloss itself is instructive. Paulus, speaking of urban praedial servitudes, is quoted in the *Digest*, at 8.2.1.pr, as follows:

> *Si intercedat solum publicum vel via publica, neque itineris actusve neque altius tollendi servitutes impedit: sed immitendi protegendi prohibendi item fluminum et stillicidiorum servitutem impedit: quia coelum quod supra id solum intercedit, liberum esse debet.*

K
The words "*quia coelum*" are then glossed in this way:

> *Quia coelum. Nota. Cujus est solum ejus debet esse usque ad coelum.*

Lyall says that in later editions of the *Glossa Ordinaria* this gloss itself is noted with the comment, at p148: "*cujus solum, ejus coelum*".

[18] I think it is significant that the glossator took as his starting point the rule that applied to the underlying strata and then applied it to what took place above the surface. The context for the annotation was the proposition that although the owner may erect structures as high as it likes on the solum of land in its ownership, its freedom to do so is restricted by the praedial servitude *non altius tollendi* that protects L
its neighbour's right to light and prospect. The owner of the dominant tenement is entitled to insist that there should be no interference with the sky over its land. The assumption appears to have been that it was generally understood that the ownership of land carried with it the right to everything that lay below the surface. The point that the glossator was making, as an explanation for the praedial servitude, was that the existing rule as to what lay below (*cuius est solum*) should be (*debet esse*) applied to the airspace above it. The rule that applied to the underlying strata appears to have been of greater antiquity.

[19] The problems that a rule in these terms might give rise to as man's understanding of the earth's structure improved, airspace began M
to be used for the passage of aircraft and means were developed to penetrate deep below the surface were not, of course, obvious in the 13th century. However, the simple notion that each landowner is the proprietor of a column or cylinder of land that stretches down to the centre of the earth and upwards indefinitely into outer space is plainly no longer tenable. The earth is not flat, as the glossator may have supposed. A greater understanding of geology has taught us that most of the earth's interior, owing to extremes of pressure and temperature, is a complex and inhospitable structure that is beyond man's capacity to enter or make use of. It has been observed that anything that is drilled below a depth of around 8.7 miles or 14km would be crushed by the earth's pressure of 50,000 pounds per square inch and vaporised by a

temperature of 1,000° Fahrenheit: see John G Sprankling, "Owning the Center of the Earth" (2008) 55 UCLA Law Review 979, at p993, fn 84. As Sprankling explains, at p994, productive human activity is possible only within the shallowest portion of the earth's crust, and humans have never penetrated below it. As for that portion of it, the development of heat mining and carbon capture, storage and sequestration technologies to reduce greenhouse gas emissions, which he discusses at pp1030-1032, would be difficult to achieve if the subsurface within which it is sought to carry out these activities in the public interest were to be broken up into columns of rock owned by the surface owners.

[20] As for the position above the surface, the development of powered flight has made it impossible to apply the brocard "*usque ad coelum*" literally. In *Bernstein* v *Skyviews & General Ltd* [1978] QB 479*, Baron Bernstein failed in his claim that the defendants, who had flown over his land to take an aerial photograph of his property that they then offered to sell to him, were guilty of trespass. Griffiths J noted, at p485, that the proposition that an owner has certain rights in the airspace above its land was well established by authority. In *Kelsen* v *Imperial Tobacco Co (of Great Britain and Northern Ireland) Ltd* [1957] 2 QB 334, for example, a mandatory injunction was granted ordering the defendants to remove a sign that projected 8in over the plaintiff's property on the ground that, applying the brocard, this was a trespass. Griffiths J was willing to accept as a sound and practical rule that any incursion into airspace at a height that may interfere with the ordinary user of land was a trespass. However, he said that wholly different considerations arise when considering the passage of aircraft at a height that in no way affects the user of the land. In his judgment, at p488, the balance was best struck by restricting the rights of the owner to such height as is necessary for the ordinary use and enjoyment of its land and the structures on it, and declaring that above that height it has no greater rights in the airspace than any other member of the public.

[21] The respondents say that this analysis should be applied to subsurface ownership too. They submit that a sensible and pragmatic solution would be for each surface owner to own directly down beneath the boundaries of its land, as far as is necessary for the use and enjoyment of the surface, the buildings on the surface and any minerals that have not been excluded from its ownership by conveyance, common law or statute that lie beneath it. Mr Driscoll was unable to point to any English authority that provided direct support for this approach to the position beneath the surface. However, there is some support for it in the US. In *Boehringer* v *Montalto* 142 Misc 560 (1931), the New York Supreme Court held that a sewer laid 150ft below the surface was not included in the surface owner's title. The judge said that title above the surface was now limited to the extent to which the owner of the soil might reasonably make use of it, and that by analogy his title was not to be extended to a depth below ground beyond which the owner might reasonably make use of it.

[22] In *United States* v *Causby* 328 US 256 (1946), the US Supreme Court held that there was a taking of the respondents' property within the meaning of the Fifth Amendment by frequent and regular flights of army and navy aircraft over their land at low altitudes. However, in para 5, at p261, of the court's opinion, Douglas J said that the doctrine expressed in the brocard has no place in the modern world:

> The air is a public highway, as Congress has declared. Were that not true, every transcontinental flight would subject the operator to countless trespass suits. Common sense revolts at the idea. To recognise such private claims to the airspace would clog these highways, seriously interfere with their control and development in the public interest, and transfer into private ownership that to which only the public has a just claim.

That was a case about limitations on the absolute rights of surface owners above the surface, as was *Willoughby Hills* v *Corrigan* 278 NE2d 658, at p664 (Ohio 1972), in which the court said that the doctrine of the common law that the ownership of land extends to the periphery of the universe has no place in the modern world. However, in *Chance* v *BP Chemicals Inc* 670 NE2d 985 (Ohio 1996), the Supreme Court of Ohio took the same approach to subsurface ownership rights. In para 8, at p992, of the court's opinion, the judge said:

> we do not accept the appellants' assertion of absolute ownership of everything below the surface of their property. Just as a property owner must accept some limitations on the ownership rights extending above the surface of the property, we find that there are also limitations on property owners' subsurface rights. We therefore extend the reasoning of *Willoughby Hills*, that absolute ownership of air rights is a doctrine which "has no place in the modern world", to apply as well to ownership of subsurface rights.

The court held that some type of physical damage or interference with the use of the land must be shown for the owner to recover for a trespass and that the use of lateral migration of injection technology to dispose of refining by products below the surface did not meet this test.

[23] Sprankling, at pp991-992, points out, however, that most modern US legal texts continue to endorse the centre of the earth theory and that almost all modern cases continue to embrace it: see, for example, *Kankakee County Board of Review* v *Property Tax Appeal Board* 871 NE2d 38 (Illinois 2007) and *Orr* v *Mortvedt* 735 NW2d 610 (Iowa 2007). Addressing himself to the question, how far below the earth's surface do property rights extend, he asserts, at p1033, that the surface owner should certainly hold property rights to a portion of the subsurface. After exploring four alternative models — ownership of the entire crust, ownership based on first-in-time exploitative use, ownership for reasonable and foreseeable uses and ownership to a specified depth — he comes down in favour of a specified depth such as 1000ft, but he acknowledges that reasonable minds may differ as to the appropriate extent. The goal of his article, he said, was to ignite that debate, not to extinguish it: see at p1039.

[24] Sprankling's article suggests that the debate as to the extent of subsurface rights remains alive in the US. In Canada, Griffiths J's approach in *Bernstein* to the right to use airspace above the land was described by the Alberta Court of Appeal as most persuasive in *Didow* v *Alberta Power Ltd* [1988] 5 WWR 606, at p613. However, we were not referred to any Canadian or Australian authority that extends that approach to ownership below the surface. In Todd *The Law of Torts in New Zealand* (5th ed) 2009, p426, it is stated that it appears to be generally accepted that any intrusion into the subsoil beneath the owner's land will constitute trespass, and that there appears to be no case in the Commonwealth where a plaintiff has failed on the basis that the area of subsoil invaded was so deep that the surface occupier's possessory rights did not extend that far. In a footnote to that passage, in which Sprankling's article is referred to, the editors note that American authority points both ways: see fn 22.

[25] Coming closer to home, Jean Howell, "Subterranean Land law: Rights below the Surface of Land" (2002) 53 NILQ 268, at p270, acknowledges that it might be argued that the same test as that which Griffiths J applied in *Baron Bernstein* should be used for land below the surface. However, as she also notes, it was implicit in that case that even above the notional height at which the landowner's usable rights stop, there is not a free-for-all in the airspace above. To characterise the surface owner's rights as following technological advances as to the depth at which land can be exploited, she says, would offend against all notions of "property" whose defining quality in land is certainty. She concludes, at p285, that any intrusion into land that is not sanctioned by some countervailing property right will be a trespass, and that although the surface owner will not usually want or be able to utilise the ground below the surface, it has rights in the land that could be valuable.

[26] In my opinion, the brocard still has value in English law as encapsulating, in simple language, a proposition of law that has commanded general acceptance. It is an imperfect guide, since it has ceased to apply to the use of airspace above a height that may interfere with the ordinary user of land: see *Bernstein*. The position in Scotland may be different: see *Stair Memorial Encyclopaedia*, vol 19, *Property*, in para 198, where the question is seen as relating to the extent of ownership rather than the balancing of rights in the airspace. However, I think that the reasons for holding that the brocard has no place in the

---

* Editor's note: Also reported at [1977] 1 EGLR 96; (1977) 241 EG 917

modern world with regard to what goes on below the surface, even in England, are not by any means as compelling as they are in respect of the use of airspace. In *United States*, the US Supreme Court regarded the airspace as a public highway to which only the public had a just claim. The same cannot be said of the strata below the surface. As Aikens LJ said in the Court of Appeal, it is not helpful to try to make analogies between the rights of an owner of land with regard to the airspace above it and its rights with regard to the strata beneath the surface: see [61]. Although modern technology has found new ways of making use of it in the public interest, there is no question of it having become a public highway. The test applied in *Chance*, that some type of physical damage or interference with the use of the land must be shown, would lead to much uncertainty. It overlooks the point that, at least so far as corporeal elements such as land and the strata beneath it are concerned, the question is essentially one about ownership. As a general rule, anything that can be touched or worked must be taken to belong to someone.

[27] The better view, as the Court of Appeal recognised, in [59], is to hold that the owner of the surface is the owner of the strata beneath it, including the minerals that are to be found there, unless there has been an alienation of them, by a conveyance, at common law or by statute, to someone else. That was the view that the Court of Appeal took in *Mitchell*. Much has happened since then as the use of technology has penetrated deeper and deeper into the earth's surface. However, I see no reason why its view should not still be regarded as good law. There must obviously be some stopping point, as one reaches the point at which physical features, such as pressure and temperature, render the concept of the strata belonging to anybody so absurd as to be not worth arguing about. However, the wells that are at issue in this case, extending from around 800-2,800ft below the surface, are far from being so deep as to reach the point of absurdity. Indeed, the fact that the strata can be worked on at those depths points to the opposite conclusion.

[28] I would hold, therefore, that the appellant's title extends down to the strata through which the three wells and their casing and tubing pass.

Possession

[29] The next question is whether possession or a right to possession is a precondition for bringing a claim in trespass. The respondents maintain that possession, not ownership, is essential and that the claim should fail because the appellant is not in possession of the substrata where the wells entered the substrata at least 800ft below the surface of its land.

[30] In *Powell* v *McFarlane* (1979) 38 P&CR 452, at p470, Slade J said:

> In the absence of evidence to the contrary, the owner of land with the paper title is deemed to be in possession of the land, as being the person with the prima facie right to possession. The law will thus, without reluctance, ascribe possession either to the paper owner or to persons who can establish a title as claiming through the paper owner.

In *JA Pye (Oxford) Ltd* v *Graham* [2002] UKHL 30; [2003] 1 AC 419, in [40], Lord Browne-Wilkinson approved of this definition, making the point that, without the requisite intention, in law there can be no possession. This is highly relevant if the law is to attribute possession of land to a party that cannot establish a paper title to possession. However, in this case, the appellant has the paper title. That, in the absence of evidence to the contrary, is sufficient for it to be deemed to be in possession of the land.

[31] As Aikens LJ said in the Court of Appeal, it is difficult to say that the appellant has actual possession of the strata below the Oxted Estate because it has done nothing to reduce those strata into its actual possession: see [66]. However, he held that the appellant, as the paper title owner to the strata and all within it (other than any gold, silver, saltpetre, coal and petroleum that belong to the Crown at common law or by statute), has the prima facie right to possession of those strata so as to be deemed to be in factual possession of them. I think that he was right to conclude that this was the effect of Slade J's *dictum*. As the paper title carries with it title to the strata below the surface, the appellant must be deemed also to be in possession of the subsurface strata. There is no one else who is claiming to be in possession of those strata through the appellant as the paper owner.

Does either common law or the 1934 Act provide a defence to the claim in trespass?

[32] There remains the question of whether the respondents have a defence to the claim of trespass, either under the common law or under the statute. I think there is nothing in the argument that there is a defence at common law. It would be different, as Aikens LJ said in the Court of Appeal, in [74], if the right to extract the petroleum had been granted to the respondents by the appellant. The principle of non-derogation from grant would prevent the appellant from doing anything that would hamper the respondents' use of the strata for the purpose that both parties contemplated at the time of the grant. However, the right to search and bore for and to get the petroleum was obtained by the respondents under licence from the Crown. I do not think there is any common law principle that the respondents can invoke in that situation to regulate their position in respect of a landowner that was not a party to that arrangement.

[33] This leaves the question of whether the matter can be said to have been regulated by the statute. Section 10(3) of the 1934 Act (now repealed and re-enacted as section 9(2) of the 1998 Act) provided:

> Nothing in this Act shall be construed as conferring, or as enabling the [Secretary of State] to confer, on any person, whether acting on behalf of His Majesty or not, any right which he does not enjoy apart from this Act to enter on or interfere with land.

The respondents say that they had a right, under the licence granted under section 2(1) of the 1934 Act, "to search and bore for and get" the petroleum, to penetrate the strata under Bocardo's land and that, as a matter of ordinary language, drilling the pipelines diagonally into the substrata would not be considered as "entering on" it or as "interfering with" Bocardo's use and enjoyment of it. They were not sunk on the surface of Bocardo's land, but were justified by the statutory right to search and bore for and get the petroleum. Moreover, there were no minerals that were capable of being enjoyed as such under the surface of Bocardo's land that were entered onto or interfered with.

[34] In the Court of Appeal, Aikens LJ, albeit with some reluctance, concluded that it was impossible to say that the 1934 Act, when read with the Mines (Working Facilities and Support) Act 1923 (the 1923 Act — later replaced by the 1966 Act) and the existing common law, granted a licensee under the 1934 Act the express or implied right to bore pipelines at depth through the land of another within the licensed area in the absence of agreement or the grant of an ancillary right under those Acts: see [80] to [83]. His reasoning was based in part on the wording of section 10(3). In his opinion, the words "enter on" land were intentionally general and sufficiently broad to include entering land beneath the surface. It was also based on the provisions of section 3(1) of that Act read with section 3(2)(*b*) of the 1923 Act. The opening words of section 3(1) of the 1934 Act provided that the 1923 Act was to apply:

> for the purpose of enabling a person holding a licence under this Act to acquire such ancillary rights as may be required for the exercise of the rights granted by the licence, and shall have effect accordingly…

Section 3(2)(*b*) of the 1923 Act provided that the expression "ancillary right" in respect of minerals was to include:

> a right of air-way, shaft-way or surface or underground wayleave.

The word "wayleave", he said, was sufficiently broad to encompass a right to bore a pipe through strata as well as create and use a passage to get to and carry minerals such as coal. The wording of section 10(3) indicated that the licensee could continue to enjoy such rights as it already has to enter onto or interfere with land, but that it was not within the power of the Secretary of State to confer on it any other right to do so.

A [35] Despite Mr Driscoll's submissions to the contrary, I have not been able to detect any flaw in this reasoning. The subsurface strata through which the wells and pipelines were sunk is Bocardo's land. There is nothing in section 10(3) or the context in which it was enacted that restricts the reference to "land" in that subsection to things that happen only on the surface. In the context of a statute that is concerned with the right to search for and bore for and get petroleum existing in its natural condition in strata below ground, the words "enter on" in that subsection are apt to apply to underground workings as well as workings on the surface. The words "interfere with" are not restricted, as was suggested, to interfering with the owner's use and enjoyment of
B the land for the time being. The owner of the subsurface is entitled to say that its land is being interfered with when it is bored into by another party. Its right to object is inherent in its right of ownership of the land. It is nothing to the point that it is not making any use of it. The fact that an underground wayleave is included in the ancillary rights referred to in section 3(2)(*b*) of the 1923 Act reinforces the conclusion that is to be drawn from the provisions of the 1934 Act that a licensee that does not already enjoy a right to enter onto someone else's land needs to acquire an ancillary right from the owner of that land if it wishes to do this.

[36] For all these reasons, I would hold, in agreement with the Court of Appeal, that the respondents have trespassed onto Bocardo's land and
C that, subject to their submissions as to the amount of the damages, they have no defence to Bocardo's claim. I would dismiss the cross-appeal.

(b) Damages

[37] The parties are agreed that if damages are to be assessed on a wayleave or user basis, their measure is the price that reasonable persons in the position of the parties would have negotiated for a grant of a contractual right for the licensee to extract the oil through the substrata below the Oxted Estate using wells PW5, PW8 and PW9: see statement of facts and issues, principal issue 2, in para 2(a). It is also agreed that in assessing the price that reasonable parties would have negotiated, the negotiation must be assumed to have taken place against the relevant
D statutory background, which at the relevant date would have included the 1934 and the 1966 Acts. I gratefully adopt Lord Clarke's description of the general background and the statutory framework.

[38] Section 8(2) of the 1966 Act provides that the compensation or consideration is to be assessed on the basis of what would be fair and reasonable between a willing grantor and a willing grantee having regard to the conditions subject to which the right is or is to be granted. The word "consideration" is included in this subsection because the rights that may be granted under section 1 of the Act include the right to search for, work and take away minerals such as coal. In the present case, however, the relevant word is "compensation". This is because
E the transaction that is in issue is the acquisition of the right to sink the wells under Bocardo's land that, as Lord Brown says in [62], Bocardo had no option but to allow the respondents to do. Had it refused to grant them a wayleave, the respondents would have been able to go to the court for an order granting them the necessary ancillary rights under section 3(2)(*d*) of the 1966 Act. I agree with Lord Walker, Lord Brown and Lord Collins, for the reasons that they give, that this must be taken to be a case of compulsory acquisition. So the general principles of compulsory acquisition law must be applied, including the "value to owner" principle and the "no scheme rule" in particular: see Lord Collins, in [101] and [102].

F [39] Accordingly, an increase in value that is consequent on the scheme for which the land is being acquired must be disregarded. The basis on which compensation is awarded is the value of the land to the owner, not its value when taken by the promoter of the scheme. However, if the land has a special value because it is the key to the development of other land, that will represent part of its value to the owner that may be taken into account in the assessment of compensation in just the same way as it would if the owner were negotiating to realise its value in the open market: *Waters* v *Welsh Development Agency* [2004] UKHL 19; [2004] 1 WLR 1304*, in [64] to [65], *per* Lord Nicholls of Birkenhead.
G It would be wrong to approach bringing this element of value into account as amounting to the exercise of a right of veto, as Harman LJ suggested in *Edwards* v *Minister of Transport* [1964] 2 QB 134*, at p156; see also *Logan* v *Scottish Water* [2005] CSIH 73; 2006 SC 178, in [102], where his approach was adopted. It is a legitimate element of the value of the land to the owner, so long as it is justified by the facts of the case.

[40] What then is one to make of the facts of this case? In respectful disagreement with Lord Walker, Lord Brown and Lord Collins, I would not attribute the key value of Bocardo's land in the hypothetical negotiation that must be imagined entirely to the scheme underlying
H the acquisition by the respondents of the right to obtain access to the petroleum. It seems to me that the key to its value lies in the geographical position that it occupies on top of the apex to the reservoir. It is of course clear that after the coming into force of the 1934 Act, only the Crown or someone holding a licence from the Crown had the right to search, bore for and get the petroleum. The market that has to be envisaged was therefore a limited one. There could be only one licence holder at any one time. However, this does not mean that the respondents must be taken to be the only possible bidder in the hypothetical market for the right to obtain access to the apex of the reservoir. The scheme that the respondents devised was dictated by the position of the drilling
J sites that they had created, but it has not been suggested that it was the only way in which access could be obtained to it. I agree with that part of Lord Clarke's judgment in which he examines this question on the assumption that the *Pointe Gourde* principle applies to the assessment of compensation under section 8(2): see [158] to [163].

[41] Support for Lord Clarke's reasoning is to be found in the decision of the Lands Tribunal in *Chapman Lowry & Puttick Ltd* v *Chichester District Council* (1984) 47 P&CR 674†, to which Lord Clarke refers in [161]. Lord Walker sees that case differently: see [55]. However, I prefer Lord Clarke's interpretation of it. The approach that the tribunal took was to ask itself whether the acquiring authority's
K need for the strip of waste land as access for the land that it owned to the rear was special or peculiar to the authority. This question could not be determined unless the needs of other possible owners of the rear land were considered and taken into account. It was reasonable to assume that such hypothetical owner or developer could expect to receive precisely the same planning permission for precisely the same residential development as that obtained by the acquiring authority and subject to the same constraints in respect of the highway. From this it followed that any owner of the rear land would have precisely the same need for the waste strip as had the acquiring authority: see pp679-680.

[42] I think that exactly the same points can be made in this case. Anyone who had obtained a licence to search, bore for and get the
L petroleum under Bocardo's land would have had precisely the same need to obtain a wayleave to obtain access to it if it were not to commit a trespass. So it was not the respondents' scheme that gave the relevant strata beneath Bocardo's land its peculiar and unusual value. It was the geographical position that its land occupies above the apex of the reservoir, coupled with the fact that it was only by drilling through Bocardo's land that any licence holder could obtain access to that part of the reservoir that gives it its key value. I agree with Lord Clarke that this case is on the side of the line identified by Mann LJ in *Batchelor* v *Kent County Council* (1989) 59 P&CR 357‡, at p361, in which the land has a key value that was pre-existent to the scheme proposed by
M the respondents for their development.

[43] I do not think it would be right to take into account what Viscount Hailsham said during the second reading of the bill that became the 1934 Act, as reported in *Hansard (HL Debates)*, 19 April 1934, cols 691-692. What he said does not fall within the limited exception to the general rule that resort to *Hansard* is inadmissible that was recognised in *Pepper (Inspector of Taxes)* v *Hart* [1993] AC 593. This is available to prevent the executive from seeking to place a

---

* Editor's note: Also reported at [2004] 2 EGLR 103

* Editor's note: Also reported at (1963) 188 EG 1081

† Editor's note: Also reported at [1984] 1 EGLR 188; (1984) 269 EG 955

‡ Editor's note: Also reported at [1990] 1 EGLR 32; [1990] 14 EG 129

A meaning on words used in legislation that is different from that which ministers attributed to those words when promoting legislation in parliament; see *R* v *Secretary of State for the Environment, Transport and the Regions, ex parte Spath Holme Ltd* [2001] 2 AC 349*, at pp407-408. Viscount Hailsham was the Secretary of State for War, but he was not the promoter of the bill. That task was being undertaken by the Secretary of State for Air, the Marquess of Londonderry, in whose support Viscount Hailsham spoke when the bill was being debated. Moreover, this is not a case in which the executive is seeking to put a different meaning on words used in the 1934 Act from that which the minister attributed to those words when promoting the bill. Nor do I
B find clear and convincing support in the wording of the 1934 Act for the argument that all that was to be compensated for was the amenity value of the land. The 10% uplift in the compensation, provided for by section 3(2)(*b*) of the 1934 Act as an allowance on account of the acquisition being compulsory, appears to me to be a neutral factor, for the reasons that Lord Clarke gives in [142].

[44] On all the remaining issues, including the way the amount of the damages ought to have been quantified, I agree with Lord Clarke. I think that the trial judge went too far in applying his figure of 9% to all the oil extracted or to be extracted during the period covered by his award until the oil and gas extraction is exhausted. In my opinion, the
C sum of £621,180 plus interest that he awarded as damages was excessive because it was not restricted to the amount that was attributable to the key value of the land. I would not be averse to using his figure of 9%, so long as it was applied only to the extra amount of oil and gas that was obtained by drilling into the apex of the reservoir. If this had been a live issue, it would have been necessary to remit the case to the High Court so that it could assess the amount of the extra value and complete the exercise of calculating, on this much more limited basis, the amount of the damages.

[45] Having dismissed the cross-appeal on the trespass issue, I would allow the appeal and remit the issue of damages to the High Court.

D Giving the second opinion, **Lord Walker** said:

[46] I agree with the judgment of Lord Hope on the trespass issue and with that of Lord Brown on the damages issue. What follows should be read as no more than footnotes to Lord Brown's judgment.

[47] It is common ground (see para 2.2 of the statement of issues) that if damages are to be assessed on a wayleave basis, the measure of damages is the price that reasonable persons in the position of the parties would have negotiated for a grant of the appropriate contractual rights, against the statutory background of the 1934 and 1966 Acts. I am inclined to think that that starting point might have been open to argument on the lines indicated in the comprehensive and scholarly
E judgment of Warren J in *Field Common Ltd* v *Elmbridge Borough Council* [2008] EWHC 2079 (Ch); [2009] 1 P&CR 1, in [55] to [99]. However, I put that aside.

[48] The starting point, therefore, is (in the words of section 8(2) of the 1966 Act):

> what would be fair and reasonable between a willing grantor and a willing grantee, having regard to the conditions subject to which the right is or is to be granted.

The statute requires the adjudicator to predict the outcome of a hypothetical negotiation between willing negotiators that reaches
F a concluded agreement. In my opinion, this statute (in conjunction with section 3 of the 1934 Act) is plainly concerned with compulsory acquisition of rights in or over land. Indeed, section 3(2) of the 1934 Act (requiring an uplift of at least 10%) says as much. I cannot accept the appellant's submission that there is a fundamental distinction between a statute that gives a public authority a right to acquire property and one that regulates property rights between private parties. The whole law of compulsory purchase began and developed with infrastructure projects (first canals, then railways) undertaken by companies in the private sector.

[49] The oldest of the statutory formulae was in section 63 of the G
Land Clauses Consolidation Act 1845 (the 1845 Act). It was also the simplest:

> … The value of the land to be purchased or taken by the promoters..

(There was also compensation for injurious affection.) There was then a long period of judicial interpretation of this simple phrase, resulting in the firmly established "value to the owner" principle. This was explained by Lord Collins in *Spirerose Ltd (in administration)* v *Transport for London* [2009] UKHL 44; [2009] 1 WLR 1797, in [119] to [129], in his exposition of the *Pointe Gourde* principle: see *Pointe Gourde Quarrying & Transport Co Ltd* v *Sub-Intendent of* H
Crown Lands [1947] AC 565. It is a principle of statutory construction, and in *Spirerose* I suggested (in [24]) that the principle's vigour is now channelled and restrained by a much more complex statutory scheme (especially sections 14 to 17 of and Schedule 1 to the Land Compensation Act 1961 (the 1961 Act)).

[50] One way of looking at the principle is to see it as an answer (at least in part) to the question: in the hypothetical negotiation called for by the statute, how far are the actual purpose and circumstances of the compulsory purchase to be taken into account? The principle tells us that compensation "cannot include an increase in value which is *entirely* due to the scheme underlying the acquisition": J
Lord MacDermott in *Pointe Gourde*, at p572, emphasis supplied. Similarly, Lord Nicholls in *Waters* v *Welsh Development Agency* [2004] UKHL 19; [2004] 1 WLR 1304, in [18], referred to the disregard of "enhancement in the value of land attributable *solely* to the particular purpose for which it is being compulsorily acquired" (emphasis supplied).

[51] This can be summarised, with some loss of precision, by saying that the hypothetical negotiation takes place in a "no-scheme world". Statutory hypotheses are notoriously troublesome. In *East End Dwellings Co Ltd* v *Finsbury Borough Council* [1952] AC 109, the issue was whether a hypothetically rebuilt block of flats would have K
been subject to the Rent Restriction Acts. Lord Asquith of Bishopstone said, at pp132-133, a passage that has since been cited in many different contexts:

> If you are bidden to treat an imaginary state of affairs as real, you must surely, unless prohibited from doing so, also imagine as real the consequences and incidents which, if the putative state of affairs had in fact existed, must inevitably have flowed from or accompanied it. One of these in this case is emancipation from the 1939 level of rents. The statute says that you must imagine a certain state of affairs; it does not say that having done so, you must cause or permit your imagination to boggle when it comes to the inevitable corollaries of that state of affairs. L

However, the need for the case to go to the House of Lords suggests that there was room for argument over what were the "inevitable corollaries" of the hypothesis.

[52] There is no difficulty about extreme cases such as *Stebbing* v *Metropolitan Board of Works* (1870) LR 6 QB 37, the graveyards case. The hypothetical sale of the graveyards was not to take place in a world in which hundreds of graves containing human remains (whose presence precluded any normal development) were to be exhumed so that a road could be built. The enhancement in value was entirely due to the road-building scheme. However, M
other cases are more difficult. One such case is the so-called *Indian* case [1939] AC 302: *Gajapatiraju* v *Revenue Divisional Officer, Vizagapatam* [1939] AC 302. It is difficult partly because of the unusual facts (the proposed anti-malarial works consisted of closing unhealthy wells that supplied several different villages, and the spring was seen as a source of fresh water both for the new harbour undertaking and for the villages) and partly because of Lord Romer's rather discursive discussion of an imaginary auction. I agree with Lord Brown that it may not be necessary or helpful to refer to the *Indian* case again.

[53] Another case that illustrates the tangles that hypothesis can lead to is *Porter* v *Secretary of State for Transport* [1996] 3 All ER

* Editor's note: Also reported at [2001] 1 EGLR 129

A 693*, considered by Lord Collins in *Spirerose*, in [115] to [118]. The no-scheme rule required the actual proposal for a bypass round Evesham to be disregarded, but because the town really did need a bypass somewhere the valuation was made on the basis that a relief road would be built on another route. So the court posited a world in which the scheme actually proposed did not exist, but an imaginary scheme did. This convoluted approach was altered for the future by amending legislation.

[54] The case now before the court is on very unusual facts, but its unusual facts do not to my mind make it more difficult. It is an extreme case, in that Star Energy's operations did not (in the words of
B Peter Smith J) interfere "one iota" with Bocardo's enjoyment of its land. Subject to a contrary argument put forward by Bocardo, the value of the right granted to Star Energy was entirely due to the scheme, which was (as Aikens LJ said in the Court of Appeal, in [111]) "the exploitation of the petroleum licence in the specified area".

[55] The contrary argument (picking up the well-known observations of Mann LJ in *Batchelor* v *Kent County Council* (1989) 59 P&CR 357, at p361) is that the value was not "entirely due to the scheme underlying the acquisition" but was pre-existent. It is true that the natural petroleum was pre-existent. It had been there no doubt for tens of thousands of years. However, the petroleum did not belong to Bocardo. It is true
C that Bocardo held a key (not, I think, the only possible key) to the most efficient exploitation of the petroleum by diagonal drilling to the apex of the oilfield. However, the key's value depended entirely on the scheme, unlike a ransom strip for which there might have been a variety of possibilities of profitable realisation, some not involving compulsory purchase, as in *Chapman Lowry & Puttick Ltd* v *Chichester District Council* (1984) 47 P&CR 674.

[56] For these reasons, and for the fuller reasons in the judgment of Lord Brown, I would dismiss this appeal.

Giving the third opinion, **Lord Brown** said:

[57] What sum would the court have assessed as being the proper
D compensation to be paid by Star to secure its right to install deviated wells or pipelines beneath Bocardo's land had Star sought to enforce that right pursuant to the 1966 Act? Agreeing as I do with Lord Hope's judgment on all the issues raised by Star's cross-appeal on liability, and concerned as I am to address only the issues arising on Bocardo's appeal as to damages, that is what I regard as the ultimate question for the court's determination. For this purpose, I shall take as my starting point the scenario described in the following four paragraphs (based partly on a somewhat simplified account of the facts and partly on what I understand to be common ground between the parties as to the proper measure of damages for trespass in this particular case given that
E Bocardo succeed on all issues of liability).

[58] Pursuant to section 2 of the 1934 Act, Star held a licence issued by the Secretary of State on behalf of the Crown giving it the exclusive right to search and bore for and get the petroleum lying underground (the property in which section 1 of the 1934 Act had vested in the Crown) in a part of Surrey including the Palmers Wood oilfield. Under the licence, Star is required to pay royalties to the Crown equal to 5% of the market value of the petroleum won (potentially rising, depending on the amount won, to 12.5%). The apex of this oilfield lies at a depth of some 2,800ft below ground within the Oxted Estate, land in Bocardo's freehold ownership.

F [59] To win the petroleum, Star needed to drill and install three pipelines, two (PW5 and PW8) down towards the apex, one (PW9) so as to inject water into a different part of the oilfield (not within the Oxted Estate) to maximise petroleum recovery. These three pipelines were drilled diagonally from a site (known as the Coney Hill well-head) outside Bocardo's Oxted Estate: PW5 entering the estate at around 1,300ft below ground level, running for around 500m and terminating at around 2,900ft below; PW8 entering at around 800ft below, running for around 700m and terminating at around 2,800ft below; PW9 entering at around 950ft below, crossing a corner of the estate and exiting deeper still after around 250m. The pipelines are variously of G 8.5in and 12.25in diameter and lined with steel casing. Their drilling and installation occasioned no harm whatsoever to the estate. It did not interfere with Bocardo's use or enjoyment of its land "one iota".

[60] Pursuant to section 3 of the 1934 Act, Part I of the 1923 Act applied, and on the replacement of the 1923 Act, as amended by the 1966 Act, the 1966 Act applied — to enable Star as licence holder to acquire such ancillary rights as it required in order to win the petroleum. The ancillary right that Star required was or was akin to that described in section 2(1)(*b*) of the 1966 Act, as amended by section 27 of the Petroleum Act 1987, as "a right of... shaft-way... or underground way-leave, or other right for the purpose of access to or conveyance H of minerals".

[61] Section 3(2)(*d*) of the 1966 Act (replacing section 4(1)(*d*) of the 1923 Act) provided that had Bocardo unreasonably refused to grant Star such ancillary right or demanded unreasonable terms for its grant, Star was entitled to ask the minister to refer the matter to the court both for it to grant the right and to assess the compensation or consideration payable for it under section 8(2) of the 1966 Act (section 9(2) of the 1923 Act). Section 8(2) provides, so far as material in the present context:

> The compensation or consideration in respect of any right... shall be assessed J by the court on the basis of what would be fair and reasonable between a willing grantor and a willing grantee...

By section 3(2)(*b*) of the 1934 Act, it was provided that:

> in determining the amount of any compensation to be paid in respect of the grant of any right, an additional allowance of not less than 10 per cent shall be made on account of the acquisition of the right being compulsory.

[62] In the light of those basic facts and those governing statutory provisions, I return to the question that I posed at the outset: what sum should the court have assessed as being proper compensation to be paid to Bocardo for having no option but to allow Star to install its pipelines K under Bocardo's land?

[63] The answer to that question — ultimately determinative of this appeal — must in turn depend on the answers to two other fundamental questions. First, do the principles ordinarily governing the approach to valuation in the field of compulsory land purchase apply equally to the construction and application of section 8(2) of the 1966 Act with regard to the compulsory acquisition of ancillary rights over (or, as here, under) land? Second, even assuming (contrary to Bocardo's argument) that compulsory purchase principles *do* apply in this context, do they operate to deny Bocardo what would otherwise be regarded as the powerful bargaining position of a landowner able to control access to a L valuable oilfield partially sited beneath its land?

[64] Bocardo's core argument is to be found in its printed case (in para 48) as follows:

> This is a classic "key" case. The second party does not own the treasure but he does own the key to the treasure chest. The key has little or no intrinsic value. Its value is what it gives access to. What the owner of the key has is purely a bargaining position. He is in the position of the owner of land which is needed to give the access necessary for the exploitation of a valuable asset.

Bocardo then contends (in para 53) that there being no direct comparables, in order to determine a fair and reasonable price "[i]t M was accordingly necessary to approach the valuation by enquiring what would be a fair share of the spoils for the landowner to receive for granting a right of access to the oil deposit".

[65] It was an acceptance of essentially this argument that led Peter Smith J at first instance to assess Bocardo's damages at £621,180 (being 9% of the gross revenue from the oil extracted during the relevant period up to trial) plus interest (together with 9% of all future revenue derived from the pipelines — the price of suspending an injunction otherwise imposed in respect of their further use). The Court of Appeal by contrast held that ordinary compulsory purchase principles apply to the assessment of compensation under section 8(2) and that pursuant to these Bocardo had no key value to exploit, was suffering no loss

* Editor's note: Also reported at [1996] 2 EGLR 101; [1996] 35 EG 89

A whatever and would be amply compensated by an award of £1,000 (to include the 10% uplift under section 3(2)(*b*) of the 1934 Act) in respect of both past and continuing trespass.

[66] I turn then to the first of the two underlying questions earlier identified: do compulsory purchase principles of valuation apply to section 8(2)? First and foremost of these principles is what is commonly known as the "no-scheme rule" or the *Pointe Gourde* rule. This rule was stated by Lord MacDermott in *Pointe Gourde Quarrying & Transport Co Ltd* v *Sub-Intendent of Crown Lands* [1947] AC 565, at p572, as follows:

> It is well settled that compensation for the compulsory acquisition of land B cannot include an increase in value which is entirely due to the scheme underlying the acquisition.

[67] The two most authoritative recent decisions on the proper approach to compensation for compulsory purchase are *Waters* v *Welsh Development Agency* [2004] UKHL 19; [2004] 1 WLR 1304 and *Spirerose Ltd (in administration)* v *Transport for London* [2009] UKHL 44; [2009] 1 WLR 1797. *Waters* was concerned principally with the correct identification of the extent of the scheme whose effect in increasing the value of the land is to be disregarded; *Spirerose* was concerned rather with the value of the acquired land's pre-existing C potential for development and more particularly with whether that has to be discounted for future uncertainties.

[68] Although it is unnecessary to traverse again most of the ground covered by those cases, it is important to note, first, Lord Nicholls of Birkenhead's description of the *Pointe Gourde* principle (in [42] of *Waters*) as "no more than the name given to one aspect of the long established 'value to the owner' principle"; second, Lord Walker's observation (in [12] of *Spirerose*) that the *Pointe Gourde* principle "is essentially concerned with statutory construction… not… with the meaning of a particular word or phrase which has appeared in a succession of statutes dealing with the same subject-matter, but with the D general attitude and expectation with which the court should approach a statute dealing with compensation for the compulsory acquisition of land [operating, as it is put in *Bennion, Statutory Interpretation* (5th ed) 2008, p442, as a "special interpretative convention"]"; and, third, as was clearly held by the majority in *Waters,* that it is the *Pointe Gourde* principle as explained in the cases, rather than the statutory rules for assessing compensation contained in section 5 of the 1961 Act (rules originally enacted in section 2 of the Acquisition of Land (Assessment of Compensation) Act 1919 (the 1919 Act)), that nowadays principally governs the approach to compensation in compulsory purchase cases, although it none the less seems to me also worth noting the terms of E section 5(3) of the 1961 Act (as amended by Schedule 15(1) of the Planning and Compensation Act 1991):

> The special suitability or adaptability of the land for any purpose shall not be taken into account if that purpose is a purpose to which it could be applied only in pursuance of statutory powers, or for which there is no market apart from the… requirements of any authority possessing compulsory purchase powers.

[69] The policy underlying the principle is of course that identified by Lord Nicholls in *Waters*, in [18]:

> When granting a power to acquire land compulsorily for a particular purpose Parliament cannot have intended thereby to *increase* the value of the subject F land. Parliament cannot have intended that the acquiring authority should pay as compensation a larger amount than the owner could reasonably have obtained for his land in the absence of the power. For the same reason there should also be disregarded the "special want" of an acquiring authority for a particular site which arises from the authority having been authorised to acquire it.

As Lord Nicholls then added, in [19]:

> This approach is encapsulated in the time-hallowed pithy, if imprecise, phrase that value in this context means value to the owner, not value to the purchaser.

[70] Bocardo contends that section 8(2) is simply not subject to the compulsory purchase principles of valuation exemplified by the *Pointe Gourde* rule. There is no mention either in the 1923 Act or the 1966 Act of the statutory rules governing the approach to compensation G to be found in the 1919 Act or the 1961 Act, in contrast to a number of other statutes similarly conferring compulsory powers to acquire subterranean land or rights: notably, the Channel Tunnel Act 1987, the Water Resources Act 1991, the Electricity Act 1989, the 1962 Act and the Gas Act 1986. Instead, there is in section 8(2) an unadorned provision for fair and reasonable compensation as between a willing grantor and a willing grantee without even an entitlement to seek the court's assessment unless and until it is shown to be not reasonably practicable to obtain the right by private arrangement because "the person with power to grant the right unreasonably refuses to grant it or demands terms which, having regard to the circumstances, are H unreasonable": section 3(2)(*d*) of the 1966 Act. Accordingly, submits Bocardo, it is in no worse negotiating position under the 1966 Act than it would be at common law — indeed better placed because, under the Act, a willing seller cannot merely hold out for a price that properly reflects the value of the right to the purchaser but can also act so as to delay the acquisition of the right and (pursuant to section 3(2)(*c*) of the 1934 Act) put the purchaser to considerable expense by way of irrecoverable costs. Moreover, of course, the seller gets a minimum 10% uplift.

[71] Ably though these arguments were presented, for my part I cannot accept them. That the present context is one of compulsory J acquisition of rights over land seems to me indisputable. How, indeed, could this be more clearly demonstrated than by the express requirement, under section 3(2)(*b*) of the 1934 Act, for a 10% or greater uplift in compensation "on account of the acquisition of the right being compulsory"? Quite why the 1923 Act (and, in turn, the 1966 Act) does not incorporate the statutory rules contained in the general land compensation legislation is unclear, but it may be because the 1923 Act (and the 1966 Act) provides not only (as is directly relevant here) for "compensation" for rights over land to win minerals not in the landowner's ownership but also for "consideration", for example for the working of coal whereby the property in the mineral passes from the grantor to the grantee and so calls for a valuation of that property right K on an ordinary commercial basis. For the life of me, however, I can think of no sound reason why parliament in 1934 should have intended an ancillary right of the kind under consideration here to be valued on a different and altogether more generous basis than comparable rights acquired under general compulsory purchase powers or, indeed, under the several Acts mentioned above. Quite the contrary. A strongly arguable case in fact arises here for saying that Star's diagonal wells are actually to be regarded as pipelines within the meaning of the 1962 Act (in which event the ordinary approach to compensation for compulsory acquisition most obviously applies). Like the Court of Appeal, I think it unnecessary to reach a conclusion on the argument. L However, the mere fact that it arises to my mind underlines the oddity of the proposition that an entirely different compensation requirement exists for the ancillary right here in question depending on whether it is enforceable under the 1923 and 1966 legislation, or under the 1962 Act.

[72] It would seem to me most odd had parliament in 1934 expropriated with no compensation whatever the property in all subterranean petroleum together with the exclusive right to search, bore for and win it, and yet at the same time intended that the landowner, beneath whose land the Crown's licence holder must necessarily bore to achieve the statutory purpose of maximising petroleum recovery in M the national interest, should thereby become entitled to "a fair share of the spoils" as the appellants contend. Moreover, to my mind such a supposition becomes more bizarre still when one recognises that parliament expressly stipulated for not less than a 10% uplift in the statutory compensation payable on account of the landowner being powerless to deny the licence holder the ancillary right it requires. Why would parliament both allow it to exploit its bargaining position for all the world as though the parties' negotiation was taking place in a routine commercial context beyond the reach of legislation and then add upwards of 10% simply because it cannot, at the end of the day, refuse to grant the ancillary right required and is unable to charge for it more than is fair and reasonable? Is the licence holder, as Peter Smith J

held, really required to pay not merely a 5% (rising to 12.5%) royalty to the Crown but in addition compensation amounting to a further 9% of the value of the petroleum won in order to be able to avail itself of its statutory entitlement to win the petroleum?

[73] This issue cannot be resolved by reference simply to the language of section 8(2): what is fair and reasonable compensation as between a willing grantor and a willing grantee must inevitably depend on whether the willing grantee is or is not entitled in the notional negotiation between the parties to exploit the position it would be in but for the grant of compulsory purchase powers to deny the licence holder access to the petroleum it is statutorily empowered to win. It depends, in short, on whether the court construing section 8(2) should approach it with the same general attitude and expectation as ordinarily it brings to the construction of statutory provisions dealing with compensation for compulsory land acquisition. If so, the *Pointe Gourde* principle applies: the landowner's compensation should not be assessed at more than it could reasonably have attained for the grant of the ancillary right had the licence holder not enjoyed a statutory power to acquire it compulsorily for a particular purpose (plus of course upwards of 10%).

[74] I recognise of course that the word "value" (which had appeared in section 63 of the 1845 Act) is not to be found in section 8(2). However, as Lord Walker observed in *Spirerose* (see [12] above), the *Pointe Gourde* approach is not dependent on "a particular word or phrase" but rather on the correct approach to statutory construction in this particular context. If the court is to construe section 8(2) consistently with other legislative provisions governing compulsory acquisition, it falls to be approached on the basis that what is fair and reasonable depends not on what the grantee is gaining but rather on what the grantor is losing. It is for this reason too that the wealth of authority concerning damages for trespass ("user damages") — damages "measured by the benefit received by the trespasser, namely by his use of the land", as Lord Nicholls put it in *Attorney-General* v *Blake* [2001] 1 AC 268, at p278F — seems to me of no assistance in the present case, Bocardo having conceded throughout that its entitlement to damages can be no more than it would have received as compensation under section 8(2).

[75] As already indicated, it seems clear to me that parliament in 1934 must have intended compensation under the 1923 Act to be assessed on similar principles to the assessment of compensation under other compulsory purchase legislation (save only that there was to be added the 10% or greater uplift payable for such ancillary rights as were required to win petroleum, notwithstanding that, by section 2(1) of the 1919 Act, parliament had abolished the 10% addition for compulsory purchase that had earlier characterised compensation awards).

[76] Were, however, parliament's intention in 1934 to be unclear and resort to be had to the *Hansard* reports of the day — as Bocardo itself prayed in aid the speech of the Marquess of Londonderry, the Secretary of State for Air and the minister promoting the bill, in support of its contention (which indeed I accept) that section 10(3) was inserted into the Act "to remove any possible doubt" as to whether a licence holder wishing to enter onto or interfere with land needed to obtain an ancillary right to do so under the 1923 Act; clearly it did — it could only serve to support my understanding of the position with regard to the intended basis of compensation. It is sufficient for this purpose to set out a passage from the speech of Viscount Hailsham, the Secretary of State for War, supporting the bill on its second reading in the House of Lords: see *Hansard (HL Debates)* 19 April 1934, cols 691-692:

> Now it is said… "You are not compensating the owners for the value of the oil which is under their land." It is quite true we are not. We are not compensating them for the value of the oil that is under their land, or, for the matter of that, for the value of oil which is under their neighbour's land. But it is fair to remember that at this moment and that is one reason why the Bill ought to be introduced and passed, as we think, at this stage there is no value in the oil under their land, or under their neighbour's land. In the three cases in which licences have been granted and are being worked [licences under the Petroleum (Production) Act 1918 pursuant to which landowners were demanding royalty payments for the right to drill for oil under their land] we have been careful to exclude those areas altogether from the provisions of the Bill, because we recognise that, in accordance with our principles it would not be right to say that, where vested interests have been created and there is a chance of land having appreciated by the possibility of oil being found there, that value should be taken away without due compensation being given.
>
> In the cases with which this Bill deals — the rest of the country — there is no value at all today in the possible oil rights, in the chances of finding oil under the soil. But we have been careful to provide that where in any particular place arrangements are made, or asked for, for the sinking of wells or for bore-holes, or in any way interfering with the actual rights that exist, interfering with the surface rights, there shall be paid not merely full compensation in the sense of the full market value, not merely full compensation for any loss of amenity value, but in addition to that it is expressly provided in the Bill that there shall be an addition of 10% because the acquisition is a compulsory one and the owner may not necessarily desire to realise that asset. So that we are careful to give full compensation in every case in which any valuable right is interfered with. All we do is to say before there is any value established, before any vested right is created, that the oil if it exists — which nobody knows — shall belong to the State in future; but that any interference with the rights of property on the soil, or with the value of the property under which the oil is situated, shall be fully compensated for to the owner whose property is interfered with.

[77] It seems to my mind perfectly clear that the compensation — and the only compensation — contemplated by parliament in enacting the 1934 Act was for "any loss of amenity value" consequent on interference "with actual rights that exist", in particular "the surface rights". For any such loss "full compensation in the sense of the full market value" was to be paid, plus 10% because "the owner may not necessarily desire to realise that asset". Compensation was to be for interference with "any valuable right". Landowners, however, had no right at all in the oil or "in the chances of finding oil under the soil". It seems clear that parliament in 1934 was not contemplating the boring of deep wells diagonally beneath land, but that had it done so, it would not have regarded that as an interference with any actual existing right or as involving any loss of amenity value or at any rate not such an interference as required more than essentially nominal compensation.

[78] I pass now to the second of the two fundamental questions arising: even supposing ordinary compulsory purchase principles apply to the assessment of compensation under section 8(2), can Bocardo nevertheless assert and benefit from the key value of the ancillary right that Star needed to acquire here? Bocardo submits that its control over the necessary right of passage of wells through its land is in principle indistinguishable from the ownership of a ransom strip of land giving necessary access to other land: in the latter case, without acquisition of the ransom strip, the second plot is landlocked; here, without acquisition of the required ancillary right of passage, the petroleum is earth-locked.

[79] For my part, I readily acknowledge the closeness of the parallel between the two situations. However, to point to the parallel is by no means to answer the question arising. It merely invites the posing of the question in another form: supposing a 1934 Act licence holder needs to acquire a strip of wasteland to be able to exploit its statutory right to search, bore for and get petroleum, would the owner of that strip be entitled to its key value?

[80] It is convenient at this stage to return to *Waters* to see what Lord Nicholls said there under the heading "Ransom value":

> 64 One last point should be noted before returning to the present case. This concerns so-called "ransom" value or, less pejoratively, "key" value. I have already mentioned that under the "value to the owner" principle or the *Pointe Gourde* principle, whichever nomenclature is preferred, the pressing need of an acquiring authority for the subject land as part of a scheme should be disregarded when assessing its value for compensation purposes. The value of the land is not the price a "driven" buyer would be prepared to pay. But a strip of land may have special value if it is the key to the development of other land. In that event this feature of the land represents part of its value as much for purposes of compensation as on an actual sale in the open market.
>
> 65. The intersection of these two principles was identified neatly by Mann LJ in *Batchelor* v *Kent County Council* (1989) 59 P&CR 357, 361:
>
> "If a premium value is 'entirely due to the scheme underlying the acquisition' then it must be disregarded. If it was pre-existent to the [scheme] it must in my judgment be regarded. To ignore the pre-existent value would be to expropriate it without compensation and would be to contravene the fundamental principle of equivalence… ."

66. In the present case the claimants contend their land had key value because of its importance as compensatory wetlands required for completion of the Cardiff Bay barrage project. Whether this contention is well-founded for compensation purposes depends, in accordance with the principle enunciated by Mann LJ, on the ambit of the scheme of which the subject land's acquisition was an integral part.

[81] In reality, Bocardo is advancing here essentially the same argument as was advanced by the owners of the wetlands in *Waters* — which I there identified (in [153]) as their second argument and (in [156]) rejected, observing that: "If correct, it would emasculate the no-scheme rule to the point of extinction." Like Lord Nicholls, I too (in [157]) cited Mann LJ's judgment in *Batchelor* and (in [158]) concluded:

> Assuming, however, that any premium value, or indeed any other particular value, of the land *were* "entirely due to the scheme underlying the acquisition" (or, if one prefers Lord Nicholls's formulation in *Director of Buildings and Lands* v *Shun Fung Ironworks Ltd* [1995] 2 AC 111, 136, due to the 'very scheme of which the [acquisition] forms an integral part'), then in my judgment, notwithstanding that it represents the land's "unrealised potentiality"..., it clearly falls to be disregarded.

[82] To my mind, there can be no doubt as to what constitutes the scheme in the present case: the Court of Appeal (in [111]) correctly identified it as: "The exploitation of the petroleum licence in the specified area." Nor can there be any doubt that, whatever particular value existed in the ancillary right here required to facilitate that exploitation (any "premium" or "key" or "ransom" value), it existed exclusively ("entirely" or "solely" are other words used in this context) because of the scheme. But for the scheme, there was no potential use or value whatever in the right being granted. It thus fell to be disregarded under the *Pointe Gourde* principle as indeed to my mind, had it been a ransom strip of land, it would no less obviously have fallen to be disregarded, whether under that principle or under section 5(3) of the 1961 Act: the purpose served by the suitability of such land for providing access could only have been achieved in pursuance of statutory powers, there being no market for such right of access apart from the requirements of the statutorily empowered licence holder.

[83] To my mind, it is impossible to characterise the key value in the ancillary right being granted here as "pre-existent" to the scheme. There is of course always the chance that a statutory body with compulsory purchase powers may need to acquire land or rights over land to accomplish a statutory purpose for which these powers have been accorded to it. However, that does not mean that on the materialisation of such a scheme, the "key" value of the land or rights that now *are* required is to be regarded as "pre-existent". This is well illustrated by Fletcher Moulton LJ's judgment in *Re Lucas & Chesterfield Gas & Water Board* [1909] 1 KB 16, where land had been compulsorily purchased for the construction of a reservoir. Having stated (at pp29-30) the "absolute rule" that the landowner "is only to receive compensation based upon the market value of his lands as they stood before the scheme was authorized by which they are put to public uses", the lord justice turned to consider the question of the "special adaptability" of land "for purposes for which lands are required only when used for works of public utility", and continued, at pp30-31:

> Ought the owner to be entitled to higher compensation by reason of the, to him, useless peculiarities which the lands possess? No better example of the problem could be found than that which we have in the present case. The land in question is by its position and conformation marked out as a favourable site for an impounding reservoir to collect water for the public supply of a district. The peculiarities which make it suitable for that purpose add nothing to its value as agricultural or grazing land, which I will assume to be the only alternative uses. A public authority obtains powers to take it for a reservoir; ought it to pay any higher price than is represented by its agricultural or grazing value? Is not any price in excess of this a violation of the canon that you are only to give that which represents its worth to the seller, and that you are to disregard all questions of its worth to the buyer?
>
> The decided cases seem to me to have hit upon the correct solution of this problem. To my mind they lay down the principle that where the special value exists only for the particular purchaser who has obtained powers of compulsory purchase it cannot be taken into consideration in fixing the price, because to do otherwise would be to allow the existence of the scheme to enhance the value of the lands to be purchased under it. But when the special value exists also for other possible purchasers, so that there is, so to speak, a market, real though limited, in which that special value goes towards fixing the market price, the owner is entitled to have this element of value taken into consideration, just as he would be entitled to have the fertility or the aspect of a piece of land capable of being used for agricultural purposes.

Towards the end of his judgment, at p35, Fletcher Moulton LJ concluded:

> The scheme which authorises the new reservoir only entitles the owner of the land to receive as compensation the value of the land unenhanced by that scheme, and, unless its situation and peculiarities create a market for it as a reservoir site for which other possible bidders exist, I do not think that the single possible purchaser that has obtained parliamentary powers can be made to pay a price based on special suitability merely by reason of the fact that it was easy to foresee that the situation of the land would lead to compulsory powers being some day obtained to purchase it.

[84] Now it is perfectly true to say that subsequently, in *Gajapatiraju* v *Revenue Divisional Officer, Vizagapatam* [1939] AC 302 (the *Indian* case) the Privy Council preferred the view expressed in *Lucas* by Vaughan Williams LJ (at p28), namely that in assessing the award for the land, the umpire "may value the possibility of the site going into the market as being required for the enlargement of the waterworks, [albeit] not on the basis of a realised possibility, or on account of the promoters having obtained from Parliament compulsory powers".

[85] However, there are two important points to be made. The first is that the *Indian* case affords no assistance at all as to how much the acquiring authority should be regarded as being willing to pay for the particular value of the land to it. Although the board was critical of Rowlatt J's view in *Sidney* v *North Eastern Railway Co* [1914] 3 KB 629, at p637, that the acquiring authority should pay for the land's existing use value or development potential no more than the highest price realisable from any competing prospective purchaser, there is certainly nothing to suggest that they disagreed with his view that compensation was plainly not to be assessed on the basis of the owner "obtaining for himself a share in [the] value [of the land to the promoter for his scheme]".

[86] As Lord Nicholls said in *Waters*, in [36], with regard to the *Indian* case:

> Potentiality is part of the market value of land and must be taken into account when assessing compensation. Potentiality should be valued even if the only likely purchaser is the acquiring authority itself. That was decided in the *Indian* case. But market value does not include enhanced value attributable solely to the particular use proposed to be made of the land under a scheme of which compulsory acquisition of the subject land is an integral part. This element of value is not part of market value because it is not an element the owner could have realised in the open market. That is the traditional view, which has long been acted upon in this country.

In practice, it appears, a more or less token increase on what otherwise would be assessed as the land's market value tends to be made in deference to the *Indian* case — for example, in *BP Petroleum Developments Ltd* v *Ryder* [1987] 2 EGLR 233, at p248K, an increase from £40 pa per acre to £45 for the rights over the additional land sought by the special purchaser there (the increase being made "for him to be certain that he will acquire the rights he seeks"); and, indeed, the increase from £50 to £75 that the Court of Appeal in [116] of the present case suggested would be made by a court assessing compensation here "to account for the fact that Star, as the holder of the petroleum licence, was a 'special purchaser'". The basic £50, I should note, is the standard compulsory purchase compensation paid for a deep tunnel.

[87] The second point to be made, to my mind more important still, is that made by Lord Nicholls in [38] of *Waters*:

> The legislation under consideration in the *Indian* case contained no equivalent of rule 3. Rule 3 is expressed in absolute terms which appear to leave no room for taking into account a potential use of the land where the acquiring authority is the only person who could turn this potentiality into an actuality. In this

> A regard rule 3 is more restrictive of compensation than the “value to the owner” principle as clarified on this point by the decision of the Privy Council in the *Indian* case.

Indeed, as Lord Nicholls had earlier noted (in [28]), r 3 in section 5 of the 1919 Act (section 5(3) of the 1961 Act — see [12] above), constituted “legislative affirmation of the approach adopted on this point by Fletcher Moulton LJ in [*Lucas*]” — an observation reiterated by Lord Walker in *Spirerose*, in [18].

[88] In my opinion, therefore, it is now to be regarded as clearly established in English law that the Fletcher Moulton (or r 3) approach is to apply to the assessment of compensation for compulsory purchase, B whether of land or rights over land, and that this approach must be recognised as an integral part of the *Pointe Gourde* principle. I go so far as to question, therefore, whether it will be necessary, or indeed helpful, ever again to refer to the *Indian* case.

[89] Lord Clarke suggests, in [140], that had the owners of the Oxted Estate before the 1934 Act been aware of the oilfield and its potential, the key value of their land as the necessary (or best) access route to the apex of the field would already have been apparent. He then asks, in [158], “whether Parliament increased the key value of the land when it enacted the 1934 Act” and, in [161], concluded not: “The key value was not created or enhanced by the scheme or the 1934 Act because the C Oxted Estate already had a key value in the market.”

[90] To my mind, however, this approach is to overlook the true effect of the 1934 Act. It must be recognised that, by this Act, parliament in terms: (i) vested the property in all petroleum in the Crown; (ii) gave the Crown “the exclusive right of searching and boring for and getting such petroleum” (a right that could be licensed to others, as here to Star); and (iii) enabled any licensee compulsorily to acquire any necessary ancillary right (as here to access the petroleum through Bocardo’s land). The correct analysis seems to me to be this: that, by these provisions, parliament was at one and the same time extinguishing whatever pre-existing key value Bocardo’s land might be thought to D have had in the open market and creating a new world in which only the Crown and its licensees had any interest in accessing the oilfield and in which they had been empowered to do so (to turn the key if one wants to persist in the metaphor) compulsorily and thus on terms subject to the *Pointe Gourde* approach to compensation.

[91] As from 1934, all exploitation of petroleum was pursuant to the new statutory licensing scheme; all, that is, save for the petroleum won pursuant to the three licences previously granted under the 1918 Act that — consistently with the first of the two paragraphs quoted above, in [76], from Viscount Hailsham’s speech promoting the 1934 Bill — was expressly excluded from the 1934 Act scheme (by the proviso E to section 1(1)). As, however, the second quoted paragraph from the speech makes plain, save for those previously licensed areas, there was to be no value at all in possible oil rights or the chances of finding oil under the soil. The 1934 Act marked the end of key values and the payment of royalties. As I have sought to explain, compensation thereafter was to be paid on the usual basis in compulsory acquisition cases (with, of course, a 10% uplift).

[92] In summary, I reject Bocardo’s contentions: first, that the principles governing the approach to valuation in compulsory purchase cases have no application to the assessment of compensation under section 8(2); second, that in any event there is here no relevant F “scheme” to be discounted under the *Pointe Gourde* principle; third, that Bocardo’s power of control over the passage by wells or pipes through its land gave a pre-existing key value to the ancillary rights that Star needed to acquire from it; and, fourth, that for purely geographical reasons the land through which access was required always had potential value so long as petroleum resources lay underground. Each contention is in reality a reformulation of the same essential argument, namely that Bocardo is entitled to some share of the value of the petroleum being accessed through its land. If it is, no doubt substantial damages such as those awarded here at first instance are appropriate (although there are detailed criticisms to be made of the precise calculation arrived at). If not, however, the £1,000 awarded by the Court of Appeal can be regarded as positively generous: compensation under section 8(2) would have been assessed at no more than £82.50, including the 10% uplift. G There is frankly no coherent basis for any intermediate award.

[93] As will already be apparent, the Court of Appeal’s approach here (following as it does Peter Gibson J’s decision in very similar circumstances in *BP Petroleum Developments*) is, to my mind, strongly to be preferred. I would dismiss this appeal.

Giving the fourth opinion, **Lord Collins** said:

[94] I agree with Lord Hope’s reasons for concluding that Star’s cross-appeal should be dismissed, and with Lord Brown’s reasoning on the quantum of damages.

[95] The principles for the award of damages in cases such as this are H fully canvassed in *Pell Frishmann Engineering Ltd* v *Bow Valley Iran Ltd* [2009] UKPC 45; [2010] BLR 73, *per* Lord Walker, in [46] to [54]. It is common ground that the measure of damages is to be assessed by reference to the amount that Bocardo would have been awarded under section 8(2) of the 1966 Act had Star obtained an order from the court granting it the necessary ancillary rights over Bocardo’s land.

[96] The statutory scheme is simple. The issues would now be regulated by the 1998 Act and the 1966 Act, but the licence to Star in the present case was issued pursuant to the 1934 Act. Section 1 of the 1934 Act vested in the Crown the property in petroleum “existing in its natural condition in strata in Great Britain” and gave the Crown J the exclusive right to search and bore for and get such petroleum. By section 2, the Crown had the power to grant licences to search and bore for and get petroleum.

[97] Star held a licence from the Crown under the 1934 Act that gave it the exclusive right to search and bore for and get the petroleum lying under (among other land) Bocardo’s land. The royalty payable by Star to the Crown is 5% of the market value of the petroleum extracted, rising, depending on the amount of petroleum, to 12.5%.

[98] By section 3 of the 1934 Act, Part I of the 1923 Act applied to enable a person holding a licence under the 1934 Act to acquire such ancillary rights as may be required for the exercise of the rights granted K by the licence. The 1923 Act as amended was replaced by the 1966 Act, a consolidating Act. By section 1 of the 1966 Act, the court may confer any ancillary right on a person having the right to work minerals, who is working or desirous of working the minerals, if the right is required in order that the minerals may be properly and conveniently worked by the licensee, and the proper and efficient working of the minerals is unduly hampered by his inability or failure to obtain that right. Among the relevant ancillary rights are “a right of… shaft-way… or underground way-leave, or other right for the purpose of access to or conveyance of minerals”: see section 2(1)(*b*).

[99] By section 3, no such right is to be granted under section 1 L unless the court is satisfied that the grant is expedient in the national interest, and it is not reasonably practicable to obtain the right by private arrangements because (among other reasons) “the person with power to grant the right unreasonably refuses to grant it or demands terms that, having regard to the circumstances, are unreasonable”: see section 3(2)(*d*). The licensee may then apply to the Secretary of State for Energy, who may refer the matter to the court: see section 4. The court may grant the right and “such compensation or consideration as in default of agreement may be determined by the court shall be paid or given by the applicant …”: see section 5(2). By section 8(1), where a right is granted under section 1, the court may determine “the amount M and nature of compensation or consideration to be paid or given…”. By section 8(2):

> The compensation or consideration in respect of any right… shall be assessed by the court on the basis of what would be fair and reasonable between a willing grantor and a willing grantee…

[100] Section 3(2)(*b*) of the 1934 Act provided that “in determining the amount of any compensation to be paid in respect of the grant of any right, an additional allowance of not less than ten per cent. shall be made on account of the acquisition of the right being compulsory”.

[101] Even without the express reference in section 3(2)(*b*) to the acquisition of the right being compulsory, there can be no doubt that

this would have been a case of compulsory acquisition and that any general principles of compulsory acquisition law are applicable. For present purposes, the most plainly relevant is the "value to the owner" principle, expressed in the first edition of *Cripps on Compensation* (1881) as follows, at p144:

> The basis on which all compensation for lands required or taken should be assessed, is their value to the owner, and not their value when taken to the promoters. The question is not, what the persons who take the land will gain by taking it; but what the person from whom it is taken will lose, by having it taken from him.

[102] One aspect of the value to the owner is the principle known as the *Pointe Gourde* rule or the "no-scheme rule", namely that "compensation for the compulsory acquisition of land cannot include an increase in value that is entirely due to the scheme underlying the acquisition": see *Pointe Gourde Quarrying & Transport Co Ltd* v *Sub-Intendent of Crown Lands* [1947] AC 565, at p572 (PC, *per* Lord McDermott), extensively discussed in *Waters* v *Welsh Development Agency* [2004] UKHL 19; [2004] 1 WLR 1304, especially in [40] *et seq* by Lord Nicholls and in [124] *et seq* by Lord Brown, and in *Spirerose Ltd (in administration)* v *Transport for London* [2009] UKHL 44; [2009] 1 WLR 1797, in [19] *et seq* by Lord Walker and in [119] *et seq* by myself. It is not necessary to repeat what is said there, except to say that it has long been recognised that "increase in value consequent on the execution of the undertaking for or in connection with which the purchase is made must be disregarded": see *South Eastern Railway Co* v *London County Council* [1915] 2 Ch 252, at p258, *per* Eve J.

[103] In this case, there can be no doubt that Bocardo will have suffered no quantifiable physical loss. It has no property rights in the petroleum. The most that it can say is that the ancillary right that Star would have asked the court to value was the "ransom value" or "key value" in the hypothetical negotiation. Plainly, a strip of land may have special value if it is the key to the development of other land, but if the premium value is entirely due to the scheme underlying the acquisition it must be disregarded: see *Waters*, in [64] to [65], *per* Lord Nicholls, approving *Batchelor* v *Kent County Council* (1989) 59 P&CR 357, at p361, *per* Mann LJ.

[104] Put differently, the question in this case is whether the legislature intended the landowner, under whose land petroleum was discovered but that did not hold a licence to exploit the petroleum and in respect of whose land the licence holder needed access, to have a share in the enterprise or in the prospective value of the petroleum. In my judgment, the whole scheme of the legislation against the background of well-established principles of compensation for compulsory acquisition demonstrates that that was not the intention. The 10% uplift in the compensation would make no sense if the landowner were entitled to rely on the ransom value of the ancillary right.

[105] In *Edwards* v *Minister of Transport* [1964] 2 QB 134, at p156 (applied in *Logan* v *Scottish Water* [2005] CSIH 73; 2006 SC 178, in [102]), Harman LJ said:

> I do not find anywhere in the textbooks or in any of the authorities any suggestion that a kind of ransom value, to which a man having a power of veto might hold the promoting authority, was the measure of his damage; for in fact he does not have a right of veto and the question, therefore, does not really arise.

In this case, Bocardo would have had no right of veto and is not entitled, on the hypothetical valuation, to compensation for a right that it would never have had. The true key to the oil is not Bocardo's ownership of the land but Star's licence, which gives it alone the right to bore for and produce petroleum: see Peter Gibson J in *BP Petroleum Developments Ltd* v *Ryder* [1987] 2 EGLR 233, at p247.

[106] I have come to this conclusion in the light of the legislation and general principles of law applicable to compulsory acquisition. It is true that what was said on the second reading of what became the 1934 Act (quoted by Lord Brown in [76]) supports the view that what was envisaged was loss of amenity value. However, I prefer not to take that into account in the light of the continuing controversy over *Pepper (Inspector of Taxes)* v *Hart* [1993] AC 593 and its limits: Aileen Kavanagh, *Pepper (Inspector of Taxes)* v *Hart and Matters of Constitutional Principle* (2005) 121 LQR 98; Sir John Baker, *Our Unwritten Constitution* (2010) 167 Proceedings of the British Academy 91, at pp99-100.

Giving the fifth opinion, **Lord Clarke** said:

I. Introduction

[107] This appeal raises two questions of some interest and perhaps importance. The first is whether the principle sometimes known by the proposition that a landowner owns the land *usque ad coelum et ad inferos* is part of English law (and, if so, to what extent) and the second is the measure of damages for trespass in circumstances in which the trespasser could have sought a licence to acquire ancillary rights under section 3 of 1934 Act and, if it had such a licence, would have had to pay compensation under section 8(2) of the 1966 Act.

[108] In a judgment given on 24 July 2008, Peter Smith J (the judge) held that the respondents had committed a trespass and awarded the appellant damages in the sum of £621,180 plus interest. The Court of Appeal (Jacob, Aikens and Sullivan LJJ) reduced the damages to £1,000. Permission to appeal against the quantum of damages was refused by the Court of Appeal but granted by this court. The respondents were subsequently granted permission to cross-appeal against the finding that they were liable in trespass.

II. Facts

[109] The appellant is and has since 1974 and 1988 been the freehold owner of Barrow Green Court and Barrow Green Farm respectively. They are near Oxted in Surrey and form the Oxted Estate. The Palmers Wood oilfield (the oilfield) is a naturally occurring reservoir of petroleum and petroleum gas, the north-eastern part of which extends beneath the Oxted Estate. The remainder of the oilfield lies under land in different ownerships.

[110] The respondents (Star) were successive holders of a petroleum production licence (the licence) issued by the Secretary of State for Energy on behalf of the Crown on 17 November 1980 under the 1934 Act. The licence granted the licensee exclusive licence and liberty to "search and bore for, and get, petroleum" in an area that included the oilfield. The licence incorporated specified clauses of the then model clauses and required the licensee to pay royalties to the Crown as percentages of the market value of the petroleum obtained as follows: 5% in respect of the first 100,000 tonnes won in any half-year, 7.5% of the next 50,000 tonnes, 10% of the next 50,000 tonnes, and 12.5% of the tonnes in excess of 200,000 tonnes, all in the same half-year.

[111] There are two drilling sites for the extraction of petroleum from the oilfield, although we are concerned only with that at Coney Hill, which is immediately next to the Oxted Estate. The three wells that are the subject of this dispute are PW5, PW8 and PW9. They were drilled by Star's predecessors from the Coney Hill site. Importantly, they were not drilled vertically down from Coney Hill but diagonally, so that they entered the substrata under the Oxted Estate at substantial depths beneath the surface.

[112] The wells were lined with steel casing with tubing inserted. The casing of PW5, PW8 and PW9 is of 8.5in and 12.25in in diameter. PW5 and PW8 begin at the Coney Hill site and then deviate, entering the substrata below the Oxted Estate at around 1,300ft and 800ft respectively below ground level and continue until termination at around 2,900ft and 2,800ft respectively below ground level. They run under the Oxted Estate for distances of around 0.5km and 0.7km respectively. Both wells are production wells to extract petroleum and petroleum gas from the reservoir that lies beneath the Oxted Estate and neighbouring lands. PW9 passes between the substrata beneath the Oxted Estate at around 950ft below ground level and ends beyond it, terminating at a point on the reservoir at around 1,400ft below ground level. PW9 is used for injecting water into the oilfield so as to maximise and speed recovery. PW5 was first drilled as an exploration well in 1986. PW8 and PW9 were drilled in July 1992. Production from PW5 began in October 1990 and from PW8 in September 1992. PW9 was used to inject water into the oilfield from August 1992.

A [113] The reason the wells were drilled diagonally under neighbouring land rather than vertically under Star's land was to maximise recovery of oil from the north-eastern part of the oilfield. Since oil is lighter than water, in order to maximise recovery it is necessary to drill the well into or close to the apex of the field, which lay under the Oxted Estate. The experts at the trial agreed that if the wells had not been drilled under the Oxted Estate, the recovery of the petroleum from the oilfield would not have been maximised.

[114] At no stage was permission sought from the appellant to drill beneath its land.

B *III. Issues*

[115] There are two principal issues in this appeal, namely whether the appellant is in principle entitled to recover damages in trespass and, if so, what is the measure of damages.

IV. Trespass

[116] Both the judge and the Court of Appeal held that the appellant is in principle entitled to recover damages in trespass. This issue has been considered in detail by Lord Hope, who has answered the question in the affirmative, as both the judge and the Court of Appeal did. I agree with his conclusion and his reasons and there is nothing I wish to add
C in this regard.

V. Limitation of action

[117] As explained by Aikens LJ (with whom Jacob and Sullivan LJJ agreed) in [9], the judge held that the appellant's claim was time-barred in respect of any trespass committed before 22 July 2000. He held that it was entitled to damages for the trespass committed by Star only from 22 July 2000 until the trial and to damages (in lieu of an injunction) for the continued trespass until the oilfield was exhausted. The judge's conclusions on limitation have been accepted both in the Court of Appeal and in this court.

VI. Damages at common law

D [118] The appellant's case is that it is entitled to damages on what has sometimes been called the user (or here the wayleave) basis. It is and has throughout been accepted by and on behalf of the appellant that it has suffered no damage or loss as a result of the trespass. As Aikens LJ stressed in [112], the drilling of three pipelines at depths of 800ft to 2,800ft below the appellant's land would not disturb or detract from its use of the land (to use the judge's phrase) "one iota". However, Mr Jonathan Gaunt QC submitted on behalf of the appellant that it does not follow from that that it is not entitled to substantial damages because it is now well settled that where a claimant cannot show loss or damage, it may be entitled to user damages.

E [119] For my part, I would accept Mr Gaunt's analysis of the position at common law. The courts have held that, in the case of trespass to land, damages may be recovered equal to the value to the defendant of the use it has made of the claimant's land even though the claimant has suffered no consequential loss and the value of its land has not been diminished. The principle originated in cases not unlike this, where the defendant trespassed by carrying coals along an underground way through the claimant's land. The damages were assessed by determining what the claimant would have received had it been paid for a wayleave: see, for example, *Stoke-on-Trent City Council* v *W&J Wass Ltd (No 1)*
F [1988] 1 WLR 1406, at pp1410G-1411E, and the cases there cited.

[120] Those principles were then applied to cases of wrongful trespass on the surface of land and wrongful retention of the possession of land in circumstances where the claimant would not otherwise have made use of the land: see *Whitwham* v *Westminster Brymbo Coal & Coke Co* [1896] 2 Ch 538, *per* Rigby LJ, at p543.

[121] Much more recently, Lord Nicholls put the principle thus in *Attorney-General* v *Blake* [2001] 1 AC 268, at p278F:

> A trespasser who enters another's land may cause the landowner no financial loss. In such a case damages are measured by the benefit received by the trespasser, namely, by his use of the land. The same principle is applied where the wrong consists of use of another's land for depositing waste, or by using a path across the land or using passages in an underground mine. In this type of case the damages recoverable will be, in short, the price a reasonable person G
> would pay for the right of user.

[122] Lord Nicholls restated the principle in *Kuwait Airways Corporation* v *Iraqi Airways Co (Nos 4 and 5)* [2002] UKHL 19; [2002] 2 AC 883, in [87] to [90] and in *Sempra Metals Ltd (formerly Metallgesellschaft Ltd)* v *Commissioners of Inland Revenue* [2008] 1 AC 561, in [116]; see also Lord Scott, in [140], and Lord Mance, in [230]. The same principle applies where a landowner is awarded mesne profits, whether or not it would have relet the property during the relevant period: *Swordheath Properties Ltd* v *Tabet* [1979] 1 WLR 285*, which was applied by the Privy Council in *Inverugie Investments*
Ltd v *Hackett* [1995] 1 WLR 713†. H

[123] The same principles have been applied in assessing damages in lieu of an injunction: see, for example, *Bracewell* v *Appleby* [1975] Ch 408‡ and *Jaggard* v *Sawyer* [1995] 1 WLR 269§, both of which were cases of obtaining access to a newly built house, and *Horsford* v *Bird* [2006] UKHL 3; [2006] 1 EGLR 75¶, *per* Lord Scott in [12] and [13]. It was in this connection that Lord Nicholls said in *Attorney-General*, at p281G, that the measure of damages is often analysed as a loss of a bargaining opportunity or, which (he said) comes to the same, "the price payable for the compulsory acquisition of a right".

[124] Many other examples could be given, including the leading
case of *Wrotham Park Estate Co Ltd* v *Parkside Homes Ltd* [1974] J
1 WLR 798**, where damages were awarded on this basis in respect of an unlawful act that had been committed and it was too late to restrain it by injunction. It was a case where land had been developed in breach of a restrictive covenant and where the existence of the new houses did not diminish the value of the benefited land "by one farthing", which is perhaps not very different from the "iota" referred to by both the judge and Aikens LJ. Other well-known examples are *Experience Hendrix LLC* v *PPX Enterprises Inc* [2003] EWCA Civ 223; [2003] 1 All ER (Comm) 830, *WWF World Wide Fund for Nature (formerly Word Wildlife Fund)* v *World Wrestling Federation Entertainment*
Inc [2007] EWCA Civ 286; [2008] 1 WLR 445 and *Lunn Poly Ltd* K
v *Liverpool & Lancashire Properties Ltd* [2006 EWCA Civ 430; [2006] 2 EGLR 29††.

[125] Finally, the most recent case in this area is the decision of the Privy Council in *Pell Frischmann Engineering Ltd* v *Bow Valley Iran Ltd* [2009] UKPC 45; [2010] BLR 73, where Lord Walker, giving the judgment of the board, reviewed the principles in detail in [46] to [54]. In [48], he set out five general principles established by the authorities. They included the following (omitting the case references):

> (1) Damages (often termed "user damage") are readily awarded at common
> law for the invasion of rights to tangible moveable or immoveable property (by L
> detinue, conversion or trespass).
>
> (2) …
>
> (3) Damages under Lord Cairns's Act are intended to provide compensation for the court's decision not to grant equitable relief in the form of an order for specific performance or an injunction in cases where the court has jurisdiction to entertain an application for such relief. Most of the recent cases are concerned with the invasion of property rights such as excessive user of a right of way. The breach of a restrictive covenant is also generally regarded as the invasion of a property right… since a restrictive covenant is akin to a negative easement. … the decision of the House of Lords in *Blake* decisively covers what their Lordships have referred to 0as a non-proprietary breach of contract.
>
> (4) Damages under this head (termed "negotiating damages" by M
> Neuberger LJ in *Lunn Poly* at para 22) represent "such a sum of money as might reasonably have been demanded by [the claimant] from [the defendant] as a *quid pro quo* for [permitting the continuation of the breach of covenant or other invasion of right]" (*Lunn Poly* at para 25).

* Editor's note: Also reported at [1979] 1 EGLR 58; (1978) 249 EG 439
† Editor's note: Also reported at [1996] 1 EGLR 149; [1996] 19 EG 124
‡ Editor's note: Also reported at [1976] 1 EGLR 119; (1974) 237 EG 731
§ Editor's note: Also reported at [1995] 1 EGLR 146; [1995] 13 EG 122
¶ Editor's note: Also reported at [2006] 15 EG 136
** Editor's note: Also reported at (1973) 229 EG 617
†† Editor's note: Also reported at [2006] 25 EG 210

(5) Although damages under Lord Cairns's Act are awarded in lieu of an injunction it is not necessary that an injunction should actually have been claimed in the proceedings, or that there should have been any prospect, on the facts, of it being granted...

[126] Lord Walker added, in [49]:

[49] Several of the recent cases have explored the nature of the hypothetical negotiation called for in the assessment of *Wrotham Park* damages. It is a negotiation between a willing buyer (the contract-breaker) and a willing seller (the party claiming damages) in which the subject-matter of the negotiation is the release of the relevant contractual obligation. Both parties are to be assumed to act reasonably. The fact that one or both parties would in practice have refused to make a deal is therefore to be ignored...

[127] I recognise that it is common ground that the measure of damages in this case must have regard to the statutory context and thus to the 1934 and 1966 Acts in particular. However, subject to that, the correct measure of damages for the trespass on the facts here would be to award the appellant user or wayleave damages and to assess them by reference to a hypothetical negotiation of the kind referred to by Lord Walker in [49] of the judgment of the board in *Pell Frischmann.* The question would be what would be a fair and reasonable figure for Star to agree to pay and for the appellant to agree to receive for the use of the part of the appellant's land that was in fact used by Star as described above. That figure would reflect the key value of the wayleave. It is inconceivable that it would be only the £1,000 awarded on the facts here by the Court of Appeal.

[128] Since both the reservoir that forms the oilfield and the Oxted Estate have been there for very many years and long before the enactment of any of the statutes relevant in this appeal, it follows that if Star or its predecessors in title had committed the trespass that was in fact committed in order to remove oil from the oilfield before, say, 1934, the correct measure of damages would have been user or wayleave damages calculated as stated above. I turn to the statutory framework.

VII. Statutory framework

[129] The property in petroleum existing in its natural condition in strata in Great Britain was originally vested in the owner of the land above it: see, for example, the decision of the Court of Appeal in Singapore in *NV de Bataafsche Petroleum Maatschappij* v *War Damage Commission* (1956) 23 ILR 810. However, by section 1(1) of the 1934 Act, the property in such petroleum was vested in the Crown, which was given the "exclusive right of searching and boring for and getting such petroleum". By section 2(1), the Board of Trade (later the Secretary of State for Energy) was given the power, on behalf of the Crown, to grant licences "to search and bore for and get petroleum". By section 2(2), any such licence was to be granted "for such consideration (whether by way of royalty or otherwise) as the Board of Trade with the consent of the Treasury may determine, and upon such other terms and conditions as the Board of Trade think fit".

[130] Section 3(1) provided that Part 1 of the 1923 Act should apply for the purposes of enabling a person holding such a licence to acquire such "ancillary rights" as might be required for the exercise of the rights granted by the licence. Those rights were stated to include "a right to enter upon land and to sink bore holes therein for the purpose of searching for and getting petroleum" and "a right to use and occupy land for the erection of such buildings, the laying and maintenance of such pipes... as may be required for the purpose of searching and boring for and getting, carrying away, storing, treating and converting petroleum".

[131] Section 3(2)(*b*) of the 1934 Act provided that, in determining the amount of compensation to be paid in respect of the grant of any right, which included any ancillary right, an additional allowance of not less than 10% was to be made on account of the acquisition of the right being compulsory. Section 10(3) provided:

Nothing in this Act shall be construed as conferring, or as enabling the Board of Trade to confer, on any person, whether acting on behalf of His Majesty or not, any right which he does not enjoy apart from this Act to enter on or interfere with land.

[132] Part 1 of the 1923 Act, which contained provisions for ancillary rights in section 3 and for compensation in section 9, was repealed and consolidated by the 1966 Act, which provided that the reference to Part 1 of the 1923 Act in section 3 of the 1934 Act was now a reference to the 1966 Act. Section 1 of the 1966 Act empowered the court to confer any rights described in a table, which included in para 5 "ancillary rights", which were defined in section 2 and included by section 2(1)(*b*):

a right of air-way, shaft-way or surface or underground wayleave, or other right for the purpose of access to or conveyance of minerals... or the ventilation or drainage of the mines.

[133] It is clear from these provisions that a grant of a petroleum licence under the 1934 Act did not itself entitle its licensee to enter land belonging to another party and that the Act did not empower the Secretary of State to grant a licence to enter such land. If the licensee wanted to drill a deviated well beneath another party's land, it needed to negotiate or apply under the 1966 Act for an ancillary right, here an underground wayleave. This is important because it shows that parliament was drawing a distinction between the oil, on the one hand, and the access to the oil, on the other. It provided for the licensee to pay a royalty to the Crown for the oil but provided a different scheme in respect of access to the oil.

[134] That scheme involved the obtaining of ancillary rights. Section 4 of the 1966 Act provided for applications for ancillary licences to be made to the minister, who is now the Secretary of State for Energy, and for him to refer the matter to the court. Section 5(1) of the 1966 Act provides that where the matter is referred to the court, the court may grant an ancillary licence. Section 5(2) provides:

Where a right is granted, such compensation or consideration as in default of agreement may be determined by the court shall be paid or given by the applicant in respect of the acquisition of the right to such persons as the court may determine to be entitled thereto.

[135] Section 8 provides, so far as relevant:

Compensation

8.—(1) Where a right is granted under section 1 of this Act,... the court may determine the amount and nature of compensation or consideration to be paid or given and the persons to whom it is to be paid or given, either at the time when it determines whether the right should be granted or the restrictions imposed or at any subsequent time.

(2) The compensation or consideration in respect of any right,... shall be assessed by the court on the basis of what would be fair and reasonable between a willing grantor and a willing grantee, having regard to the conditions subject to which the right is or is to be granted.

Section 8 thus replaced section 9 of the 1923 Act, which was in similar terms.

VIII. Correct approach to section 8(2)

[136] It is, to my mind, striking that the negotiation contemplated by section 8(2) is, at any rate on the face of it, essentially the same as is deployed by the common law in assessing wayleave damages. Its purpose is, again on the face of it, the same, namely to ascertain what would be a fair and reasonable figure for Star to agree to pay and for the appellant to agree to receive for the use of the part of the appellant's land that was in fact used by Star as described above. There is nothing in the language of the subsection concerning the value of the land used or taken, and there is no suggestion that the purpose for which the right is to be acquired is to be ignored during the postulated negotiation.

[137] Nor is there any reference in section 8(2) or elsewhere in the 1934 or 1966 Acts to the basis on which compensation is to be assessed, as, for example, in section 63 of the 1845 Act, section 2 of the 1919 Act or section 5 of the 1961 Act. See also the 1962 Act, the Gas Act 1986, the Channel Tunnel Act 1987, the Electricity Act 1989 and the Water Resources Act 1991. In particular, section 2 of the 1919 Act provided detailed rules for the assessment of compensation in respect of land acquired compulsorily. It is to my mind striking that those rules were not incorporated into the 1923 Act. Moreover, section 5 of the 1961 Act

A in turn set out detailed rules for the assessment of compensation, which again were not incorporated into the 1966 Act.

[138] There are many different types of compulsory acquisition legislation. The 1934 and 1966 Acts are one example and the 1919 and 1961 Acts are another. I can see no principled basis for applying the provisions of the latter Acts to the assessment of compensation under the former. If parliament had intended those provisions to apply, it would have so provided in 1923 and 1966. In this regard, I agree with the reasoning of Judge Hague QC in *Mercury Communications Ltd* v *London & India Dock Investments Ltd* [1994] 1 EGLR 229.

[139] If section 8(2) of the 1966 Act is given its ordinary and natural
B meaning, it postulates a negotiation in which it is assumed that both parties are willing to reach agreement and that they both act reasonably. In such a negotiation, the seller will naturally stress the value of the right being sold (here, the wayleave) to the purchaser. On the facts of this case, that value is the key to unlocking the oil in the reservoir because it was necessary to dig diagonal tunnels in order to maximise the oil recovered from the reservoir. On this approach, the figure agreed at the postulated negotiation would be the same as it would be at common law. Moreover, it would, as I see it, be the same whether it is treated or described as compensation or consideration. That is because what is being negotiated is a fair price for the wayleave.

C [140] In the absence of authority to the contrary, I would approach the matter in that way. In the particular context of the 1934 and 1966 Acts, I see no reason not to do so. The position can be tested by a comparison between the position immediately before and after the 1934 Act came into force. Before the Act, assuming that the parties were aware of the reservoir of petroleum and its potential, and the owner of the reservoir and its oil wanted to exploit it, it would need to obtain a wayleave through the Oxted Estate. If it acted lawfully, it would have to seek a wayleave from the owner and would (it is assumed) pay a fair price. If it acted unlawfully and committed trespass, the measure of damages at common law would be the notional price of the wayleave, which is to
D be a fair and reasonable price arrived at after a postulated negotiation between a willing seller and a willing buyer. The strength of the seller's position (and the price or damages arrived at) would depend on the existence and physical position of the oil, on the one hand, and the existence and position of the land through which it was necessary to obtain a wayleave in order to be able to exploit it, on the other.

[141] Why, so far as the wayleave is concerned, should the position be any different after the 1934 Act? The effect of the Act is to transfer the oil to the Crown without compensation. That was no doubt a political decision. After the Act, the Crown is in the same position as the owner of the reservoir was in before the Act. It now has the right to
E exploit the oil, either itself or by granting a licence. In order to exploit the oil to its full extent, it (or its licensee) needs to obtain a wayleave. It is entitled to do so, but only if it pays compensation under section 8(2). That compensation is to be assessed by postulating a negotiation as described above. As I see it, the seller of the wayleave is in the same position before and after the Act. The reason it is in that position is not because of the Act, and certainly not solely because of the Act, but because of the physical position of the seller's land in respect of the reservoir.

[142] In these circumstances, absent authority to the contrary, I would hold that the measure of compensation under the Act is the
F same as at common law. It is true that section 3(2)(*b*) provides for an additional allowance of 10%, but that is expressed to be because the acquisition of the right was compulsory. It was at one time common for such a provision to be included in statutory provisions that provided for compensation for compulsory purchase. By section 2(1) of the 1919 Act, the 10% allowance was abolished in respect of cases to which the 1919 Act applied. That very fact, coupled with the enactment of section 3(2)(*b*) of the 1934 Act, demonstrates clearly that the 1919 regime does not apply to the new provisions of the 1934 Act. It appears that parliament took the view that in this different regime an additional payment should be made over and above the amount arrived at in the postulated negotiation, which of course assumes that there is a willing seller and a willing buyer and thus that the seller (or more accurately the grantor) cannot refuse to sell. This factor tells us nothing G
about what would be a fair and reasonable figure to arrive at as a result of the negotiation.

[143] It is said that the approach that I have described is the wrong approach to the construction of the subsection because of the *Pointe Gourde* principle, to which I now turn.

IX. Pointe Gourde principle

[144] This principle, which is also known as the "no scheme rule", takes its name from the decision of the Privy Council in *Pointe Gourde Quarrying & Transport Co Ltd* v *Sub-Intendent of Crown Lands* [1947] AC 565. The principle has been considered in some detail in two comparatively recent cases in the House of Lords. They are *Waters* H
v *Welsh Development Agency* [2004] UKHL 19; [2004] 1 WLR 1304, and *Spirerose Ltd (in administration)* v *Transport for London* [2009] UKHL 44; [2009] 1 WLR 1797. In both *Waters* and *Spirerose*, the compensation fell to be calculated in accordance with the 1961 Act.

[145] Lord Brown summarised the principle thus in one sentence in *Waters*, in [125]:

> *Pointe Gourde* has long been regarded as authority for the principle that compensation for compulsory purchase "cannot include any increase in value which is entirely due to the scheme underlying the acquisition"…

There are two questions that arise under this head. The first is whether J
this principle applies to compensation assessed under section 8(2) of the 1966 Act. The second is, if so, whether there has been an increase in the value of the wayleave entirely or solely due to the scheme.

[146] There has been some discussion in the cases as to the juridical basis of the principle. However, it is, in my opinion, now clear that it is a principle of statutory construction. This is clear from [127] and [128] of the speech of Lord Collins in *Spirerose* with which the other members of the appellate committee agreed. He put it thus:

> 127. What is the juridical basis of the *Pointe Gourde* principle? Lord Nicholls said in *Waters*... para 42 that the principle is no more than the name given to one K
> aspect of the long established "value to the owner" principle.
>
> 128. In my opinion it is a principle of statutory interpretation, mainly designed and used to explain and amplify the expression "value." It is in this sense that it has sometimes been referred to as a common law principle: see eg *Fletcher Estates (Harlescott) Ltd* v *Secretary of State for the Environment* [2000] 2 AC 307, 315, *per* Lord Hope of Craighead; *Waters*... para 142, *per* Lord Brown of Eaton-under-Heywood. In *Rugby Joint Water Board* v *Shaw-Fox* [1973] AC 202, 213-215 Lord Pearson reviewed the authorities and concluded, at p214, that although the *Pointe Gourde* principle had been described as a "common law principle", it could not be such a principle "because compulsory acquisition and compensation for it are entirely creations of statute". He went on, at pp214-215: "The *Pointe Gourde* principle in my opinion involves an interpretation of the word 'value' in those statutory provisions which require L
> the compensation for compulsory acquisition to include the value of the lands taken". I am satisfied that this is the right approach and that there is nothing in Lord Nicholls' speech in *Waters* which is inconsistent with this view.

It is clear from those conclusions that the question is whether the *Pointe Gourde* principle applies to compensation under section 8(2) of the 1966 Act as a matter of construction of that subsection.

[147] In *Waters*, Lord Nicholls gave the leading speech, in which he analysed in some detail the approach to compensation under the various different statutes referred to above, although he did not analyse the position under the 1934 or 1966 Acts. All the statutes he analysed M
either expressly referred to value or contained a compensation code. Thus, from [15], he discussed the meaning of "value" in section 63 of the 1845 Act. He recognised, in [17], that land may have a special value and gave an example based on *Inland Revenue Commissioners* v *Clay* [1914] 3 KB 466:

> Thus a house, worth £750 as a house but £1,000 as an annex to an adjoining nursing home, has a market value of £1,000.

Lord Nicholls then said, in [18]:

> [18] In principle, subject to one qualification, this approach is equally applicable when assessing value for the purposes of compensation. It is this qualification which has given rise to difficulty. The qualification is that

A enhancement in the value of the land attributable solely to the particular purpose for which it is being compulsorily acquired, and an acquiring authority's pressing need of the land for that purpose, are to be disregarded. If statute authorises an authority to acquire some ancient graveyards in the City of London and use the land for new buildings and a new street from Blackfriars to the Mansion House, the increased value the land will have when applied to these more profitable secular purposes should be left out of account. This is implicit in the yardstick of "value" in the Lands Clauses Consolidation Act 1845. When granting a power to acquire land compulsorily for a particular purpose Parliament cannot have intended thereby to *increase* the value of the subject land. Parliament cannot have intended that the acquiring authority should pay as compensation a larger amount than the owner could reasonably have obtained for his land in the
B absence of the power. For the same reason there should also be disregarded the "special want" of an acquiring authority for a particular site which arises from the authority having been authorised to acquire it.

[148] Lord Nicholls added, in [19], that it was in this context that the cases distinguish between value to the owner and value to the purchaser, which he further described in [21]:

> Drawing a distinction between value to the owner and value to the purchaser makes it necessary to distinguish the one from the other. It is necessary to separate from the market value of land any enhancement in value attributable solely to the presence of the acquiring authority in the market as a purchaser of the land in exercise of its statutory powers. It is important to recognise that,
> C for this purpose, it is not the existence of a power of compulsory acquisition which increases the value of land. What is relevant, because this may affect the value of the land, is the use the acquiring authority proposes to make of the land it is acquiring. Accordingly, in identifying any enhanced value which must be disregarded it is always necessary to look beyond the mere existence of the power of compulsory purchase. It is necessary to identify the use proposed to be made of the land under the scheme for which the land is being taken. Hence the introduction of the concept of the 'scheme' or equivalent expressions such as project or undertaking.

[149] Lord Nicholls then considered the 1919 Act, the *Indian* case, namely *Gajapatiraju* v *Revenue Divisional Officer, Vizagapatam*
D [1939] AC 302, the *Pointe Gourde* principle and the 1961 Act. In the context of the *Pointe Gourde* principle, he said it was one aspect of the "value to the owner" principle.

[150] In *Spirerose*, Lord Collins summarised the position thus in part of a section of his speech enitled "*Principles of valuation*":

> [88] … "It is well settled that compensation for the compulsory acquisition of land cannot include an increase in value which is entirely due to the scheme underlying the acquisition": *Pointe Gourde…* [1947] AC 565, 572, *per* Lord MacDermott. …
>
> [89] Some elementary principles of the law of compensation for compulsory acquisition provide a starting point. First, the underlying principle is that fair
> E compensation should be given to the owner claimant whose land has been compulsorily taken. The aim of compensation is to provide a fair financial equivalent for the land taken. The owner is entitled to be compensated fairly and fully for his loss, but the owner is not entitled to receive more than fair compensation: *Director of Buildings and Lands* v *Shun Fung Ironworks Ltd* [1995] 2 AC 111, 125; *Waters…* para 4.
>
> [90] Second, the basis of compensation is the value to the owner, and not its value to the public authority. In the first edition of *Cripps* (later Lord Parmoor), *Principles of the Law of Compensation* (1881) it was said, at p 144:
>
> "The basis on which all compensation for lands required or taken should be assessed, is their value to the owner, and not their value when taken to the promoters. The question is not, what the persons who take the land will gain
> F by taking it; but what the person from whom it is taken will lose, by having it taken from him."
>
> [91] The classic example mentioned by *Cripps* is *Stebbing* v *Metropolitan Board of Works* (1870) LR 6 QB 37, 42 where Cockburn CJ said that it was intended that the landowner should be compensated to the extent of his loss and "his loss shall be tested by what was the value of the thing to him, not by what will be its value to the persons acquiring it."
>
> [92] Third, and directly in point on this appeal, one plainly relevant element in the value to the owner is the prospect of exploiting the property. I have already mentioned *R* v *Brown* [(1867)] LR 2 QB 630, in which Cockburn CJ said at p 631, that the jury assessing compensation under the 1845 Act had to consider
>
> "the real value of the land, and may take into account not only the present purpose to which the land is applied, but also any other more beneficial purpose to which in the course of events at no remote period it may be applied, just as an owner might do if he were bargaining with a purchaser in the market." G

[151] In cases where those principles apply, the critical question is what was the value to the owner, which is arrived at by excluding the value to the acquirer. This is not, however, to say that the value to the owner may not have regard to the use that will be made of the land. In a case such as this, where the value of the land is that it has a ransom or key value because it is needed by the buyer to exploit its land, as I see it the value to the owner is (or includes) the ransom or key value. Both this principle and its relation to the *Pointe Gourde* principle can be seen from [64] and [65] of Lord Nicholls speech in *Waters*.

[152] Lord Nicholls said, in [64] and [65]: H

> [64] One last point should be noted before returning to the present case. This concerns so-called "ransom" value or, less pejoratively, "key" value. I have already mentioned that under the "value to the owner" principle or the *Pointe Gourde* principle, whichever nomenclature is preferred, the pressing need of an acquiring authority for the subject land as part of a scheme should be disregarded when assessing its value for compensation purposes. The value of the land is not the price a "driven" buyer would be prepared to pay. *But a strip of land may have special value if it is the key to the development of other land. In that event this feature of the land represents part of its value as much for purposes of compensation as on an actual sale in the open market.*
>
> [65] The intersection of these two principles was identified neatly by Mann LJ in *Batchelor* v *Kent County Council* (1989) 59 P&CR 357, 361: J
> "If a premium value is 'entirely due to the scheme underlying the acquisition' then it must be disregarded. If it was pre-existent to the [scheme] it must in my judgment be regarded. To ignore the pre-existent value would be to expropriate it without compensation and would be to contravene the fundamental principle of equivalence. …"

[153] The part of [64] that I have italicised shows that the key value is part of the value to the owner. This view was also expressed by Lord Brown in *Waters*, in [140], where he said:

> [140] True it is that in the *Indian* case [1939] AC 302, 312, Lord Romer said: K
> "The disinclination of the vendor to part with his land and the urgent necessity of the purchaser to buy must alike be disregarded. Neither must be considered as acting under compulsion." It by no means follows, however, that the open market value to the seller will exclude whatever key value the land may have. On the contrary, any such value properly falls to be taken into account, as it was in *Stokes* v *Cambridge Corpn* (1961) 13 P&CR 77, …

Land is not to be valued merely by reference to the use to which it is being put at the valuation date but by reference to any unusual features or potentialities it may have: see the *Indian case*, *per* Lord Romer, giving the judgment of the Privy Council, at pp312-313.

[154] Thus, value to owner includes or potentially includes key L
value, but is to be disregarded in the circumstances identified if, as Mann LJ put it in *Batchelor*, at p361, the premium or key value is "entirely due to the scheme underlying the acquisition". As I see it, that approach encapsulates the *Pointe Gourde* principle and is a gloss on or modification (or perhaps explanation) of the value to owner principle. On the other hand, if, on the facts of the particular case, the premium value or key value was pre-existent to the scheme, it is taken into account and not disregarded.

[155] The question is whether these principles apply to compensation under section 8(2) of the 1966 Act. It is difficult to see how they do as a matter of construction of the Act. I have already expressed the M
view that the codes in the 1919 and 1961 Acts do not form part of the approach identified in section 8(2) of the 1966 Act and thus in the 1934 Act. There is, moreover, no reference to "value" in that subsection. In these circumstances, although it is a compensation provision, as Lord Pearson put it in the passage approved by Lord Collins in [128] of his speech in *Spirerose* quoted above (and thus by the House), the *Pointe Gourde* principle involves an interpretation of the word "value". Since the word "value" does not appear in section 8(2), it is difficult to see why it should be construed as though it did.

[156] For these reasons, I would hold that the subsection should be construed by the application of the language used in it, as explained above, without reference either to the codes or to the *Pointe Gourde*

principle. In this regard, I prefer the reasoning of Judge Hague in *Mercury Communications* to that of Peter Gibson J in *BP Petroleum Development Ltd* v *Ryder* [1987] 2 EGLR 233.

[157] I should, however, say a word concerning the *Hansard* materials relied on by the respondents. I agree with Lord Hope, for the reasons that he gives in [43], that there is no legitimate basis for using *Hansard* as an aid to construction of the simple terms of the 1923, the 1934 or the 1966 Acts. Like section 9 of the 1923 Act, section 8(2) of the 1966 Act is in simple terms. It can readily be applied to the facts of this case. It is true that there is no reference to key value in the *Hansard* materials; nor is there any reference to the *Pointe Gourde* principle. There is, accordingly, no reference to the distinction drawn in *Batchelor*, so that no assistance can be obtained from *Hansard* as to the principles the court must apply to the postulated negotiation other than those stated in the subsection. It appears to me that no consideration was given at all to the problems that have arisen in this connection and that the correct approach is for the court to construe the statute in accordance with its language and having regard to its statutory purpose. This can readily be done without the assistance of *Hansard*.

[158] I recognise that others do not take the same view of the subsection. I therefore turn to the question of what, on the assumption that the *Pointe Gourde* principle applies to the assessment of compensation under section 8(2), is the correct approach to the key value. Again, in a case of this kind the starting value is the key value, which must be disregarded only if it represents an increase in value that is entirely due to the scheme underlying the acquisition. I take this to refer to the acquisition of the wayleave. The contrast is that identified by Mann LJ in *Batchelor* as approved by Lord Nicholls in [65] of *Waters*. The question is whether the key value was entirely due to the scheme underlying the acquisition or whether it was pre-existent to the scheme. Another way of putting what seems to me to be essentially the same question is to ask, as Lord Nicholls does in [18] of *Waters*, whether parliament increased the key value of the land when it enacted the 1934 Act.

[159] It is important to note that this approach does not disregard the key or ransom value but encapsulates it. It is expressly accepted as relevant by Lord Nicholls in [64] of *Waters* set out above. In this connection, I should mention two cases that are referred to by Lord Collins in [105] but that were not, I think, relied on in argument. They are *Edwards* v *Minister of Transport* [1964] 2 QB 134, *per* Harman LJ, at p156, and *Logan* v *Scottish Water* [2005] CSIH 73; 2006 SC 178, which applied his analysis. Harman LJ there said that the possibility of assessing damages by reference to a "ransom value" did not really arise because such a value would be relevant only if the postulated seller had a right of veto, which it does not because the rights are being compulsorily purchased.

[160] In my opinion, that approach cannot be correct in the light of *Waters*, at any rate under section 8(2). The figure to be arrived at as a result of the postulated negotiation in a case of this kind is the key or ransom value of the wayleave, that is, a fair and reasonable price for the access to the oil reservoir, but on the assumption of a willing seller or grantor and a willing buyer or grantee. Thus, the value reflects the importance of the particular access to the particular oil. I note in passing that, in these circumstances, it seems to me that the 10% uplift is also understandable as reflecting the fact that the seller was compelled to sell.

[161] The critical question is thus that stated above, namely whether the key value was entirely due to the scheme underlying the acquisition or whether it was pre-existent to the scheme. In my opinion, the key value was not created or enhanced by the scheme or the 1934 Act because the Oxted Estate already had a key value in the market. As I see it, this is a case like *Chapman Lowry & Puttick Ltd* v *Chichester District Council* (1984) 47 P&CR 674, where the acquiring authority owned a plot of housing land to the rear of a small quantity of waste land that was necessary to gain access to the housing land. The question was whether the purpose for which the land was required was to be taken into account. It was held that it was. The tribunal (Mr VG Wellings QC) said, at p680:

> It appears to me that the reference land is the key which unlocks the development value of the rear land in whosoever's hands the rear land happens to be. By reason of that fact the reference land has acquired naturally a value in excess of its existing use value. The matter can be tested by analogy with the *Pointe Gourde* principle as was done by the Court of Appeal in *Lambe* v *Secretary of State for War* [[1955] 2 QB 612]. It is not the scheme underlying the acquisition which gives value to the reference land in excess of its existing use value; it is its geographical position, coupled with the fact that there is no other suitable access for residential development on the rear land.

[162] I would accept the appellant's submission that that is essentially the position on the facts here. This is a case that is on the side of the line identified by Mann LJ in *Batchelor* in which the land had a key value that was pre-existent to the Act and the scheme. That key value depended on the juxtaposition of two physical features of the land, namely the existence of the reservoir of oil and the existence of the land that was part of the Oxted Estate and that was above the apex of the oil deposit and, given the absence of any other suitable or optimal access to that apex, would be needed by anyone who exploited the oil in the reservoir in order to maximise the recovery of the oil. Moreover, it had and has this key value even if, as is likely, there is only one owner or licensee that wishes to exploit the oil.

[163] In these circumstances, the key value cannot, in my opinion, fairly be described as solely or entirely due to the scheme because it pre-existed it. Whenever the owner of the oil chose to exploit it fully it would need a wayleave over the Oxted Estate. That is so whether the owner was a private landowner before the oil was nationalised or was the Crown after the 1934 Act and, in that case, whether the Crown exploited the oil itself or granted a petroleum licence to another to do so. In short, the key value was not created by the 1934 Act or the grant of the petroleum licence to Star. It was pre-existing key value.

[164] For my part, save perhaps for *BP*, I would not accept the submission that if that is the case here, it must follow that any of the decided cases was wrongly decided. There was no equivalent of the existing reservoir in *Waters* or *Spirerose*. In these circumstances, assuming (contrary to my view) that the no-scheme rule derived from *Pointe Gourde* applies to the assessment of compensation under section 8(2) of the 1966 Act and, thus, to the 1934 Act, I do not accept that this approach would emasculate the no-scheme rule to the point of extinction. Whether the relevant value falls to be disregarded depends on which side of the line drawn by Mann J in *Batchelor* and approved in *Waters* the facts of a particular case falls. If this case does not fall on what may be called the appellant's side of the line, it is difficult to see what case would. For these reasons, I have reached a different conclusion from the Court of Appeal on this part of the case.

[165] Since writing the above paragraphs, I have seen Lord Brown's response in [89] to [91]. I remain unpersuaded. The key value attaches to the access land and not to the oil. Both before and after the 1934 Act it was or would have been necessary for the party seeking to exploit the oil to obtain access through the Oxted Estate by acquiring ancillary rights. It would thus have been necessary for that party to obtain such rights, whether as the owner of the oil or the licensee from the Crown. In both cases, it would have needed a wayleave and in both cases it would have had to pay a price that reflected the key value of the access land (not the oil). In these circumstances, I remain of the view that the key value was not solely attributable to the scheme. It was at least in part attributable to the physical juxtaposition of the access land and the reservoir.

X. Pipe-lines Act 1962

[166] The respondents rely in the alternative on the 1962 Act. They say that it is part of the background of statutory legislation against which the hypothetical negotiation would have taken place and that they could have obtained a right to drill or maintain oil wells and pipelines through the appellant's land under sections 11 and 12 of the 1962 Act, which provided for compensation under the 1961 Act.

[167] The appellant's response is twofold. The first is that, since neither the respondents nor their experts or lawyers thought of this point when they were considering the matter in detail before the judge or when they prepared the grounds of appeal to the Court of Appeal on which

permission to appeal was granted, it is hardly likely to have played any part in hypothetical negotiations in the 1980s. It was certainly never suggested in evidence that it would have done, although that is hardly surprising since nobody thought of the point until it appeared in the respondents' revised skeleton argument in the Court of Appeal. I would be reluctant to rest a decision on this analysis.

[168] The appellant's second response is that the 1962 Act does not apply. Section 65 provides, so far as relevant:

> (1) In this Act "pipe-line" (except where the context otherwise requires) means a pipe… for the conveyance of any thing other than air, water, water vapour or steam…

It is submitted that the pipes in this case were not for the "conveyance of substances" and that this can be seen from section 57 of the 1962 Act, which amended section 3(2)(*b*) of the 1923 Act. In its unamended form section 3(2)(*b*) provided that "ancillary rights" included:

> (*b*) [A] right of… underground wayleave, or other right for the purpose of access to or conveyance of minerals or the ventilation or drainage of the mines;…

Section 57 of the 1962 Act amended section 3(2)(*b*) of the 1923 Act by excluding from it rights for the conveyance of minerals "by means of a pipe", thus (as Mr Gaunt put it) leaving all other ancillary rights, including the right to bore wells, to be covered by the 1923 Act. Section 57 was repealed in 1966 when the amendment to the 1923 Act was reflected in the definition of ancillary rights in section 2(1)(*b*) of the 1966, as quoted above. He submitted that, in these circumstances, the correct conclusion is that the 1962 Act was not intended to apply to pipes constituting the oil well itself. That seems to me to be correct.

[169] A third point has occurred to me under this head. Where it applies, the effect of the 1962 Act is to apply the compensation scheme under the 1961 Act; so that if it applies here, those principles, including the *Pointe Gourde* principles, would apply. It seems to me to follow that compensation would be payable for the reasons given in section IX above, namely that the key value was not solely or entirely due to the scheme.

XI. Human Rights Act

[170] The appellant sought permission to rely on the Human Rights Act 1998 on the basis that compensation of so little as was awarded by the Court of Appeal would infringe its rights under Article 1 of Protocol 1 of the European Convention on Human Rights. However, since, if the conclusions that I have already expressed are correct, the appellant would be no better off in terms of damages than it would be under section 8(2), there is no need for me to express a view under this head.

XII. Measure of damages

[171] This topic would of course arise only if I were right so far. Since the majority of the court takes a different view on that question, the measure of damages is not relevant. It seems to me that if it were relevant, the correct approach would be to assess a fair and reasonable amount to reflect the key value of the wayleave, in the words of section 8(2), as between a willing grantor and a willing grantee, and to add 10% in accordance with the statute. However, as I see it, that key value would not reflect the value of the access to all the oil in the reservoir. Its particular value was to provide access to the apex (or attic) oil. I am not, at least at present, persuaded that that was the basis on which the figure was arrived at by the judge. In these circumstances, if this were a live issue, I would remit it to the High Court for determination.

XII. Conclusion

[172] For the reasons that I have given, I would allow the appeal and would remit the issue of damages to the High Court.

Appeal and cross-appeal dismissed.

Eminence Property Developments
A Ltd *v* Heaney

COURT OF APPEAL

[2010] EWCA Civ 1168

21 October 2010

MUMMERY, ETHERTON AND SULLIVAN LJJ

Sale of land — Contract — Rescission — Repudiatory breach — Vendor serving notices to complete by reference to standard* B *conditions 6.8 and 7.5 — Notices incorrectly calculating date for expiry of notices — Vendor accordingly serving rescission notices prematurely — Whether this amounting to repudiatory breach of contract entitling purchaser to return of deposit

The respondent contracted to purchase long leases of 13 flats in a block that the appellant freeholder was constructing. He paid a deposit of £36,000 towards the purchase price of £1.47m. The 13 sale contracts, incorporating the standard conditions of sale (4th ed), were exchanged in December 2007. The contractual completion date fell in December 2008, by which time the property market had suffered a significant downturn.

C The respondent did not complete on the contractual completion date. The appellant served notices to complete, under standard condition 6.8, in respect of each contract, drawing attention to the consequences prescribed by standard condition 7.5 should the respondent fail to complete within 10 working days thereafter. A covering letter stated that the respondent had calculated the final date for completion as being 15 December 2008; this was erroneous since the 10 working days expired on 19 December. On 17 December, the appellant gave notice to rescind the contract and forfeit the deposit under standard condition 7.5. Those notices were premature owing to the mistake over the expiry date of the notices to complete.

D The respondent claimed that the premature rescission notice amounted to a repudiatory breach of contract, which he accepted, thereby discharging him from further performance of the contracts and entitling him to the return of his deposit. The appellant sought a declaration that the deposit had been validly forfeited owing to the respondent's failure to complete by 19 December. Dismissing the claim, the recorder held that the appellant had repudiated the contract because, regardless of the reasons for giving the premature rescission notices, they showed a clear intention to terminate the contracts immediately. The appellant appealed.

***Held*: The appeal was allowed. (1) The legal test of repudiatory** E **conduct is whether, looking at the circumstances objectively from the perspective of a reasonable party in the position of the innocent party, the contract-breaker has shown an intention to abandon and refuse to perform the contract. Whether that is so will be fact-sensitive, such that it is dangerous to attempt to clarify the application of the legal principle by a series of propositions derived from cases decided on their own facts. All the circumstances must be taken into account so far as they bear on an objective assessment of the contract-breaker's intention. The contract-breaker's motive, although irrelevant if relied on solely to show its subjective intention, may be relevant as reflecting something of which the innocent party or a reasonable party in its position would have been aware, so as to throw light on the way in which such a reasonable** F **party would have viewed the alleged repudiatory breach. (2) The appellant's notice of rescission was not a repudiatory breach of the sale contracts. The appellant had been ready, willing and able to complete, as confirmed by its notice to complete and the covering letter thereto. Although that covering letter had stated an erroneous date for the expiry of the notice to complete, its wording made it clear that the calculation of the date was based on the contractual provisions, such that it was an obviously mistaken application of those provisions rather than an attempt to vary them, an abandonment of them or a refusal to comply with them. No reasonable party would have viewed it in any other way.** G **Neither the respondent nor his solicitor had any reason to think that had the error been pointed out to the appellant, it would not immediately have acknowledged its mistake. The same reasoning applied to the notice of rescission. A reasonable party would have realised that the premature service of the notice was due to a simple error of calculation, analogous to a clerical error, which, if pointed out, would have resulted in the notice being treated as ineffective and in the appellant waiting until the true date of expiry of the notices to complete before deciding whether to treat the contracts as having been terminated. It was impossible, on those facts, to find an intention on the part of the appellant to abandon and refuse to perform the contracts.** H

The following cases are referred to in this report.

Bridgen v *Lancashire County Council* [1987] IRLR 58, CA
CDV Software Entertainment AG v *Gamecock Media Europe Ltd* [2009] EWHC 2965 (Ch)
Chilean Nitrate Sales Corporation v *Marine Transportation Co Ltd (The Hermosa)*; *Nitrate Corporation of Chile Ltd* v *Pansuiza Compania de Navegacion SA*; *Marine Transportation Co Ltd* v *Pansuiza Compania de Navegacion SA* [1982] 1 Lloyd's Rep 570, CA
Dalkia Utilities Services plc v *Celtech International Ltd* [2006] EWHC 63 (Comm); [2006] 1 Lloyd's Rep 599; [2006] 2 P&CR 9
Federal Commerce & Navigation Co Ltd v *Molena Alpha Inc (The Nanfri)*; J *Federal Commerce & Navigation Co Ltd* v *Molena Gamma Inc (The Lorfri)*; *Federal Commerce & Navigation Co Ltd* v *Molena Beta Inc (The Benfri)* [1979] AC 757; [1978] 3 WLR 991; [1979] 1 All ER 307; [1979] 1 Lloyd's Rep 201, HL, [1978] QB 927; [1978] 3 WLR 309; [1978] 3 All ER 1066, [1978] 2 Lloyd's Rep 132, CA
James Shaffer Ltd v *Findlay Durham & Brodie* [1953] 1 WLR 106, CA
Mannai Investment Co Ltd v *Eagle Star Life Assurance Co Ltd* [1997] AC 749; [1997] 2 WLR 945; [1997] 3 All ER 352; [1997] 1 EGLR 57; [1997] 24 EG 122; [1997] 25 EG 138, HL
Ross T Smyth & Co Ltd v *TD Bailey Son & Co*; *sub nom TD Bailey Son & Co* v *Ross T Smyth & Co Ltd* [1940] 3 All ER 60; (1940) 56 TLR 825; 164 LT 102, HL
Spettabile Consorzio Veneziano di Armamento e Navigazione v *Northumberland Shipbuilding Co* (1919) 121 LT 628, CA K
Sweet & Maxwell Ltd v *Universal News Services Ltd* [1964] 2 QB 699; [1964] 3 WLR 356; [1964] 3 All ER 30, CA
Woodar Investment Development Ltd v *Wimpey Construction UK Ltd* [1980] 1 WLR 277; [1980] 1 All ER 571, HL

This was an appeal by the appellant, Eminence Property Developments Ltd, from a decision of Mr Recorder Blohm QC, sitting in the Bristol District Registry of the Chancery Division, dismissing a claim against the respondent, Kevin Heaney, for declaratory relief as to its entitlement to forfeit a deposit under a contract for the sale of land.

A Bernard Livesey QC and Paul Mitchell (instructed by Solicitor Direct, of Leyland) appeared for the appellant; Andrew Clarke QC and Charles Auld (instructed by Foot Anstey Solicitors, of Exeter) represented the respondent.

Giving judgment, **ETHERTON LJ** said:

Introduction

[1] The issue on this appeal is whether a vendor of land that served a notice to complete making the time for completion of the essence of the sale contract and then, mistakenly, treating the contract as at an end prior to the expiry of the notice, was thereby in repudiatory breach of B the contract, entitling the purchaser to terminate the contract.

[2] Mr Recorder Blohm QC, sitting as a judge of the High Court, held that vendor in the present case, Eminence Property Developments Ltd (Eminence), was, in those circumstances, in repudiatory breach of the contract of sale, and the purchaser, Mr Kevin Heaney, lawfully terminated the contract by accepting that repudiation. This is an appeal by Eminence from the recorder's order dated 15 January 2010.

Factual background

[3] The material facts, largely taken from the recorder's judgment, may be simply stated. Eminence is the freehold owner of a block of C one- and two-bedroom flats called Carpenters Place, in Bristol, which it constructed. Mr Heaney is a property developer. Prior to completion of the flats, Mr Heaney negotiated and agreed with Eminence the purchase by him off-plan of 13 flats for the aggregate price of £1.47m and a deposit of 2.5% of the purchase price of each flat, amounting to £36,750 in total. On 10 December 2007, 13 separate contracts were exchanged in identical terms, each contract being for the sale of a flat on a long lease.

[4] The contracts provided for the contractual completion date to be fixed by reference to the date on which the relevant flat under construction was ready for occupation. It was common ground before D the recorder that, in the circumstances that occurred, the contractual completion date was 4 December 2008.

[5] The contracts incorporated the standard conditions of sale (4th ed) (the conditions), subject to minor amendments. Condition 1.1 contains the following material definitions:

> 1.1.1(m) "Working day" means any day from Monday to Friday (inclusive) which is not Christmas Day, Good Friday or a statutory Bank Holiday
>
> 1.1.3 A party is ready, able and willing to complete:
>
> (a) if he could be, but for the default of the other party, and…

[6] Condition 6.1.1 provides:

> E 6.1.1… time is not of the essence of the contract unless a notice to complete has been served.

[7] Condition 6.8 concerns the service of a notice to complete and provides:

> 6.8. Notice to complete
>
> 6.8.1 At any time on or after the completion date, a party who is ready, able and willing to complete may give the other a notice to complete.
>
> 6.8.2 The parties are to complete, the contract within ten working days of giving a notice to complete, excluding the day on which the notice is given. For this purpose, time is of the essence of the contract.
>
> 6.8.3 On receipt of a notice to complete:
>
> F (a)…
>
> (b) if the buyer paid a deposit of less than 10 per cent (no less than £500), he is forthwith to pay a further deposit equal to the balance of that 10 per cent deposit

[8] Condition 7.5 concerns the consequences of the buyer's failure to comply with a notice to complete, and provides:

> 7.5 Buyer's failure to comply with notice to complete.
>
> 7.5.1 If the buyer fails to complete in accordance with a notice to complete, the following terms apply:
>
> 7.5.2 The seller may rescind the contract, and if he does so:
>
> (a) he may
>
> (i) forfeit and keep any deposit and accrued interest
>
> (ii) resell the property and any chattels included in the contract G
>
> (iii) claim damages,
>
> (b) The buyer is to return any documents he received from the seller and is to cancel any registration of the contract.
>
> 7.5.3 The seller retains his other rights and remedies.

[9] Following the making of the contracts, there was a significant downturn in the property market. By late 2008, Mr Heaney was having difficulty raising the funds necessary to complete at the contract price. There were negotiations between Eminence and Mr Heaney for a lower purchase price and higher specification for the flats, but no concluded agreement was ever reached. H

[10] Eminence wanted to complete the contracts on 4 December 2008, the contractual completion date, but Mr Heaney did not do so. On 5 December 2008, Eminence, by its solicitor, Solicitor Direct, served a notice to complete in respect of each contract pursuant to condition 6.8. Each notice referred to the relevant contract for the flat in question, stated that the sale had not been completed on the date fixed and that the seller was ready, willing and able to complete, gave "notice under condition 6.8 of the Standard Conditions of Sale 4th edition (to which the contract was made subject) to complete the contract in accordance with that condition" and drew "attention to the consequences set out in condition 7.5 of the Standard Conditions" should Mr Healey "fail J to complete the transaction within 10 working days of giving [the] notice, exclusive of the date on which it was given". It is not disputed that the terms of the notices were in accordance with the contracts and the conditions.

[11] The notices were sent under cover of a letter dated 5 December 2008 from Mr Matthew Jones, a solicitor with Solicitor Direct, to Foot Anstey, Mr Heaney's solicitor. In the letter, Mr Jones referred to the flats that Mr Heaney had contracted to purchase, and said:

> We are disappointed to note that we have not completed the above and we have had no explanation as to why. K

[12] Mr Jones referred to the enclosed notices to complete. He then set out "the contractual daily compensation rate for each flat", said that Mr Heaney was liable for certain legal costs and concluded, in the last paragraph:

> We calculate the final date for completion under the notice is 15th December 2008.

[13] That statement as to the final date for completion was incorrect. It is common ground on this appeal that the combined effect of condition 6.8.2 and the definition of "working day" in condition 1.1.1(m) was that the notices to complete expired on L 19 December 2008. Mr Jones's evidence before the recorder, in para 28 of his witness statement, as to why he had said in the letter that the final date for completion under the completion notices was 15 December 2008, was as follows:

> I am not sure why I did this. I presume I calculated the "days" rather than the "working days". I can only put this down to human error on my part.

[14] Mr Heaney took no steps to complete the contracts. Eminence decided to rescind them pursuant to condition 7.5 and to exercise the other remedies specified in that condition, including forfeiting the M deposits. However, it did so prematurely. On 17 December 2008, under cover of a letter from Mr Jones to Foot Anstey, Solicitor Direct sent notices of rescission in respect of each contract. Each notice said as follows:

> Further to the expiry of the notice to complete served under condition 6.8 of the Standard Conditions of Sale 4th edition (The Standard Conditions) you have failed to complete the purchase of the above property. Pursuant to 7.5 of the Standard Conditions please treat this letter as formal notice of rescission of the Agreement for sale dated 10th December 2007. Your clients deposit has been forfeited along with accrued interest and our client is entitled to damages for breach of contract, including but not limited to the remainder of the deposit payable under the contract.

A [15] In the covering letter, Mr Jones asked Foot Anstey to "confirm whether you will be dealing with the claim against your client as well".

[16] Foot Anstey replied by letter dated 18 December 2008. The recorder described it as "almost by return". It referred to "each of your letters of 17 December 2008" and to the notices to complete and the provisions of condition 6.8, and continued as follows:

> 1.3. The Notices to Complete that were served by you required completion of the contracts on or before 18 December 2008, which is the date 10 working days after service of the Notices to Complete (assuming it was deemed served on 5 December rather than 8 December for the purpose of standard
> B condition 1.3.
>
> 1.4. The act of rescinding contracts under cover of your letters of 17th December 2008 constitutes a repudiatory breach of contract. Our client accepts your clients' repudiatory breach of contract and elects to rescind the contracts and is discharged from them.
>
> 1.5. Please remit, by return, our client's deposits together with any interest accrued on those deposits.

[17] That letter from Foot Anstey was written by Mr Richard Bagwell, who was a litigator and not involved in the conveyancing. The promptness of the response to the letter from Solicitor Direct enclosing
C the rescission notices and the fact that the letter was written by Mr Bagwell led the recorder to speculate that Foot Anstey was already aware, on receipt of Mr Jones's letter of 5 December 2008 enclosing the notices to complete, that Mr Jones was mistaken in his stated calculation of the final date for completion as being 15 December 2008, and that (as the recorder said in [15] of his judgment) "they were prepared and awaiting Eminence to fall into the elephant trap that it had dug for itself".

[18] Solicitor Direct replied to Foot Anstey's letter of 18 December 2008 on the same day. It said:

> Further to your letter of the 18th December 2008 we note your comments.
> D As you will note the rescission notice stipulates "Further to the expiry of the notice to complete…" Your client is still entitled to the full 10 working days to complete under the notice. The contents letter of the 17th was not intended to come into effect until after the expiry of the notice. If your client is willing and able to complete we look forward to receiving the purchase prices.

[19] Mr Bagwell's response by e-mail on the same day was that:

> you cannot seek to retrospectively alter the fact of rescission… Your notice of rescission was unambiguous and immediate, as were the consequences. Your client has committed a repudiatory breach of contract, which our client has accepted. Please send the deposits by return.

E [20] It is frankly accepted on behalf of Mr Heaney that he was pleased to have the opportunity to extricate himself from the contracts.

These proceedings

[21] On 9 February 2009, Eminence commenced these proceedings against Mr Heaney for a declaration that the deposits had been forfeited as at 19 December 2008 and for damages for breach of contract. The particulars of claim proceed on the basis that Mr Heaney's default was in failing to complete on or before 19 December 2008 or at all.

[22] By his defence and counterclaim, Mr Heaney alleged that it was not possible for him to complete the purchase on 18 or
F 19 December 2008 because Eminence "had by then wrongfully purported to rescind the contract and to have forfeited the Defendant's deposits". He counterclaimed for a declaration that the parties were discharged from future performance of their obligations under the contracts and for the repayment of the deposits.

[23] On 10 July 2009, District Judge Watson ordered the trial of the following preliminary issue:

> whether the defendant has rescinded the contract in relation to the claimant's repudiatory breach and that, as a result, the parties are discharged from future performance of their obligations under the 13 contracts, all dated 10th December 2007, for the sale and purchase of 999 year leasehold interests in each of plots 1, 2, 3, 4, 5, 6, 7, 8, 9, 16, 19 and 24 Carpenters Place, Melvin Square, Bristol.

Judgment G

[24] That issue was tried before the recorder on 14 and 15 January 2010.

[25] He reached the following conclusions of fact, against which there is no appeal or cross appeal:

> [30]…
>
> (1) The claimant's solicitors intended to serve the notice of rescission after the expiry of the notice to complete;
>
> (2) They miscalculated the period of the notice to complete, probably by counting ten days rather than ten working days, but I do not think that the mechanism for the mistake matters too much; H
>
> (3) On receipt of the notice Foot Anstey were aware that the claimants could only serve a valid notice of rescission after 19th December 2008;
>
> (4) Foot Anstey believed that the claimants had made a mistake in calculation. I reach that conclusion because, as I say, it is screamingly obvious;
>
> (5) A reasonable recipient of the notice to complete would have been aware of the claimant's intention and would also have been aware that the letter had been sent when it was sent by reason of a mistake and would have believed that had the claimants been aware of the true position the letter would not have been sent, but would have been sent after 19th December 2008 had the contract remained uncompleted.

[26] The recorder noted that Eminence, by its then counsel, J
Mr Steven Ball, advanced two arguments. The first was that, on the basis of the approach to interpretation in *Mannai Investment Co Ltd* v *Eagle Star Assurance Co Ltd* [1997] AC 749*, the notices to complete should be interpreted as determining the contracts on 19 December 2008. The second argument was that, in the light of the decision of the House of Lords in *Woodar Investment Development Ltd* v *Wimpey Construction UK Ltd* [1980] 1 WLR 277 and of Gloster J in *CDV Software Entertainment AG* v *Gamecock Media Europe Ltd* [2009] EWHC 2965 (Ch), the letter of 17 December 2008 was, on any view, not an unequivocal repudiation of the contracts.

[27] The recorder rejected both arguments. He gave his brief K
conclusion on the first argument as follows:

> [32] I agree with Mr Auld [counsel for Mr Heaney] that the only meaning that can be given to the letter of 17th December 2008 is that Eminence intended to bring the contract to an end immediately and that they did so, and said they did so, on the assumption that the time for completion of the notice to complete had expired. In my view the other possible meanings of the letter put forward by Mr Ball were not realistic meanings.

[28] There is no appeal against that finding.

[29] As to the second argument, the recorder referred to *Woodar*, *CDV Software Entertainment*, *Chilean Nitrate Sales Corporation* v *Marine Transportation Co Ltd (The Hermosa)* [1982] 1 Lloyd's L
Rep 570, *Dalkia Utilities Services plc* v *Celtech International Ltd* [2006] EWHC 63 (Comm); [2006] 1 Lloyd's Rep 599 and *Spettabile Consorzio Veneziano di Armamento e Navigazione* v *Northumberland Shipping Co Ltd* (1919) 121 LT 628. His analysis and conclusions on the second argument were as follows:

> [41] In the present case I conclude that the letter of 17th December 2008 comprised a clear case of refusal by Eminence to perform their future obligations under the contract, those obligations going to the very root of the contract. That, after all, is the very purpose of a letter terminating a contract after a failure to comply with a notice to complete. This letter made that doubly clear by referring to the remedies that the vendor had and was exercising, the forfeiture M
> of the deposit and the claim for damages. These are remedies which were only available on the termination of the contract by the vendor (see the general conditions of sale, which are the standard conditions, I think, paragraph 7.5).
>
> [42] Next, in my view, it is abundantly plain that this letter was not intended to be conditional. It set out the basis for termination, which was the failure on the part of Mr Heaney to complete, and it set out its entitlement to repudiate, which was based on the notice to complete having made time of the essence. These were the basic assumptions underlying the letter and the termination. Had Eminence, or more realistically their solicitors, been aware of the true position,

---

* Editor's note: Also reported at [1997] 1 EGLR 57; [1997] 24 EG 122; [1997] 25 EG 138

they would undoubtedly not have served that letter, or indeed any letter, at that time. That seems to me to be wise with the benefit of hindsight.

[43] The letter does not seek to reserve their position should these assumptions be incorrect. One can understand why it does not do so. There may be difficulties for a party who wishes to accept a repudiatory breach if his letter of acceptance is couched in equivocal terms and therefore conditions or reservations may not be easy to express without risk to the basic effectiveness of the letter. Be that as it may, the letter was, in my view, clear in its terms. Having all due regard to the circumstances surrounding it, it is simply not possible to construe it as being conditional upon the assumptions that it makes being correct. In my view the letter has the effect set out in the words of Lord Justice Atkin at *Spettabile Consorzio Veneziano di Armamento e Navigazione* v *Northumberland Shipbuilding Co Ltd* [1918-1919] All ER Rep 963 in page 968 that: "The party to the contract made quite plain his own intention not to perform the contract."

[44] I am fortified in this view by the reference to the vendor's remedies contained in the letter and in particular the reference to forfeiture of the deposit. I do not agree with Mr Auld's submission that this changed the status of the deposit. The deposit was paid to Eminence's solicitors as agents for Eminence and therefore became their property. Eminence was subject to the contractual obligation to repay that sum in certain circumstances. The announcement of forfeiture simply purported to affect Eminence's remedy on the contract coming to an end. However, it did underline the assertion that the contract was at an end and that Eminence would not perform any of its obligations but would, henceforth, act on the basis that the contract had been determined.

[45]…

[46] Although I do not base my decision on the more general consequences of finding in Eminence's favour, it does seem to me that a finding that Eminence's letter of termination was only conditional would be contrary not only to the true construction of the letter but also to the understood practice of conveyancers, certainly as I understand it, and more generally would be damaging of the reasonable expectations and requirements of certainty as regards contractual matters.

[47] In those circumstances my conclusion is that Eminence committed a repudiatory breach of the contract by their letter of 17th December 2008 and that that breach was accepted by the letter dated 18th December 2008 sent by Foot Anstey and that the contract has therefore been determined, and I answer the preliminary issue accordingly.

Appeal

[30] Eminence's case on this appeal, as presented by its leading counsel, Mr Bernard Livesey QC, may be summarised briefly as follows. Eminence will have repudiated the contracts if, but only if, by its conduct it clearly intimated an intention to abandon and altogether to refuse to perform them. That is a matter of fact, to be considered objectively, that is to say from the perspective of a reasonable person in the position of Mr Heaney, looking at all the circumstances and Eminence's entire conduct. In support of those propositions, Mr Livesey relied on the speeches of the majority of the appellate committee of the House of Lords in *Woodar* and on the judgment of Donaldson LJ in *Chilean Nitrate Sales*.

[31] Mr Livesey submitted that, having regard to those principles, the recorder could not properly have found that the notices of rescission sent on 17 December 2008 constituted a repudiatory breach of the contracts. He said that, first and critically, the notices, far from being clear, were internally inconsistent. Each of them referred to the notice to complete, which had been properly served and was properly worded in accordance with the conditions, and stated that the notice to complete was served under condition 6.8 and that notice of rescission was being given pursuant to condition 7.5. In effect, the rescission notice incorporated those conditions by reference, and so disclosed an obvious inconsistency between the requirement of the conditions that 10 working days must expire before the remedies under condition 7.5 arose and the statement in the notice that the contract was thereby rescinded and the deposit forfeit. Second, a reasonable person would have appreciated, and as the recorder found as a fact Mr Heaney, by his solicitor, on receipt of the notices of rescission, was perfectly aware that a mistake had been made by Mr Jones in prematurely giving notice of rescission. Mr Heaney knew, or at any event a reasonable person in Mr Heaney's position would have known, that had Mr Jones been aware of the correct position, Eminence would have waited until after the expiry of the 10 working days specified in condition 6.8 before rescinding the contracts. Third, there was never any doubt, objectively or in the mind of Mr Heaney, that Eminence wanted to enforce the contracts rather than abandon them. That was apparent from the fact that the property market had fallen since the contracts had been made, as well as the fact that Eminence was manifestly intending to follow the contractual procedure in serving the notices to complete under condition 6.8 and then purportedly exercising the remedies conferred by condition 7.5.

[32] Mr Andrew Clarke QC, for Mr Heaney, advanced the following six key propositions in support of the decision of the recorder. First, a clear and absolute declaration that a contract is at an end, the making of which is later found to be a breach of contract, is repudiatory of that contract. Second, the mere fact that the party making the declaration asserts and believes that it is acting in accordance with the terms of the contract does not make what would otherwise be a repudiatory breach non-repudiatory. Third, one party can declare to the other that a contract should be interpreted or performed in a particular way without being thereby in repudiatory breach of contract if it is wrong. Parties often make such an assertion while performing the contract (as they see it) and, in such cases, the court must determine whether their conduct was in all the circumstances repudiatory. To declare the contract to be at an end in reliance on that erroneous view is altogether different and repudiatory of the contract: it is conduct inconsistent with the continuance of the contract. Fourth, that last proposition is subject to qualification where the declaration is merely a vehicle to achieve a court resolution of a dispute *and* the parties anticipate that the one making the declaration will perform if wrong (and such performance is possible). In such a case, what would otherwise be a declaration that the contract is at an end does not have the quality because of the parties' agreement. Fifth, the fact that a party is mistaken in its grounds for declaring a contract to be at an end (even where that the mistake is known to the other party) does not deprive an otherwise repudiatory breach of that quality. Sixth, the present case is one in which there was a clear, absolute and immediate repudiation of the contract by the notices of rescission — a declaration of termination coupled with forfeiture and a claim for damages: that repudiation could be and was accepted. The fact that Eminence was mistaken and that was known to Mr Heaney is of no assistance to the Eminence since it declared the contract to be at an end and acted accordingly. In support of those propositions, Mr Clarke relied particularly on *Federal Commerce & Navigation Co Ltd* v *Molena Alpha Inc (The Nanfri, Benfri and Lorfri)* [1978] QB 927 (CA), [1979] AC 757 (HL) (*The Nanfri*), *Woodar, Chilean Nitrate Sales,* and *Bridgen* v *Lancashire County Council* [1987] IRLR 58. Mr Clarke submitted that, in the circumstances, the recorder's decision was a proper allocation of risk as between Eminence and Mr Heaney.

Discussion

[33] Notwithstanding the carefully crafted and attractive submissions of Mr Clarke and the recorder's detailed analysis, I consider that it is clear that the notices of rescission did not constitute a repudiatory breach of the contracts. I reach that conclusion essentially for the reasons advanced by Mr Livesey.

[34] It is convenient to refer, first, to *The Nanfri* because reference is made to that case in *Woodar. The Nanfri* concerned time charters by which ships were let for periods of around six years. Most of the cargoes were carried on cif terms, the shippers paying the freight in advance and receiving "freight pre-paid" bills of lading. The charters provided for the payment of hire twice monthly in advance; the owners to have the right to withdraw the vessel in default of payment. The charters also contained provision for permissible deductions from hire and stipulated that the masters were to be under the orders of the charterers with regard to employment, agency or other arrangements. Accordingly, the bills of lading were issued and signed by the charterers on behalf of the master and freight was paid to the charterers or their agents. The charterers made deductions from hire that the owners did not accept were permissible. The owners informed the charterers that the authority of the charterers or their agents to sign any bill of lading was being withdrawn. The owners also instructed the masters not to

sign any bill of lading endorsed "freight pre-paid" or not bearing an indorsement incorporating a lien in favour of the owners on all cargoes and sub-freights belonging to the charterers and any bill of lading freight. If the order to the masters had been implemented, it would have had disastrous consequences for the charterers. The charterers treated the totality of the owners' conduct as a repudiation of the charters, which they accepted. On a case stated in respect of each ship from arbitrators, it was held by the Court of Appeal and the House of Lords that the charterers were entitled to make the deductions and that the owners had, by their conduct, repudiated the charters, which repudiation had been accepted by the charterers.

[35] The owners argued that their conduct was not repudiatory because they had acted on the advice of their lawyers in New York and London, and they were under an honest misapprehension as to their rights. They relied on *James Shaffer Ltd* v *Findlay Durham & Brodie* [1953] 1 WLR 106, *Sweet & Maxwell Ltd* v *Universal News Services Ltd* [1964] 2 QB 699 and *Ross T Smyth & Co Ltd* v *TD Bailey Son & Co* (1940) 56 TLR 825. In the Court of Appeal, Lord Denning, rejecting that argument and distinguishing those three cases, said, at pp979E-980A:

I have yet to learn that a party who breaks a contract can excuse himself by saying that he did it on the advice of his lawyers: or that he was under an honest misapprehension. Nor can he excuse himself on those grounds from the consequences of a repudiation. In those three cases the conduct of the party concerned was entirely innocent. It did not evince any intention to break his contractual obligations. I would go by the principle as I have always understood it that if the party's [conduct] — objectively considered in its impact on the other party — is such as to evince an intention no longer to be bound by his contractual obligations, then it is open to the other party to accept his repudiation and treat the contract as discharged from that time onwards. A most important point here is that the conduct of the owner was such as to lead the charterers reasonably to believe that the owners would issue such orders again in the future whenever they, the owners, wished to force the charterers to comply with the owners' demands in similar circumstances. In short, the owners were determined to give orders to the masters — in flat contradiction of the charterparty — time and time again so long as the contract continued so as to enforce their demand that hire should be paid in full without any deductions unless the owners agreed. To my mind such conduct amounted to a repudiation of the contract within the principles laid down by Lord Blackburn in *Mersey Steel and Iron Co Ltd* v *Naylor, Benzon & Co* (1884) 9 App Cas 434, 442-444 and followed ever since. It was open to the charterers to accept the repudiation, as they did, and to treat themselves as discharged from any further performance. It was then at an end for the future. It cannot be revived by any subsequent "without prejudice" agreement.

[36] Goff LJ emphasised the need for an objective assessment. He said, at p991D:

The question is not what the owners wanted or wished in the recesses of their minds, but did they by their conduct evince an intention no longer to be bound by the contract or to perform it only in a way inconsistent with their obligations under the charter?

[37] Goff LJ quoted the following findings of fact by the arbitrators, which he described, at p991E, as "entirely destructive of the owners' case":

The issuing of the order of October 4, 1977, was not to secure their claim for disputed deductions (the offer for an escrow deposit having been rejected by owners as "meritless") but rather to compel the charterers to pay over all sums deducted from hire by the charterers which the owners disputed, irrespective of whether such deductions should ultimately be determined to be valid or invalid, in whole or in part, and to ensure that in the future the charterers made no deductions whatever from hire unless they had previously been expressly approved by the owners or they were supported by vouchers signed by the master or a proper tribunal had pronounced on their validity. At the same time, the owners knew that part of the amounts, payment of which they were demanding as a condition of the withdrawal of their orders, were due to the charterers.

[38] He concluded, at p992E, that:

the irresistible inference is that they were saying, "We will only perform the contract upon our terms and not on yours" and, as Lord Denning MR has said, they were holding a pistol to the heads of the charterers and we have the finding as to their true purpose which I have read from the award.

[39] On appeal to the House of Lords, Lord Wilberforce described the law in this area as "perspicuous". He said the following, at pp778F-779C:

Was this breach, or threatened breach, repudiatory or not? I shall not set out at any length the numerous authorities on anticipatory breach: this is one of the more perspicuous branches of the law of contract and the modern position is clear. The form of the critical question may differ slightly as it is put in relation to varying situations:

"... an intimation of an intention to abandon and altogether to refuse performance of the contract..." or "evince an intention no longer to be bound by the contract... ." (*Freeth* v *Burr* (1874) LR 9 CP 208, 213, *per* Lord Coleridge CJ)

"I do not say that it is necessary to show that the party alleged to have repudiated should have an actual intention not to fulfil the contract. He may intend in fact to fulfil it, but may be determined to do so only in a manner substantially inconsistent with his obligations, and not in any other way" (*Ross T Smyth & Co Ltd* v *TD Bailey, Son & Co* [1940] 3 All ER 60, 72, *per* Lord Wright) such as to deprive "the charterers of substantially the whole benefit which it was the intention of the parties. ... that the charterers should obtain from the further performance of their own contractual undertakings" (*Hongkong Fir Shipping Co Ltd* v *Kawasaki Kisen Kaisha Ltd* [1962] 2 QB 26, 72, *per* Diplock LJ).

"To constitute repudiation, the threatened breach must be such as to deprive the injured party of a substantial part of the benefit to which he is entitled under the contract. ... Will the consequences of the breach be such that it would be unfair to the injured party to hold him to the contract and leave him to his remedy in damages...?" *Decro-Wall International SA* v *Practitioners in Marketing Ltd* [1971] 1 WLR 361, 380, *per* Buckley LJ).

The difference in expression between these two last formulations does not, in my opinion, reflect a divergence of principle, but arises from and is related to the particular contract under consideration: they represent, in other words, applications to different contracts, of the common principle that, to amount to repudiation a breach must go to the root of the contract.

[40] Lord Wilberforce, like the Court of Appeal, dismissed as irrelevant any subjective intention or desire of the party in breach not to abandon the contract. He said, at p780E-F:

It is thirdly irrelevant that it was in the owners' real interest to continue the charters rather than to put an end to them. If a party's conduct is such as to amount to a threatened repudiatory breach, his subjective desire to maintain the contract cannot prevent the other party from drawing the consequences of his actions. The two cases relied on by the appellants (*James Shaffer Ltd* v *Findlay Durham & Brodie* [1953] 1 WLR 106 and *Sweet & Maxwell Ltd* v *Universal News Services Ltd* [1964] 2 QB 699) do not support a contrary proposition, and would only be relevant here if the owners' action had been confined to asserting their own view — possibly erroneous — as to the effect of the contract. They went, in fact, far beyond this when they threatened a breach of the contract with serious consequences.

[41] That case is the principal authority on which Mr Clarke relied for his submissions that: it is irrelevant that the rescission letter of 17 December 2008 was served under a mistake on the part of Eminence's lawyer; had Eminence appreciated the mistake, it would not have sent the letter; and Eminence's preference was to maintain the contracts. Mr Clarke further submitted that the present case is an even stronger and clearer one of repudiatory conduct since the owners' conduct in *The Nanfri* concerned the manner in which they intended to perform the contract, whereas the present case concerns Eminence's act of bringing the contracts finally to an end.

[42] *Woodar* concerned a contract for the sale of development land. A special condition provided for the purchasers to have the following right to rescind:

if prior to the date of completion... (iii) any authority having a statutory power of compulsory purchase shall have commenced to negotiate for the acquisition by agreement or shall have commenced the procedure required by law for the compulsory acquisition of the property or any part thereof.

[43] The purchasers sent to the vendors a notice purporting to rescind the contract under that provision on the ground that the Secretary of State

A for the Environment had commenced the procedure for compulsory acquisition of the land. The vendors brought proceedings against the purchasers for a declaration that the condition gave the purchasers no right to rescind and for damages for breach of contract in serving the notice. The House of Lords, allowing the purchasers' appeal, held that the service of the notice of rescission was not a repudiatory breach of the contract.

[44] Lord Wilberforce emphasised that, in considering whether the purchasers had repudiated the contract, it was necessary to consider all the circumstances and, in particular, the purchasers' conduct as a whole. He said, at pp280F-281A:

> B This gives rise to the first issue in this appeal: whether, by invoking special condition E(a)(iii), and in the circumstances, the appellants are to be taken as having repudiated the contract. The respondents so claim, and assert that they have accepted the repudiation and are entitled to sue the appellants for damages.
>
> My Lords, I have used the words "in the circumstances" to indicate, as I think both sides accept, that in considering whether there has been a repudiation by one party, it is necessary to look at his conduct as a whole. Does this indicate an intention to abandon and to refuse performance of the contract? In the present case, without taking the appellants' conduct generally into account, the respondents' contention, that the appellants had repudiated, would be a difficult one. So far from repudiating the contract, the appellants were
> C relying on it and invoking one of its provisions, to which both parties had given their consent. And unless the invocation of that provision were totally abusive, or lacking in good faith, (neither of which is contended for), the fact that it has proved to be wrong in law cannot turn it into a repudiation. At the lowest, the notice of rescission was a neutral document consistent either with an intention to preserve or with an intention to abandon the contract, and I will deal with it on this basis — more favourable to the respondents. In order to decide which is correct the appellants' conduct has to be examined.

[45] Lord Wilberforce stated that the purchasers' subjective intention was inconclusive. He said, at p281D:

> It is clear in the first place that, subjectively, the appellants, in 1974, wanted
> D to get out of the contract. Land prices had fallen, and they thought that if the contract were dissolved, they could probably acquire it at a much lower price. But subjective intention is not decisive: it supplied the motive for serving the notice of rescission: there remains the question whether, objectively regarded, their conduct showed an intention to abandon the contract.

[46] Lord Wilberforce then set out the relevant facts. He referred to a meeting between the purchasers' advisers and a Mr Cornwall, acting for the vendors, before service of the notice of rescission, in which Mr Cornwall was told of the intention to serve the notice, and Mr Cornwall said that, if that were done, the vendors would take the purchasers to court and would let the judge decide whether the contract
E could be rescinded. Mr Cornwall also said that if a notice to rescind were served, he would accept it as intending to protect the purchasers fully and prudently and would not regard it as a hostile act. Following the service of the notice, the vendors' solicitor wrote that the vendors did not accept its validity and they instituted proceedings. Mr Cornwall then wrote to the purchasers' president, in which reference was made to the proceedings, stating that both parties would have to await the decision of the court as to the validity of the purchasers' claim to be entitled to rescind on the grounds stated. He wrote a further letter to the effect that the vendors and, doubtless, the purchasers would abide by the result of the court proceedings. Lord Wilberforce summarised as
F follows the relevance of that evidence, at pp282D and 282H-283B:

> My Lords, I cannot find anything which carries the matter one inch beyond, on Wimpey's part, an expressed reliance on the contract (condition E(a)(iii)), on Woodar's side an intention to take the issue of the validity of the notice (nothing else) to the courts, and an assumption, not disputed by Wimpey, that both sides would abide by the decision of the court. This is quite insufficient to support the case for repudiation.
>
> …
>
> My Lords, in my opinion, it follows, as a clear conclusion of fact, that the appellants manifested no intention to abandon, or to refuse future performance of or to repudiate the contract. And the issue being one of fact, citation of other decided cases on the other facts is hardly necessary. I shall simply state that the proposition that a party who takes action relying simply on the terms of the contract, and not manifesting by his conduct an ulterior intention to abandon it, G
> is not to be treated as repudiating it is supported by *James Shaffer Ltd* v *Findlay Durham & Brodie* [1953] 1 WLR 106 and *Sweet & Maxwell Ltd* v *Universal News Services Ltd* [1964] 2 QB 699.

[47] Lord Wilberforce contrasted *The Nanfri.* He said, at p283B-C:

> In contrast to these is the case in this House of *Federal Commerce & Navigation Co Ltd* v *Molena Alpha Inc* [1979] AC 757 which fell on the other side of the line. Of that I said at p780:
>
> The two cases relied on by the appellants (*James Shaffer Ltd* v *Findlay Durham & Brodie* [1953] 1 WLR 106 *and Sweet & Maxwell Ltd* v *Universal News Services Ltd* [1964] 2 QB 699)… would only be relevant here if the owners' action had been confined to asserting their own view — possibly H
> erroneous — as to the effect of the contract. They went, in fact, far beyond this when they threatened a breach of the contract with serious consequences.

[48] He concluded on this aspect as follows, at p283D-E:

> In my opinion therefore the appellants are entitled to succeed on the repudiation issue, and I would only add that it would be a regrettable development of the law of contract to hold that a party who bona fide relies upon an express stipulation in a contract in order to rescind or terminate a contract should, by that fact alone, be treated as having repudiated his contractual obligations if he turns out to be mistaken as to his rights. Repudiation is a drastic conclusion which should only be held to arise in clear cases of a refusal, in a matter going to the root of the contract, to perform contractual obligations. To uphold the J
> respondents' contentions in this case would represent an undesirable extension of the doctrine.

[49] Lord Keith, agreeing with Lord Wilberforce in the result, also emphasised, at p294G-H, the need to consider the matter objectively and to look at all the circumstances. He referred, at p295A, to "a tract of authority which vouches the proposition that the assertion by one party to the other of a genuinely held but erroneous view as to the validity or effect of a contract does not constitute repudiation". He cited, in that regard, *Spettabile*, *James Shaffer* and *Sweet & Maxwell,* and he quoted, at p295H, the following statement of Lord Wright in *Ross T Smyth*, at p107: K

> a mere honest misapprehension, especially if open to correction, will not justify a charge of repudiation.

[50] Lord Keith continued as follows, at p296A:

> So in the present case the question comes to be whether, having regard to all the circumstances, the conduct of the appellants in relation to their invocation of special condition E(a)(iii) of the contract was such that a reasonable person in the position of the respondents would properly infer an intention "in any event," to use the expression employed by Warrington and Atkin LJJ in the *Spettabile* case, 121 LT 628, to refuse to perform the contract when the time came for performance. L

[51] He concluded that, on the facts, he was unable to regard the purchasers' conduct as evincing an intention altogether to refuse performance of the contract. He distinguished the comments of Lord Denning MR in *The Nanfri*, quoted above, on the basis that the time for performance had not yet passed. He said, at pp296H-297C:

> I would accept without hesitation the statement of Lord Denning MR in *Federal Commerce & Navigation Co Ltd* v *Molena Alpha Inc* [1978] 1 QB 927, 979 that a party who breaks a contract cannot excuse himself by saying that he did it on the advice of his lawyers, or that he was under an honest misapprehension. If in the present case the time for performance had passed M
> while the appellants were still maintaining their position based on the erroneous interpretation of special condition E(a)(iii), they would have been in breach of contract and liable in damages accordingly. Lord Denning goes on to say: "Nor can he excuse himself on those grounds from the consequences of a repudiation." That may be so, but it is first necessary to determine whether or not there has been a repudiation.
>
> The doctrine of repudiatory breach is largely founded upon considerations of convenience and the opportunities which it affords for mitigating loss, as observed by Cockburn CJ in *Frost* v *Knight* (1872) LR 7 Ex 111, 114. It enables one party to a contract, when faced with a clear indication by the other that he does not intend to perform his obligations under it when the time for performance arrives, to treat the contract, if he so chooses, as there and then at an end and to claim damages as for actual breach. Where one party, honestly but

A erroneously, intimates to the other reliance upon a term of the contract which, if properly applicable, would entitle him lawfully to rescind the contract, in circumstances which do not and are not reasonably understood to infer that he will refuse to perform his obligations even if it should be established that he is not so entitled, legal proceedings to decide that issue being in contemplation, I do not consider it in accordance with ordinary concepts of justice that the other party should be allowed to treat such conduct as a repudiation. Nor, in my opinion, are there any considerations of convenience which favour that course.

[52] Lord Scarman agreed with the reasons given by Lord Wilberforce. He said that the error of the majority of the Court of Appeal was in concentrating on the notice of rescission and its B accompanying letter and in failing to take into account all the acts and conduct of the purchasers in their dealings with Mr Cornwall. He elaborated as follows, at p299C-H:

> My Lords, as I see it, the error of the majority of the Court of Appeal in the instant case was, notwithstanding some *dicta* to the contrary, to concentrate attention on one act, ie the notice of rescission with its accompanying letter. They failed to give the consideration which the law requires of all the acts and conduct of the defendants in their dealings with Mr Cornwell — the "alter ego" of the plaintiff company. The law requires that there be assessed not only the party's conduct but also, "objectively considered," its impact on the other party. The error is neatly exposed in Goff LJ's terse conclusion: "In my C judgment rescission is repudiation, and if it cannot be justified by the terms of the contract it is wrongful and a breach." The learned Lord Justice was, with respect, concentrating too much attention on one act isolated from its surrounding circumstances and failing to pay proper regard to the impact of the party's conduct upon the other party.
>
> In this case the contract provided for the possibility of rescission by the defendants. But the notice of rescission, which the defendants gave, was not, in the circumstances which existed when it was given, one which the defendants had any contractual right to give. But they honestly believed the contract did give them the right. When one examines the totality of their conduct and its impact upon Mr Cornwell it is plain, as shown by my noble and learned friend's analysis of the facts, that the defendants, though claiming mistakenly to exercise a power given them by the contract to bring it to an end, were not evincing an D intention not to be bound by the contract. On the contrary, they believed they were acting pursuant to the contract. And Mr Cornwell well understood the situation… It never occurred to Mr Cornwell that the defendants, if held not to have been entitled to give notice of rescission, would refuse to perform the contract. In fact, it would seem that he believed exactly the contrary. Such was the impact upon him of the defendants' conduct.

[53] Lord Salmon and Lord Russell dissented. Mr Clarke relied on the following passage in Lord Salmon's speech, at p288E-G:

> I do not recall that any of these definitions of a repudiation of a contract have ever until now, been questioned. The fact that a party to a contract mistakenly believes that he has the right to refuse to perform it cannot avail him. Nor is E there any authority for the proposition that if a party to a contract totally refuses to perform it, this refusal is any the less a repudiation of the contract because he honestly but mistakenly believes that he is entitled by a condition of the contract to refuse to perform it.
>
> It would indeed be unfortunate if the law were otherwise. A mistake in the construction of a contractual condition, even such a glaringly obvious mistake as the present can apparently easily be made especially perhaps when the market price has fallen far below the contract price. It is acknowledged in this case that the mistake was an honest one. If, however, a case arose in which a mistake of this kind was alleged to be an honest mistake, but not acknowledged to be so, it would be extremely difficult, if not impossible to prove the contrary.

F [54] Mr Clarke distinguished *Woodar* from the present case as turning on its own special facts. It is the basis of his fourth key proposition. He submitted that it is authority that an act or declaration wrongly terminating a contract will not be a repudiatory breach only where the parties have agreed in advance that if and when the act is done or the declaration made, the question of the validity and effectiveness of the act or declaration will be brought before the courts and the parties will abide by the court's decision. Mr Clarke relied on *Woodar* as further authority that subjective intention is irrelevant and he relied on the passage in Lord Salmon's speech, which I have quoted, as illustrating the practical difficulty of distinguishing between honest mistakes precluding repudiation and other mistakes consistent with repudiation.

[55] In *Chilean Nitrate Sales*, Donaldson LJ, giving the judgment of G the court, said, at p572, that the following propositions could be taken from *Woodar* and elsewhere:

> (a) Dissolution of a contract upon the basis of renunciation is a drastic conclusion which should only be held to arise in clear cases of a refusal to perform contractual obligations in a respect or respects going to the root of the contract.
>
> (b) The refusal must not only be clear, but must be absolute. Where a party declares his intention to act or refrain from acting in a particular way on the basis of a particular appreciation of his obligations, either as a matter of fact or of law, the declaration gives rise to a right of dissolution only if in all the circumstances it is clear that it is not conditional upon his present appreciation H of his obligations proving correct when the time for performance arrives.
>
> (c) What does or does not amount to a sufficient refusal is to be judged in the light of whether a reasonable person in the position of the party claiming to be freed from the contract would regard the refusal as being clear and absolute?
>
> … (d) [T]he conduct relied upon is to be considered as at the time when it is treated as terminating the contract, in the light of the then existing circumstances. These circumstances will include the history of the transaction or relationship. Later events are irrelevant, save to the extent that they may point to matters which the parties should have considered as hypothetical possibilities at the relevant time.

[56] Mr Clarke submitted that Eminence's conduct in serving the J notices of rescission was, objectively, a clear and absolute refusal to perform its contractual obligations going to the root of the contracts, within the first two propositions of Donaldson LJ; further, in accordance with Donaldson LJ's fourth proposition, what occurred after the service of the notices, in particular Eminence's realisation of its mistake and its change of heart, is irrelevant.

[57] Mr Clarke referred us to *Bridgen*. That case concerned an unsuccessful claim for unfair dismissal by an employee who claimed that she had been constructively dismissed. It does not seem to me to have any useful bearing on the present appeal, and so I do not propose to consider it further. K

[58] Finally, in *Dalkia*, Christopher Clarke J had to consider the different approaches and outcomes in *The Nanfri* and *Woodar*. One of the issues in *Dalkia* was whether the defendant, Celtech, was in repudiatory breach of a contract under which it was obliged to pay the claimant, Dalkia, charges under agreements by which Dalkia would construct a paper mill for Celtech and would provide electricity and steam to the mill. Celtech failed to pay three instalments of the charges, and it informed Dalkia that it was not within its power and cash resources to make the payments and that it was facing insolvency. Dalkia served a notice of termination on Celtech pursuant to a contractual provision that Dalkia should have the right to terminate the L principal agreement immediately if Celtech was in material breach of its obligations to pay the charges. Celtech contended that it was not in material breach, having merely failed to pay three instalments, and that the notice of termination was a wrongful repudiation of the contract, which Celtech accepted as bringing the contract to an end.

[59] Christopher Clarke J held that Celtech was in material breach of its obligations to pay the charges and that Dalkia was entitled to serve the termination notice. He went on to consider whether, if he were wrong in his conclusion that Dalkia was entitled to terminate the contract, Dalkia's notice of termination was, as Celtech contended, a repudiation that Celtech was entitled to accept. Dalkia relied on M *Woodar.* Christopher Clarke J contrasted that case with *The Nanfri,* gave an explanation for the different outcomes, and distinguished *Woodar* on the facts as follows:

> [148] The decision of the majority [in *Woodar*] must be contrasted with the proposition most powerfully enunciated in *The Nanfri* [1979] 1 Lloyd's Rep 201; AC 757 — that it is no defence to a party who has repudiated a contract to say that he acted in good faith under a mistaken understanding of the law. A reconciliation may lie in the fact that in *The Nanfri* the repudiation consisted of an act the instruction of the master not to sign pre-paid bills of lading which had the immediate effect of substantially depriving the charterers of virtually the whole benefit of the charter since the issue of such bills was essential to the maintenance of the charterers' trade. By contrast the notice of termination in

Woodar did not have that or a similar consequence. In *Vaswani* v *Italian Motors Sales and Services* [1996] 1 WLR 270, the Privy Council said:

> "Whilst… the request for the payment of an excessive price would not in itself amount to a repudiation, if the conduct relied on went beyond the assertion of a genuinely held view of the effect of the contract the conduct could amount to a repudiation. This is the position if the conduct is inconsistent with the continuance of the contract."

and referred to Lord Wilberforce's *dictum* in *The Nanfri* that two cases relied on by the then appellants:

> "would only be relevant here if the owners' action had been confined to asserting their own view — possibly erroneous — as to the effect of the contract. They went, in fact, far beyond this when they threatened a breach of the contract with serious consequences.
>
> [149] It seems to me that *Woodar* is distinguishable. On the facts of that case the majority felt able to conclude that, despite the unqualified terms of the notice, the circumstances in which it was given did not manifest an intention to refuse further performance. The time for performance had not arisen, *Woodar* needed to serve a notice in order to reserve its position, and the discussions between the parties had proceeded on the basis that the service of a notice was not to be regarded as a hostile act, and that the entitlement or otherwise of *Woodar* to serve the notice would be determined by the court, to which *Woodar* would apply, by whose decision both parties would abide.

[60] Mr Clarke submitted that the way in which Christopher Clarke J distinguished *Woodar* in [149] is precisely the same basis on which Mr Clarke seeks to distinguish it from the present case.

[61] I would make the following general observations on all those cases. First, in this area of the law, as in many others, there is a danger in attempts to clarify the application of a legal principle by a series of propositions derived from cases decided on their own particular facts. Instead of concentrating on the application of the principle to the facts of the case in hand, argument tends to revolve around the application of those propositions, which, if stated by the court in an attempt to assist in future cases, often become regarded as prescriptive. So far as concerns repudiatory conduct, the legal test is simply stated, or, as Lord Wilberforce put it, "perspicuous". It is whether, looking at all the circumstances objectively, that is from the perspective of a reasonable person in the position of the innocent party, the contract-breaker has clearly shown an intention to abandon and altogether refuse to perform the contact.

[62] Second, whether there has been a repudiatory breach is highly fact-sensitive. That is why comparison with other cases is of limited value. The innocent and obvious mistake of Mr Jones in the present case has no comparison whatever with, for example, the cynical and manipulative conduct of the shipowner in *The Nanfri.*

[63] Third, all the circumstances must be taken into account in so far as they bear on an objective assessment of the intention of the contract-breaker. This means that motive, although irrelevant if relied on solely to show the subjective intention of the contract-breaker, may be relevant if it is something or it reflects something of which the innocent party was, or a reasonable person in its position would have been, aware and throws light on the way in which the alleged repudiatory act would be viewed by such a reasonable person. So, Lord Wilberforce in *Woodar*, at p281D, expressed himself in qualified terms on motive, not by saying that it will always be irrelevant, but that it is not of itself decisive.

[64] Fourth, although the test is simply stated, its application to the facts of a particular case may not always be easy to apply, as is well illustrated by the division of view among the members of the appellate committee in *Woodar*.

[65] On the facts of the present case that are agreed and were found by the recorder, I consider it clear that Eminence's rescission notices sent under cover of Mr Jones's letter of 17 December 2008 did not constitute a repudiatory breach of the contracts, for the following reasons.

(1) It is common ground that Eminence was, as required by condition 6.8.1, "ready, able and willing to complete" and so entitled to service the notices to complete. The covering letter dated 5 December 2008 confirms that that was the case and that completion did not take place on the contractual completion date of 4 December 2008 in consequence of Mr Heaney's default. That is further supported by the economic reality that, owing to the downturn in the property market, the contractual sale prices of the flats were above market values, and Mr Heaney was attempting to negotiate lower prices and a higher specification.

(2) By para 3 of the notices to complete, Eminence gave "notice under condition 6.8… to complete the contract in accordance with that condition". Condition 6.8.2 required completion of each contract "within 10 working days". By virtue of the definition of "working day" in condition 1.1.1(m), completion was therefore required to take place by close of business on 19 December 2008. The covering letter of 5 December 2008 from Eminence's solicitor stated that it "calculate[s] the final date for completion under the notice is 15th December 2008". That was, as its wording plainly indicates, a calculation based on the contractual provisions. It was not an attempt to vary the contractual provisions or an abandonment of them or a refusal to comply with them, but an obviously mistaken application of them. No reasonable party could have viewed it in any other way.

(3) Mr Heaney's solicitor did not point out that obvious error. It either did not notice it or, as the recorder speculated, it lay in wait for Eminence to act on Mr Jones's mistake, so that it could fortuitously extricate Mr Heaney from the contracts by claiming that premature termination of the contracts by Eminence was itself a repudiatory breach. What is clear is that neither it nor Mr Heaney had any reason whatever to think that if the error was pointed out to Mr Jones, he would not immediately have acknowledged his mistake.

(4) The notices of rescission were served prematurely, before the expiry of the 10 working days specified in condition 6.8 and, accordingly, before Eminence was in a position lawfully to exercise the remedies specified in condition 7.5, including rescission and forfeiture of the deposits. That was due to the same mistake on the part of Mr Jones in calculating the contractual date of expiry of the notices to complete. It was obvious that a mistake had been made because, as the recorder said, it was "screamingly obvious". Again, on the basis of the recorder's findings, a reasonable recipient of the notices of rescission would have appreciated that had the error been pointed out, it would immediately have been acknowledged by Mr Jones. Indeed, that is precisely what happened on 18 December 2008, when [Mr Jones] received Foot Anstey's letter of the same date in response to the rescission notices. This was not a case such as those cited by counsel and mentioned earlier in this judgment, in all of which the parties were maintaining in litigation different interpretations of contractual provisions. In the present case, as a reasonable party in Mr Heaney's position would have realised, there was a simple error of calculation by Eminence's solicitor, analogous to a clerical error, that, once pointed out, would have been (as it was) conceded immediately. The consequence of pointing it out, as the reasonable party would have appreciated, was that the rescission notices would have been treated, as they were in law, as ineffective, and Eminence would have waited until the expiry of the notices to complete before deciding whether to treat the contracts as being at an end. No question of allocation of risk, as advanced by Mr Clarke, arises.

(5) It is, in my judgment, impossible clearly to find on those facts an intention by Eminence to abandon and altogether to refuse to perform the contacts that, in view of the state of the market, had become highly advantageous to Eminence and onerous to Mr Heaney. On the contrary, the obvious inference from those facts is that Mr Heaney and his solicitor were only too well aware that Eminence very much wanted to enforce the contracts, either by completing them or by rescinding them and exercising the other remedies conferred by them, in either case in accordance with the contractual terms.

(6) The error of the recorder was to concentrate on the rescission notices without taking into account all the circumstances. To do so results in a lack of reality. Mr Clarke's submissions reflected that same approach as he sought to emphasise the distinction between, on the one hand, cases such as *The Nanfri* and *Bridgen*, in which the court was considering wrongful performance of an ongoing contract, and, on the other hand, cases such as the present, which concern a wrongful termination of the contract. The essence of Mr Clarke's approach was that, in the latter case, if the act of termination — here, the notice of

A rescission — is clear in its intended effect, that is the end of the matter, save for a limited exception reflecting the precise facts of *Woodar*. For the reasons that I have given, that is not a proper application of the legal test. It was precisely that approach that was criticised by Lord Scarman in *Woodar*, at p299C.

(7) In any event, as Mr Livesey emphasised and, indeed, placed at the forefront of his submissions, that even if the rescission notices are looked at in isolation, they do not clearly show an intention by Eminence to abandon and altogether to refuse to perform the contact. They are, as he submitted, internally inconsistent and, on their face, throw up a query as to Eminence's intention. By referring to the notices to complete and
B conditions 6.8 and 7.5, the rescission notices, far from clearly indicating an intention to abandon the contracts, show an intention to implement the contractual procedure for bringing the contracts to an end and exercising the remedies specified in the contracts. Their service was, however, inconsistent with those contracts in that they were premature. From the perspective of a reasonable party in Mr Heaney's position, that left unclear whether the intention of Eminence was to insist on the effectiveness of the notices of rescission, notwithstanding the terms of C
the contracts or, if the error as to the contractual terms was pointed out, to abide by those terms.

[66] It follows that I do not accept Mr Clarke's fourth key proposition that the factual scenario in *Woodar* is the only one in which a wrongful declaration that a contract is at an end will not be repudiatory. Moreover, I do not believe that Christopher Clarke J went so far as that proposition in *Dalkia*.

Conclusion

[67] For those reasons, I would allow this appeal.

Sullivan LJ said: D

[68] I agree.

Mummery LJ said:

[69] I also agree.

Appeal allowed.

TABLE OF CASES

Titles of cases shown in bold type.
Names of parties reversed shown in ordinary type.

References are to year, volume number and page,
e.g **[2010]** 1 **100** refers to vol 1 of 2010 at p 100. Cases beginning with Figures,
e.g 24, are sorted as if spelled out, e.g Twenty four.

INDEX OF SUBJECT MATTER

References are to year, volume number and page,
eg **[2009]** 1 **89** refers to vol 1 of 2009 at p89

ADVERSE POSSESSION — *continued*
Intention to possess
Permission
Agricultural land — Occupation of land originally pursuant to licence — Occupation continuing after termination of licence on sale of land — Whether occupant acquiring title by adverse possession — Whether having necessary intention to possess where acknowledging need to give up possession should owner require land — Whether occupation by implied permission of owner — Held, intention to possess existing where occupant intending to hold land indefinitely on own account — No requirement of intention to dispossess and infringe rights of paper owner — Recognition of need to give up possession should owner so require not negating intention to possess — Continuation of occupation without objection of owner following expiry of licence not sufficient to show implied permission — Evidence of permission required to be probative not merely consistent with grant of permission
***J Alston & Sons Ltd* v *BOCM Pauls Ltd (Ch),* [2009]** 1 **93**

Limitation Act 1980
Acts constituting "adverse possession"
Squatter holding on after expiry of grazing licence — Whether quality of possession "adverse" — Interruption by highway works — Whether acquiescence by squatter in highway works stopping time running — Whether true owner entitled to possession of land — Held, possession following expiry of licence capable of being adverse — Running of time not interrupted by highway works — True owner not entitled to possession
***Topplan Estates Ltd* v *Townley (CA),* [2005]** 1 **89**

Mortgage
Limitation Act 1980
Mortgagor falling behind with repayments — Mortgagee making demand for repayment then taking no further action — Date when right of action for possession accruing — Whether action time-barred under s15 such that charge extinguished under s17 — Para 8 of Part I of Sched 1 — Applicability of para 8 to action for possession by mortgagee — Whether mortgagee's possession required to be adverse for time to run — Whether mortgagee's possession permissive — Held, para 8 of Part I of Sched 1 applying to actions by mortgagee for possession of mortgaged land — Adverse possession by mortgagor required for time to run — Requirement met by ordinary possession by person in whose favour time can run — Possession of mortgagor not permissive — Legal charge extinguished
***Ashe* v *National Westminster Bank plc (CA),* [2008]** 1 **123**

Permission
Evidence
Application for registration as proprietors of land — Adjudicator allowing application on grounds of adverse possession — Whether sufficient reasoning on issue of possession — Whether occupation with permission of registered owner — Held, reasons sufficient where adjudicator implicitly accepting evidence of respondent as to acts of possession — No overt acts or demonstrable circumstances sufficient to prove permission from landowner
***Hicks Developments Ltd* v *Chaplin (Ch),* [2007]** 1 **1**

Limitation Act 1980
Implied permission — Tenant remaining in occupation of house without paying rent — Para 8(4) of Sched 1 — Whether adverse possession — Whether squatter having implied permission — Whether additional areas occupied as part of implied permission — Held, adverse possession not established — Para 8(4) applying — Squatter having implied permission — Additional areas occupied as part of implied permission
***Batsford Estates (1983) Co Ltd* v *Taylor (CA),* [2005]** 2 **12**

Possession
Temporary exclusion
Squatter in possession of land for limitation period — Self-exclusion by squatter from small area for temporary period — Whether period of self-exclusion defeating claim to title by adverse possession — Held, period of self-exclusion by squatter capable of defeating claim to title by adverse possession
***Generay Ltd* v *The Containerised Storage Co Ltd (CA),* [2005]** 2 **7**

ADVERTISEMENTS — See **Town and country planning**

AGREEMENT FOR LEASE
Specific performance
Minimum net internal area
RICS *Code of Measuring Practice* — Agreement to take lease for £1m premium — Agreement that premises should be of specified minimum net internal area (NIA) — Determination of NIA — Whether NIA referring to usable area or floor areas — Whether sloping areas within NIA — Whether area subject to fire-escape route within NIA — Held, NIA referring to usable areas — Sloping areas capable of use — Area subject to fire-escape route incapable of practical use
***Kilmartin SCI (Hulton House) Ltd* v *Safeway Stores plc (Ch),* [2006]** 1 **59**

Pre-emption or option
Construction — Agreement to sell lease — Agreement providing that contract to purchase lease coming into effect on service of counternotice to offer price — Whether agreement creating pre-emption or option — Law of Property (Miscellaneous Provisions) Act 1989 — Whether order of specific performance — Held, agreement creating option — Order of specific performance
***Coaten* v *PBS Corporation (Ch),* [2006]** 3 **43**

AGRICULTURAL HOLDINGS
Agricultural Holdings Act 1986
Succession
Holding farmed by family for generations — Consensual successions by grant of new tenancy to different family members in 1972 and 1998 — Whether exhausting maximum two statutory successions permitted by 1986 Act — Whether s37(2) of 1986 Act applying to consensual tenancy granted prior to coming into force of original statutory succession scheme in 1976 — Held, tenancy granted by consent prior to 1976 not taking case out of statutory succession scheme — S37(2) not applying to transactions occurring prior to 1976 — Further statutory succession possible
Kemp* v *Fisher (Ch), 9

Licence
Agricultural Holdings Act 1986, s2
Licence granted for use of agricultural land — Whether gratuitous licence — Whether consideration for licence — Whether s2 of 1986 Act having application and creating agricultural tenancy — Rent memorandum including disputed land — Whether surrender and regrant by operation of law — Held, licence granted for a consideration — S2 having application and creating tenancy — No surrender by operation of law
***Well Barn Farming Ltd* v *Backhouse (Ch),* [2005]** 3 **109**

Notice to quit
Counternotice
Agricultural land tribunal — Case stated — Tenant allegedly failing to comply with notice to remedy — Landlord serving notice to quit under Case D — Tenant serving counternotice and seeking reference to arbitrator — S28 of Agricultural Holdings Act 1986 — Whether arbitration request rendering counternotice ineffective — Whether tribunal having jurisdiction — Whether tribunal entitled to conclude fair and reasonable landlord would not insist on possession — Whether tribunal required to state a case — Held, tribunal not required to state a case — Simultaneous service of counternotice and request for arbitration not rendering counternotice ineffective — Tribunal having jurisdiction — Tribunal entitled to conclude fair and reasonable landlord would not insist on possession
***William Smith (Wakefield) Ltd* v *Parisride Ltd (Admin),* [2005]** 2 **22**

Notice to remedy repairs
Arbitration
Agricultural Holdings Act 1986 — Notice to remedy breaches — Legislation distinguishing between repair and non-repair breaches — Legislative discrimination in favour of repair notices over other breach notices — Human Rights Act 1998 — Art 14 of European Convention on Human Rights — Whether unjustifiable discrimination under Art 14 — Whether grounds for declaration of incompatibility — Whether appellant entitled to apply for declaration of incompatibility — Whether 1986 Act discriminating on grounds of property or status — Whether any discrimination objectively justified — Whether 1986 Act compatible with ECHR — Held, tenant not entitled to apply for declaration of incompatibility — Legislation not discriminatory under Art 14 — Legislation compatible with ECHR
***Lancashire County Council* v *Taylor (CA),* [2005]** 2 **17**

Service
Service on husband of one of joint owners — S93 of Agricultural Holdings Act 1986 — Whether service on landlord — Whether service on agent of landlord — Whether valid service — Held, notice not validly served — Husband not agent of landlord — Service not effective for purposes of s93(5)
***Lodgepower Ltd* v *Taylor (CA),* [2005]** 1 **1**

Succession
"Eligible person"
S36(3)(*a*) of Agricultural Holdings Act 1986 — Requirements of eligibility — S41(1)(*b*) "material extent" requirement — Agricultural Land Tribunal deciding that applicant for succession tenancy not satisfying s36(3)(*a*) requirement to "material extent" — Application to state a case — Whether question of law that was fairly arguable — Meaning of "material extent" — Whether matter decided by *Littlewood* v *Rolfe* [1980] 1 EGLR 7 — Held, no arguable question of law — "Material extent" having wide interpretation — Matter not decided by *Littlewood*
***Thomson* v *Church Commissioners for England (Admin),* [2006]** 2 **1**

Tenancy
Surrender by operation of law
Gratuitous licence — Whether rent memorandum adding land to tenancy — Whether surrender and regrant — Whether Agricultural Tenancies Act

BURDEN OF PROOF — See **Sale of land** (Agency)

BUSINESS TENANCY RENEWAL — See **Landlord and tenant** (New tenancy)

CAUSE OF ACTION — See **Negligence**

CAVEAT LESSEE PRINCIPLE — See **Landlord and tenant** (Nuisance)

CHARGE — See **Land registration** (Alteration of registered title)

CHARGING ORDER — See also **Sale of land**

Enforcement

Limitation Act 1980

Charging order against property to secure judgment debt — Charging order not enforced — Owner of property applying to set aside order — Whether s20(1) of 1980 Act applying — Whether charging order ceasing to be enforceable after 12 years — Whether appropriate to discharge order pursuant to s3(5) of Charging Orders Act 1979 — Held, s20 of 1980 Act no bar to enforcement of charging order after 12 years — No adverse possession by owner of property so as to extinguish chargee's title pursuant to ss15 and 17 — Discharge of order under s3(5) of 1979 Act not appropriate where no statutory provision preventing steps to enforce order in future

***Yorkshire Bank Finance Ltd* v *Mulhall (CA),* [2008]** 3 **7**

CHARGING ORDERS ACT 1979 — See **Charging order**; **Insolvency**; **Sale of land** (Charging order)

CIVIL PRACTICE AND PROCEDURE

Jurisdiction

EC Regulation 44/2001

Claim against guarantor of lessee's obligations — Guarantor domiciled in France — Arts 2, 5(1)(a) and 22 of regulation — Whether English courts having jurisdiction to hear claim — Whether principal subject matter of claim relating to rights *in rem* in property situated in England — Whether lease impliedly specifying England as country for performance of relevant obligation — Held, rights *in rem* principal subject matter of claim against guarantor of obligations under lease — English court best forum for establishing factual basis of claim and deciding questions of English law of landlord and tenant — England also place for performance of guarantor's obligations

***Wellington Pub Co plc* v *Hancock (CC),* [2009]** 3 **45**

CIVIL PROCEDURE RULES — See also **Civil Practice and Procedure; Evidence** (Expert report); **Landlord and tenant** (Housing Act 1985) (Secure tenancy); **Party wall**; **Sale of land**

Costs

Fixed costs

CPR 45 — Admission by defendant of full amount claimed — Claim for purpose of enforcing decision of adjudicator — Whether only fixed costs payable by defendant — Whether failure of claimant to specify fixed commencement cost in claim form precluding application of CPR 45.3 — Whether appropriate for court to award substantial costs in exercise of discretion — Held, omission of fixed commencement cost from claim form not precluding application of fixed costs — Departure from fixed costs regime appropriate where claimant justified in issuing proceedings to enforce adjudicator's decision and following procedure in para 9.2 of *TCC Guide* — Claimant entitled to substantial costs

***Amber Construction Services Ltd* v *London Interspace HG Ltd (TCC),* [2008]** 1 **1**

Part 36

Whether rejection of Part 36 offer to settle proceedings rendering offer incapable of acceptance at later date — Whether implied withdrawal of Part 36 offer possible — Whether later offer superseding earlier offer — Held, CPR 36 operating as self-contained code — Code not subject to rules of general law relating to formation of contracts save where that clearly intended — Part 36 offer remaining open until trial or formal withdrawal in accordance with CPR 36.3(7) — Offer not lapsing on being rejected by offeree — Offer incapable of implied withdrawal — Earlier offer not superseded by later offer not made in accordance with CPR 36 requirements

Gibbon* v *Manchester City Council; LG Blower Specialist Bricklayer Ltd* v *Reeves (CA), 85

Protective costs order

Judicial review

Planning permission — Directive 83/337/EEC (EIA Directive) — Application for protective costs order (PCO) in connection with application for judicial review of planning decision — Whether claimant required to show issues raised of general public importance and public interest requiring their resolution in accordance with principles in *R (on the application of Corner House Research)* v *Secretary of State for Trade and Industry* [2005] EWCA Civ 192; [2005] 1 WLR 2600 — Whether PCO necessary to comply with requirement in art 10a of directive that procedure to challenge legality of decision to which directive applying not to be prohibitively expensive — Whether issue of prohibitive expense depending on financial means of individual claimant — Held, *Corner House* principles to be modified to comply with directive in cases to which directive applying — Directive imposing no requirement for issues raised to be of general public importance or that public interest requiring their resolution — Issue of prohibitive expense to be considered with regard to whether potential costs would be prohibitive for ordinary member of the public — PCO granted to claimant – Reciprocal cap imposed on liability of respondent — Such cap not necessarily inconsistent with art 10a

R (on the application of Garner)* v *Elmbridge Borough Council (CA), 137

COLLATERAL CONTRACT — See also **Landlord and tenant** (Assignment)

COLLECTIVE ENFRANCHISEMENT — See **Land registration**; **Leasehold enfranchisement** (Leasehold Reform, Housing and Urban Development Act 1993)

COMMISSION — See **Estate agents**

COMMON LAND — See **Easements**

COMMON PARTS — See **Leasehold enfranchisement** (Leasehold Reform, Housing and Urban Development Act 1993)

COMMONHOLD AND LEASEHOLD REFORM ACT 2002 — See **Landlord and tenant** (Right to manage) (Service charges); **Right to manage**

COMMONS

Law of Property Act 1925, s194

Land subject to rights of common

Council proposing to inclose common — Objection from pressure group — Secretary of State granting consent to fence common without holding public inquiry — Whether s194 requiring public inquiry prior to grant of consent — Held, no need to hold public local inquiry for every application under s194(1)

***R (on the application of Ashbrook)* v *Secretary of State for the Environment, Food and Rural Affairs (Admin),* [2005]** 1 **99**

COMMONS ACT 1876 — See **Town and village greens**

COMMONS ACT 1879 — See **Town and village greens**

COMMONS ACT 2006 — See Town and village greens

COMMONS REGISTRATION ACT 1965 — See **Town and village greens**

COMMONS REGISTRATION (NEW LAND) REGULATIONS 1969 — See **Town and village greens**

COMPANIES ACT 1985 — See Leasehold enfranchisement (Leasehold Reform, Housing and Urban Development Act 1993)

COMPANY VOLUNTARY ARRANGEMENT — See **Insolvency**

COMPULSORY PURCHASE — See also **Easements** (Rights of way)

Compensation

Assumptions

Planning permission — Land Compensation Act 1961 — Whether planning permission for mixed-use development to be assumed on *Pointe Gourde* principle — Whether assumptions confined to those required by ss14 to 17 of 1961 Act — Whether appropriate to award full value of land with planning permission or only percentage reflecting hope value — Held, statutory code for determining planning status not exhaustive — Planning permission to be assumed where tribunal finding on balance of probability that such would have existed at valuation date in no-scheme world — Compensation to be awarded by reference to full value of land with planning permission

***Spirerose Ltd (in administration)* v *Transport for London (CA),* [2009] 1 107**

Notice to treat

Compulsory Purchase Act 1965 — Compensation claim alleging interest in land compulsorily acquired — No notice to treat served on owners of alleged interest — Acquiring authority disputing existence of interest — Whether acquiring authority obliged to serve notice to treat on owners of all interests in land acquired — Whether omission to serve notice to treat precluding claim for compensation — Whether entitlement to have compensation determined arising under ss22 or 10 of 1965 Act — Held, acquiring authority obliged to act reasonably in determining which interests in land "required" for scheme — Where freehold required for scheme any option to purchase held by third parties acquired with it — Any omission to be remedied by acquiring authority under s22 — Holder of omitted interest entitled to initiate s22 process where acquiring authority failing to do so and unwilling to give up possession of land in question

***Union Railways (North) Ltd* v *Kent County Council (CA),* [2009] 2 1**

Compulsory Purchase Act 1965 — Jurisdiction of Lands Tribunal (LT) — Compensation claim alleging interest in land compulsorily acquired — No notice to treat served on owners of alleged interests — Acquiring authority disputing existence of interest — Whether acquiring authority obliged to serve notice to treat on owners of all interests in land acquired — Whether omission to serve notice to treat precluding claim for compensation by owner — Whether LT having jurisdiction to determine disputed questions as to title – Held, acquiring authority obliged to serve notice to treat on owners of all interests acquired — Compensation claim to be determined under s10 of 1965 Act where no notice served — LT able to determine disputed issues of title where necessary to determine question of entitlement to compensation

***Union Railways (North) Ltd* v *Kent County Council (LT),* [2008] 2 183**

Valuation method

Development potential — Agricultural land — Landowners seeking to have development potential of land taken into account in valuation — No direct comparables sufficiently similar to land compulsorily acquired — Whether residual method of valuation applicable — Whether spot valuation appropriate — Held, residual method to be applied only in exceptional cases — Residual method appropriate only where prospects of successful development such that variables few and effect on valuation readily assessable — Spot valuation based on existing use plus element of hope value — Hope value taking into account likelihood of obtaining planning permission and other matters

***Mon Tresor & Mon Desert Ltd* v *Ministry of Housing and Lands (PC),* [2008] 3 13**

Value of lease

Compulsory acquisition of underlease — Property in disrepair — Whether compensation to be reduced to take account of dilapidations liability from which tenant relieved as result of CPO — Whether appropriate to set off value of dilapidations liabilities against claimed compensation figure — Held, any liability of tenant for dilapidations to be taken into account in assessing compensation for compulsory acquisition of lease — Effect on compensation a matter for valuation — Inappropriate simply to deduct cost of remedying dilapidations from compensation otherwise payable

***Richard Parsons Ltd* v *Bristol City Council (LT),* [2007] 3 73**

Material considerations

Town and Country Planning Act 1990

Local planning authority granting planning permission to rival parties for development on same site — Resolution to exercise compulsory purchase powers under s226(1)(a) of 1990 Act in favour of one party — That party promising to develop another site funded by CPO site development — Whether local planning authority entitled to take into account prospect that development of CPO site would cross-fund other development either as s226(1A) well-being benefit or other material consideration — Held, impermissible under s226(1A) to take into account commitment by developer of CPO site to secure development of unconnected site to achieve further well-being benefits for the area — Such matters not otherwise material considerations relevant to exercise of s226(1)(a) power

R (on the application of Sainsbury's Supermarkets Ltd)* v *Wolverhampton City Council and another (SC), 103

COMPULSORY PURCHASE ACT 1965 — See **Compulsory purchase**

CONSENT — See **Restrictive covenants**

CONSENT ORDER — See **Boundaries**

CONSERVATION (NATURAL HABITATS ETC) REGULATIONS 1994 — See **Sale of land (Environment)**

CONSTRUCTION

Adjudication

Appointment of adjudicator

Jurisdiction — Building contract — Dispute — Contract providing for either party to request nomination of adjudicator by RIBA — Contractor wanting adjudicator with legal qualifications — Contractor contacting particular adjudicator to check availability and suggesting name in application to RIBA — Adjudicator appointed and giving decision — Whether adjudicator's appointment and decision valid — Whether implied term that party not seeking unilaterally to influence nominator's decision — Whether apparent bias — Held, appointment and decision valid — Unnecessary to imply term — No apparent bias

***Makers UK Ltd* v *Camden London Borough Council (TCC),* [2008] 3 1**

Enforcement

Construction — Adjudication — Enforcement — S107 of Housing Grants, Construction and Regeneration Act 1996 — Clause 9A.7.1 of JCT Management Contract — Basis for adjudicator's jurisdiction where no written contract existing between parties — Whether parties reaching final determination by agreement on disputed item so as to render adjudicator's decision non-binding — Whether enforceability of decision affected by mistakes — Whether appropriate to adjourn enforcement proceedings pending final account — Held, adjudicator's jurisdiction in absence of clear written contract founded on exchange of written submissions in which parties not disputing jurisdiction — Final determination by agreement possible only by agreement reached after adjudicator giving decision and not by agreement during adjudication process — Mistakes not affecting enforceability of decision provided adjudicator answering questions referred to him and not proceeding in obviously unfair manner — Inappropriate to adjourn enforcement proceedings pending outcome of final account

SG South Ltd* v *Swan Yard (Cirencester) Ltd (TCC), 7

Set-off — Stay of execution — First adjudicator deciding contractor entitled to sums under terms of contract — Second adjudicator deciding employer entitled to claim liquidated damages — Contractor seeking to enforce first adjudicator's decision — Whether employer entitled to set off entitlement to liquidated damages — Whether stay to be granted by reason of contractor's parlous financial position and contractual obligation to repay sums due — Held, sums found due to contractor in adjudication decision subject to right of set-off where clear and unequivocal decision in further adjudication as to employer's entitlement to recover specific sum by way of liquidated and ascertained damages — Stay of execution granted where contractor not in financial position to repay sums awarded and this representing a change from position when relevant contract made and not due to employer's failure to pay sums awarded by the adjudicator — Further grounds for stay where terms of contract requiring imminent repayment of sum claimed

***JPA Design & Build Ltd* v *Sentosa (UK) Ltd (TCC),* [2009] 3 15**

Liquidated and ascertained damages

Adjudicator making award to contractor in building dispute — Summary judgment sought on claim to enforce award — Whether principle in *Bramall & Ogden Ltd* v *Sheffield City Council* (1983) 29 BLR 73 applying — Whether employer entitled to set off claim for liquidated and ascertained damages (LADs) against sum due — Whether entitled to stay pending second adjudication in which employer referring party — Held, contractor entitled to rely upon *Bramall* principle in respect of LADs claim where partial possession granted and contract not providing for sectional completion — No entitlement to set off LADs claim against adjudicator's award — No stay to await outcome of second adjudication

***Avoncroft Construction Ltd* v *Sharba Homes (CN) Ltd (TCC),* [2008] 2 7**

Easement of necessity

Implication

Transfer of land without express reservation of easement of way — Transferred land adjoining public highway — Exercise of common law right of access to public highway dependent upon planning permission — Planning permission for access unlikely — Whether easement of way of necessity implied — Scope of any easement of necessity — Held, no implication of easement of necessity — Common law right of access benefiting transferred land at date of transfer — Unlikelihood of planning permission to exercise common law right of access not a relevant circumstance — No necessity

***Adealon International Corporation Proprietary Ltd* v *Merton London Borough Council (Ch),* [2006] 2 1**

Environmental Protection Act 1990

Remediation notice

Remediation works to remove contamination from land formerly used as gasworks — Remediation notice served on current gas undertaker in respect of cost of works — Whether current undertaker "appropriate person" to pay for works under s78F — Whether s78F imposing liability on successors of actual polluter — Effect of s49(1) of Gas Act 1986 — Held, liability confined to actual polluter — Transfer of assets and liabilities under 1986 Act not extending to liabilities not in existence at time of transfer

***R (on the application of National Grid Gas plc (formerly Transco plc))* v *Environment Agency (HL),* [2007] 3 5**

Excessive use

Right of way

Access to rear of terraced property by footpath across property of neighbour — Right exercisable for "all proper purposes" connected with reasonable enjoyment of dominant tenement — Whether right of way restricted to purposes for which use of front door not reasonably practicable — Whether use of way excessive — Held, unrestricted right to use footpath for access and egress to dominant tenement in connection with ordinary domestic activities of household — No excessive use

***Brooks* v *Young (CA),* [2008] 3 27**

Extinguishment

Exercise of statutory powers

Easement of access benefiting flat — Building in multiple occupation — Notice under s352 of Housing Act 1985 requiring works — Works obstructing easement — Whether easement extinguished — Whether remedy for nuisance where easement not extinguished — Held, statutory duty and power to carry out works under s352 notice — Easement no longer capable of benefiting flat — Easement extinguished — No remedy in nuisance if easement continuing

***Jones* v *Cleanthi (QB),* [2006] 1 1**

Statutory obligation

Building in multiple occupation — Housing Act 1985 — S352 notice requiring landlord to carry out works — Works blocking easement of access in favour of lessee — Whether statutory power to carry out works — Whether easement extinguished — Held, statutory obligation to perform works giving rise to statutory power to do so — Easement not extinguished once and for all — No actionable interference with lessee's rights

***Jones* v *Cleanthi (CA),* [2007] 1 7**

Implied term

S106 agreement

Agreement by estate owners A and B to construct bypass — B holding land required for bypass — Whether normal canons of construction of contracts applying to s106 agreements — Whether rights of access to construct bypass in favour of A necessarily implicit or granted by implied term — Whether proprietary rights — Whether rights capable of protection by notice on registers — Held, normal canons of construction applying to s106 agreements — A having rights of access under s106 agreement as necessarily implicit or by implied term — Rights proprietary and capable of protection under Land Registration Acts

***Stroude* v *Beazer Homes Ltd (Ch),* [2006] 2 115**

Prescription

Acquiescence

Right of way — Long user — Freehold owner granting tenancy of servient land after beginning of period of long user — Whether grant of tenancy preventing finding of acquiescence in user — Whether right over secondary route abandoned — Held, rights of way acquired — Grant of tenancy not affecting finding of acquiescence where freeholder having knowledge of user at or before date of grant — No intention shown to abandon secondary route

***Williams* v *Sandy Lane (Chester) Ltd (CA),* [2007] 1 10**

Right of way

S2 of Prescription Act 1832 — Claim to prescriptive right of way over common land — Wimbledon and Putney Commons Act 1871 — Whether owners of common capable of granting private right of way – Whether possible to acquire prescriptive right on basis of 40 years' user in absence of capable grantor — Held, owners of common able lawfully to grant easement not interfering with open and unenclosed nature of commons — Capable grantor required for acquisition of easement by 40 years' user under second part of s2 of 1832 Act

***Housden* v *Conservators of Wimbledon and Putney Commons (CA),* [2008] 2 107**

Right of access

Parking

Dwelling transferred with right of access over defined land to front and rear — Garage constructed at rear by agreement with transferor — Whether right of access to rear to be construed purposively as extending over land of sufficient width to permit access by motor vehicles to garage — Whether visitor parking rights to be implied into right of vehicular access to dwelling — Held, right of access to rear not extended by agreement regarding construction of garage — Right defined in transfer by reference to lane of fixed location not open to purposive construction regarding width — Right to park implied into right of vehicular access only where reasonably necessary for exercise or enjoyment of right — No right implied where transfer making express provision for parking rights and right of access capable of substantial enjoyment without further rights

***Waterman* v *Boyle (CA),* [2009] 2 7**

Right of entry

Rebuilding or renewal

Property benefiting from right of entry to adjoining yard for purposes of rebuilding or renewal — Planning permission to demolish single-storey industrial unit and build five-storey commercial and residential block — Whether right of entry to yard for that purpose — Whether redevelopment "rebuilding or renewal" — Held, "rebuilding" including demolition of existing buildings on property and replacement with either similar or different buildings or none at all — Right of entry extending to proposed redevelopment works

***Risegold Ltd* v *Escala Ltd (CA),* [2009] 1 5**

Property benefiting from right of entry to adjoining yard for purposes of rebuilding or renewal — Planning permission to demolish single-storey industrial unit and build five-storey commercial and residential block — Whether right of entry to yard for that purpose — Whether redevelopment "rebuilding or renewal" — Held, grant to be construed strictly — No right of entry for purposes of implementing planning permission — "Rebuilding or renewal" not extending to complete redevelopment by construction of something different

***Risegold Ltd* v *Escala Ltd (Ch),* [2008] 1 13**

Right to light

Damages

Development infringing right to light of adjoining building — Damages awarded in lieu of injunction — Applicable principles for calculating damages by reference to loss of ability to prevent infringement — Held, damages to reflect outcome of hypothetical negotiation between parties — Damages to include a fair percentage of the profit from the development

***Tamares (Vincent Square) Ltd* v *Fairpoint Properties (Vincent Square) Ltd (Ch),* [2007] 1 26**

Residential development — Owner of neighbouring hotel claiming infringement of right to light — Claim in nuisance — Basis upon which damages to be calculated — Whether remedy of account of profits available — Held, damages awarded upon basis of actual loss to hotel owner — Remedy of account of profits available in nuisance claims only in exceptional circumstances if at all

***Forsyth-Grant* v *Allen (CA),* [2008] 2 16**

Prescription

S3 of Prescription Act 1832 — Office building previously let on terms reserving to landlord full right to build on adjoining land — Whether reservation in lease amounting to "consent or agreement" preventing prescriptive right to light from arising — Whether such consent required to refer expressly to light — Held, reservation making clear that right to light not absolute and indefeasible — reservation amounting to agreement or consent — Express reference to light unnecessary

***RHJ Ltd* v *FT Patten (Holdings) Ltd (Ch),* [2007] 3 1; (CA), [2008] 2 11**

INSOLVENCY — *continued*

Company voluntary arrangement

Guarantees

Commercial leases — Insolvent company's liabilities under leases guaranteed by parent company — CVA purporting to provide for payment in satisfaction of all liabilities including those guaranteed — Ss 5(2) and 6(1)(*a*) of Insolvency Act 1986 — Whether CVA effective to release guarantors' liability to landlords — Whether landlords otherwise precluded from enforcing guarantees — Whether terms of CVA unfairly prejudicial to landlords — Held, CVA not directly effective to release guarantor from guarantees — Insolvent company able to enforce implied agreement not to claim under guarantees — VCA unfairly prejudicial to landlords

***Prudential Assurance Co Ltd* v *PRG Powerhouse Ltd; Luctor Ltd* v *PRG Powerhouse Ltd (Ch),* [2007]** 3 **131**

Unfair prejudice — Insolvency Act 1986 — Tenant's obligations under leases of retail units guaranteed by parent company — Tenant becoming insolvent and entering into company voluntary arrangement (CVA) — Administrators structuring CVA so as to deprive guaranteed landlords of benefit of guarantees on payment of compensation — Application to set aside CVA under s6(1) of Insolvency Act 1986 — Whether CVA unfairly prejudicial to guaranteed landlords — Held, unfairness assessed by vertical comparison with position on liquidation and horizontal comparison with position of other creditors — CVA unfair on vertical comparison where guaranteed landlords losing rights that would be retained on liquidation — Payment for loss of guarantees inadequate in time of market uncertainty where impossible accurately to calculate appropriate amount of compensation — CVA unfair on horizontal comparison where guaranteed landlords receiving less favourable treatment than unsecured creditors and tenant's associated companies

Mourant & Co Trustees Ltd* v *Sixty UK Ltd (in administration) (Ch), 125

Insolvency Rules 1986

Value of landlord's claim

Company in administration — Landlord voting at creditors' meeting on CVA — Landlord relying upon dilapidation schedule and future loss of rent totalling £1.175m — Chairman of creditors' meeting valuing landlord's claim at £1 — Whether chairman entitled, under rr 2.38(5) and 1.17(3), to value at £1 — Held, consideration of r 2.38(5) unnecessary since proposal would have been carried even if landlord's claim included in voting — Chairman entitled to decide £1 value under r 1.17(3) — Landlord's claim unliquidated and unascertained

***In the matter of Newlands (Seaford) Educational Trust; Chittenden* v *Pepper (Ch),* [2007]** 2 **7**

Sale of property

Sale at undervalue

Insolvency Act 1986 — Sale and leaseback entered into with purpose of putting assets beyond reach of judgment creditors — Sale price less than unencumbered freehold value owing to grant of lease — S423 of 1986 Act — Whether lease forming part of consideration — Whether transaction at undervalue — Held, sale and leaseback at price less than unencumbered freehold value not at undervalue provided tenancy having sufficient value to vendor — Premium value of tenancy making up for any shortfall in purchase price — Irrelevant that tenancy unassignable

Chen* v *Delaney (Ch), 21

INSOLVENCY ACT 1986 — See **Insolvency; Landlord and tenant** (Disclaimer of lease) (Forfeiture); **Title to land**

INSOLVENCY RULES 1986 — See **Construction** (Insolvency); **Insolvency** (Administration)

INSURANCE

Joint names

Negligence

Liability to employer — Construction of manufacturing plant — Contract to install sprinkler system — Flooding causing damage to plant and contents — Claim against contractor responsible for installing sprinkler system — Whether contractual provision requiring employer to take out joint-names insurance applying to contractor — Whether policy excluding liability of contractor for negligence — Held, provision in question not providing contractors with separate liability insurance in respect of structures outside area of own works — Obligation to insure in joint names extending only to matters not caused by negligence in absence of plain language to contrary

***Tyco Fire & Integrated Solutions (UK) Ltd (formerly Wormald Ansul (UK) Ltd)* v *Rolls-Royce Motor Cars Ltd (formerly Hireus Ltd) (CA),* [2008]** 2 **113**

INSURANCE RENT — See **Landlord and tenant** (Insurance)

INTERMEDIATE LEASEHOLD INTERESTS — See **Leasehold enfranchisement** (Leasehold Reform, Housing and Urban Development Act 1993)

INTERNAL FLOOR AREA — See **Leasehold enfranchisement** (Leasehold Reform, Housing and Urban Development Act 1993)

INTERPRETATION — See **Boundaries**

JCT MANAGEMENT CONTRACT — See **Construction**

JCT STANDARD FORM — See **Construction** (Building contract)

JOINT-NAMES INSURANCE — See **Insurance**

JOINT OWNERSHIP

Beneficial interests

Presumptions

Secure tenant purchasing property under right-to-buy provisions of Housing Act 1985 — Daughter sharing in purchase under s123(3) and contributing to purchase price — Purchase investment for letting to tenants — Joint legal ownership — Whether parties owning property in equal beneficial shares — Whether shares to be determined according to contribution to purchase price — Whether daughter's contribution including part of discount obtained by secure tenant on right-to-buy purchase — Whether including half of mortgage amount — Whether daughter entitled to account of rental income — Held, presumption of equal beneficial shares not applying where property purchased by family members as investment — Discount attributable to secure tenant only — Daughter's contribution including half of mortgage amount — No account of rental income

***Laskar* v *Laskar (CA),* [2008]** 2 **70**

JOINT TENANCY — See **Landlord and tenant** (Human rights)

JUDICIAL REVIEW — See also **Lands Tribunal; Tort**

Airports

Consultations

White paper on future of air transport — Airport expansion — Proposed new runway at Stansted — Proposed extension of runway at Luton — Whether proposals within scope of earlier consultation exercises — Whether decision to adopt policies lawful — Held, adoption of certain policies unlawful — Statement in white paper regarding wide-spaced runway option going further than basis of consultation exercise — White paper unfairly and wrongly managing to convey impression that proposal to extend existing runway at Luton had been consulted upon and was supported — Declaratory relief granted

***R (on the application of Wandsworth London Borough Council)* v *Secretary of State for Transport; R (on the application of Essex County Council)* v *Secretary of State for Transport (Admin),* [2006]** 1 **91**

JURISDICTION — See **Agricultural holdings** (Notice to quit); **Civil practice and procedure; Compulsory purchase; Construction; Landlord and tenant** (Assured tenancy) (Housing Act 1988) (Possession order) (Service charges); **Leasehold enfranchisement** (Jurisdiction of leasehold valuation tribunal) (Leasehold Reform, Housing and Urban Development Act 1993 — Leasehold valuation tribunal)

LAND COMPENSATION ACT 1961 — See **Compulsory purchase**

LAND COMPENSATION ACT 1973 — See **Compensation; Compensation for injurious affection; Compulsory acquisition; Highways Act 1980**

LAND REGISTRATION

Alteration of registered title

Rectification for mistake

Charge — Transfer of land — Transferee registered as proprietor — Transferee granting charge in favour of bank to secure indebtedness — Previous owner claiming transfer obtained by fraud or forgery — Whether charge invalid — Whether rectification available to set aside charge on grounds of mistake — Held, charge valid where executed in proper form by registered proprietor — Registration of charge not a mistake unless mortgagee having notice of defect in mortgagor's title — Lodging of caution appropriate means of putting prospective mortgagee on notice

***Barclays Bank plc* v *Guy (CA),* [2008]** 2 **74**

Assignment — *continued*

Virtual assignment — continued

Lease of commercial premises — Premises occupied by underlessee — Tenant transferring all economic benefits and burdens and management responsibilities to third party — Consent of landlord not obtained for assignment — Whether breach of covenant against alienation without landlord's consent — Held, covenant against assignment applying only to legal assignment unless context showing otherwise — Virtual assignment not amounting to underletting or declaration of trust within meaning of covenant — Virtual assignment amounting to parting with possession of demised premises contrary to covenant where demised premises underlet and tenant relinquishing power to deal with property to assignee

Clarence House Ltd **v** ***National Westminster Bank plc (Ch),*** **[2009]** 2 **19**

Assured shorthold tenancy

Insured tenancy deposit scheme

Landlord and tenant — Assured shorthold tenancy — Insured tenancy deposit scheme — Prescribed information — Housing Act 2004 – Tenant paying deposit – Deposit lodged in insured tenancy deposit scheme under s212 of 2004 Act — Landlord failing to provide tenant with prescribed information regarding scheme within required period – Information subsequently provided — Ss 213(6), 214(3) and (4) of 2004 Act — Whether tenant entitled to refund of deposit — Whether landlord liable to make penalty payment — Held, deposit refundable and penalty payable only where prescribed information not given — Landlord not so liable where information given in writing albeit outside 14-day time period

Harvey **v** ***Bamforth (CC),*** **[2008]** 3 **66**

Termination

Fixed-term tenancy for one year less one day — Tenancy providing for rent as a yearly sum payable quarterly — Statutory periodic tenancy — Termination — S21 notice served by reference to quarterly statutory periodic tenancy — Whether statutory periodic tenancy quarterly or yearly — Proper construction of s5 of Housing Act 1988 — Held, statutory periodic tenancy quarterly — S5(3)(d) requiring ascertainment of period for which rent last payable

Church Commissioners for England **v** ***Meya (CA),*** **[2006]** 2 **39**

Assured tenancy

Possession

Adjournment — Ground 8 claim for possession — Mandatory ground of possession — S7 of Housing Act 1988 — Claim for possession — Application for possession to enable tenant to reduce rent arrears below ground 8 threshold — Whether county court having jurisdiction to grant adjournment — Held, court cannot normally grant adjournment for purpose of enabling tenant to reduce rent arrears below ground 8 threshold limits

North British Housing Association Ltd **v** ***Matthews; London & Quadrant Housing Trust*** **v** ***Morgan (CA),*** **[2005]** 1 **31**

Variation

Registered social landlord — Standard form of tenancy agreement — Unfair Terms in Consumer Contracts Regulations 1999 — Tenancy agreement containing contradictory provisions for variation of terms — Whether conferring unilateral power on landlord to vary terms without approval of tenants — Whether such provision "not binding" on tenants as unfair term — Held, no right of unilateral variation of terms — Doubt raised by contradictory terms to be resolved in favour of tenant as consumer — Unilateral variation clause unfair and not binding unless Office of Fair Trading guidelines taken into account

Governors of Peabody Trust **v** ***Reeve (Ch),*** **[2008]** 3 **61**

Breach of covenant

Damages

Damages assessed on negotiating basis — Date of breach — Lease liable to forfeiture on date of breach — Whether hypothetical negotiations should assume that lease liable to forfeiture at breach date — Held, presumption that damages under negotiating basis assessed at date of breach — Lord Cairns' Act raising quasi-equity considerations — Court entitled to depart from usual presumption in appropriate cases

Lunn Poly Ltd **v** ***Liverpool & Lancashire Properties Ltd (CA),*** **[2006]** 2 **29**

Break clause

Agency

Commercial underlease containing break clause permitting tenant to end lease by giving six months' notice — Notice given in name of other company in tenant's group — Landlord asserting notice invalid and underlease continuing — Whether notice valid — Whether notice given by tenant as required by underlease — Whether given by third party acting as agent for tenant — Held, notice not given by tenant invalid even if reasonable recipient would think it was so given — Subsequent ratification by tenant of notice served by third party not effective to save notice — Notice given by third party as principal not as agent for tenant

Hexstone Holdings Ltd **v** ***AHC Westlink Ltd (Ch),*** 13

Assignment

Commercial lease — Lease assigned with licence of landlord — Licence to assign containing break clause — Right to break personal to defendant as assignee — Defendant assigning to associated company with consent of landlord – Subsequent reassignment without landlord's consent — Landlord refusing consent on grounds relating to risk of defendant exercising break clause — Whether defendant able to exercise break clause while not the tenant — Whether right to break reviving on reassignment — Whether consent to reassignment unreasonably withheld — Held, break clause a right incident to lease passing to assignee on assignment — No commercial sense in construing licence to assign so as to permit termination of lease by party after ceasing to be tenant — Right not reviving on reassignment — Claimant not acting unreasonably in refusing consent to avoid perceived possibility of loss of significant rental income

Norwich Union Life & Pensions Ltd **v** ***Linpac Mouldings Ltd (Ch),*** **[2009]** 3 **147**

Material compliance

Lease containing break clause subject to material compliance by tenant of all obligations — Tenant carrying out large programme of repairs — Landlord declining to inspect or comment on specification during repair works — Whether material compliance requiring strict compliance — Whether tenant having materially complied with repair obligations — Whether lease terminating on termination date — Held, material compliance not requiring strict compliance — Tenant having materially complied – Lease terminating on termination date

Fitzroy House Epworth Street (No 1) Ltd **v** ***The Financial Times Ltd (TCC),*** **[2006]** 1 **19**

Lease containing break clause subject to material compliance by tenant of all obligations — Tenant undertaking large programme of repairs — Landlords declining to inspect or comment upon specification during repair works — Whether material compliance requiring strict compliance — Whether tenant having materially complied with repair obligations — Whether lease terminating on termination date — Held, material compliance not requiring strict compliance — Tenant having materially complied — Lease terminating on termination date

Fitzroy House Epworth Street (No 1) Ltd **v** ***The Financial Times Ltd (CA),*** **[2006]** 2 **13**

Notice

Acknowledgment of receipt — Commercial lease containing tenant's break clause exercisable on six months' prior notice — Notice to be valid only where given by hand or posted to appropriate address unless receipt by other means acknowledged by landlord — Tenant giving notice by fax — Landlord bringing claim for rent for period after break date — Acknowledgment of receipt of break notices — Whether acknowledgment retrospectively validating break notices — Whether lease continuing — Held, faxed notices invalid unless and until receipt acknowledged by landlord — Notice invalid at critical moment six months prior to break date incapable of validation by later acknowledgment — Lease continuing — Tenant liable for rent

Orchard (Developments) Holdings plc **v** ***Reuters Ltd (CA),*** **[2009]** 1 **13**

Collateral contract

Dilapidations

Agreement for lease — Whether collateral contract that no terminal schedule of dilapidations would be served upon expiry of lease — Whether position affected by conduct of negotiations "subject to formal deed" — Held, labelling of negotiations as "subject to contract" not precluding formation of collateral contract at time of main contract though not before — No contract concluded between parties save that resulting from grant of lease — Collateral contract not providing defence to claim by assignee of landlord for dilapidations

Business Environment Bow Lane Ltd **v** ***Deanwater Estates Ltd (CA),*** **[2007]** 2 **51**

Dilapidations — Whether tenant entering into lease in return for assurance that no terminal schedule of dilapidations would be served at its end — Whether proviso in lease agreed as substitute for assurance — Whether position affected by conduct of negotiations "subject to formal deed" — Held, landlord bound by assurance — Proviso in lease no substitute for

LANDLORD AND TENANT — *continued*

Forfeiture

Eviction

Landlord and tenant — Forfeiture — S2 of Protection from Eviction Act 1977 — Tenancy of mixed business and residential premises — Peaceable re-entry of mixed premises — Whether s2 of 1977 Act applying — Whether phrase "let as a dwelling" means "let wholly or partly as a dwelling" — Held, s2 of the 1977 Act having application to mixed premises — "Let as a dwelling" means "let *wholly or partly* as a dwelling"

***Pirabakaran* v *Patel (CA),* [2006]** 3 **23**

Registered land

Lease — Reversionary interest assigned by Land Registry TR1 form with all rights in property — Assignee forfeiting lease prior to registration of reversionary interest at Land Registry — Whether assignee holding only equitable interest entitled to forfeit — Whether s141(2) of Law of Property Act 1925 applying — Whether forfeiture effective — Held, s141(2) applying — Equitable assignee of reversionary interest entitled to forfeit lease — Forfeiture effective

***Scribes West Ltd* v *Relsa Anstalt (No 3) (CA),* [2005]** 1 **22**

Relief

Terms of relief — Whether equitable assignee entitled to claim relief from forfeiture — Whether exception to *Gill* v *Lewis* [1956] 2 QB 1 that forfeiture security only for rent — Whether relief from forfeiture — Held, equitable assignee entitled to apply for relief from forfeiture — Principles in *Gill* v *Lewis* departed from in very exceptional case — Relief would have been refused on grounds other than security for rent and costs

***Test Valley Borough Council* v *Minilec Engineering Ltd (in liquidation) (Ch),* [2005]** 2 **113**

Waiver

Arrears of rent — Bankruptcy petition by landlord — Landlord seeking to forfeit lease for breaches of covenant — One tenant sending cheque paying bankruptcy sum and other rent arrears — Landlord processing cheque, retaining bankruptcy sum but repaying remainder — Whether acceptance of rent waiving right to forfeit for breaches of covenant — Whether bankruptcy proceedings constituting waiver — Held, processing cheque not acceptance of rent — No acceptance where landlord retaining only part of sum for purpose other than rent — No waiver of forfeiture by pursuance of bankruptcy proceedings

***Seahive Investments Ltd* v *Osibanjo (CA),* [2009]** 1 **32**

Proviso for forfeiture of lease in event of late payment of rent or insolvency — Insolvency Act 1986 — Tenant entering into company voluntary arrangement — Tenant appropriating moneys to specific rent arreas — Landlord receiving moneys but refusing to accept appropriation — Whether landlord bound by tenant's appropriation — Whether acceptance of moneys amonting to waiver of right to forfeit lease — Held, tenant's appropriation valid — Landlord's acceptance of payments amounting to waiver of right to forfeit

***Thomas* v *Ken Thomas Ltd (CA),* [2007]** 1 **31**

Housing Act 1985

Succession

Possession order — Revival of tenancy — Tenant under secure tenancy failing to comply with terms of suspended possession order — Tenant remaining in occupation until death — Brother of tenant applying to postpone possession date in possession order so as to revive tenancy with view to exercising succession rights under s89 of 1985 Act — Whether tenancy continuing until execution of possession order — Whether possible to vary terms of possession order under s85 after death of former tenant — Held, tenancy not continuing until execution of possession order — Inappropriate to reverse established case law indicating that tenancy determining on breach of terms of order — Death of former secure tenant not precluding application under s85(2) to postpone the date of possession under order — Brother of former tenant able to apply under CPR 19.8 to represent deceased's estate for purposes of such application

Austin* v *Southwark London Borough Council (SC), 45

Possession order — Secure tenancy determining on failure to comply with terms of order — Former tenant remaining in property until death as tolerated trespasser — Brother of former tenant entitled to succeed to tenancy had it subsisted at time of death — Brother seeking to be joined to possession proceedings as representative of former tenant's estate under CPR 19.8 — Whether possible to vary terms of possession order under s85 of Housing Act 1985 after death of former tenant — Whether tenancy capable of being revived in order to permit succession — Held, CPR 19.8 not applying — Right to apply under s85 terminating on death of former tenant and not surviving for benefit of estate — Person who would succeed to secure tenancy if subsisting at death of former tenant having no right to apply to revive tenancy in own name

***Austin* v *Southwark London Borough Council (CA),* [2009]** 2 **41**

Secure tenancy — Joint tenancy — Married couple holding secure tenancy of property as joint tenants — Husband leaving but wife and daughter continuing to live in property until wife's death — Landlords serving notice to quit on husband – Whether daughter entitled to succeed to secure tenancy — Whether tenancy continuing to be held by husband by survivorship — Whether landlords entitled to possession — Held, succession provisions of 1985 Act not displacing doctrine of survivorship — Secure tenancy vesting in surviving joint tenant on death of other joint tenant — Husband becoming sole tenant by survivorship — Daughter not succeeding to tenancy — Tenancy ceasing to be secure where surviving tenant not residing in property — Landlords entitled to possession

Solihull Metropolitan Borough Council* v *Hickin (CA), 147

Housing Act 1988

Assured tenancy

Determination of rent — Local Government and Housing Act 1989 — Former long residential tenancy coming to an end — Tenant becoming assured tenant by virtue of Sched 10 to 1989 Act — Rent determined by rent assessment committee disregarding tenant's improvements — Improvements not disregarded on subsequent reference under s13(4) of 1988 Act — Whether words to be read into statutory provisions so as to permit disregard of improvements — Held, tenant not meeting conditions for disregard of improvements where no assured tenancy at date of improvements since such tenancies not yet existing — No provision in 1989 Act for deeming long residential tenancy to have been assured tenancy — Legislation not unintelligible, absurd or totally unreasonable such as to justify reading in words so as to permit application of disregard

***Hughes* v *Borodex Ltd (Admin),* [2009]** 2 **47**

Assured tenancy

Determination of rent — Local Government and Housing Act 1989 — Former long residential tenancy ending — Tenant becoming assured tenant by virtue of Sch 10 to 1989 Act — Initial rent determined by rent assessment committee disregarding tenant's improvements — Improvements not disregarded on subsequent reference under s13(4) of Housing Act 1988 — Whether words to be read into statutory provisions so as to permit disregard of improvements – Held, paras 9 and 10 of Sch 10 applying only to fixing of initial rent at outset of assured tenancy — No disregard of improvements permitted on subsequent determinations under ss 13 and 14 — No justification for reading words to contrary into statute

Hughes* v *Borodex Ltd (CA), 25

Rent review clause — Assured tenancy providing for annual rent increase both during term and any subsequent statutory periodic tenancy — Tenants holding over after end of contractual term as statutory periodic tenants — Managing agents proposing substantial rent increase – S13(1) of 1988 Act — Whether rent review clause in original assured tenancy continuing in effect — Whether rent assessment panel having jurisdiction to determine rent — Held, statutory periodic tenancy governed by s13(1)(a) not (b) — Statutory scheme governing increase of rent — Rent review clause in original assured tenancy ceasing to have effect — Rent assessment panel having jurisdiction to determine rent

***London District Properties Management Ltd* v *Goolamy (Admin),* [2009]** 3 **67**

Possession order

Assured tenancy — Position of assured tenant after making of possession order with enforcement suspended on terms — S9(2) of 1988 Act — Standard form N28 — Whether assured tenancy continuing after date for possession specified in order — Whether tenant becoming tolerated trespasser — Held, result depending upon correct construction of county court order — Order on form N28 having effect of ending assured tenancy on date for possession specified in order — Former tenant thereafter having status of tolerated trespasser

***White* v *Knowsley Housing Trust (CA),* [2007]** 2 **61**

Protected and statutory tenants

Transitional protection — Tenant originally protected by Rent Act 1977 — Joint tenancies — S34(1)(b) of 1988 Act — Series of joint tenancies with fluctuating membership — Original tenant protected by 1977 Act ceasing to be a joint tenant — Whether remaining joint tenants under successor joint tenancies entitled to protection of s34(1)(b) — Held, protection continuing for joint tenancies only where joint tenant originally protected prior to enactment of 1988 Act

***Secretarial & Nominee Co Ltd* v *Thomas (CA),* [2005]** 3 **37**

Rent assessment

Practice and procedure — Rent assessment committee taking into account supplemental submission from tenant — Submission not copied to landlord — Whether breach of natural justice — Correct approach to valuation of security of tenure — Held, fairness requiring that all parties see all relevant evidence — Further material submitted to committee after conclusion of

Leasehold Reform Act 1967 — *continued*

Enfranchisement price

Improvements — Assumption under s9(1A)(d) — Whether value of potential to make improvements should be subtracted — Whether value of planning permission for improvements should be disregarded — Held, value of potential or any planning permission for improvements should not be disregarded

***Fattal* v *Keepers and Governors of the Possessions, Revenues and Goods of the Free Grammar School of John Lyon (CA),* [2005]** 1 **51**

Hope value

Valuation under s9(1A) — Whether permissible to include hope value in valuation of landlord's interest — Application of *Earl Cadogan* v *Sportelli* — Held, valuation of landlord's interest not to include hope value — Decision on point in *Sportelli* given *per incuriam*

***Pitts* v *Cadogan; Cadogan* v *Atlantic Telecasters Ltd (LT),* [2007]** 3 **86**

Low rent test

S4A test that initial rent less than two-thirds of rateable value — At grant of tenancy property consisting of two cottages with rateable values — Tenancy envisaged conversion to single dwelling-house — Dwelling-house later entered in valuation list with rateable value — Initial rent more than two-thirds of rateable values of two cottages but less than two-thirds of rateable value of single house — Whether property having nil rateable value at date of grant of tenancy — Whether two cottages or single dwelling-house "the property" for purposes of s4A — Whether low rent test satisfied — Held, low rent test not satisfied — "The property" under s4A being two cottages — Two cottages having rateable values — Initial rent more than two-thirds of sum of rateable values of two cottages

***Neville* v *The Cowdray Trust Ltd (CA),* [2006]** 3 **47**

Meaning of "house"

Notice of desire to acquire freehold of property — Property one in parade of shops with residential accommodation above — Whether "house" for purposes of 1967 Act — Whether tenant entitled to exercise enfranchisement rights — Held, property a "house" within 1967 Act — Removal of residence test irrelevant — Test of whether a house reasonably so called objective

***Hareford Ltd* v *Barnet London Borough Council (CC),* [2005]** 2 **72**

S2(1) test that house "designed or adapted for living in" — Whether test must be satisfied at date of tenant's desire notice — Whether at date of desire notice premises must be capable of occupation as residence — Held, premises must be viewed as at the moment of the desire notice — Question was whether premises were designed or adapted for living in — Premises that were dilapidated or adapted for a different purpose such as commercial were not designed or adapted for living in

***Boss Holdings Ltd* v *Grosvenor West End Properties; Mallet & Sons (Antiques) Ltd* v *Grosvenor West End Properties (CA),* [2006]** 2 **43**

Leasehold Reform, Housing and Urban Development Act 1993

Claim for new lease

S45 counternotice — Validity — Mistake as to identity of landlord — Substantial estate with primary and subsidiary trusts — Whether tenant in position of actual tenant would have been misled by mistake — Whether counternotice valid — Held, counternotice valid — Tenant in position of actual tenant would not have been misled by mistake as to identity of landlord — Counternotice valid

***Lay* v *Ackerman (CA),* [2005]** 1 **139**

Collective enfranchisement

Anti-avoidance provisions — S13 initial notice served and registered on behalf of qualifying tenants — Landlord subsequently granting long lease of basement caretaker's flat and light well to associated company — Whether lease void as severance of freehold interest under s19(1)(i) — Whether void as disposal of acquirable common parts under s19(1)(ii) and s2(1)(b) and (3) — Whether interest under lease liable to acquisition on behalf of participating tenants under s19(2) and (3) — Held, no severance of reversion where long lease granted rather than freehold interest — Insurance rent and service charges under lease amounting to "rent" such that lease not capable of enlargement into freehold under s153 of Law of Property Act 1925 — Lease void under s19(1)(a)(ii) since caretaker's flat a "common part" acquirable on behalf of participating tenants — Light well a common part but not necessary to acquire it on behalf of participating tenants — Grant of lease not a disposal of freeholder's interest within s19(2) and (3) so as to render interest acquirable on behalf of participating tenants

Panagopoulos* v *Earl Cadogan (Ch), 151

Application to acquire part of building containing 20 flats — That part divisible into two sets of 10 flats — Whether property self-contained part of a building within s3 — Whether that term denoting a part incapable of further subdivision into smaller self-contained parts — Held, property qualifying for enfranchisement where comprising self-contained part of building and none of statutory exclusions in s4 applying — S3 imposing no further requirement that property indivisible into smaller self-contained parts

41-60 Albert Palace Mansions (Freehold) Ltd* v *Craftrule Ltd (Ch), 45

Charitable Housing Trust — Whether flats on long leases held for charitable purposes — Whether tenants of charitable housing trust entitled to exercise rights of collective enfranchisement — Held, only flats let as social housing held for charitable purposes — Tenants holding long leases entitled to seek collective enfranchisement

***Brick Farm Management Ltd* v *Richmond Housing Partnership Ltd (QB),* [2005]** 3 **57**

Counternotice — Counternotice held invalid in earlier proceedings — Landlord ordered to serve further counternotice admitting claim within time limit — Second counternotice not admitting entitlement to enfranchise — Whether landlord able to serve second counternoice denying entitlement — Whether nominee purchaser entitled to order under s25 that premises be acquired on terms of initial notice — Whether landlord entitled to order varying time limit to serve third counternotice admitting entitlement — Held, where court determining issue as to entitlement of tenant to collective enfranchisement, landlord cannot serve further counternotice denying entitlement — Nominee purchaser entitled to acquire on terms of initial notice — No jurisdiction to vary time limit to permit landlord to serve third counternotice admitting entitlement

***Brick Farm Management Ltd* v *Richmond Housing Partnership Ltd (No 2) (Ch),* [2006]** 2 **46**

Deferment rates — Extended lease — Whether appropriate to apply same percentage for capitalisation rate as for deferment rate — Whether deferment rate for central London properties applicable to properties elsewhere — Held, no valuation rationale for applying same capitalisation rate — Same deferment rate applied as for central London properties

***Nicholson* v *Goff (LT),* [2007]** 1 **83**

Initial notice — Four out of seven qualifying tenants seeking to enfranchise — S13 initial notice signed by three individual tenants and director of fourth corporate tenant — Whether notice validly signed "by" corporate tenant within s99(5) of 1993 Act — Whether signature of two directors or one director and company secretary required under s36A of Companies Act 1985 — Whether notice invalid on ground that not given by majority of qualifying tenants — Company signing notice under s13 of 1993 Act required to provide signature of two directors or director and company secretary or affix common seal under s36 of 1985 Act — Notice not validly signed by corporate tenant where signature of only one director provided — Notice invalid since not signed by majority of qualifying tenants

Hilmi & Associates Ltd* v *20 Pembridge Villas Freehold Ltd (CA), 41

Internal floor area — Common parts — Building containing both residential and commercial units — Residential lessees serving notice to acquire freehold — S4(1) of 1993 Act — Landlord disputing right to acquire freehold on ground that more than 25% of the internal floor area neither residential nor comprised in the common parts — Whether common parts including those for use of commercial occupants only — Held, all common parts qualifying as such regardless of whether used by residential lessees — Internal floor area including balconies let with flats

***Marine Court (St Leonards on Sea) Freeholders Ltd* v *Rother District Investments Ltd (CC),* [2008]** 1 **39**

Landlord's counternotice — Validity — Failure to state if premises within area of estate-management scheme — Requirements of landlord's counternotice — Requirements of s21 — Whether additional requirements of reg 4 of Leasehold Reform (Collective Enfranchisement)(Counternotices) (England) Regulations 2002 — Whether reg 4 requirements mandatory or directory — Whether landlord's counternotice valid — Held, requirement in reg 4 to serve negative statement specifying premises not within EMS area directory — Landlord's counternotice valid

***7 Strathray Gardens Ltd* v *Pointstar Shipping & Finance Ltd (CA),* [2005]** 1 **53**

Purchase price — Block of flats — Whether positive value to be applied to caretaker's flat — Whether service charge payable by underlessees to include loss of rack-rent from caretaker's flat in circumstances where head lessee obliged by headlease to provide such flat rent-free — Held, service charge to include loss of rack-rent from caretaker's flat where expressly defined to come within cost of outgoings for such accommodation — Headlease to be valued on assumption that headlessee entitled to be compensated in service charge for loss of rack-rent from caretaker's flat

McHale* v *Earl Cadogan (CA), 51

Purchase price — Deferment rate — Flats — Enfranchisement of block of flats in Hove — Whether uplift to generic 5% *Sportelli* rate for flats justified in light of greater obsolescence and management risks and lower capital growth rates than in prime central London (PCL) area — Held, 0.25% addition justified for obsolescence to reflect fact that property values lower than in PCL while building costs comparable — Addition of 0.5% justified to reflect lower capital growth rates in Brighton and Hove area than in PCL — Further addition of 0.25% justified to reflect greater management problems of flats over houses — Deferment rate of 6% applied

Ashdown Hove Ltd* v *Remstar Properties Ltd (LVT), 61

Purchase price — Deferment rate — Hope value — Enfranchisement of 1930s block of flats by 50% of qualifying tenants — Leasehold valuation tribunal applying 5% *Sportelli* deferment rate and making no allowance for hope value — Whether higher deferment rate justified for increased risk of obsolescence and variation in growth rates in non-prime areas — Held, appropriate percentage of hope value to be awarded for non-participating flats — No evidence that non-prime property prices more volatile than in prime areas so as to justify different deferment rate — Obsolescence concerned with risk of deterioration to building not increase in value as site for redevelopment — Property having no features increasing risk of obsolescence or differentiating from generality of properties in that regard — Hope value for non-participating flats assessed at 10% of marriage value where 50% of flats not participating and leases having 65 years to run

***Culley* v *Daejan Properties Ltd (Ch),* [2009]** 3 **165**

Right to amend initial notice — Whether claim including property that no right to acquire — Whether self-contained part of a building — Provision of "relevant services" — Held, purposive approach to amendment provisions in para 15(2) of Sched 3 to Act not permissible — Amendment allowed since claim including property that claimant not entitled to acquire — Specified premises not self-contained part of building where not possible to provide relevant services independently by reconfiguring existing services

***Oakwood Court (Holland Park) Ltd* v *Daejan Properties Ltd (CLCC),* [2007]** 1 **121**

Tenant's notice proposing purchase price of £210 — Landlord's counternotice proposing £130,000 — Whether counternotice valid — Whether *Cadogan* v *Morris* test applying to landlord's counternotice — Whether *Cadogan* v *Morris* test of unrealistic price invoking objective or subjective condition — Held, counternotice valid — *Cadogan* v *Morris* test not applying to landlord's counternotice — *Cadogan* v *Morris* test invoking subjective condition — Landlord entitled to reply upon price proposed in good faith

***9 Cornwall Crescent London Ltd* v *Kensington and Chelsea Royal London Borough Council(CA),* [2005]** 2 **131**

Terms of acquisition — S1(4)(a) of 1993 Act — Land proposed to be retained by freeholder — Whether rights offered over retained land sufficient to comply with s1(4)(a) — Test to be applied — Enfranchisement premium — Deferment rate — Held, rights offered to be considered in light of freeholder's proposals as a whole — Offered rights insufficient where effectively taken away by rights retained under s21(3)(d) — Deferment rate of 7% applied

***Ulterra Ltd* v *Glenbarr (RTE) Co Ltd (LT),* [2008]** 1 **103**

Vesting order — S24 of 1993 Act — LVT determining price payable for freehold — Terms of transfer subsequently agreed — Nominee purchaser applying for vesting order — Whether application made out of time — Whether time running from date of LVT decision or subsequent agreement of terms of transfer — Held, LVT having power to issue more that one decision dealing with issues in dispute in stages — Terms of transfer not taken to be determined at time of determination of price — Time running from date of agreement of terms of transfer — Application for vesting order made in time

***Goldeagle Properties Ltd* v *Thornbury Court Ltd (CA),* [2008]** 3 **69**

Counternotice

Initial notice of collective enfranchisement served on then freeholder — Purchaser of freehold reversion serving counternotice before purchase registered — Tenants applying to acquire freehold on terms of initial notice — Effect of s19 of 1993 Act — Whether valid counternotice served — Whether validated retrospectively upon registration of purchaser's title — Whether expert evidence relevant to bona fides of price proposal in initial notice — Held, no valid counternotice — S19 not retrospectively validating acts of purchaser done without legal right — Expert valuation evidence not probative of issue of good faith in relation to tenants' proposals

***Renshaw and others* v *Magnet Properties South East LLP (CC),* [2008]** 1 **42**

Determination of price

Applications for enfranchisement or extension of leases — Appropriate deferment rate — Whether hope value to be excluded from valuation — Method of valuation for hope value where included — Held, setting of general deferment rate appropriate — Deferment rate to be calculated as risk-free rate minus housing market growth rate plus risk premium — Deferment rate 4.5% for houses and 4.75% for flats in prime central London area — Departure from generic deferment rate justified only if compelling evidence indicating necessary owing to particular features of individual case — Hope value to be excluded from valuation on collective enfranchisement claim

***Earl Cadogan v Sportelli (LT),* [2007]** 1 **153**

Collective enfranchisement — Block of flats — Purchase price — Deferment rate — Whether departure from general rate of 5% for flats justified — Location — Obsolescence — Management risks — Valuation date — Whether date of landlord's counternotice or date of decision of leasehold valuation tribunal — Held, general deferment rate for flats applicable — No adjustment for location — No adjustment for obsolescence where fully reflected in vacant possession value and risk premium — No difference between flats managed by head lessee and those managed directly by freeholder — Valuation date that of counternotice

***Hildron Finance Ltd* v *Greenhill Hampstead Ltd (LT),* [2008]** 1 **179**

Deferment rate — Evidence — Whether freeholder should be excluded from adducing evidence challenging generic deferment rate laid down in *Earl Cadogan* v *Sportelli* — Held, leave to amend required for freeholder to contend for deferment rate outside range contended for in pleadings on appeal — Evidence directed to challenging guidance of Lands Tribunal of kind given in *Sportelli* to be excluded save in exceptional circumstances

***Earl Cadogan* v *Erkman (LT),* [2009]** 1 **87**

Deferment rates — Absence of market evidence on deferment rates — Conventional 6% — Settlement evidence — Financial markets — Property-specific factors — RICS *Red Book* valuation manual — Relevance of *Red Book* guidance to statutory valuations — Held, deferment rates determined from evidence — Settlement evidence unreliable without agreed analysis — In absence of market sales evidence, financial market evidence of index-linked gilts acceptable — 4½% appropriate deferment rate on Cadogan Estate — No adjustments between individual flats and collective enfranchisement — One-quarter per cent adjustment between houses and flats on Cadogan Estate — Regard to be had to statutory valuation definitions as interpreted by case law and not to *Red Book*

***Arbib* v *Earl Cadogan; Earl Cadogan* v *55/57 Cadogan Square Freehold Ltd; Earl Cadogan* v *Moussaieff; Day* v *32 Rosary Gardens (Freehold) Ltd; In re Earl Cadogan's Appeal (LT),* [2005]** 3 **139**

Hope value — Deferment rate — Whether hope value to be excluded from purchase price — Whether general deferment rate laid down by Lands Tribunal to be applied — Held, hope value to be excluded from lease extension and collective enfranchisement claims under 1993 Act — No error in Lands Tribunal's approach to deferment rate

***Earl Cadogan* v *Sportelli; Earl Cadogan* v *27-29 Sloane Gardens Ltd; Earl Cadogan* v *Grandeden Property Management Ltd; Howard de Walden Estates Ltd* v *Maybury Court Freehold Co Ltd (CA),* [2008]** 1 **137**

Head lessees

Collective enfranchisement — Whether landlord's leaseback proposals required to be specified in s21 counternotice —Whether possible to serve leaseback notice at any time up to acquisition by nominee purchaser — Effect of definition of "appropriate time" in Sched 9 — Held, not possible to require leaseback unless specified in counternotice — Definition of "appropriate time" imposing condition subsequent preventing leaseback specified in counternotice if flat having qualifying tenant by date of acquisition by nominee purchaser

***Cawthorne* v *Hamdan (CA),* [2007]** 1 **67**

Head lessee — Chapter II of 1993 Act — Buildings incorporating flats and common parts — Whether head lessee entitled to individual lease extension in respect of flat within the building — Held, lessee under lease including flat together with other premises qualifying as "tenant of a flat" for purposes of Chapter II — Head lessee of such premises entitled to exercise right of individual lease extension

***Howard de Walden Estates Ltd* v *Aggio; Earl Cadogan* v *26 Cadogan Square Ltd (HL),* [2008]** 2 **57**

MEANING OF "PREMISES" — See **Leasehold enfranchisement** (Leasehold Reform, Housing and Urban Development Act 1990 — Lease extension)

MESNE PROFITS — See **Landlord and tenant** (Tolerated trespasser)

MINES (WORKING FACILITIES AND SUPPORT) ACT 1966 — See **Trespass**

MISREPRESENTATION — See **Estate agents** (Property Misdescriptions Act 1991); **Landlord and tenant**

MISTAKE — See **Land registration** (Alteration of registered title — Rectification for mistake)

MORTGAGE — See also **Adverse possession**

Consumer Protection Act 1974

Agreement in parts

Mortgage loan made to customer on condition that existing mortgage paid off out of advance — Customer defaulting on payments — Possession orders made — Whether mortgage unenforceable for failure to comply with requirements of s127(3) of 1974 Act — Whether amount of advance within threshold for application of Act — Whether agreement in parts within s18(1)(*a*) such that each part to be considered as separate agreement — Held, agreement in parts within s18(1)(*a*) where agreement by its terms containing two or more parts in different categories — S18(1)(*a*) not applying since terms creating single agreement falling as a whole within two disparate categories — S18(1)(*b*) applying — Credit exceeding threshold for application of Act — Mortgage unregulated by Act and enforceable

Southern Pacific Mortgage Ltd* v *Heath (CA), 55

Enforcement

Undue influence

Wife persuaded by husband to agree to remortgage of matrimonial home to refinance his debts — Husband having clandestine affair and later obtaining divorce — Mortgagee seeking possession — Whether mortgage enforceable against wife – Whether vitiated by husband's undue influence — Held, obligation of candour and fairness owed by husband where wife reposing trust and confidence in him — Requisite trust and confidence capable of arising in connection with impugned transaction — Obligation of candour and fairness requiring disclosure of material facts necessary to enable wife to make informed decision — Obligation breached where husband failing to disclose affair while asking wife to agree to charge to preserve family home

First Plus Financial Group plc* v *Hewett (CA), 51

Equitable mistake

All-moneys charge

Deed of release — Company in administration — Bank releasing charge over property in erroneous belief that all secured indebtedness discharged — Administrator wishing to sell property — Whether bank to be treated as secured or unsecured creditor — Whether bank entitled to rescind deed of release for mistake — Held, relief from consequences of mistake available where party affected would have acted differently had it known the true facts — No bar to relief that legal effect of document correctly understood by parties — Bank to be treated as secured creditor

***Fender (administrator of FG Collier & Sons Ltd)* v *National Westminster Bank plc (Ch),* [2008]** 3 **80**

Overriding interest

Possession

Land Registration Act 2002 — Mortgagee holding charge over property — Property transferred to mortgagor by person suffering from mental illness — Transferor resident in care institution at time of registration of mortgage — S29 of 2002 Act — Whether transfer to mortgagor voidable for incapacity or undue influence — Whether transferor in actual occupation at date of mortgage such that right to set aside transfer taking priority over charge — Held, transferor entitled to set aside transfer — Transferor in actual occupation at date of mortgage so that right to set aside transfer having priority over charge — Sufficient evidence of continuity and permanence of occupation where involuntary residence elsewhere explained by objective reasons and where persistent intention to return home manifested

Link Lending Ltd* v *Bustard (CA), 55

Tenancy

Proprietary estoppel — Fraud — Land Registration Act 2002 — Agreement for sale and leaseback of property on understanding that purchaser to pay off vendors' mortgage liabilities and permit them to remain in the property for life so long as terms of lease complied with — Purchaser mortgaging property – Mortgagee seeking possession — Housing Act 1988 — Whether — having assured tenancy — Whether vendors' tenancy and equitable rights against purchaser binding on mortgagee — Whether amounting to overriding interests with priority over mortgagee's rights — Held, assured tenancy held — Description as "Short Assured Tenancy" a "provision to the effect" that tenancy not assured shorthold tenancy for purposes of para 3 of Sched 2A to1988 Act — Vendors' rights under tenancy and those arising by proprietary estoppel and through fraud of purchaser binding on mortgagee — Rights amounting to overriding interests taking priority over mortgagee's rights — Substance of transaction that purchaser's title always subject to vendors' rights

***Redstone Mortgages plc* v *Welch (CC),* [2009]** 3 **71**

Possession

Limitation Act 1980

Mortgagor falling behind on repayments — Mortgagee making demand for repayment then taking no further action — Date when right of action for possession accruing — Whether action time-barred under s15 of 1980 Act such that charge extinguished under s17 — Whether adverse possession necessary for limitation period to run — Whether mortgagor acknowledging mortgagee's title — Held, right of action for possession not depending upon adverse possession by mortgagor — Right accruing on date mortgagee becoming entitled to possession — Mortgagor not acknowledging mortgage — Action time-barred

***Ashe* v *National Westminster Bank plc (Ch),* [2007]** 2 **137**

Power of sale

Human rights

Law of Property Act 1925 — Mortgagors falling into arrears with mortgage repayments — Mortgagee appointing receivers and exercising power of sale under s101 of 1925 Act — Sale overreaching mortgagors' rights — New registered proprietor seeking possession of premises — Whether statutory power of sale without court order compatible with Art 1 of First Protocol to European Convention on Human Rights — Held, s101 implementing rather than overriding private bargain of parties — No deprivation of possessions by state intervention — No incompatibility with Art 1

***Horsham Properties Group Ltd* v *Clark (Ch),* [2008]** 3 **75**

Security

Costs

Mortgagee appointing receivers in respect of mortgaged property — Legal proceedings by mortgagor against mortgagee — Whether mortgagor entitled to add legal costs to security — Whether costs reasonably and properly incurred — Ss101 and 109(6) of Law of Property Act 1925 — Receivers' fees and costs — Whether power to add receivers' costs to security — Held, mortgagee entitled to legal costs as reasonably and properly incurred in preserving or enforcing security — No power to charge mortgagor in respect of receivers' costs beyond those provided for by statute

***Vedalease Ltd* v *Averti Developments Ltd (CLCC),* [2007]** 2 **125**

Subrogation

Unjust enrichment

Land Registration Act 2002 — Bank advancing loan by way of remortgage — Registrable charge granted to bank but not registered — Further charge subsequently registered by another lender — Whether that lender unjustly enriched by bank's failure to register its charge — Whether bank entitled to be subrogated to earlier charge securing loan repaid from bank's advance so as to obtain priority over subsequent charge — Held, carelessness of bank in failing to register own charge not precluding remedy of subrogation — Remedy depending on whether subsequent chargee unjustly enriched — Justice of position as between bank and subsequent chargee to be examined — Subsequent chargee unjustly enriched where enrichment due to failure of bank to obtain legal charge on which decision to lend premised — Irrelevant that borrower performing all terms of bargain with bank

Anfield (UK) Ltd* v *Bank of Scotland plc (Ch), 75

NEGLIGENCE — See also **Insurance**

Architect

Duty of care

Chain of causation — Latent defect — Building designed with inadequate capacity for rainwater drainage — Flood damage to tenant's property — Whether architect under continuing duty of care — Whether reasonable to expect defect to be discovered and remedied following earlier flood — Whether defect ceasing to be latent upon earlier discovery of which tenant not informed – S14B of Limitation Act 1980 — Whether claim time-barred — Possibility of defect being discovered upon intermediate inspection not affecting liability if claimant not discovering it — Flood damage foreseeable — Discovery of defect by third party not rendering defect patent to others who remained unaware — Claim not time-barred

***Pearson Education Ltd* v *The Charter Partnership Ltd (CA)* [2007]** 2 **89**

NEGLIGENCE — *continued*

Licensed conveyancer

Damages

Property advertised for sale with option to purchase additional land — Purchaser's conveyancer failing to include option in conveyance — Claim for negligence and breach of contract — Quantum of damages — Whether award to be made for loss of chance to develop both areas of land together — Held, damages recoverable on claimed basis where conveyancer knowing about special circumstances that purchaser intending to develop both areas of land together for profit — Damages calculated as percentage of anticipated profit of development less profit obtainable from development of property without additional land — Overall percentage derived from multiplying together percentage chances of grant and exercise of option and of obtaining planning permission and funding for development

Joyce* v *Bowman Law Ltd (Ch), 129

Solicitor

Assessment of damages

Notice under s25 of Landlord and Tenant Act 1954 wrongly relying upon ground *(f)* — Liability to pay statutory compensation — Identity of tenant — Whether chance that party in occupation would be found to be tenant — Whether chance that under correct s25 notice no statutory compensation payable — Assessment of chance — Whether trial judge failing properly to assess evidence relating to speculative chance — Whether conclusions on hypothetical facts conclusions on facts — Held, trial judge failing to reach proper conclusions on evidence of speculative chance — Damages increased

***Talisman Property Co (UK) Ltd* v *Norton Rose (CA),* [2006]** 3 **59**

Notice under s25 of Landlord and Tenant Act 1954 wrongly relying upon ground *(f)* — Admitted liability for incorrect notice — Liability to pay statutory compensation — Identity of tenant — Whether party in occupation found to be tenant — Whether chance that under correct s25 notice no statutory compensation would have been payable — Assessment of chance — Held, by serving incorrect notice on occupier, landlord missing opportunity to argue that occupying party was the tenant and no compensation would then have been payable — Assessment of that chance at 30% — Damages assessed at 30% of statutory disturbance compensation

***Talisman Property Co (UK) Ltd* v *Norton Rose (Ch),* [2006]** 1 **145**

Limitation period

Cause of action — Negligent advice alleged in respect of entry into building agreement and subsequent loss of rights by waiver of certain contractual provisions — Date upon which cause of action accruing — Whether damage resulting at later date than entry into agreement or waiver of provisions — Whether loss contingent — Whether claims time-barred — Held, damage consisting of loss of chance of renegotiating agreement occurring at date agreement entered into — Damage immediate not contingent where benefit of contractual provision lost — Claims time-barred

***Watkins* v *Jones Maidment Wilson (CA),* [2008]** 1 **149**

Limitation Act 1980 — Party purporting to be owner of property instructing solicitor to act on transfer of it — Transfer fraudulent — Actual owner alleging breach of duty of care by solicitor in failing to verify identity of client — Whether claim statute-barred under section 2 of 1980 Act — Whether cause of action accruing on completion of fraudulent transfer or on later registration of title by fraudulent transferee — Whether solicitor owing continuing duty of care after completion — Held, cause of action accruing on completion of fraudulent transfer — Forged transfer exposing any purchaser of property to risk of litigation and therefore reducing price willing to pay — Consequent diminution in open market value of property at date of completion of forger transfer — This constituting actual loss to owner — No special facts creating continuing duty of care after completion of transfer for which solicitor retained

Nouri* v *Marvi (CA), 79

S14A(5) of Limitation Act 1980 — Three-year limitation period — Knowledge required for bringing action — Leases containing options to determine personal to original tenant — Assignments to subsidiary company — Negligent failure to advise consequence of assignments — Options ceasing to have effect — Whether claimant entitled to rely upon s14A(5) three-year limitation period — Whether appellants having requisite knowledge for bringing action — Whether appellants having knowledge of material facts — Whether claimant knowing enough to investigate further — Held, claimant having sufficient knowledge to investigate further with view to instigating proceedings before writ — S14A(5) three-year period having expired

***3M United Kingdom plc* v *Linklaters & Paines (CA),* [2006]** 2 **53**

Surveyors

Damages

Supervision of building contract — Extent of duties of supervising building surveyors/architects — Defects — Alternative remedial schemes — Measure of damages — Defective Premises Act 1972 — Whether building surveyors negligent — Whether any claim under 1972 Act — Whether alternative measure of damages to reinstatement costs — Determination of alternative remedial schemes — Measure of damages — Held, building surveyors/architects negligent — No claim under 1972 Act — Cost of repair appropriate measure of damages — Less expensive remediation scheme appropriate — Damages awarded

***Catlin Estates Ltd* v *Carter Jonas (a firm) (TCC),* [2006]** *2* **139**

Surveyor negotiating grant of leases on behalf of landowners — Surveyor failing to negotiate turnover rent — Whether breach of duty by surveyor — Whether damages to be assessed on valuation basis or loss of earnings basis — Date at which loss to be assessed — Held, surveyor in breach of duty — Chance lost to obtain turnover rent — Damages to be assessed on valuation basis by reference to comparative value of landlord's interest under lease actually granted and that which should have been obtained — Valuation by reference to values as at date of transaction

***Earl of Malmesbury* v *Strutt & Parker (QB),* [2007]** 3 **153**

Valuation

Identification of property — Surveyor agreeing to value property — Surveyor misled by owner into valuing different property on nearby plot — Mortgage lender advancing loan in reliance upon valuation — Whether appellant liable to lender for breach of duty — No allegation of failure to exercise reasonable care and skill — Whether unqualified obligation on surveyor to value property to which instructions relate — Held, normal retainer of surveyor to inspect and value property containing absolute obligation to inspect and value correct property — Valuation of different property amounting to breach of duty notwithstanding exercise of reasonable care and skill

***Platform Funding Ltd* v *Bank of Scotland plc (CA),* [2008]** 3 **83**

NET INTERNAL AREA — See **Agreement for lease** (Specific performance)

NOISE — See **Nuisance**

NON-COMPLETION CERTIFICATE — See **Construction** (Building contract)

NON-COMPLIANCE — See **Town and country planning**

NO-SCHEME WORLD — See **Compulsory purchase**

NOTICE TO COMPLETE — See **Sale of land**

NOTICE TO QUIT — See Agricultural holdings

NOTICE TO TREAT — See **Compulsory acquisition; Compulsory Purchase**

NOTICES — See also **Agricultural holdings** (Notice to quit) (Notice to remedy repairs — Service); **Compulsory acquisition; Landlord and tenant** (Landlord and Tenant Act 1987) (Landlord and Tenant (Covenants) Act 1995 — Lease) (Landlord and Tenant (Covenants) Act 1995 — Yield up and reinstatement obligations) (Option to determine — Tenant's break option) (Rent review — Trigger notice); **Leasehold enfranchisement** (Leasehold Reform, Housing and Urban Development Act 1993 — Claim for new lease) (Leasehold Reform, Housing and Urban Development Act 1993 —Collective enfranchisement); **Option agreement**

Option

Validity

Option in lease to acquire freehold — Exercise of option requiring at least three months' notice expiring within specified period — Date of expiry of notice relevant to valuation and completion — Notice expressed to expire before three months — *Mannai Investment Co Ltd* v *Eagle Star Life Assurance Co Ltd* [1997] 1 EGLR 57 —Whether notice valid — Held, notice invalid

***Peaceform Ltd* v *Cussens (Ch),* [2006]** 3 **67**

RIGHT TO MANAGE — See also **Landlord and tenant**

Commonhold and Leasehold Reform Act 2002

Part of building

Vertical division — Tenants of flats seeking to exercise right to manage part of building in which flats located — S72(3)(a) of 2002 Act — Whether part of building self-contained so as to qualify for right to manage — Whether departure from vertical such as to exclude right to manage — Held, no departure from vertical permitted unless *de minimis* — Relative floor area of horizontal part not relevant — Tenants' part of building not self-contained

***Holding & Management (Solitaire) Ltd* v *1-16 Finland Street RTM Co Ltd (LT),* [2008]** 1 **107**

Qualifying premises

Premises in mixed use — Para 1 of Schedule 6 to 2002 Act — Extent of non-residential part — Whether exceeding 25% of internal floor area — Whether vaults to be excluded in determining floor area — Whether porter's flat residential where converted into offices without planning permission and in breach of covenant in headlease owed by landlord to freeholder — Whether wishes of residential tenants relevant — Held, vaults part of demise — Porter's flat not residential where landlord having no intention to use for residential purposes and residents having no right under terms of leases to insist upon such use

***Connaught Court RTM Co Ltd* v *Abouzaki Holdings Ltd (LT),* [2008]** 3 **175**

SALE OF LAND

Auction

Seller as trustee

Title containing obligation to impose covenants against competition in sales of other properties — Seller not imposing such covenants — Whether breach of seller's duty as trustee for the buyer — Contract — Completion — Whether specific performance — Agent — Whether bidder acting as agent personally liable — Held, seller's duty as trustee not requiring imposition of anti-competition covenants in other sales — Agent personally liable — Specific performance granted on summary judgment

***Englewood Properties Ltd* v *Patel (Ch),* [2005]** 2 **77**

Charging order

Trustee

Contract for sale and purchase of land — Charging Orders Act 1979 — Contracts exchanged — Purchaser acting as nominee for another — Charging order on property in respect of costs orders made against purchaser — Effect of trust and sub-trust on interest of intermediate trustee — Whether purchaser having interest as trustee upon which order under s2(1)(b)(i) of 1979 Act able to bite — Whether non-party costs order appropriate against purchaser's principal — Held, interest of intermediate trustee not extinguished where trust and sub-trust existing — Vendor holding property on bare trust for purchaser while purchaser holding on bare trust for principal — Purchaser retaining interest as trustee for purposes of charging order — Non-party costs order appropriate where principal supporting litigation

***Nelson* v *Greening & Sykes (Builders) Ltd (CA),* [2008]** 1 **59**

Completion

Deed or escrow

Legal charge — Mortgage lender advancing funds — Funds remitted to vendors' solicitor and released to vendors — Contracts exchanged — Transfer executed and delivered by vendor — Purchaser granting legal charge in favour of mortgagee — Purchaser failing to pay balance of purchase price — Whether execution and delivery of transfer amounting to effective transfer or transfer made in escrow pending payment of balance — Whether mortgagee having valid charge taking priority over unpaid vendors' lien — Held, where part of purchase price unpaid inference arising that transfer made in escrow pending payment of balance — Inference rebuttable by evidence to contrary — Transfer unconditional where funds released to and used by vendors — Mortgagee's charge valid

***Bank of Scotland plc* v *King (Ch),* [2008]** 1 **65**

Notice to complete

Compliance — Contract providing for completion by 1pm on completion date failing which purchaser to be treated as having completed on next working day — Whether failure of purchaser to pay purchase moneys by 1pm on last day permissible under notice to complete entitling vendors to rescind — Whether failure to include vendors' legal costs in sum paid having that effect — Held, 1pm cut-off point relevant only to contractual completion date — Breach placing appellant in breach of contract but not giving right to rescind — Payment after 1pm on last day of notice to complete sufficient to comply with notice — Vendor's legal costs not payable as condition of completion such that failure to include them in purchase moneys not giving right to rescind — Specific performance ordered

***Chinnock* v *Hocaoglu (CA),* [2009]** 1 **61**

Compliance — Contract providing for completion by 1pm on completion date failing which purchaser to be treated as having completed on next working day — Whether completion date in contract to be rectified on ground of common mistake — Whether failure of purchaser to pay purchase moneys by 1pm on last day permissible under completion notices entitling vendors to rescind — Held, purchaser entitled to rectification to amend completion date — Vendor serving valid notice to complete — Tendering of purchase moneys after 1pm on final day sufficient to comply with notice — Absence of vendor's legal costs from moneys tendered amounting to failure to comply with notice such that vendor entitled to rescind

***Chinnock* v *Hocaoglu (Ch),* [2008]** 2 **77**

Law of Property Act 1925 — Agreement for sale of shares in company owning commercial property — Contractual term that purchaser not obliged to complete in absence of specified documentation – Notice to complete served by vendor — Vendor rescinding contract and forfeiting deposit — Whether vendor ready, willing and able to complete – Whether appropriate to order return of deposit under s49(2) of 1925 Act — Held, vendor obliged to call evidence to support assertion that ready, willing and able to complete where this disputed — Vendor ready, willing and able to complete where able to set up necessary administrative arrangements to enable completion to take place within reasonable time — No return of deposit in case of default by purchaser save in exceptional circumstances — Later sale by vendor to third party at profit not sufficient reason to order return of deposit

***Midill (97PL) Ltd* v *Park Lane Estates Ltd and another (CA),* [2009]** 1 **65**

Contract

Benefit and burden

Land in ownership of third parties — Vendor contracting to buy from third parties and assigning rights under contracts to purchaser — Purchaser retaining part of sale price pending completion of works under terms of assigned contract — Purchaser subsequently failing to pay balance — Whether third party entitled to recover retention moneys — Whether purchaser bound to observe vendor's obligations under assigned contract as burden appurtenant to benefit received — Held, burden imposed with benefit only where both conferred by same legal transaction and benefit conditional on or reciprocal to burden — Assignment of benefit of contract not made conditional on purchaser undertaking burden — Completion of sale not so conditional — Burden of assigned contract remaining obligation of vendor alone

Davies* v *Jones (CA), 82

Frustration

Sale of block of flats — Purchaser agreeing to sell long lease of porter's flat if no longer needed as such and to pay overage to vendor — Sale of porter's flat prevented by enfranchisement of block under Leasehold Reform, Housing and Urban Development Act 1993 — Whether agreement to be construed so as to require overage payment — Whether agreement frustrated — Held, sale of freehold in porter's flat as part of statutory enfranchisement of entire block falling outside contemplation of parties at date of agreement — Differences between agreed steps under agreement and events leading to enfranchisement so numerous and significant that agreement incapable of applying to such events — Unreasonable to impose absolute liability even where performance of agreement impossible — Agreement discharged by frustration

Hildron Finance Ltd* v *Sunley Holdings Ltd (Ch), 1

Purchase price

Sale of property for development — Price to be fixed sum less costs of obtaining release or variation of restrictive covenants — No planning permission or agreement with covenantees at completion date — Whether deduction to be made from purchase price — Whether implied term that cost of obtaining release to be estimated subject to later correction once agreement reached — Whether court able to reach interpretation different from those advanced by parties — Vendor's proposed term not making commercial sense — No grounds for implying term proposed by purchaser — Declaration in contractual dispute to be made only on basis of submission advanced by a party and tested by adversarial argument

***Anglo-Continental Educational Group (GB) Ltd* v *Capital Homes (Southern) Ltd (Ch),* [2008]** 3 **96**

Rescission

Land Registration Act 2002 — Contract for sale of registered freehold land on full title basis under general condition 4.3(a) — Special condition requiring purchaser to accept leasehold interest shown in schedule to freehold title previously determined by surrender — Notice and restriction of bankruptcy of lessee shown against leasehold title in land register and bankruptcy petition and order registered in land charges register — S86 of 2002 Act — Whether surrender void by reason of bankruptcy — Whether vendor having notice of bankruptcy — Whether purchaser entitled to rescind on ground that freehold title encumbered by lease — Held, surrender void by reason of bankruptcy proceedings of which landlord having notice by reason of s86 notice on land register — Notice not depending on search of land charges register that landlord under no obligation to make — Purchaser entitled to rescind for failure of vendor to comply with full title guarantee

Weir* v *Area Estates Ltd (Ch), 77

Land Registration Act 2002 — Contract for sale of registered freehold land on full title basis under general condition 4.3(a) — Special condition requiring purchaser to accept leasehold interest shown in schedule to freehold title previously determined by surrender — Surrender in fact ineffective due to pending bankruptcy petition against lessee at relevant time — Whether special condition precluding reliance on continued existence of lease as ground for rescission of contract by purchaser — Whether existence of lease a purely technical conveyancing defect insufficient to justify rescission — Held, purchaser entitled to rescind notwithstanding terms of special condition — Such condition not protecting vendor in respect of latent defect in title of which properly advised vendor knew or should have known — Existence of lease not purely technical defect where termination of lease not a foregone conclusion at date for completion

Weir* v *Area Estates Ltd (CA), 91

Rescission

Purchaser contracting to buy off-plan underlease of property in development — Contract conditional on vendor obtaining headlease by specified date — Vendor failing to obtain headlease within stipulated time limit but not informing purchaser — Headlease later obtained — Purchaser giving notice to rescind contract – Proper construction of sale contract — Whether rescission possible once headlease obtained — Held, sale contract giving option to rescind if headlease not granted by specified date but not preventing contract from becoming unconditional thereafter – Rescission no longer possible once headlease granted — Notice to rescind served after grant of headlease ineffective

McGahon* v *Crest Nicholson Regeneration Ltd (CA), 84

Repudiatory breach — Vendor serving notices to complete by reference to standard conditions 6.8 and 7.5 — Notices incorrectly calculating date for expiry of notices — Vendor accordingly serving rescission notices prematurely — Whether this amounting to repudiatory breach of contract entitling purchaser to return of deposit — Held, test of repudiatory conduct is whether contract-breaker showing intention to abandon and refuse to perform the contract in all circumstances as viewed objectively from the perspective of reasonable person in position of innocent party — Motive of contract-breaker relevant if something of which innocent party would be aware so as to throw light on way such party would view breach — No repudiatory breach where premature rescission indicates a mistaken application of contractual provisions not refusal to comply with them

Eminence Property Developments Ltd* v *Heaney (CA), 165

Termination

Obligation of vendor to use all reasonable endeavours to secure planning agreement — Purchaser not obliged to complete until agreement secured — Whether contract determining if vendor unable to obtain agreement within reasonable time — Implication of term as to termination — Whether vendor making all reasonable endeavours — Whether purchaser entitled to specific performance — Held, contract not determining automatically where all reasonable endeavours failing to produce planning agreement in reasonable time — Purchaser having option to terminate contract or proceed without planning agreement — Claimant not making all reasonable endeavours — Defendant entitled to specific performance

***Yewbelle Ltd* v *London Green Developments Ltd (Ch),* [2007]** 1 **137**

Contractual term

Construction

Development agreement — Whether term lacking commercial sense — Whether possible to construe term to give effect to intentions of parties without need for rectification — Whether parties' pre-contractual negotiations admissible — Held, court able to correct mistake in contract by process of construction where clear both that something wrong with language of contract and what reasonable person would understand parties to have meant — Case made out where interpretation of term in accordance with ordinary rules of syntax rendering provisions arbitrary and irrational — Not possible to have regard to pre-contractual negotiations as aid to construction — Rectification available if written instrument not reflecting prior consensus of parties

***Chartbrook Ltd* v *Persimmon Homes Ltd (HL),* [2009]** 3 **119**

Deposit

Forfeiture

S49(2) of Law of Property Act 1925 — Sale contracts purporting to exclude application of s49(2) — Vendor purporting to forfeit deposit — Action by purchaser for return of deposit under s49(2) — Vendor seeking summary dismissal of claim — Whether possible to exclude s49(2) by agreement — Held, not possible to contract out of s49(2) — Contractual term purporting to exclude s49(2) void as attempt to oust jurisdiction of court

***Aribisala* v *St James Homes (Grosvenor Dock) Ltd (Ch),* [2007]** 3 **39**

S49(2) of Law of Property Act 1925 — Contracts for sale — Purchaser unable to complete — Vendor purporting to forfeit deposit — Whether just in all circumstances to order return of deposit under s49(2) — Held, deposit exceptionally returnable only where purchaser unable either to perform contract or offer alternative — Return of deposit not justified in ordinary case where purchaser unable to find purchase money

***Aribisala* v *St James' Homes (Grosvenor Dock) Ltd (No 2) (Ch),* [2008]** 2 **65**

Leaseback

Possession

Sale and leaseback of residential property — Final payment of 30% of purchase price not payable to vendor in event of termination of tenancy by purchaser pursuant to rights under tenancy agreement — Termination for arrears of rent — Whether vendor entitled to final payment — Whether repayable as deposit not held by landlord on scheme authorised by Ch 4 of Pt 6 of Housing Act 2004 — Whether provision for retention of final payment unenforceable as penalty or unfair contract term — Held, Ch 4 of Part 6 applying where tenant paid money to landlord and seeking repayment — No application to sale and leaseback where tenant paid nothing but instead seeking balance of purchase price — Relevant contractual provision not a penalty and not a case of relief from forfeiture where no proprietary right to final payment but merely contingent contractual right to payment of debt — No significant imbalance between parties' rights and obligations so as to give rise to unfairness under Unfair Terms in Consumer Contracts Regulations 1999.

UK Housing Alliance (North West) Ltd* v *Francis (CA), 81

Notice to complete

Specific performance

Law of Property (Miscellaneous Provisions) Act 1989 — Vendor contracting to construct flats with due dispatch and sell them to purchasers at discounted price — Discount including finder's fee not reflected in sale contracts — Contracts containing entire agreement clause — Purchasers serving notice to complete when flats unfinished – Vendor subsequently completing flats and serving notice to complete on purchasers — Claim for specific performance — S2 of 1989 Act — Whether contracts void for failure to set out all agreed terms — Whether purchasers entitled to rescind — Held, s2 requiring land contract to contain all expressly agreed terms of land sale — Terms of other simultaneous agreement forming part of same composite transaction not required to be set out in land contract provided performance of land contract not conditional thereon — Finder's fee agreement not part of consideration for land contracts or condition of their performance — S2 of 1989 Act complied with — Specific performance available — Purchasers not entitled to rescind in reliance on own notice to complete owing to failure to specify reasonable time for compliance

North Eastern Properties Ltd* v *Coleman (CA), 161

Part 36 settlement

Enforcement

Law of Property (Miscellaneous Provisions) Act 1989 — Whether CPR 36 settlement creating contractual or *sui generis* obligations — Whether settlement involving disposition of land invalid for failure to comply with s2 of 1989 Act — Held, CPR 36 regime creating *sui generis* obligations enforceable by court under inherent jurisdiction — S2 of 1989 Act not to be applied to defeat settlement reached by parties — Power to order parties to sign document complying with s2

***Orton* v *Collins (Ch),* [2007]** 2 **147**

Positive covenants

Enforcement

House purchased from registered social landlord — Vendor's predecessor obtaining funding from local housing authority under written agreement — Agreement containing positive covenant by vendor to carry out essential repairs — S609 of Housing Act 1985 — Whether covenant enforceable against purchaser — Whether section 609 permitting enforcement of positive covenants against successors in title of covenantor — Held, s609 permitting enforcement of restrictive covenants by local housing authority in same manner as if covenant benefiting retained land — Positive covenants not enforceable against covenantor's successors in title pursuant to s609

***Cantrell* v *Wycombe District Council (CA),* [2008]** 3 **93**

Solicitor's undertakings

Release of charges

Properties subject to all-moneys charges granted by vendors to mortgagees — Indebtedness of vendors exceeding value of properties — Vendors' solicitor giving undertakings to secure release of charges — No redemption statement obtained — Charges not released upon completion — Whether solicitor for purchasers entitled to summary enforcement of undertakings — Whether redemption figure calculated by reference to sum mortgagees would have accepted at time of sale or sum currently required — Held, summary enforcement granted — No relevant dispute over redemption figure requiring resolution by inquiry — Redemption figure ascertained by reference to sum currently required by mortgagees not figure mortgagees might have accepted had undertakings been complied with promptly

***Angel Solicitors* v *Jenkins O'Dowd & Barth (Ch),* [2009]** 1 **71**

TABLE OF STATUTES

References are to year, number and volume,

eg **[2010]** 2 **127** refers to vol 2 of 2010 at p127

TABLE OF STATUTORY INSTRUMENTS

References are to year, number and volume

eg **[2010]** 2 **127** refers to vol 2 of 2010 at p127

TABLE OF EU LEGISLATION

References are to year, number and volume,
eg **[2010] 2 127** refers to vol 2 of 2010 at p127

For Product Safety Concerns and Information please contact our EU representative GPSR@taylorandfrancis.com Taylor & Francis Verlag GmbH, Kaufingerstraße 24, 80331 München, Germany

Batch number: 10706093

Printed by Printforce, the Netherlands